D0731569

Dr. Robert E. Smith

THE COMPLETE BOOK OF BIORHYTHM LIFE CYCLES

AARDVARK PUBLISHERS, Inc.

NEW YORK

AARDVARK PUBLISHERS, INC.
257 CENTRAL PARK WEST
NEW YORK, NEW YORK 10024

Library of Congress Cataloging in Publication Data
Smith, Robert Elijah, 1911-
    The complete book of biorhythm life cycles.

    Bibliography: p. 195.
    1. Biological rhythms. I. Title.
QH527.S56   612   76-40329
ISBN 0-917384-01-6
ISBN 0-917384-00-8 pbk.

Printed in the United States of America

In this book, Dr. Smith has applied his knowledge and experience in scientific studies to the field of biorhythm theory. The numerical relationships and calculations in this book are based upon tables that are not only original in content and design but also accurate to the best of his knowledge. Results have been cross-checked innumerable times to eliminate inaccuracies or misprints. In the course of this work, Dr. Smith makes or infers certain interpretations of results. These interpretations are based solely upon his own knowledge or upon historical knowledge within the field of biorhythm research. Their acceptance or rejection is solely the responsibility of the reader.

Photographs on pages 58, 60, 61, 63, 64, 66, 70, 71, 75, 76, 77, 85, 86, 87, 88, 90 (top), and 91 are courtesy of United Press International.

Photographs on pages 67, 72, 78, and 90 (bottom) are courtesy of Wide World Photos, Inc.

Simple-to-Follow Instructions
for the Use of the Tables
Are on Pages 18 and 19.

CONTENTS

PREFACE

WE ALL HAVE ups and downs, days when we're dreary and days when we're full of energy after only six hours of sleep. What makes us feel so fresh or so droopy for no apparent reason? Mounting evidence suggests that such swings are due to natural rhythms of the body that occur at measurable intervals. If the rhythms can be measured, then the swings can be anticipated. If you could know in advance when your physical energies would be low, you probably wouldn't schedule an extra workout on that day. Or days when your brain will perk along at top speed: a good time to tackle that demanding assignment. Or a so-called "critical" day, when the rhythms are shifting from high to low or low to high and your general condition is likely to be unstable: time for extra caution driving a car or working with machinery. Few people manage to cash in on their extra-energy days or to forgive themselves a little for their lows, though both are in a sense natural resources and can be used to advantage.

This book is designed to help you put these resources at your service. At the heart of the book is one of the world's largest and fastest computers. Its size and speed are important because they made this book unique—and of unique value in attempting to understand the rhythmic swings in human behavior, in order to maximize the times of high potential and compensate for the lows. The computer makes available at a glance the periodic changes everyone experiences in the physical, emotional, and intellectual life cycles. A computer is only a machine, but it can point the way to a better understanding of that complex, individual, changeable organism—the human being.

Hopefully, the theory of biorhythms will help us understand ourselves a little better. Scientists have recently begun to investigate the rhythmic nature of all life; many studies have been done in the specific area of biorhythms, but more rigorous research is called for to determine exactly how they fit into human behavior patterns. Some of this research must be carried forward by persons such as yourself. One way to do this is to apply the theory to your own life experiences. Another possibility is to apply the theory to lives other than your own. This book provides both opportunities, as it is applicable to everyone, yet everyone can apply it differently to unique events and circumstances.

The physical, emotional, and intellectual rhythms are thought to run in cycles of unchanging length from birth throughout life. This is the first book that allows an element of flexibility in calculating the cycles. If consistent experience indicates your biorhythmic curves have been offset in either direction by one to three days, an adjustment table is provided that allows you to incorporate this change.

This book also covers a wider range of years than any previous book on biorhythms. It includes tables for birthdates from 1800 through the futuristic, Stanley Kubrick year of 2001. The reader has the opportunity to research and test the theory of biorhythms for exciting lives and events in this two-century time span.

Although there is not yet hard and fast scientific proof for the theory of the three periodic life cycles, the studies and empirical evidence to date, and the everyday life experiences of many people, lend it support. When you begin testing the theory for yourself, you may discover some very interesting things about your day-to-day behavior and that of others.

Remember the day you won that golf tournament, or successfully auditioned for your first part, or ran the car into a lamppost? You can check your biorhythm cycles to see if your mind, emotions, and body might have given you a special boost that day, or whether, if you'd only known you wouldn't be up to par on that Tuesday, you could have been super cautious and avoided the accident. Perhaps you're curious about Grandpa's emotional state the day he fell off the roof and broke his leg, or John Wilkes Booth's the night he shot President Lincoln. Or whether you're likely to be in top form next Monday when you'd planned to ask for a raise.

The tables in the back of the book will help you find the biorhythm information for these days. And if math has always been one big mystery, don't despair! The computer has done all the work. It calculated three cycles every day for 202 years, a total of 221,190 operational procedures, each of which involved a lot of arithmetic. If each of those calculations had required fifteen minutes of your time to do by hand, you would have been working eight hours a day for more than nineteen years to turn out all these tables. A single speedy computer reduced this to a matter of minutes.

The tables work for anyone, with any birthdate, and for any date and event. This means the book can be passed from husband to wife, from mother to son, from you to a friend, and in every case the tables are immediately valid to the new reader. And because the tables have been worked out so far into the future, they continue to be applicable year after year.

With these versatile tables, the reader can find personal biorhythm data for a single day of a particular year, for a single month, or for a complete year. Other information is available too. Would you like to know your exact location on the physical, emotional, or intellectual curve on any day of any year? Your critical, high, and low days for any month of any year? Would you like help in predicting the sex of your unborn child? Its birthdate? How compatible you're likely to be with a new lover or co-worker? Would you like to know when the next "triple-critical" day in your life will most likely occur? The tables in the back of the book will give the answers. And most important, you will want to know how to interpret your biorhythm results once you see them. A complete chapter on interpretations is provided.

A biorhythm table is not a crystal ball. For example, it cannot be used to predict future events. What it can do is provide you with information on what your physical, emotional, and intellectual state should be on a particular day, which ought to make your life a lot easier. Thus, if the table indicates a critical day for next Monday, it would be a good idea to drive with extra caution on that date, as studies indicate automobile accidents frequently occur on the driver's critical day. But don't cancel your plane reservations. Your mood can't influence the chances of a plane crash. However, if it should be the *pilot's* critical day . . . well, maybe you should have checked his biorhythms too.

If you knew his birthdate, you could do it with the tables. Or anyone's. All because of the computer. Sure, there is an author, editor, publisher, printer, and many others who helped put this book together. Yet the most crucial link —to which this work is dedicated—is the computer which made it possible. Once you begin using this book, you'll agree.

—Dr. Robert E. Smith
June, 1976

iii

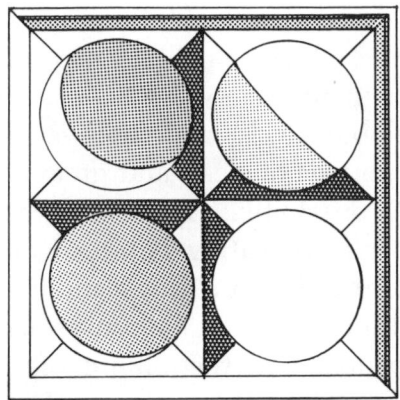

CHAPTER ONE

THE DEVELOPMENT OF BIORHYTHM THEORY

Rhythms in Man and Nature

MAN AND ALL living things are rooted in time—and no one fully understands time. Each of us questions the unknown forces of our ancient origin, and the all too sudden passage of our present lives. Every creature seems to have a biological clock, or set of clocks, that ticks off rhythms regulating its life and sense of self.

Man has always been fascinated by the rhythms of Nature. Markings in caves and on ivory from as far back as the Paleolithic period indicate that men were already observing the cycles of the moon. Stonehenge in England was an ancient observatory enabling priests to calculate the rhythms of changing seasons and movements of the stars. Over 2,400 years ago, Hippocrates, father of medicine, advised his associates that regularity was a healthy sign and irregular habits led to illness. In the Middle Ages, health, strength, sexual power, and menstruation were tied to the periodic waxing and waning of the moon; bloodletting and death associated with full and low tides.

The rhythmic influences of Nature and of internal biological clocks have continued to preoccupy man right up to the present time. Today, scientists are investigating many different kinds of rhythms in man: internal monthly rhythms, such as the physical changes in hormone levels and body temperature; the circadian rhythms, which coincide approximately with a twenty-four-hour day and regulate the daily changes in body temperature, blood pressure, blood sugar and hemoglobin levels; and the even shorter rhythms, such as those that occur during periods of sleep and dreaming.

Other recent studies on rhythms have involved mood changes and depression, jet lag and its effects, eating habits related to the enzyme cycle, worker efficiency cycles, reactions to drugs and therapy at different times, allergic sensitivity, breathing rhythms, the onset and severity of certain diseases, immunity to infection, and the timing of drug-taking and surgical procedures. Many current studies on rhythms are also being conducted with plants and lower forms of animal life.

But what stamps all these rhythms on man and his environment? How did the human mind and body become so interwoven with time? No one is certain. Some say such rhythms evolved in prehistory in response to external events like the daily turning of the earth, and that they are passed on from generation to generation. Some lean to the idea that man's rhythms are imprinted early in life, perhaps at birth, by parents or environment. Others believe that living things constantly respond to external events, such as phases of the moon, shifts from light to darkness, and changes in temperature and barometric pressure.

Whatever the explanation, it is clear that in almost every aspect of our lives we are part and parcel of that strange concept we call time, and modern man is only beginning to understand the details of his rhythmic life.

Biorhythms

In the nineteenth century, studies first began on certain life rhythms or cycles that were later termed "biorhythms." The word *biorhythm* is a compound of two Greek words, *bios* and *rhythmos*, which mean *life* and *a constant or periodic beat*. The theory of biorhythms defines and measures three basic and important life cycles in man: the physical, emotional, and intellectual.

Wilhelm Fliess, a highly respected and prominent doctor in Berlin, did pioneer work on biorhythms in the 1890s. Fliess, who had observed 23- and 28-day rhythms in many of his patients, began to collect statistics on the periodic occurrence of fevers, childhood disease, and the susceptibility to disease and death. With these statistics in hand, Fliess believed he had detected rhythms which were fundamental to man's life.

Dr. Fliess later developed two major biorhythm theories: first, that Nature bestows on man "master internal clocks" which begin counting time at birth and continue throughout life; and second, that one of these clocks regulates a 23-day cycle influencing man's physical condition and another regulates a 28-day cycle influencing emotions or degree of sensitivity.

A widely read man, Fliess speculated on why these two rhythms should prevail. He believed, much as we do today, that man is essentially bisexual in nature, composed of both male and female elements. Fliess called the 23-day physical cycle the male cycle, since it influenced strength, endurance, and vitality. He considered the 28-day cycle to be representative of the female element in all human beings; it governed sensitivity, intuition, love, and creativity—the entire emotional spectrum.

Subsequent research has reinforced the idea of the 23-day physical and 28-day emotional cycles. Of course, today few would agree with the premise that all physical components are male and all emotional matters female. Instead, both are now considered to be essential characteristics of each sex.

Wilhelm Fliess wrote extensively about the biorhythm theory, but the mathematics and statistics he used to support it were so massive and confusing that few people bothered to closely examine or to understand them. Still, the basic premise of the theory caught on. The idea of periodic rhythms in man created a considerable controversy among his colleagues, one which still exists today. Most scientists have accepted the fact that man's physical and emotional states are in constant flux, but many do not agree that these changes are influenced by regular biological cycles that start at birth.

One of Fliess' contemporaries who kept an open mind to his ideas was Sigmund Freud, a man with extremely revolutionary ideas of his own at the time. Early in his career, Freud showed extreme interest in and admiration for Fliess' theories, and they soon became very close friends. One hundred and eighty-four letters from Freud to Fliess have been published; unfortunately, the replies from Fliess have been lost.

Important ideas tend to spread rapidly in the scientific community. Dr. Hermann Swoboda, professor of psychology at the University of Vienna, read Fliess' work while still a young man, and by the turn of the century was himself researching, lecturing, and writing on biorhythms. Swoboda, who detected a periodicity in the occurrence of dreams and thinking processes, and in fevers, asthma, heart attacks, and the outbreak of illness, believed his own investigations confirmed Fliess' observations on the 23-day and 28-day cycles. Swoboda contributed to the theory the notion of the "critical" day, when the cycle shifts from high to low or low to high; a day of instability and usually of some stress for most people.

He was still working with biorhythms in 1963, when he died at the age of ninety. By that time, his ideas, largely ignored when he began working, had been widely accepted in many European countries. In the 1960s, certain Swiss hospitals planned operations to synchronize with the patients' most favorable days in their biorhythm cycles, and some German cities directed traffic safety programs based on drivers' biorhythm charts.

The third biorhythm cycle, the one associated with the intellect, was not determined until the 1920s. Alfred Teltscher, a doctor of engineering and professor at Innsbruck, Austria, grew curious about the fluctuations in his students' intellectual performance and capacity. Analyzing the academic and examination records of a large number of students, Teltscher discovered a regular rhythmic cycle that repeated every 33 days. His study suggested that mental powers, ability to concentrate on or to apply new ideas, might also be subject to internal clocks.

In the late twenties and early thirties, Drs. Rexford B. Hersey and Michael J. Bennett at the University of Pennsylvania stumbled onto a similar rhythm. Studying the behavior patterns of workers in a railroad shop over a period of months, Hersey and Bennett uncovered a 33- to 36-day cycle. It has been assumed by some later studies that they too had actually detected the intellectual cycle, so vital to concentration and decision-making, that Teltscher had first proposed.

Today, the idea that these three basic rhythms exist and repeat throughout life remains a matter of some debate. Numerous studies made since the pioneering days of biorhythms seem to indicate that these cycles exert a strong influence on all our lives and actions. Much, however, is still unknown. Scientists have not yet studied biorhythms extensively or rigorously enough either to prove or disprove their existence and effect. Support for the theory of biorhythms comes chiefly from successful accident prevention programs, from examination of correlations between biorhythm data and outstanding accomplishments, especially in the field of sports, and from the testimony of many individuals who insist they have lived more productively, aware of these underlying rhythms in their lives.

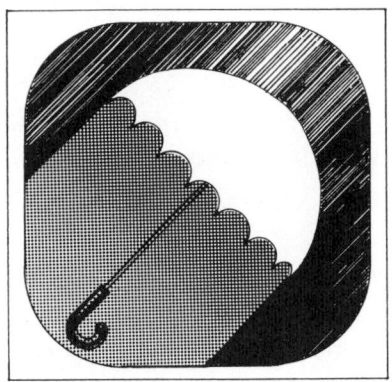

CHAPTER TWO

PROVING AND APPLYING THE THEORY

Can You Prove the Biorhythm Theory?

INVARIABLY, the question of proof arises with respect to the biorhythm theory. Frankly, it is difficult to prove. One reason lies in the fact that science, to date, has had little success with proof in areas where human values are involved. In fact, scientists are generally annoyed by the biorhythm theory or any theory that resists proof by clinical or "scientific procedures." On the one hand, valid mathematics are used to arrive at its results; on the other hand, one must interpret those results in psychological terms. Consequently it is possible to agree with the validity of the mathematics, but disagree with the interpretations.

Some people also find it difficult to accept the simplicity of the theory. Although people, as a rule, do not like complications, at the same time they tend to be suspicious of anything that appears too simple. The biorhythm theory falls in this latter category. It *is* difficult to admit that one's life can be simply divided or categorized into "critical," "high," or "low" days. Yet day-to-day experiences tell us that such may truly be the case! There are days when we seem to have more energy, vitality, and emotional control. There are days when these same feelings are at low ebb. And there are also those days when we react to situations in a totally unexpected way.

There are many people who support the biorhythm theory. Bertram Brown, Director of the National Institute of Mental Health, has said, "These biorhythms have a lot of validity. They help explain in part everything from having a bad week to exciting scientific things like the varied effects medications have when administered at different times."

Douglas Kelley, a statistician with the National Safety Council, is quoted as saying: "When chemistry was at the state where biorhythm is today, it was called alchemy. But alchemy became chemistry, and within fifty years research may do the same for biorhythm."

On the other side is Colin Pittendrigh, an expert on biological rhythms at Stanford University. *The Washington Post* quoted him as saying, "I consider this stuff an utter, total, unadulterated fraud. I really know nothing about it because we've been unable to track it down. But I consider anyone who offers to explain my life in terms of 23-day rhythms a numerological nut, just like somebody who wants to explore the rhythms of pig iron price to 11 decimal places."

Against these pros and cons and lacking sufficient clinical methods to prove the theory, an alternative procedure is to apply it to numerous situations and to carefully note the results, rather than to constantly criticize its assumptions. This alternative is recommended to the reader. Numerous opportunities are provided throughout this book for the reader to test the theory. Actually, the situation is similar to accepting or rejecting the daily weather forecast. The forecast can't be proved. But it is too useful and important in the life of an individual to neglect or refuse to accept. Nor is one too concerned if the weather forecast is not completely reliable. I may carry my raincoat tomorrow when, in fact, the sun will shine brightly. However, I still feel rewarded in that I was prepared for the event of rain. I also know the next forecast is quite likely to be valid.

These are not unique circumstances for man. He has always had to choose between the objective and the subjective, that which he can feel versus that which he can sense, fact versus fancy if one pleases. Economic and social men are perfect examples. They cannot act through certainty because proof does not exist for the many actions they follow. Economic man —like biorhythmic man—must be completely informed. Being completely informed is to know all courses of action that are open to him.

Against this criteria it is foolhardy, indeed, to completely ignore or refuse to examine the biorhythm theory. "Too stupid to come in out of the rain" is often a result of refusing to observe the forecast of rain. A hasty decision, made now, without regard to another time when mental capabilities may be supposedly keener, is the mark of insensitivity; and irrationality is often the inability or reluctance to observe all factors and possible courses of action available.

Applying the Theory

One of the major areas in which the biorhythm theory has been applied is in accident prevention; it is also the easiest in which to collect statistics and analyze results. As early as the 1930s, biorhythm researchers were studying the relationship of accidents to the "critical" or "cautionary" days in the cycles. These are the days when a cycle switches from high to low or low to high, often creating a state of instability for a few days each month. It is during this period that most people seem to have a greater propensity toward accidents, illness, and stress. By making workers aware of this period in their cycles, they can either take on easier tasks for those days, or compensate by just being a little extra-cautious and aware.

In 1939, Hans Schwing, a student at the Swiss Federal Institute of Technology in Zurich, Switzerland, studied 700 random accident cases of all types and found that 322 fell on a single critical day, 74 on a double-critical day, and 5 on a triple-critical day. The balance, 299 accidents, happened on non-critical, or "normal," days. Pretty impressive numbers: 401 accidents out of 700—almost 60 percent—took place on critical days, which only occupy 20 percent of a person's life. Obviously, something besides chance was at work—possibly slow reflexes, impaired coordination, poor judgment, all of which add up to human error.

Recent studies tend to confirm these findings, and today biorhythm application has improved the safety records for factory workers, taxi drivers, truck drivers, and airline pilots. The *National Safety News* reported that Jacob Sanhein of the Naval Weapons Support Center in Crane, Indiana, did a study of 776 industrial and motor vehicle accidents—of the total, 41.75 percent took place on critical days. Pfizer, Inc., tried rotating production workers in accordance with their biorhythms; their accident rate dropped nearly 60 percent. Exxon Chemical in Dayton, Texas, has recently begun to send safety reminders to approximately 900 employees on their critical days. A transport insurance company in Los Angeles is planning to start biorhythm programs for as many trucking companies as will accept the idea.

Many other industrial companies, businesses, and airlines have been reported to be experimenting with the biorhythm theory, but when questioned are reluctant to say so. Perhaps a good indication of the present "biorhythm controversy."

During the last several years, newspaper and magazine articles have reported on the application of biorhythms in Japan. There they have incorporated the biorhythm theory into traffic safety programs, work efficiency studies, and insurance plans. The Japanese statistics look so good one would think they were trying to prove a point instead of working to save lives! In his book *Is This Your Day?*, George Thommen includes a letter sent to him from Mr. Yujiro Shirai of the Japan Biorhythm Association, in which Shirai states that over five thousand firms have begun to use the biorhythm theory in that country.

In the United States, interest is now being shown in the field of health and medicine as well. In his book, *Occult Medicine Can Save Your Life*, Dr. C. Norman Shealy, a neurosurgeon who has introduced a microscopic brain surgery technique for victims of pituitary disturbance, has written, "My own feeling is that physicians particularly should be aware of their own critical days and should then avoid serious decisions or actions. Surgeons, for example, should not perform operations on critical days. Patients, too, should be protected by not having any surgery or potentially dangerous tests or procedures performed on them during these days."

Doctors in the field of mental health are studying the correlation between biorhythms and sudden changes in behavior. Nursing associations in different cities are becoming interested in learning about biorhythms and their use and importance to the nursing profession.

Recent and past studies have shown that death tends to fall on or near critical days. Hans Schwing, in addition to conducting accident studies, also analyzed 300 death dates and came to the amazing conclusion that death was nearly eleven times more likely on a critical day than on any other. Other studies are ongoing: The Life Cycles Research Foundation in Peoria, Illinois, is commencing a study on the relationship between critical days and sudden death cases. Between 500 and 1,000 cases will be analyzed.

The biorhythm theory is also being applied to more positive areas. Correlations between biorhythm patterns and achievement in athletic or sporting events, for example, are most interesting. Studies have shown that some athletes perform best on "high" days; others gather all strength available to excel on the unstable "critical" or "cautionary" days; extremely strong individuals often perform best when the physical cycle is not at its absolute peak. And, of course, different sports and skills require different emphasis in the individual cycles.

Creative endeavors—writing, acting, painting, craftwork—can be easier, more productive, when carried through in tune with the individual's biorhythm cycles. Barbara Bigham, in an article for *Writer's Digest*, suggested that "writer's block" could be avoided by dividing the phases of a writing project to concur with the biorhythm cycles: the intellectual phase of researching the topic, taking notes, organizing and preparing an outline could be done during the peak days of the individual's intellectual cycle; the actual writing could be done during the "high" days of the emotional cycle, which "influences our moods, sensitivity, and, most important to writers, our creativity." The physical chores necessary, such as retyping, mailing, returning library books, could best be done at another time.

Biorhythm may also help to improve our relationships with others; at the very least, give us greater insight and understanding. But this is the area in which human values and interpretations in psychological terms present the greatest problems in reaching definitive conclusions. Early studies done by Fliess and Swoboda as well as current research seem to indicate that people with similar cycles will be more compatible than those who are opposite. It seems reasonable that if your moods, "high" and "low" days coincide, you're more likely to get along. But then there are those individuals who may function best when interacting with others whose cycles complement rather than coincide with their own. What we *can* do is be more tolerant and understanding if we know why others around us do not always act and react in a consistent way. Reduction of family stress, work-situation conflicts, and improvement in group or "team" efforts through biorhythm application are all possibilities for the future.

In the meantime, the proof remains with you. The following chapters describe the three cycles and their monthly effects and explain how to use the tables to trace your own biorhythms. In the beginning, you might try keeping a diary for a month or two, noting down how you feel each day—whether you're sharp, optimistic, energetic, irritable. After a couple of months, check your cycles for this period to see if your observations and your biorhythms jibe. Nothing will ever convince you like your own experience.

Bio-Quiz 1

Farmers in some regions believe severe droughts follow a 17-year cycle. The year 1976 was thus predicted as a drought year. Can you prove or disprove the 17-year theory? (Maybe not, but the farmers who prepared for 1976 are harvesting bumper crops.)

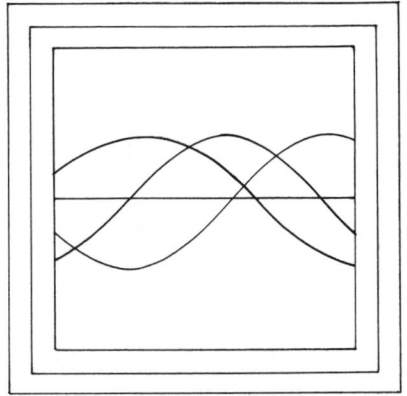

CHAPTER THREE

THE THREE CYCLES

The Physical Cycle

THE FLUCTUATIONS of the 23-day physical cycle are thought to influence man's strength, endurance, energy, and general physical well-being. We're all aware that our energy levels vary. Some mornings we can scarcely pull ourselves out of bed; others, though we don't quite wake up somersaulting, we're ready to go even before taking a cup of coffee. The biorhythm theory suggests that we can calculate days we might be more likely to zip through and days when we'll drag along.

Let's use the sample physical curve below to become acquainted with the theory. During the first half of the 23-day cycle (11½ days), one's physical well-being is increasing. This portion of the cycle is sometimes referred to as a "discharge period," analogous to the time when a battery is discharging electrical power, using up its stored energy. During this first half of the cycle, a person is quite vigorous and appears to command a powerful source of energy. Some surgeons who follow the biorhythm theory favor this period, usually the 2nd to 9th or 3rd to 8th days, as the most favorable for elective surgery.

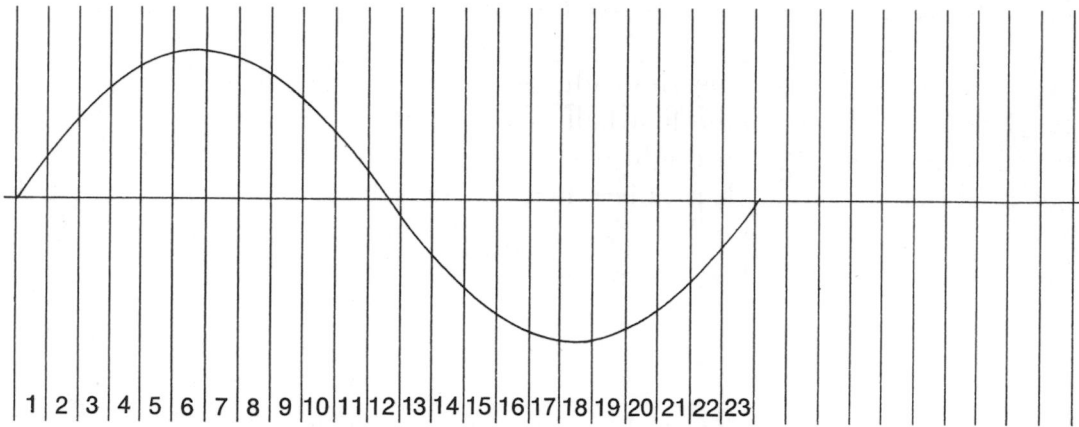

1 2 3 4 5 6 7 8 9 10 11 12 13 14 15 16 17 18 19 20 21 22 23

THE 23-DAY PHYSICAL CURVE

Critical Days occur at the *start* and *middle* of each 23-day cycle. In this sample cycle, days 1 and 12.

The High Day is at the top of the curve, above the base line. In this cycle, day 7.

The Low Day is at the bottom of the curve, below the base line. In this cycle, day 18.

The second half of the 23-day cycle (also 11½ days) is a time of reduced vigor, recuperation, and storage of new energy, as when a battery is recharging. These are the days when the curve on the chart is below the base line.

To an athlete, this is the time when he is in a slump, and the stay-at-home experiences his own kind of slump as well. During this time, man is more content to rest and regain his strength. That's not to say that work, tennis, and spring cleaning simply cease every other 11 days, but the drive slows down, the time-outs become a little more frequent, a good book and a catnap begin to look more attractive. Doctors who follow the theory find this time ideal for a patient's recuperation or therapy.

It is important to emphasize that the curve does not divide into a "good" half and a "bad" half. Neither portion is necessarily better or worse than the other—fortunately, since each adds up to half a lifetime! The theory of biorhythms aims to alert you to your capacities and potentials, the days when your basic drives (in this case physical) are at high, low, or critical tide. A low period is no more evil than a low gas tank in your car. If you read the gauge properly, you won't plan to drive 500 miles that day. You'll add more fuel—for the human machine, this means a little rest, a little bit of being good to yourself. A low period that is observed and used wisely can nourish the body. An athlete, for instance, might adjust his training schedule to provide for more rest or less intense concentration during the second half of his cycle.

Properly used, a low period can give the same benefit as sleep to an exhausted man.

A high day may not be an unadulterated blessing, either. Although more can be accomplished in the first half of the cycle, the physical plant can be tuned so high that a man might over-exert or try to go beyond his physical potential, ending up with a pulled muscle on the fifteenth hole of the golf course. A lot depends on individual condition. Professional athletes have often hit home runs, caught long passes, and broken records at the peak of their physical cycles.

There are two "critical" or "cautionary" days in every complete 23-day physical cycle. These are the *first* day, when each new cycle begins, and the *halfway* mark, between the 11th and 12th days, when energy switches into the recharge period. The body is relatively unstable and less resistant to stress on these days; heart attacks, for example, apparently tend to fall on critical days of this curve. Again, it is important to note that the days in themselves are not critical. The person's condition on that day may bear watching; he might react badly to strain imposed by the outside world or by his own body. There is no magical hex or voodoo spell that will make a heart fail at the "critical" or "cautionary" point of the physical cycle, but a man who has recently had a heart attack might take extra precautions on such a day. A taxi driver might be extra careful too. Or someone working in a machine shop. Or someone shepherding fifteen pre-school children out to the playground.

The Emotional Cycle

The 28-day emotional cycle has a curve similar to that of the physical cycle, with the curve rising in the first half (14 days), or discharge phase, and falling in the second half (also 14 days), or recharge phase. The sample below will help you to fix its main characteristics in mind. Since this cycle probably influences sensitivity and creativity as well as feelings of love and cooperativeness, artists and writers may find their muses visiting them more often and their tempers sweeter in the first two weeks of this cycle. During the first 14 days, one is likely to be cheerier and more optimistic; during the second 14 days a little less open, friendly and hopeful; and on critical days, decidedly grumpy and irritable.

The individual pattern will no doubt vary with individual temperament. Someone with a sunny and very calm disposition may seem a bit cloudy at the low point of the cycle but is unlikely to storm a lot at any time. A more passionate and erratic type can swing through periods of bliss and then astonish us with an outburst of anger three days later. It seems quite likely that some people would experience their biorthythmic fluctuations more strongly than others.

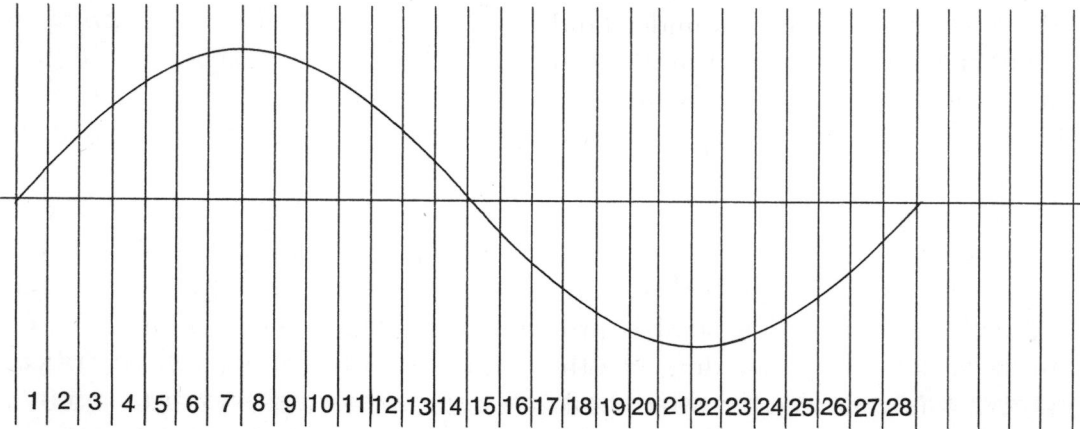

1 | 2 | 3 | 4 | 5 | 6 | 7 | 8 | 9 |10|11|12|13|14|15|16|17|18|19|20|21|22|23|24|25|26|27|28

THE 28-DAY EMOTIONAL CURVE

Critical Days occur at the *start* and *middle* of each 28-day cycle. In this sample cycle, days 1 and 15.

The High Day is at the top of the curve, above the base line. In this cycle, day 8.

The Low Day is at the bottom of the curve, below the base line. In this cycle, day 22.

An interesting condition exists with respect to the 28-day cycle which is not true of the other two. Since 28 days make up four seven-day weeks, the important days of this cycle always fall on the same day of the week, the day of the week on which one was born. A person born on a Monday will know that every other Monday will be a critical day. Whereas "blue Mondays" are traditional, Monday's child will have a critical day one Monday, a high day the next Monday, another critical, and then a low, and so on throughout the year. This same-day syndrome provides a means of testing the biorhythm theory for yourself. Does experience tell you that your emotionally shaky days are almost always on the same days of the week? It's best to test this out over a period of several months; there are always other elements to account for. Bad news can bring you down even on a high day, but you'll probably be able to handle it better.

The Intellectual Cycle

The 33-day cycle is representative of man's intellectual fluctuations. To date, researchers have focused somewhat less on the intellectual than on the other two cycles, which is understandable since much remains to be done in aiding man to understand and use his mental powers. Some doctors claim this cycle is closely related to secretions of the thyroid. Independent medical studies have pointed out correlations between variations in intellectual performance and hormone levels; again, this is an area which has not been thoroughly explored.

The first half of this cycle (16½ days) is the time when students and others engaged in intellectual pursuits are more able to absorb new concepts, be more creative, make notable progress. We all know that studying is a breeze at some times, a drag at others. The first half of the cycle is considered prime time for creative thinking, for progress in new subject areas, for mental successes when memory is at a premium, and for work which demands accurate and immediate mental responses. Since the mind responds most rapidly to new challenges during this period, it would probably be the best time to begin a new assignment or a new job.

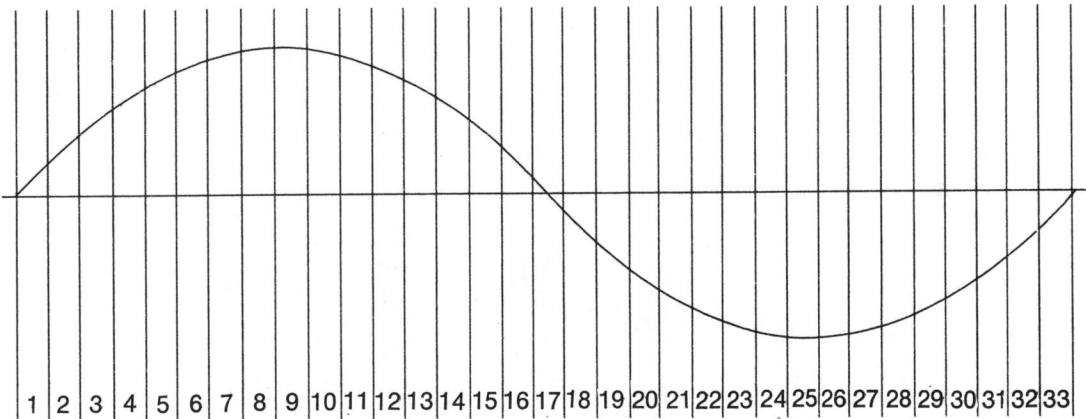

1 2 3 4 5 6 7 8 9 10 11 12 13 14 15 16 17 18 19 20 21 22 23 24 25 26 27 28 29 30 31 32 33

THE 33-DAY INTELLECTUAL CURVE

Critical Days occur at the *start* and *middle* of each 33-day cycle. In this sample cycle, days 1 and 17.

The High Day is at the top of the curve, above the base line. In this cycle, day 9.

The Low Day is at the bottom of the curve, below the base line. In this cycle, day 25.

The second half of this 33-day cycle (also 16½ days) is a time when one's ability to think logically is somewhat reduced. The I.Q. doesn't automatically lose 10 points; the brain just says it isn't quite so happy with a heavy load of new stimulation. Remember that energies have been discharged in the first half of the cycle. Now the mind wants a little rest, in order to store up energy again. During this time, it is somewhat more difficult to absorb new ideas, to do creative thinking, or to perform mental exercises where concentration, memory, and quick mental response are required. This time is probably best suited to review of previously learned concepts, to practice of lessons which should be learned by rote, to absorption and consolidation of prior gains. The second half of the intellectual cycle seems to be the best time to practice your lines, go over your French verb endings, and edit the paper you wrote.

The critical days in the 33-day intellectual cycle are the 1st and 17th days. These may be days when important decisions could best be delayed. If you know that a major matter must be resolved on one of those days, it might be a good idea to try to see all sides of the question beforehand. But we all have to face some problems without any warning. Take a little extra time to think it over; it could be helpful. Also, as you will discover in a later chapter, the circumstances under which these critical days occur will make a difference.

No part of life needs to grind to a halt just because it's a critical day or the low part of the cycle. A big exam or a big interview can be handled well even on the least promising day; most of us have taken one or the other with a cold and still scored high. We're a little cautious, or we work extra hard beforehand, or we make a special effort to get the adrenalin flowing. If we know in advance that a task will be more difficult than usual, we can be properly prepared, and this is where the biorhythm theory can be most useful.

These are the three basic biorhythm cycles—their monthly cyclical lengths and variations. But keep in mind that each person's cycles will be individual. Though everyone born on the same day in the same year will have identical cycles, they will obviously not have identical lives. Not only will their circumstances be different, but their inborn and learned reactions to events will also differ sharply. All three cycles will change with age: a young person's are likely to peak more sharply, an older person's to flatten out. Health, temperament, character, and probably heredity will influence the steepness of the curve and the way one responds to fluctuations in his potential. Awareness of where you stand in each of your life cycles at a particular time, and how you function while being at various points in each of the three cycles will help you determine the best coping mechanisms for your particular life style.

The following chapter explains how you can quickly find any biorhythm information you need by using the three tables in the back of the book.

CHAPTER FOUR

HOW TO USE THIS BOOK: YOUR OWN BIORHYTHM LIFE CYCLES

YOUR NEW understanding and knowledge of the basic principles of biorhythms will enable you to fully appreciate the simplicity of obtaining biorhythm data by using the tables in Appendix A. They are three in number and are entitled: Birth Code Tables, Inquiry Year Tables, and Biorhythm Data Tables.

The Biorhythm Data Tables consist of a set of coded data groups for each of the three biorhythm cycles: the physical, emotional, and intellectual. Each of these groups contains complete biorhythm data for one year, month by month. In order to obtain biorhythm data for any date or time period between 1800 and 2001, for yourself or for anyone else, it is necessary to know only the proper code to identify the appropriate data group. It is the use of the first two tables—the Birth Code Tables and the Inquiry Year Tables—that allows you to determine the proper code.

The practical application of the biorhythm theory is dependent on two items: first, a subject's birthdate; second, an inquiry date or time period for which we wish to learn the status of a subject's biorhythm cycles. Having the subject's birthdate and the inquiry date, it is a simple matter to find the biorhythm data by the following procedure:

How to Use the Tables

There are three steps—as simple as A, B, C.

A. *Birth Code Tables*—First, find the Birth Codes under the birthdate. The Birth Code Tables begin on page 102, and include the years 1800 to 2001. In the example on the opposite page, look up the Birth Codes for October 14, 1890, the birthdate of Dwight D. Eisenhower. Eisenhower's complete Birth Code is E27AJ:
  - The first letter = the Physical Birth Code = E in this example.
  - The middle two numbers = the Emotional/Sensitivity Birth Code = 27 in this example.
  - The last two letters = the Intellectual Birth Code = AJ in this example.

B. *Inquiry Year Tables*—Now, change the Birth Codes to Inquiry Year Codes. The Inquiry Year Tables begin on page 170. For this example, let's use November 26, 1957, as the Inquiry Date—the day on which Eisenhower suffered a heart attack. So we will change the Birth Codes to 1957 Inquiry Year Codes.
  - Opposite the Inquiry Year 1957 and under the Physical Birth Code E, we find the Inquiry Year Code P13.
  - Opposite the Inquiry Year 1957 and under the Emotional Birth Code 27, we find the Inquiry Year Code S18.
  - Opposite the Inquiry Year 1957 and under the Intellectual Birth Code AJ, we find the Inquiry Year Code M15.

C. *Biorhythm Data Tables*—Finally, we find the biorhythm data for 1957 by looking under the Inquiry Year Codes in the Biorhythm Data Tables. This third set of tables begins on page 176.
  - The Inquiry Year Code P13 gives the physical biorhythm data.
  - The Inquiry Year Code S18 gives the emotional biorhythm data.
  - The Inquiry Year Code M15 gives the intellectual biorhythm data.

By checking the dates of the criticals, highs, and lows in each of the cycles for the month of November, we find that President Eisenhower had a critical day on November 26th in his emotional cycle. He was also at the low points in the other two cycles on this date.

A. *Birth Code Tables*

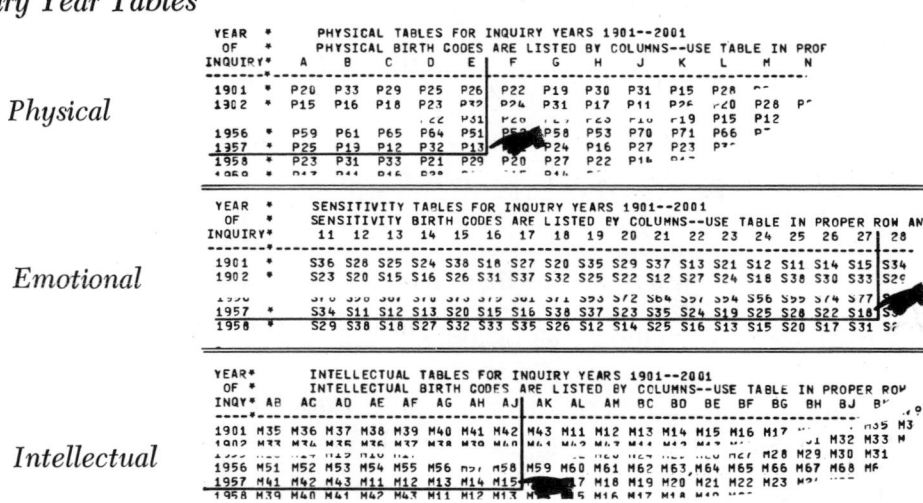

1890 DAYS	FIRST LETTER = PHYSICAL CODE, MIDDLE NUMBER = SENSITIVITY CODE, LAST TWO LETTERS = INTELLECTUAL CODE											
	JAN	FEB	MAR	APR	MAY	JUN	JUL	AUG	SEP	OCT	NOV	DEC
1	R38BK	F18BH	S18BC	J27AL	Z19AH	S23AF	D32AC	Z33CS	S35CP	D29CK	Z22CH	H34CD
2	X11BL	V12BJ	P12BD	U13AM	C30AJ	P36AG	J20AD	C15AB	P16CR	J21CL	C29CJ	S26CE
3	B31BM	E37BK	W37BE	R19BC	K14AK	W32AH	U28AE	K25AC	W24CS	U17CM	K34CK	P34CF
4	N18BP	M27BL	Y27BF	X30BD	F23AL	Y20AJ	R33AF	F35AD	Y21AB	R22CP	F26CL	W11CH
5	A12CD	H13BM	T13BG	B14BE	V36AM	T28AK	X15AG	V16AE	T17AC	X29CR	V38CM	Y31CJ
6	Z37CE	S19BP	G19BH	N23EF	E328C	G33AL	B25AH	E24AF	G22AD	B34CS	E11CP	T18CK
7	C27CF	P30CD	L30BJ	A36BG	M20BD	L15AM	N35AJ	M21AG	L29AE	N26AB	M31CR	G12CL
8	K13CH	W14CE	D14BK	Z32BH	H28BE	D25BC	A16AK	H17AH	D34AF	A38AC	H18CS	L37CM
9	F19CJ	Y23CF	J23BL	C20BJ	S33BF	J35BD	Z24AL	S22AJ	J26AG	Z11AD	S12AB	D27CP
10	V30CK	T36CH	U36BM	K28EK	P158G	U16BE	C21AM	P29AK	U38AH	C31AE	P37AC	J13CR
11	E14CL	G32CJ	R32BP	F33BL	W25BH	R24BF	K17BC	W34AL	R11AJ	K18AF	W27AD	U19CS
12	M23CM	L20CK	X20CD	V15BM	Y35BJ	X21BG	F22BD	Y26AM	X31AK	F12AG	Y13AE	R30AB
13	H36CP	D28CL	B28CE	E25BP	T16BK	B17BH	V29BE	T38BC	B18AL	V37AH	T16AF	X14AC
14	S32CR	J33CM	N33CF	M35CD	G24BL	N22BJ	E34BF	G11BD	N12AM	E27AJ	S37AG	B23AD
15	P20CS	U15CP	A15CH	H16CE	L21BM	A29BK	M26BG	L31BE	A37BC	M13AK	L14AH	N36AE
16	W28AB	R25CR	Z25CJ	S24CF	D17BP	Z34BL	H38BH	D18BF	Z27BD	H19AL	D23AJ	A32AF
17	Y33AC	X35CS	C35CK	P21CH	J22CD	C26BM	S11BJ	J12BG	C13BE	S30AM	J36AK	Z29AG
18	T15AD	B16AB	K16CL	W17CJ	U29CE	K38BP	P31BH	U37BH	K19BF	P14BC	U32AL	C28AH
19	G25AE	N24AC	F24CM	Y22CK	R34CF	F11CD	W18BL	R27BJ	F30BG	W23BD	R20AM	K33AJ
20	L35AF	A21AD	V21CP	T29CL	X26CH	V31CE	Y12BM	X13BK	V14BH	Y36BE	X28BC	F15AK
21	D16AG	Z17AE	E17CR	G34CM	B38CJ	E18CF	T37BP	B19BL	E23BJ	T32BF	B33BD	V25AL
22	J24AH	C22AF	M22CS	L26CP	N11CK	M12CH	G27CD	N30BM	M36BK	G20BG	N15BE	E35AM
23	U21AJ	K29AG	H29AB	D38CR	A31CL	H37CJ	L13CE	A14BP	H32BL	L28BH	A25BF	M16BC
24	R17AK	F34AH	S34AC	J11CS	Z18CM	S27CK	D19CF	Z23CD	S20BM	D33BJ	Z35BG	H24BD
25	X22AL	V26AJ	P26AD	U31AB	C12CP	P13CL	J30CH	C36CE	P28BP	J15BK	C16BH	S21BE
26	B29AM	E38AK	W38AE	R18AC	K37CR	W19CM	U14CJ	K32CF	W33CD	U25BL	K24BJ	P17BF
27	N34BC	M11AL	Y11AF	X12AD	F27CS	Y30CP	R23CK	F20CH	Y15CE	R35BM	F21BK	W22BG
28	A26BD	H31AM	T31AG	B37AE	V13AB	T14CR	X36CL	V28CJ	T25CF	X16BP	V17BL	Y29BH
29	Z38BE		G18AH	N27AF	E19AC	G23CS	B32CM	E33CK	G35CH	B24CD	E22BM	T34BJ
30	C11BF		L12AJ	A13AG	M30AD	L36AB	N20CP	M15CL	L16CJ	N21CE	M29BP	G26BK
31	K31BG		D37AK		H14AE		A28CR	H25CM		A17CF		L38BL

B. *Inquiry Year Tables*

Physical

YEAR OF INQUIRY	PHYSICAL TABLES FOR INQUIRY YEARS 1901--2001 PHYSICAL BIRTH CODES ARE LISTED BY COLUMNS--USE TABLE IN PROP												
	A	B	C	D	E	F	G	H	J	K	L	M	N
1901	P20	P33	P29	P25	P26	P22	P19	P30	P31	P15	P28		
1902	P15	P16	P18	P23	P32	P24	P31	P17	P11	P25	P20	P28	P
				P31	P28			P23	P10	P19	P15	P12	
1956	P59	P61	P65	P64	P51	P58	P53	P70	P71	P66			
1957	P25	P19	P12	P32	P13	P24	P16	P27	P23	P			
1958	P23	P31	P33	P21	P29	P20	P27	P22	P14				
1959													

Emotional

YEAR OF INQUIRY	SENSITIVITY TABLES FOR INQUIRY YEARS 1901--2001 SENSITIVITY BIRTH CODES ARE LISTED BY COLUMNS--USE TABLE IN PROPER ROW AN																	
	11	12	13	14	15	16	17	18	19	20	21	22	23	24	25	26	27	28
1901	S36	S28	S25	S24	S38	S18	S27	S20	S35	S29	S37	S13	S21	S12	S11	S14	S15	S34
1902	S23	S20	S15	S16	S26	S31	S37	S32	S25	S22	S12	S27	S24	S18	S38	S30	S33	S29
1957	S34	S11	S12	S13	S20	S15	S16	S38	S37	S23	S35	S24	S19	S25	S28	S22	S18	S3
1958	S29	S38	S18	S27	S32	S33	S35	S26	S12	S14	S25	S16	S13	S15	S20	S17	S31	S

Intellectual

YEAR OF INQY	INTELLECTUAL TABLES FOR INQUIRY YEARS 1901--2001 INTELLECTUAL BIRTH CODES ARE LISTED BY COLUMNS--USE TABLE IN PROPER ROW																
	AB	AC	AD	AE	AF	AG	AH	AJ	AK	AL	AM	BC	BD	BE	BF	BG	BH	BJ	B
1901	M35	M36	M37	M38	M39	M40	M41	M42	M43	M11	M12	M13	M14	M15	M16	M17			M35 M3
1902																M32	M33	M	
1956	M51	M52	M53	M54	M55	M56	M57	M58	M59	M60	M61	M62	M63	M64	M65	M66	M67	M68	MF
1957	M41	M42	M43	M11	M12	M13	M14	M15		M17	M18	M19	M20	M21	M22	M23	M		
1958	M39	M40	M41	M42	M43	M11	M12	M13	M	M16	M17	M18	M19						

C. *Biorhythm Data Tables*

TABLE P13 *Physical*

	CRITICALS		HIGHS	LOWS	
JAN	12	23	18	6	29
FEB	4	15 27	10	21	
MAR	10	22	5 28	16	
APR	2	14 25	20	8	
MAY	7	18 30	13	1 24	
JUN	10	22	5 28	16	
JUL	3	15 26	21	9	
AUG	7	18 30	13	1 24	
SEP	10	22	5 28	16	
OCT	3	15 26	21	9	
NOV	7	18 30	13	1 24	
DEC	11	23	6 29	17	

TABLE S18 *Emotional*

	CRITICALS		HIGHS	LOWS	
JAN	8	22	15	1	29
FEB	5	19	12	26	
MAR	5	19	12	26	
APR	2	16 30	9	23	
MAY	14	28	7	21	
JUN	11	25	4	18	
JUL	9	23	2 30	16	
AUG	6	20	27	13	
SEP	3	17	24	10	
OCT	1	15 29	22	8	
NOV	12	26	19	5	
DEC	10	24	17	3 31	

TABLE M15 *Intellectual*

	CRITICALS		HIGHS	LOWS
JAN	5	21	13	29
FEB	7	23	15	
MAR	12	28	20	3
APR	14	30	22	5
MAY	17		25	8
JUN	2	19	27	10
JUL	5	22	30	13
AUG	7	24		15
SEP	9	26	1	17
OCT	12	29	4	20
NOV	14		6	22
DEC	1	17	9	25

Following are some additional examples to help you get acquainted with using the tables.

Example 1. James Dean, the actor in *Rebel Without a Cause*, was killed in an automobile head-on collision on September 30, 1955. He was born on February 8, 1931. Examine his biorhythm data on or near the date of the accident.

	Physical	Emotional	Intellectual
A. Birth Codes for February 8, 1931	Y	12	BG
B. Inquiry Year Codes for 1955	P20	S12	M28
C. Biorhythm Data for September, 1955	high 29th	crit. Oct. 2nd	low 30th

James Dean was at his high point physically but at his lowest point intellectually. In addition, his emotional cycle indicates a critical two days from the day of the accident. This combination of a high physical with a low or critical in the intellectual cycle is often found in high-risk situations, where the individual is physically "charged up" and the intellectual controls for self-protection and survival are either shaky or lessened.

Example 2. New York baseball fans will never forget Joe DiMaggio. Nor are they likely to forget the date of December 11, 1951, when Joe announced that he was "quitting baseball" for personal and other reasons that he found hard to explain. Is there any indication that his decision may have been influenced by biorhythmic factors? Joe's birthdate is November 25, 1914.

	Physical	Emotional	Intellectual
A. Birth Codes for November 25, 1914	F	24	AE
B. Inquiry Year Codes for 1951	P22	S34	M25
C. Biorhythm Data for December, 1951	high 11th	high 12th	crit. 11th

Joe stood at the high points of his physical and emotional cycles. In addition, he had a critical point, intellectually, on that day. On a critical day, the individual is usually in an unstable state within the particular cycle, and has the potential for acting either in a positive or negative way. On the intellectual cycle, a critical day could mean the possibility for a wise decision or an unexplained, even hasty judgment or action. There is no doubt that Joe felt he had good reasons to quit baseball when he did. On the other hand, his announced decision to do so on this particular day may have also been partially the result of unseen biorhythmic factors.

Example 3. Ulysses S. Grant, born April 27, 1822, accepted the surrender of General Robert E. Lee on April 9, 1865. Examine the status of General Grant's biorhythm cycles on the date of the surrender.

	Physical	Emotional	Intellectual
A. Birth Codes for April 27, 1822	A	18	BF
B. Inquiry Year Codes for 1865	P28	S31	M30
C. Biorhythm Data for April, 1865	crit. 7th	high 8th	crit. 12th

Grant's physical data indicates a critical two days previously. His emotional or sensitivity data indicates he was at the high point of that cycle on the surrender day.

Example 4. Some of the most difficult personal experiences of history were those faced by King Edward VIII of England on the days between the 10th and 14th of December, 1936. On December 10, against the advice of many of his supporters, he reached the decision to abdicate his throne. On December 11, he announced to his nation that he had decided to marry Bessie Wallis Warfield (formerly Mrs. Simpson). On December 12, 13, 14, he prepared to leave his homeland with "the woman I love." Did his biorhythm cycles reveal the stress and strain of these days? King Edward's birthdate was June 23, 1894. For Inquiry Dates, use December 10–14, 1936.

	Physical	Emotional	Intellectual
A. Birth Codes for June 23, 1894	X	14	AD
B. Inquiry Year Codes for 1936	P69	S67	M65
C. Biorhythm Data for December, 1936	crit. 13th	crit. 12th	crit. 10th

All three cycles indicate critical days surrounding the day of his decision.

Example 5. Richard M. Nixon, born on January 9, 1913, resigned the Presidency on August 9, 1974. Does his biorhythm data on that date reveal any of the physical, emotional, or mental stress of those "final days?"

	Physical	Emotional	Intellectual
A. Birth Codes for January 9, 1913	S	19	BC
B. Inquiry Year Codes for 1974	P15	S36	M14
C. Biorhythm Data for August, 1974	crit. 11th	high 8th	crit. 6th low 14th

(21)

We do know Mr. Nixon faced a critical breakdown physically soon after his resignation. The biorhythm data points to August 11 as this time. The high emotional day (the twenty-four-hour period between the 8th and the 9th) was reflected and supported by the emotional tempo of his resignation message to the Nation. The downward trend of his intellectual cycle, with a low near the 14th (five days after leaving office), also seems to presage events which occurred later.

Example 6. President Gerald Ford, born July 14, 1913, announced the decision to grant Richard M. Nixon an unconditional pardon on September 8, 1974. Let us assume that this decision was made in the early days of that month. Do President Ford's biorhythm cycles in early September of 1974 indicate any of the stress or strain of that decision?

	Physical	*Emotional*	*Intellectual*
A. Birth Codes for July 14, 1913	W	34	CS
B. Inquiry Year Codes for 1974	P18	S31	M35
C. Biorhythm Data Codes for September, 1974	crit. 5th	crit. 2nd low 9th	low 4th

President Ford went through *two* critical days in early September—one each in his physical and emotional cycles. Both criticals can be interpreted as times of unexpected or unusual moments in his physical and emotional attributes. The low intellectually on September 4, when he may have made his pardon decision, and the low emotionally when he granted the pardon, are left for the reader to interpret.

And that's all there is to it! Now that you know how to use the tables in Appendix A to find your biorhythm data, you may find it helpful to keep a personal biorhythm record. An annual form for this purpose is provided on the following page. It is not necessary for you to fill out the form—all the information indicated is already available in the tables. The "copy lesson" is simply to accustom you to the procedures involved when you first use the tables. If you find that you prefer to make up and keep an annual biorhythm record in advance, additional forms are provided in the back of the book in Appendix C.

Your Own Step-by-Step
Biorhythm Record for the Current Year

Step 1.	Your Date of Birth _____		Look up your own Birth Codes; insert each in the appropriate box.
	Birth Codes		
	Physical *Emotional* *Intellectual*		

Step 2.	Select Inquiry Year _____			Select the current year as your Inquiry Year. Look opposite the Inquiry Year and under your Birth Codes to find the Inquiry Year Codes; insert each in the appropriate box.
Inquiry Year	*Inquiry Year Codes*			
	Physical	*Emotional*	*Intellectual*	

Step 3. Copy your biorhythm data for the year.

Month	Physical Crit.	High	Low	Emotional Crit.	High	Low	Intellectual Crit.	High	Low	
Jan.										
Feb.										Look under the Inquiry Year Codes in the Biorhythm Data Tables and insert your biorhythm data for each month.
March										
April										
May										
June										
July										
Aug.										
Sept.										
Oct.										
Nov.										
Dec.										

Bio-Quiz 2

Frank Shorter, the American Gold Medal marathoner in 1972, ran second in the same event on July 31, 1976. When interviewed on that date, Frank said he had not felt his usual self all through the race. Did his biorhythm data indicate unstable physical or emotional forces during that time period? Frank's birthdate is October 31, 1947.

CHAPTER FIVE

INTERPRETING YOUR BIORHYTHM DATA

NOW THAT YOU have learned what the biorhythm theory is and have been instructed in how to find your biorhythm data quickly and easily by using the tables in this book, there remains an even more important consideration. How does one interpret the results? What do critical, high, and low days mean for each individual who applies the theory to his daily activities and events?

Interpretations of biorhythm data center on the three most important days in the monthly rhythms: the critical, high, and low points in each of the physical, emotional, and intellectual cycles. The general characteristics associated with the high and low days are more straightforward and consistent than is the case with the critical or cautionary days. Generally, high days indicate full tides of: vitality, endurance, and strength from a physical view;

excitement, intense feelings, and vigorous sensibilities from an emotional view; and knowledge, reason, and thought from the intellectual view. There are gradations of these dependent upon the situations and circumstances under which they occur, and on the individual personality involved. Thus, a high in the emotional cycle can signal a time of high production for the actor or artist, similar to a high in the physical cycle of an athlete. On the other hand, a high emotional state under circumstances requiring coolness during stress is a different matter, and could signal an "off" day for the professional race driver, or the man at bat with three men on.

The general characteristics of the low days in the cycles are not necessarily directly opposed to those of the highs, and should not be thought of as the "bad" or negative days in the cycles. They do indicate a lower level of potential or energy, and should be thought of as the recharge, rest, or recovery phase of the cycles.

In general, interpretations of critical days revolve around two key words: *deviation* and *instability*. The word "deviation," for example, means "a departure from a standard." Biorhythm research seems to indicate that critical days are often times when a person's behavior and actions differ from the norm for that individual. Examples are the usually polite, easy-going individual who, on his emotional critical day, finds himself short-tempered and edgy with everyone around him; or the person who is normally charged-up, energetic, and full of drive, who finds himself feeling listless, tired, and withdrawn on the critical day in his physical cycle.

The second key word, "instability," has similar connotations, but it brings into focus the concept of motion. In fact, the word itself infers a "wobbly" or shaky condition, and aptly describes the time during which the cycle is changing direction—either from high to low or low to high. Although there is a potential for mishap at this point in a cycle, there is at the same time a potential for very positive results. If one is cautious and aware of the unstable state, and able to harness and direct the energy, one can often excel on a critical day. By taking full advantage of the positive aspects of one's deviant behavior or feelings during this time, there is the potential for outstanding achievement—the possibility for a physical, emotional, or intellectual breakthrough in one's life. As an example of this, many professional athletes, mountain climbers, and others who depend upon physical attributes, have been able to perform admirably on biorhythmic critical days in their physical cycles.

With these general characteristics as a background, the following is a guidelist of more specific interpretations of the biorhythmic high, low, and critical days.

Interpreting Your High, Low Days

Physical High:

Your energy, vitality, and strength are at full tide. Your physical system is able to withstand shock, injury, or stress at this time. Sexual drives are above normal. At the same time, physical forces may be overly strong with a danger of overexertion or susceptibility to reckless, aggressive activities.

Emotional High:

Your emotions are keyed to their highest levels. This is the time when you will feel most outwardly directed to others, able to enjoy their company, as well as to give them support, encouragement, and understanding if needed. It is the optimum time for strengthening your relationships with your lover or friends. Your high emotions can also push you to great performances, especially in creative endeavors. On the other hand, depending upon circumstances and your personality, be aware of possible runaways, emotional outbursts or over-doing. Your high emotions could also lead you to impulsive, rash behavior.

Intellectual High:

Analytical, logistical, and concentrative abilities are at upper levels. This is a prime time for creative thinking, composing, writing, decision-making. This is a good time for abstract reasoning, verbal fluency, understanding new concepts, number ability, memory manipulation, and for facing new challenges, taking on new work assignments. This is also a time of possible frustrations if opportunities for intellectual expansion are limited or missing.

Physical Low:

Your physical forces are at low tide with vitality and energy resources possibly drained. Fatigue is common. Rest, recuperation are important at this time. Your interest in or inclinations to physical, sexual activities are on the wane. In athletic activities, endurance is usually lessened. There is a lack of zest or spirit and a feeling of near exhaustion may follow performance. You may have "worn out" feelings in general during this time. Depressive tendencies may be present and must be resisted. There is a danger of overexertion if highs in the other cycles tend to override your feelings of physical low energy. Watch eating, drinking, as any overindulgence at this time is potentially dangerous.

Emotional Low:

Your emotional state is relaxed or below normal. In situations where extreme calm or lack of emotions is required, this may be helpful. In other circumstances, your feelings, sensitivities, and awareness of your environment may be low. It is a time when you may feel more inwardly directed, less able to extend yourself to others. You may even experience feelings of depression or loneliness. In general, it is a time of passivity rather than engagement.

Intellectual Low:

Your ability to move into new intellectual fields is lessened at this time. Decision-making, judgments, and concentration on tasks may all be frustrating or difficult during this period. Impulsive, spur-of-the-moment actions with regrettable consequences are possible. Loss or relinquishment of cautions or protective inclinations may occur. Depressed or exhausted mental states are also possible.

Interpreting Your Critical Days

Physical Critical:

In general, you can tire more easily, quickly. There may be some impairment of your physical control or strength with a potential for accidents or mishaps in the home, office, or car. A lessening of the body's protective senses increases the dangers of exertion, exhaustion, and possible injury. Relax, get added rest. Give extra attention to physical comforts and aids, such as proper lighting, frequent relaxation periods, that will make your day easier. Avoid activities where large amounts of energy are required—for most people, not a good day to start spring cleaning, mowing the lawn, or playing a strenuous game of tennis.

During health crises, a physical critical can signal a recurrence of problems—a possible heart attack, stroke, etc. If you have a particular health problem, it is a good time to curtail physical activities that require a large expenditure of energy.

In extreme cases, sensory deprivation may occur, with a consequent neglect for personal safety as exhibited in acts of either extreme bravery or foolhardiness.

For those who participate in sports frequently or for the professional athlete, performance hangs in the balance—it can be your best or worst day. Try to be extremely aware of your physical potential during this unstable time; intensify mental concentration to control and extend your physical capabilities for maximum performance and possible outstanding achievement.

Emotional Critical:

Either emotional exhaustion or euphoria is possible—from "the skids" (depression) to great heights (elation, outbursts). You may have a "let-down" or tired feeling, since exhaustion is often based on an imbalanced emotional state. You may also experience emotional "flair-ups" or extreme agitation, anxiety, or frustration. In extreme cases, suicides or attempted suicides have occurred on or very close to emotional critical days.

Overexertion and exhaustion due to emotional stress are also potential dangers to health and can trigger a physical crisis such as a heart attack or a stroke. Be particularly alert to health side effects in the day's activities.

Give extra care and attention to situations where emotions play a key role: driving your automobile in heavy traffic; stress situations at work; dealing with family problems; disciplining and working with young children. In general, forced calmness and mental concentration on control of emotions is advised. Keep in mind the unique opportunity of an emotional critical day— by harnessing your emotional energies at this time you could develop and strengthen a model of emotional self-control and expression which is so important to overall happiness and well-being. This day could give you the opportunity to reach important breakthroughs in your emotional life, as well as present possible new avenues of creativity.

Intellectual Critical:

On the critical day in the intellectual cycle, there is a tendency to avoid making necessary decisions and to hastily improvise acceptable answers to situations. Mental carelessness can often cause a disregard for personal safety and survival. Mental depression and a lower threshold to health hazards are also potential dangers.

Verbal fluency—an attribute of intelligence—may be decreased or restricted. The faculties of abstract reasoning, number ability, and memory appear to be lessened. Misplacement of valuables and forgetfulness are common. Often, there is a disregard for consequences: thus, the man who can't swim goes boating; the homemaker with a sore back decides to move the piano to another location. If it is possible to "*think* before leaping," this is the time to do it.

Despite these more frequent negative factors, either success or failure in academic or intellectual pursuits is possible on this day, since one's intellectual acuity is alterable in either direction by the imbalanced state. Mental control through awareness and alertness is important at these times. The opportunity to reach new heights, to have a day of intellectual, creative brilliance—to have that one stroke of genius that could change your life, are all entirely possible.

(29)

The Standard Biorhythm Curves

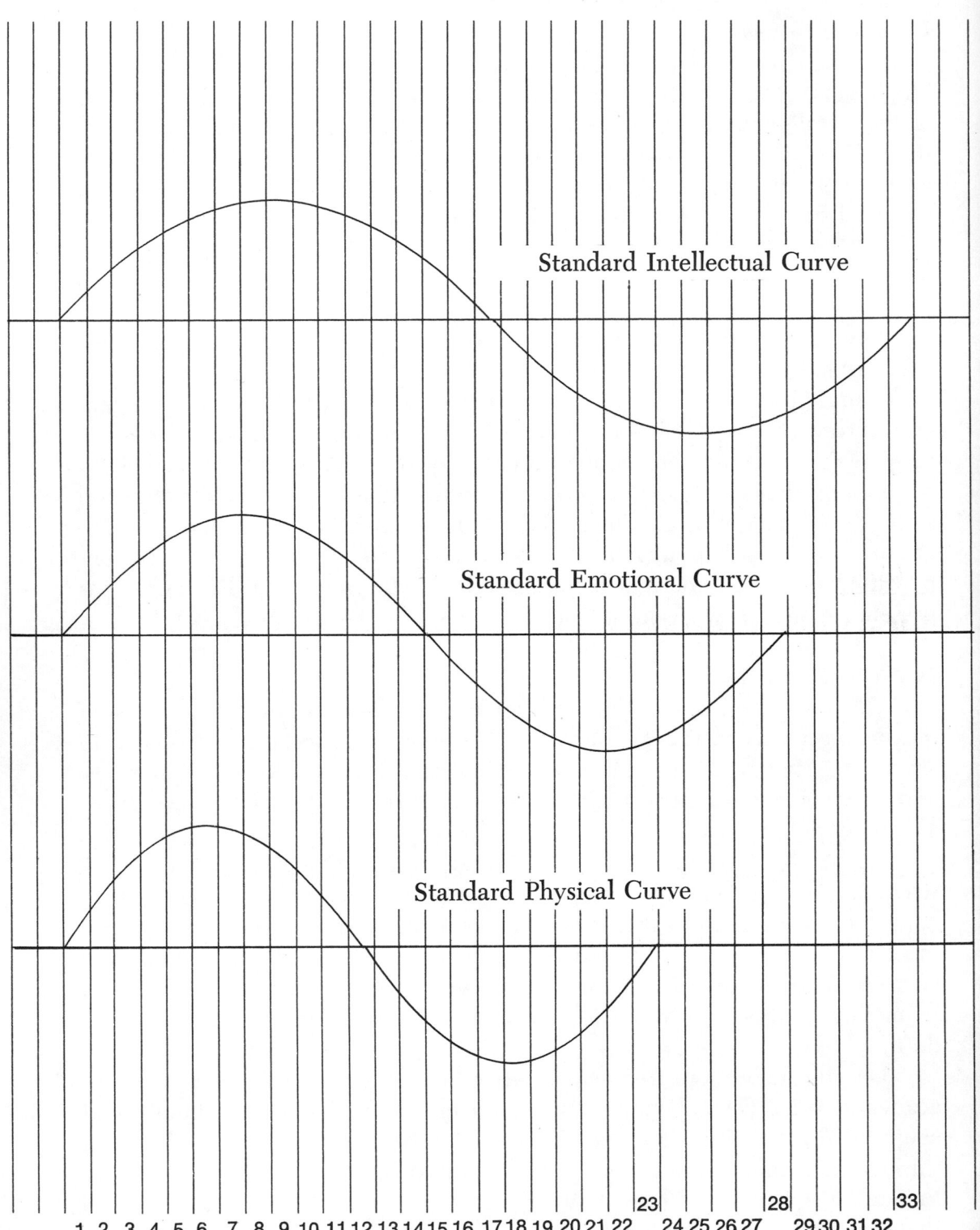

Standard Intellectual Curve

Standard Emotional Curve

Standard Physical Curve

1 2 3 4 5 6 7 8 9 10 11 12 13 14 15 16 17 18 19 20 21 22 23 24 25 26 27 28 29 30 31 32 33

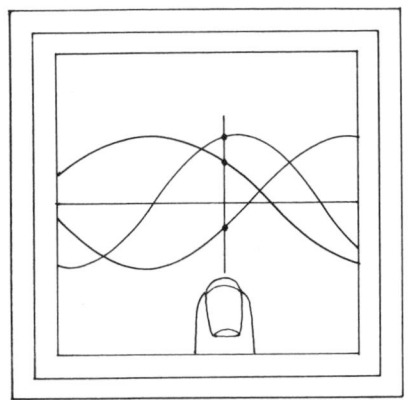

HOW TO PINPOINT
YOUR POSITION
ON EACH CYCLE

THE THREE tables in Appendix A enable you to accurately determine your critical, high, and low days in any month, in any year, for each of the three cycles. One normally seeks such information when looking at the biorhythm record for a month or more, as shown in the example below from the Biorhythm Data Tables.

```
TABLE M83
---------
        CRITICALS    HIGHS    LOWS
        ---------    -----    ----
JAN   16                8      24
FEB    2  18           10      26
MAR    6  22           14      30
APR    8  24           16
MAY   11  27           19       2
JUN   13  29           21       4
JUL   16               24       7
AUG    1  18           26       9
SEP    3  20           28      11
OCT    6  23           31      14
NOV    8  25                   16
DEC   11  28            3      19
```

Although you can easily approximate where you stand on any day of the month in relationship to these given critical, high, and low days, you may want to know your exact position on each of the three cycles on a particular day. Another set of tables is provided in this chapter for this option.

To find your cycle positions for a particular day, it is necessary to pinpoint your positions to a *standard curve* for each cycle. The three reference or standard curves are: the Standard Physical Curve, the Standard Emotional Curve, and the Standard Intellectual Curve, shown on page 30. Three corresponding tables are used with these (see pages 32–35).

Three table look-ups enable you to pinpoint your location for any specific date. Suppose, for example, you want to pinpoint your position on the physical cycle for May 18, 1977; you simply do the following:

A. *Birth Code Tables*—Look up your Physical Birth Code under your birth-date. For this example, assume it is W.

B. *Inquiry Year Tables*—Look under the Physical Birth Code W for the year 1977 to get the Physical Inquiry Year Code; in this example it is P14.

C. *Table 1: The Standard Physical Curve*—Look opposite P14 and under May to get the Index Value; in this case it is 3.

Table 1
Location for Any Given Date on the Standard Physical Curve

The Index Value is opposite the code number (left column) and under the desired month. The position on the curve = sum of the index and day number. If the sum exceeds 23, subtract 23. For codes in parentheses, add 1 to Index Value for any month after February.

TABLE NUMBERS	JAN	FEB	MAR	APR	MAY	JUN	JUL	AUG	SEP	OCT	NOV	DEC
P11 (P51)	14	22	4	12	19	4	11	19	4	11	19	3
P12 (P52)	16	1	6	14	21	6	13	21	6	13	21	5
P13 (P53)	12	20	2	10	17	2	9	17	2	9	17	1
P14 (P54)	21	6	11	19	3	11	18	3	11	18	3	10
P15 (P55)	8	16	21	6	13	21	5	13	21	5	13	20
P16 (P56)	10	18	0	8	15	0	7	15	0	7	15	22
P17 (P57)	0	8	13	21	5	13	20	5	13	20	5	12
P18 (P58)	6	14	19	4	11	19	3	11	19	3	11	18
P19 (P59)	20	5	10	18	2	10	17	2	10	17	2	9
P20 (P60)	11	19	1	9	16	1	8	16	1	8	16	0
P21 (P61)	22	7	12	20	4	12	19	4	12	19	4	11
P22 (P62)	7	15	20	5	12	20	4	12	20	4	12	19
P23 (P63)	15	0	5	13	20	5	12	20	5	12	20	4
P24 (P64)	4	12	17	2	9	17	1	9	17	1	9	16
P25 (P65)	18	3	8	16	0	8	15	0	8	15	0	7
P26 (P66)	5	13	18	3	10	18	2	10	18	2	10	17
P27 (P67)	1	9	14	22	6	14	21	6	14	21	6	13
P28 (P68)	19	4	9	17	1	9	16	1	9	16	1	8
P29 (P69)	9	17	22	7	14	22	6	14	22	6	14	21
P30 (P70)	3	11	16	1	8	16	0	8	16	0	8	15
P31 (P71)	17	2	7	15	22	7	14	22	7	14	22	6
P32 (P72)	2	10	15	0	7	15	22	7	15	22	7	14
P33 (P73)	13	21	3	11	18	3	10	18	3	10	18	2

Now, simply follow the directions in Table 1; your pinpoint location is the Index Value added to the date of the inquiry month. In this example, you add the Index Value to 18 (May 18). Using the assumed values above, $18 + 3 = 21$. By referring to the Standard Physical Curve, it is easy to pinpoint the *21st day* in the cycle—for May 18, 1977.

The same procedure is used to pinpoint locations on the emotional and intellectual cycles, using the respective standard curves and Tables 2 and 3.

Using Table 1, locate the exact position on the Standard Physical Curve for the following example.

The nation was shocked on May 15, 1972, with the news that Governor George Wallace had been shot and gravely wounded while campaigning. The immediate concern of everyone was whether he would be able to survive the assassin's bullet. Did his biorhythmic position on the physical cycle indicate that he might have the physical strength and endurance to recover? His birthdate is August 25, 1919.

A. *Birth Code Tables*—Governor Wallace's Physical Birth Code is T.

B. *Inquiry Year Tables*—Under the Physical Birth Code T for the year 1972, we find the Physical Inquiry Year Code is P69.

C. *Table 1: The Standard Physical Curve*—Look opposite (P69) and under May for the Index Value, which is 15. (According to Table 1 directions, for codes in parentheses, you add 1 to the Index Value for months after February.)

The Index Value 15 is added to the Inquiry Date (May 15) to give the position on the Standard Physical Curve. (According to Table 1 directions, if the sum exceeds 23, you subtract 23.) So in this example: $15 + 15 = 30 - 23 = 7$. *Day 7* on the Standard Physical Curve is the *high* day for that cycle, indicating the time when a person is at his peak physical strength. For Governor Wallace, it certainly may have been a factor in his ability to survive.

Table 2 will help you to find your exact location on the Standard Emotional Curve, and is used in exactly the same way as Table 1.

As an example, suppose you wanted to find out about Carol Channing's exact position on her emotional cycle on two memorable days in her stage career: December 8, 1949, when she received rave notices for her opening performance in *Gentlemen Prefer Blondes*; and January 16, 1964, for her first-night performance in *Hello, Dolly!* Carol Channing was born on January 31, 1923.

A. *Birth Code Tables*—Carol Channing's complete Birth Code is K20CD. In this case, we are only interested in the emotional or sensitivity part of the code, which is 20.

B. *Inquiry Year Tables*—Change the Birth Code 20 to its respective Inquiry Year Codes for the years 1949 and 1964. They are: S24 and S58.

C. *Table 2: The Standard Emotional Curve*—Look opposite S24 and S58 and under the respective months (December and January) of Carol's opening nights to get the Index Values 1 and 21.

Adding the respective dates and Index Values gives the exact positions on the cycle:

For December 8, 1949: $1 + 8 = 9$.   The *9th day* is the high point of the emotional cycle.

For January 16, 1964: $21 + 16 = 37 - 28 = 9$. Again, the high point of the emotional cycle.

Table 2
Location for Any Given Date on the Standard Emotional Curve

The Index Value is opposite the code number (left column) and under the desired month. The position on the curve = sum of the index and day number. If the sum exceeds 28, subtract 28. For codes in parentheses, add 1 to Index Value for any month after February.

TABLE NUMBERS	JAN	FEB	MAR	APR	MAY	JUN	JUL	AUG	SEP	OCT	NOV	DEC
S11 (S51)	23	26	26	1	3	6	8	11	14	16	19	21
S12 (S52)	20	23	23	26	0	3	5	8	11	13	16	18
S13 (S53)	17	20	20	23	25	0	2	5	8	10	13	15
S14 (S54)	14	17	17	20	22	25	27	2	5	7	10	12
S15 (S55)	7	10	10	13	15	18	20	23	26	0	3	5
S16 (S56)	4	7	7	10	12	15	17	20	23	25	0	2
S17 (S57)	1	4	4	7	9	12	14	17	20	22	25	27
S18 (S58)	21	24	24	27	1	4	6	9	12	14	17	19
S19 (S59)	16	19	19	22	24	27	1	4	7	9	12	14
S20 (S60)	10	13	13	16	18	21	23	26	1	3	6	8
S21 (S61)	2	5	5	8	10	13	15	18	21	23	26	0
S22 (S62)	0	3	3	6	8	11	13	16	19	21	24	26
S23 (S63)	13	16	16	19	21	24	26	1	4	6	9	11
S24 (S64)	3	6	6	9	11	14	16	19	22	24	27	1
S25 (S65)	6	9	9	12	14	17	19	22	25	27	2	4
S26 (S66)	25	0	0	3	5	8	10	13	16	18	21	23
S27 (S67)	18	21	21	24	26	1	3	6	9	11	14	16
S28 (S68)	9	12	12	15	17	20	22	25	0	2	5	7
S29 (S69)	27	2	2	5	7	10	12	15	18	20	23	25
S30 (S70)	15	18	18	21	23	26	0	3	6	8	11	13
S31 (S71)	22	25	25	0	2	5	7	10	13	15	18	20
S32 (S72)	11	14	14	17	19	22	24	27	2	4	7	9
S33 (S73)	8	11	11	14	16	19	21	24	27	1	4	6
S34 (S74)	26	1	1	4	6	9	11	14	17	19	22	24
S35 (S75)	5	8	8	11	13	16	18	21	24	26	1	3
S36 (S76)	12	15	15	18	20	23	25	0	3	5	8	10
S37 (S77)	19	22	22	25	27	2	4	7	10	12	15	17
S38 (S78)	24	27	27	2	4	7	9	12	15	17	20	22

Table 3 will help you to find your exact location on the Standard Intellectual Curve, using the same procedure as with Tables 1 and 2. Using Table 3, locate the exact positions on the Standard Intellectual Curve for the following examples:

Two characters out of the Old West who still live in the minds of many Americans are Wild Bill Hickok, born May 27, 1837, and Jesse James, born September 5, 1847. Both lived and died by the gun. Both were killed because of mental lapses that caused them to disregard the caution that usually ruled their lives: one while straightening a picture on a wall (James, April 3, 1882); the other who had his back to his slayer (Hickok, August 2, 1876). What were the exact positions of James and Hickok on their intellectual cycle on their last days?

Table 3
Location for Any Given Date on the Standard Intellectual Curve

The Index Value is opposite the code number (left column) and under the desired month. The position on the curve = sum of the index and day number. If the sum exceeds 33, subtract 33. For codes in parentheses, add 1 to Index Value for any month after February.

TABLE NUMBERS		INDEX VALUES FOR EACH OF TWELVE MONTHS										
	JAN	FEB	MAR	APR	MAY	JUN	JUL	AUG	SEP	OCT	NOV	DEC
M11 (M51)	0	31	26	24	21	19	16	14	12	9	7	4
M12 (M52)	32	30	25	23	20	18	15	13	11	8	6	3
M13 (M53)	31	29	24	22	19	17	14	12	10	7	5	2
M14 (M54)	30	28	23	21	18	16	13	11	9	6	4	1
M15 (M55)	29	27	22	20	17	15	12	10	8	5	3	0
M16 (M56)	28	26	21	19	16	14	11	9	7	4	2	32
M17 (M57)	27	25	20	18	15	13	10	8	6	3	1	31
M18 (M58)	26	24	19	17	14	12	9	7	5	2	0	30
M19 (M59)	25	23	18	16	13	11	8	6	4	1	32	29
M20 (M60)	24	22	17	15	12	10	7	5	3	0	31	28
M21 (M61)	23	21	16	14	11	9	6	4	2	32	30	27
M22 (M62)	22	20	15	13	10	8	5	3	1	31	29	26
M23 (M63)	21	19	14	12	9	7	4	2	0	30	28	25
M24 (M64)	20	18	13	11	8	6	3	1	32	29	27	24
M25 (M65)	19	17	12	10	7	5	2	0	31	28	26	23
M26 (M66)	18	16	11	9	6	4	1	32	30	27	25	22
M27 (M67)	17	15	10	8	5	3	0	31	29	26	24	21
M28 (M68)	16	14	9	7	4	2	32	30	28	25	23	20
M29 (M69)	15	13	8	6	3	1	31	29	27	24	22	19
M30 (M70)	14	12	7	5	2	0	30	28	26	23	21	18
M31 (M71)	13	11	6	4	1	32	29	27	25	22	20	17
M32 (M72)	12	10	5	3	0	31	28	26	24	21	19	16
M33 (M73)	11	9	4	2	32	30	27	25	23	20	18	15
M34 (M74)	10	8	3	1	31	29	26	24	22	19	17	14
M35 (M75)	9	7	2	0	30	28	25	23	21	18	16	13
M36 (M76)	8	6	1	32	29	27	24	22	20	17	15	12
M37 (M77)	7	5	0	31	28	26	23	21	19	16	14	11
M38 (M78)	6	4	32	30	27	25	22	20	18	15	13	10
M39 (M79)	5	3	31	29	26	24	21	19	17	14	12	9
M40 (M80)	4	2	30	28	25	23	20	18	16	13	11	8
M41 (M81)	3	1	29	27	24	22	19	17	15	12	10	7
M42 (M82)	2	0	28	26	23	21	18	16	14	11	9	6
M43 (M83)	1	32	27	25	22	20	17	15	13	10	8	5

A. *Birth Code Tables*—Look up their Intellectual Birth Codes: Hickok—BD; James—AE.

B. *Inquiry Year Tables*—Change the Intellectual Birth Codes to the respective Inquiry Year Codes for the years 1876, 1882: Hickok—M77; James—M14.

C. *Table 3: The Standard Intellectual Curve*—Following table directions, we arrive at the following:

	Index Value	Pinpoint Position on the Intellectual Cycle
Wild Bill Hickok (August 2, 1876)	22	$22 + 2 = 24$
Jesse James (April 3, 1882)	21	$21 + 3 = 24$

Both met death when at the *lowest point* in their intellectual cycles.

Although you can approximate your location on each cycle by simply knowing on what days your criticals, highs, and lows occur, pinpointing will give you a more precise location for a particular day. Pinpointing is especially useful if you want to compare your exact position with that of another person on the same day. This type of biorhythmic information can be valuable for many different kinds of situations: consider the athlete who will be competing against one other person on a specific future date; or the doctor and patient selecting a day for surgery; or someone planning to ask his boss for a raise; or the couple selecting a wedding date. In each of these instances, it would be very helpful to know where the other person stands in one or more of his cycles in relation to your own point in the cycles. Pinpointing is also useful for determining the exact cycle position of a woman who is expecting a child—see Chapter Eight, Predicting the Day of Birth and the Sex of an Unborn Child. In addition, as shown in this chapter, pinpointing can add another dimension of understanding when examining the cycle positions for only one individual—either for yourself, for others you know personally, or for well-known people or historical figures—is desired.

There will be many dates and events for which you will want to know your exact cycle positions. Use the pinpoint technique as one more valuable tool in learning about you and your biorhythms.

CHAPTER SEVEN

DO BIORHYTHMS START AT BIRTH?

FROM ITS pioneering days to the present, the biorhythm theory has been based on the assumption that the three cycles start at birth and proceed continuously throughout life. Most biorhythm researchers have accepted this proposition without question.

It is logical to assume that the fundamental rhythms of life begin with life itself. Birth is probably the most profound and traumatic event the human organism ever experiences. At birth, the infant is suddenly and abruptly cast from a comfortable, warm, dark, nearly silent state out into a world of discomfort or pain, a world of cold, of noise, and of harsh light.

Before birth, the unborn baby's food, water, and oxygen are supplied from the mother's bloodstream via the placenta and umbilical cord. Its waste products, including carbon dioxide, are removed by the same route. At birth, the umbilical cord is severed and all the body functions, which were previously taken care of by the mother, are suddenly the sole responsibility of the infant's own organs. The infant for the first time has to function on its own in many respects, and it follows that this is a reasonable moment to assume the beginning of the life cycles.

In the face of all these facts, one may still question the reliability of birth as the starting point for the biorhythm cycles. Or, having accepted birth as the start, one may still doubt the inflexibility of the starting time. In many cases, not just the date of birth but the hour of birth is crucial. A child born a few minutes after midnight may have a life rhythm that seems to have started on the previous day; one born just before midnight may seem to belong to the following day.

Some argue that life begins at conception, and that the life cycles should begin then, too. Although the exact conception date is extremely difficult to determine, if you prefer to agree with this premise, or wish to test it against your own biorhythm data, you can approximate the conception date by subtracting 280 days (normal gestation time) from your birthdate. Table 5 in Chapter Eight will help you to quickly determine this date. Of course, there is always the rare possibility of knowing a more exact date from records furnished by your family doctor or your parents. In either case, it is possible to readjust your biorhythm cycles to begin at the approximate time of conception if you prefer, and if that seems to synchronize better with your experience.

Other people argue that a later traumatic event may be able to reset the cycles at some point. A very serious physical, emotional, or mental illness may be cause for a readjustment in any one of the cycles. Or a significant change in the daily routines or rhythms, such as a radical change in work schedule, climate, or environment, may make it necessary to slightly readjust the longer monthly rhythms. With this book, it is possible to test the theory and to shift the curves a few days in either direction if, for whatever reason, the cycles are continually off by a certain number of days. Chapter Nine: Adjusting the Cycles to Experience, explains how to do this.

Until enough data is fed into enough computers, we are not going to have scientific confirmation of the idea that everyone's biorhythms start at birth and continue uninterrupted throughout life. To date, we have a considerable amount of information which indicates this is so. Perhaps we will find that the rules do not hold for everyone. Life is incredibly complex, and there are exceptions to every theory.

Many in the scientific community remain skeptical about the constancy of the starting times and period lengths of the cycles. The *National Safety News* has quoted Andrew Ahlgren of the University of Minnesota as saying, "Research has turned up a number of biological cycles . . . but all the cycles vary from individual to individual. They don't start at exactly the same moment. And they can shift over the period of a lifetime as a result of illness. age, or some other factor."

Harold Willis, director of the Biorhythm Clinic in Joplin, Missouri, responded: "The cycles don't necessarily start in every case at the exact moment of birth. They may vary from twelve to twenty-four hours. And there may be additional variation due to psychological or physical trauma. But those variations affect only a minority. I'd say sixty to seventy-five percent of the population conforms to the normal pattern."

Once again the proof remains with you. There is a high probability that your biorhythms began on the day of your birth, as they seem to for most people. But if your experience tells you that your critical day is always two days later than the biorhythm data says it should be, follow your experience. In your case, some event may have reset the cycles. It is easy enough to adjust them and still get the full benefit of the biorhythm tables. The biorhythm theory should always be checked out against the evidence of your own experience; properly interpreted and understood, it should enable you to live a more enjoyable, more productive life.

Bio-Quiz 3

Gamal Abdel Nasser, the Egyptian leader, was born on January 15, 1918. He died of a heart attack on September 28, 1970, after having suffered minor attacks in the weeks preceding his death. Using the biorhythm data, what critical days and what interpretations would you have made for President Nasser in September of 1970?

CHAPTER EIGHT

PREDICTING THE DAY OF BIRTH AND THE SEX OF AN UNBORN CHILD

Predicting the Birthdate

DOCTORS GENERALLY approximate the date of birth of an unborn child by adding 280 days, the normal gestation time, to the supposed date of conception. Subsequent blood tests, the mother's progress through pregnancy, and other factors often lead to revisions of this initial date. Even after revision, the date is rarely accurate. Unless a woman has been recording her temperature and pinpointing the time of ovulation, the date of conception is often a kind of average date at best. Knowledge of biorhythms adds another piece of evidence that can help prospective parents figure out when their baby will be born. So many factors are involved that no one method, including this one, is ever 100 percent correct, but it can be helpful.

The method described here uses the physical and emotional biorhythm cycles in conjunction with the 280-day advancement technique to try to predict the day of birth more firmly. The method is based on the observation that birth often occurs on or very close to the *physical* and/or the *emotional* critical days of the mother. Many biorhythm users can confirm that birth frequently takes place on a critical day. Not a surprising fact: if the mother's system is wobbly and unbalanced, if she is more subject to stress than usual, and if her body is already gearing up for labor, then it would seem natural for labor to begin at such a time.

Critical days seem to be extremely important in our life experience: a baby is likely to be born on the mother's critical day; the day of birth itself is a triple-critical day for the infant; the day of death is also likely to be a critical day in the life cycle.

The biorhythm method used to predict the date of birth consists of three steps:

Step 1. Determine the approximate conception date.

Step 2. Use Table 4 and the conception date of Step 1 to determine the "tentative birthdate."

Step 3. Using the mother's birthdate, along with the tentative birthdate of Step 2 as the Inquiry Date, find the mother's closest critical day to the tentative birthdate. This closest critical day is the predicted birthdate.

Step 1 can be found in two ways: a more exact date can be furnished by the doctor or parents-to-be; or, if this is not possible, you can approximate the date by subtracting 280 days from the estimated birthdate. If this latter method is used, Table 5 (later in this chapter) can be used to quickly determine the conception date from an estimated birthdate.

For the first example, let's assume the conception date is known from information furnished by the doctor.

Step 1. The conception date is March 30, 1976.

Step 2. Using Table 4, look opposite March 27–March 31—the tentative birthdate is January 4, 1977.

Step 3. Use the mother's birthdate (October 15, 1952) and the tentative birthdate of Step 2 (January 4, 1977) as the Inquiry Date to find the mother's nearest critical day.

	Physical	*Emotional*
Birth Codes for October 15, 1952	X	38
Inquiry Year Codes for 1977	P13	S38
Biorhythm Data critical days for January, 1977	12, 23	5 , 19

(42)

The critical day of the mother that is closest to the tentative birthdate is January 5. Therefore, the predicted birthdate is January 5, 1977.

Table 4—Birthdates Based on Conception Dates

IF CONCEPTION DATE IS	THEN MONTH OF BIRTH IS	THE DAY OF BIRTH WILL EQUAL THE	AND THE YEAR OF BIRTH WILL EQUAL THE
JAN 1 - JAN 24	OCT	CONCEPTION DAY + 7	CONCEPTION YEAR
JAN 25 - JAN 31	NOV	CONCEPTION DAY - 24	CONCEPTION YEAR
FEB 1 - FEB 23	NOV	CONCEPTION DAY + 7	CONCEPTION YEAR
FEB 24 - FEB 28	DEC	CONCEPTION DAY - 23	CONCEPTION YEAR
MAR 1 - MAR 26	DEC	CONCEPTION DAY + 5	CONCEPTION YEAR
MAR 27 - MAR 31	JAN	CONCEPTION DAY - 26	CONCEPTION YEAR + 1
APR 1 - APR 26	JAN	CONCEPTION DAY + 5	CONCEPTION YEAR + 1
APR 27 - APR 30	FEB	CONCEPTION DAY - 26	CONCEPTION YEAR + 1
MAY 1 - MAY 24	FEB	CONCEPTION DAY + 4	CONCEPTION YEAR + 1
MAY 25 - MAY 31	MAR	CONCEPTION DAY - 24	CONCEPTION YEAR + 1
JUN 1 - JUN 24	MAR	CONCEPTION DAY + 7	CONCEPTION YEAR + 1
JUN 25 - JUN 30	APR	CONCEPTION DAY - 24	CONCEPTION YEAR + 1
JUL 1 - JUL 24	APR	CONCEPTION DAY + 6	CONCEPTION YEAR + 1
JUL 25 - JUL 31	MAY	CONCEPTION DAY - 24	CONCEPTION YEAR + 1
AUG 1 - AUG 24	MAY	CONCEPTION DAY + 7	CONCEPTION YEAR + 1
AUG 25 - AUG 31	JUN	CONCEPTION DAY - 24	CONCEPTION YEAR + 1
SEP 1 - SEP 23	JUN	CONCEPTION DAY + 7	CONCEPTION YEAR + 1
SEP 24 - SEP 30	JUL	CONCEPTION DAY - 23	CONCEPTION YEAR + 1
OCT 1 - OCT 24	JUL	CONCEPTION DAY + 7	CONCEPTION YEAR + 1
OCT 25 - OCT 31	AUG	CONCEPTION DAY - 24	CONCEPTION YEAR + 1
NOV 1 - NOV 24	AUG	CONCEPTION DAY + 7	CONCEPTION YEAR + 1
NOV 25 - NOV 30	SEP	CONCEPTION DAY - 24	CONCEPTION YEAR + 1
DEC 1 - DEC 24	SEP	CONCEPTION DAY + 6	CONCEPTION YEAR + 1
DEC 25 - DEC 31	OCT	CONCEPTION DAY - 24	CONCEPTION YEAR + 1

Let's try another example, using an actual case. Robert Kennedy was born on November 20, 1925. His mother, Mrs. Rose Kennedy, was born on July 22, 1890. Assuming the conception date as February 15, 1925, predict the birthdate of Robert Kennedy.

Step 1. The conception date is February 15, 1925.

Step 2. Using Table 4, look opposite February 1–February 23—the tentative birthdate is November 22, 1925.

Step 3. Use the mother's birthdate (July 22, 1890) and the tentative birthdate (November 22, 1925) as the Inquiry Date to find the mother's nearest critical day.

	Physical	Emotional
Birth Codes for July 22, 1890	G	27
Inquiry Year Codes for 1925	P17	S28
Biorhythm Data critical days for November, 1925	7, 19 , 30	10, 24

/

(43)

The critical day of the mother that is closest to the tentative birthdate is November 19. Therefore, the predicted birthdate using the biorhythm method would have been November 19, 1925—just twenty-four hours before the actual date of birth.

But let us assume that an approximate conception date is not available either from the doctor or the parents. The next best procedure is to inquire when the baby is expected. This information is usually easier to determine and more readily available than the conception date anyway. Knowing the expected date of birth, one can eliminate Steps 1 and 2 and use the predicted birthdate as the Inquiry Date, along with the mother's birthdate, to find the nearest critical day in the mother's physical or emotional cycle.

As an example, Mary Whyte, born May 24, 1954, announces she is "expecting" on June 9, 1977. What is your prediction for the birthdate of her unborn child, using the biorhythm method?

	Physical	Emotional
Mary's Birth Codes	H	34
Inquiry Year Codes for 1977	P27	S34
Biorhythm Data critical days for June, 1977	10 , 21	6, 20

The predicted birthdate according to the biorhythm method is June 10, 1977.

Predicting the Sex

If the mother's physical and emotional cycles can determine the time of birth of the unborn child, it is logical to wonder if her biorhythms also influence her baby's sex. The limited amount of evidence available suggests that it may.

The child's sex is determined at conception and depends entirely upon whether an X or Y sperm fertilizes the egg. Sex is thus governed by the male factor rather than by the female. However, there is evidence that some women are predisposed to bear more children of one gender than the other; there may be components in the female reproductive tract that favor one type of sperm over the other. If these components change over a period of time, if they are influenced by the woman's physical or emotional condition, then her biorhythms would indeed help to determine the sex of her unborn child.

Medical opinion holds that a boy is more likely to be born if intercourse takes place when the egg is in position to be fertilized, but a girl is more likely if intercourse occurs before the egg is ready. This is because the male sperm moves faster but does not live as long. The slower moving female sperm can live until the egg is in position. If this is actually the sole determinant, the biorhythmic influence would not be strong.

But it is also thought that changes in the alkaline or acid content of the blood influence the sex of the child: if a woman's blood is more alkaline, she is more likely to have boys. Since emotional stress can cause changes in the alkaline-acid content of the blood, the emotional cycle could be important to this theory. Biorhythm researchers claim that the high point in the physical cycle is said to favor a condition of alkalinity in the blood, the high point in the emotional rhythm to favor acidity.

Thus, the biorhythm theory states that when the physical biorhythm is high in the female at the time of conception, the egg cell is more likely to accept the male sperm cell and produce a boy. When the emotional biorhythm is high in the female at the time of conception, the egg cell is more likely to accept the female sperm cell and produce a girl.

The biorhythm theory can be used to try to predict the sex of an unborn child. Its accuracy depends partly upon establishing an accurate conception date, difficult at best, so that the position of the mother's cycles at the moment of conception can be ascertained. In any event, there is a 50 percent chance of being right, so it ought to be worth trying.

The theory can also be used in trying to plan the sex of an unborn child. However, even doctors haven't managed perfect records in this area. If the sex of a child is very important, it would probably be best to consult medical experts and examine other techniques in addition to using the biorhythm method.

In the following examples, well-known people and their birthdates are used to illustrate and test the biorhythm method of predicting the sex of a child.

A baby was born on August 3, 1922, to Mamie Eisenhower. Assuming a normal gestation period of 280 days, the approximate conception date was October 27, 1921. Mamie Doud Eisenhower was born on November 14, 1896. Using Mamie's Birth Codes and the conception year 1921 as the Inquiry Year, we find the following:

	Physical	Emotional
Birth Codes for November 14, 1896	B	15
Inquiry Year Codes for 1921	P24	S18

Now, find Mamie's relative positions on the standard physical and emotional biorhythm cycles for October 27, 1921. These pinpoint positions can be located quickly in Tables 1 and 2 in Chapter Six, How to Pinpoint Your Position on Each Cycle:

Day in Cycle

Position on the standard physical cycle for October 27, 1921 $5 \ (28 - 23 = 5)$

Position on the standard emotional cycle for October 27, 1921 $3 \ (31 - 28 = 3)$

The physical position is practically at the high point of the cycle. The emotional position is lower. According to the biorhythm theory, the baby should have been a boy. It was—the baby was John Sheldon Doud Eisenhower.

The following table makes it quite easy to determine an approximate conception date from any expected birthdate. Having determined the conception date, one can then easily find the relative positions of the mother on her physical and emotional cycles on that date to predict the sex of the unborn child.

Table 5— Conception Dates Based on Expected or Actual Birthdates

TIME OF BIRTH	CONCEPTION MONTH	CONCEPTION DAY EQUALS	CONCEPTION YEAR EQUALS
JAN 1 - JAN 5	MAR	BIRTH DAY + 26	BIRTH YEAR - 1
JAN 6 - JAN 31	APR	BIRTH DAY - 5	BIRTH YEAR - 1
FEB 1 - FEB 4	APR	BIRTH DAY + 26	BIRTH YEAR - 1
FEB 5 - FEB 28	MAY	BIRTH DAY - 4	BIRTH YEAR - 1
MAR 1 - MAR 7	MAY	BIRTH DAY + 24	BIRTH YEAR - 1
MAR 8 - MAR 31	JUN	BIRTH DAY - 7	BIRTH YEAR - 1
APR 1 - APR 6	JUN	BIRTH DAY + 24	BIRTH YEAR - 1
APR 7 - APR 30	JUL	BIRTH DAY - 6	BIRTH YEAR - 1
MAY 1 - MAY 7	JUL	BIRTH DAY + 24	BIRTH YEAR - 1
MAY 8 - MAY 31	AUG	BIRTH DAY - 7	BIRTH YEAR - 1
JUN 1 - JUN 7	AUG	BIRTH DAY + 24	BIRTH YEAR - 1
JUN 8 - JUN 30	SEP	BIRTH DAY - 7	BIRTH YEAR - 1
JUL 1 - JUL 7	SEP	BIRTH DAY + 23	BIRTH YEAR - 1
JUL 8 - JUL 31	OCT	BIRTH DAY - 7	BIRTH YEAR - 1
AUG 1 - AUG 7	OCT	BIRTH DAY + 24	BIRTH YEAR - 1
AUG 8 - AUG 31	NOV	BIRTH DAY - 7	BIRTH YEAR - 1
SEP 1 - SEP 6	NOV	BIRTH DAY + 24	BIRTH YEAR - 1
SEP 7 - SEP 30	DEC	BIRTH DAY - 6	BIRTH YEAR - 1
OCT 1 - OCT 7	DEC	BIRTH DAY + 24	BIRTH YEAR - 1
OCT 8 - OCT 31	JAN	BIRTH DAY - 7	BIRTH YEAR
NOV 1 - NOV 7	JAN	BIRTH DAY + 24	BIRTH YEAR
NOV 8 - NOV 30	FEB	BIRTH DAY - 7	BIRTH YEAR
DEC 1 - DEC 5	FEB	BIRTH DAY + 23	BIRTH YEAR
DEC 6 - DEC 31	MAR	BIRTH DAY - 5	BIRTH YEAR

To use another actual example:

Mrs. Lyndon B. Johnson, born December 22, 1912, gave birth to Lynda Bird Johnson (Mrs. Charles Robb) on March 19, 1944.

Step 1. Use Lynda's birthdate to determine the approximate conception date in Table 5. This would be June 12, 1943.

Step 2. Using her mother's Birth Codes and the conception year as the Inquiry Year, we find the following:

	Physical	Emotional
Birth Codes for December 22, 1912	G	25
Inquiry Year Codes for 1943	P28	S37

Step 3. Now, find the relative positions of Mrs. Johnson on the standard physical and emotional biorhythm cycles for June 12, 1943. (Use Tables 1 and 2 in Chapter Six.)

	Day in Cycle
Position on the standard physical cycle for June 12, 1943	21
Position on the standard emotional cycle for June 12, 1943	14

The position on the emotional cycle (day 14) is slightly higher than the position on the physical cycle (day 21). The indication is a slight trend toward a girl.

The above procedure is only approximate in its predictions. One must remember it is based upon the conception date being 280 days prior to birth. In many cases it is extremely difficult to pinpoint conception, and there are many factors that may decrease or increase the 280-day gestation period.

Bio-Quiz 4

Know a mother-to-be?

Her birthdate................................

Expected date of arrival................................

Now use the Bio Tables and procedures of this last chapter to predict the baby's birthdate as................................

CHAPTER NINE

ADJUSTING THE CYCLES TO EXPERIENCE

ONE OF the more versatile features of this book is that it gives you the opportunity to adjust your biorhythm cycles forward or backward a number of days. In addition, it is possible to make a more radical adjustment to the cycles by using the date of conception rather than birth as the date from which the cycles start. For those who prefer to use the conception date, the process is quite easy and straightforward. Suppose, for example, that John Sills, born August 17, 1943, prefers to start his biorhythm cycles from the conception date. First, he establishes an approximate conception date by subtracting 280 days (normal gestation time) from his birthdate. Table 5 in Chapter Eight makes this very easy to do. The conception date in this case is November 10, 1942. Now, all John need do is use November 10, 1942, as his biorhythmic birthdate.

However, instead of a new car, you may be only interested in shifting gears—for example, to shift your physical cycle a couple of days forward or backward. This can be done by changing your Birth Code for the particular cycle, which will change the days on which your critical, high, and low days occur. Table 6, at the end of this chapter, will help you to make these minor adjustments.

There may be a number of reasons for shifting the cycles. For example, your birthdate covers a full twenty-four-hour period. If you were born at five minutes past midnight on May 20, it is just possible that your biorhythm cycles began closer to May 19 than to your official birthdate. Or, if you were born at five minutes before midnight on the 20th, you just might belong on the biorhythm calendar for the following day. If you know your exact time of birth, it will be easy to figure out if this may be causing a discrepancy in your cycles. If you notice that you are always a day or so ahead or behind the tables on your high, low, or critical days, the hour of your birth could be the reason.

Work habits may affect the times we have feelings of being "up" or "down." Daytime workers generally conform to sixteen or more hours of day-time activities to around eight hours of sleep at night. Night workers have a different schedule. Many hurry home in the early hours of morning and spend only four to six hours in sleep. Several research studies indicate that varying periods of light and darkness affect the rhythmic cycles of some plant and animal life. It is reasonable to suspect there are also some effects upon man and that some adjustment of the biorhythm cycles might have to be made to compensate for various life styles and work habits. Similarly, radical changes in environment or climate may also call for some cycle adjustment.

Those who travel frequently may want to shift the cycles one day when crossing a date line—not so rare an occurrence in this era of jet travel. In this case, one could shift his cycles a day in either direction to correspond with the new time schedule.

Although no one has seriously investigated whether a major trauma, such as an automobile accident or near-fatal illness, might cause the cycles to be reset, it would seem possible as it has been proven to occur with other bio-logical rhythms. Many claim that a serious trauma can reset the cycles, and it seems logical that a serious physical, emotional, or mental illness may call for a readjustment in one or more of the cycles.

There may be other factors that can alter your schedule that we do not yet know about. Research on biorhythms has not been nearly extensive enough to give us all the answers. Whatever the reason, if experience tells you that your biorhythm tables are consistently out of step with your life by a day or so, it is possible to shift your cycles.

As an example, let's assume that Claire Walsh, born September 17, 1942, has a case of "blue Mondays" every other Monday. (Remember that all emotional or sensitivity critical days fall on the same day of the week.) Although Claire knows from personal observation that she's down on alternate Mondays, her biorhythm record indicates that those periods of minor depression are supposed to fall on Tuesdays. If Claire's critical days are a day behind, her entire cycle must be one day off. It would be very useful to shift the curves back a day. This is easily done with Table 6, on page 53.

First, Claire finds her Birth Codes based on her birthdate, September 17, 1942. The complete code is R28CP. To switch her Emotional Code, she uses Table 10, opposite code 28 and under the column headed "1 Day." Since she is shifting *behind* one day, from Tuesday to Monday, she uses the code shown in parentheses. She finds

Opposite the original Emotional Code 28   1 Day
                                             (20)

So in the future, Claire will use 20 instead of 28 for her Emotional Code. Her "blue Mondays" are now synchronous with her biorhythmic readings.

Although the lengths of the three biorhythm cycles are generally thought to be unvarying, it is possible that a few people will find one of their cycles consistently a day longer or shorter than the norm. We know, for example, that some women have menstrual cycles longer or shorter than the norm, but we don't yet have such ample documentation in the field of biorhythms. The standard opinion is that the length of the cycles is inflexible, but for those whose experience contradicts this assumption, this chapter also offers a means of adjusting the cycle lengths to make them slightly longer or shorter than the norm.

As an example, Nancy Fawcett, born March 11, 1951, decides to change the length of her emotional biorhythm cycle from 28 days to 26 days. To shorten the cycle by two days, everything proceeds as before up to the Biorhythm Data Tables—then you have to do a little arithmetic on your own. This is because you are now changing the cycle *length* rather than merely shifting the dates when the critical, high, and low days occur. First, Nancy finds her complete Birth Code—W12CH. Then she looks at Table 10 opposite her Emotional Code 12 and under the column headed "2 Days." Since she is shifting the cycle *behind* two days, she uses the code shown in parentheses. This adjusts her code from 12 to 31.

Original Emotional Code   *New Emotional Code (2 Days)*
         12   (31)

Nancy now uses Emotional Code 31 instead of 12. Suppose she wants to use this new code for the Inquiry Year of 1976. Using Inquiry Year 1976, under Code 31 you will find Biorhythm Data Code S52. However, Code S52 is only valid now for the *first critical day* in January. The other readings must be found by successively adding one-half the new period length (26 divided by 2 = 13 days). The first three months of the *original* and *newly calculated* S52 tables are shown below:

| | *Original S52* | | | | *New S52* | | |
	Criticals	*Highs*	*Lows*		*Criticals*	*Highs*	*Lows*
Jan.	9 23	16	2 30	Jan.	9 22	15	3 28
Feb.	6 20	13	27	Feb.	4 17	10	23
March	5 19	12	26	March	1 14 27	7	20

Here, criticals occur every 14 days (28 divided by 2). Highs, lows are found by adding 7 (14 divided by 2) to each successive critical.

Thus, 9 + 14 = 23rd of January;
23 + 14 = 37 − 31 = 6th of February;
6 + 14 = 20th of February;
20 + 14 = 34 − 29 = 5th of March, etc.

Here, criticals occur every 13 days (26 divided by 2). Highs, lows are found by adding 6 (13 divided by 2) to each successive critical.

Thus, 9 + 13 = 22nd of January;
22 + 13 = 35 − 31 = 4th of February;
4 + 13 = 17th of February;
17 + 13 = 30 − 29 = 1st of March, etc.

Although the cycles should not fluctuate or deviate from the norm by more than one to three days, it is possible to use Table 10 to shift the cycles either forward or backward by more than three days, if necessary, by using a combination of codes. To shift the cycle four days, for example, first use the reading for 1 Day, then 3 Days, or use the 2-Day reading twice. But remember, if adjusting the cycle *length*, you have to add half the new cycle length to the first date in January to determine your other new critical, high, and low days, as explained in the example above.

Table 6—Adjusting the Cycles
For Persons with High or Low Days Seemingly off by 1 to 3 Days.
Ahead Codes Come First, Behind Day Codes Are in Parentheses.

OLD PHYS. CODES	NEW PHYSICAL CODES FOR HIGHS, LOWS OFF BY			OLD SENS. CODES	NEW SENSITIVITY CODES FOR HIGHS, LOWS OFF BY		
	1 DAY	2 DAYS	3 DAYS		1 DAY	2 DAYS	3 DAYS
A	Z (N)	C (B)	K (X)	11	31 (38)	18 (26)	12 (34)
B	N (X)	A (R)	Z (U)	12	37 (18)	27 (31)	13 (11)
C	K (Z)	F (A)	V (N)	13	19 (27)	3 (37)	14 (12)
D	J (L)	U (G)	R (T)	14	23 (30)	36 (19)	32 (13)
E	M (V)	H (F)	S (K)	15	25 (33)	35 (28)	16 (20)
F	V (K)	E (G)	M (Z)	16	24 (35)	21 (25)	17 (15)
G	L (T)	D (Y)	J (W)	17	22 (21)	29 (24)	34 (16)
H	S (M)	P (E)	W (V)	18	12 (31)	37 (11)	27 (38)
J	U (D)	R (L)	X (G)	19	30 (13)	14 (27)	23 (37)
K	F (C)	V (Z)	E (A)	20	28 (32)	33 (36)	15 (23)
L	D (G)	J (T)	U (Y)	21	17 (24)	22 (16)	29 (35)
M	H (E)	S (V)	P (F)	22	29 (17)	34 (21)	26 (24)
N	A (B)	Z (X)	C (R)	23	36 (14)	32 (30)	20 (19)
P	W (S)	Y (H)	T (M)	24	21 (16)	17 (35)	22 (25)
R	X (U)	B (J)	N (D)	25	35 (15)	16 (33)	24 (28)
S	P (H)	W (M)	Y (E)	26	38 (34)	11 (29)	31 (22)
T	G (Y)	L (W)	D (P)	27	13 (37)	19 (12)	30 (18)
U	R (J)	X (D)	B (L)	28	33 (20)	15 (32)	25 (36)
V	E (F)	M (K)	H (C)	29	34 (22)	26 (17)	38 (21)
W	Y (P)	T (S)	G (H)	30	14 (19)	23 (13)	36 (27)
X	B (R)	N (U)	A (J)	31	18 (11)	12 (38)	37 (26)
Y	I (W)	G (P)	L (S)	32	20 (36)	28 (23)	33 (14)
Z	C (A)	K (N)	F (B)	33	15 (28)	25 (20)	35 (32)
				34	26 (29)	38 (22)	11 (17)
				35	16 (25)	24 (15)	21 (33)
				36	32 (23)	20 (14)	28 (30)
				37	27 (12)	13 (18)	19 (31)
				38	11 (26)	31 (34)	18 (29)

INTELLECTUAL ADJUSTMENT CODES

OLD INTL. CODES	NEW INTELLECTUAL CODES FOR HIGHS, LOWS OFF BY			OLD INTL. CODES	NEW INTELLECTUAL CODES FOR HIGHS, LOWS OFF BY		
	1 DAY	2 DAYS	3 DAYS		1 DAY	2 DAYS	3 DAYS
AB	AC (CS)	AD (CR)	AE (CP)	BJ	BK (BH)	BL (BG)	BM (BF)
AC	AD (AB)	AE (CS)	AF (CR)	BK	BL (BJ)	BM (BH)	BP (BG)
AD	AE (AC)	AF (AB)	AG (CS)	BL	BM (BK)	BP (BJ)	CD (BH)
AE	AF (AD)	AG (AC)	AH (AB)	BM	BP (BL)	CD (BK)	CE (BJ)
AF	AG (AE)	AH (AD)	AJ (AC)	BP	CD (BM)	CE (BL)	CF (BK)
AG	AH (AF)	AJ (AE)	AK (AD)	CD	CE (BP)	CF (BM)	CH (BL)
AH	AJ (AG)	AK (AF)	AL (AE)	CE	CF (CD)	CH (BP)	CJ (BM)
AJ	AK (AH)	AL (AG)	AM (AF)	CF	CH (CE)	CJ (CD)	CK (BP)
AK	AL (AJ)	AM (AH)	BC (AG)	CH	CJ (CF)	CK (CE)	CL (CD)
AL	AM (AK)	BC (AJ)	BD (AH)	CJ	CK (CH)	CL (CF)	CM (CE)
AM	BC (AL)	BD (AK)	BE (AJ)	CK	CL (CJ)	CM (CH)	CP (CF)
BC	BD (AM)	BE (AL)	BF (AK)	CL	CM (CK)	CP (CJ)	CR (CH)
BD	BE (BC)	BF (AM)	BG (AL)	CM	CP (CL)	CR (CK)	CS (CJ)
BE	BF (BD)	BG (BC)	BH (AM)	CP	CR (CM)	CS (CL)	AB (CK)
BF	BG (BE)	BH (BD)	BJ (BC)	CR	CS (CP)	AB (CM)	AC (CL)
BG	BH (BF)	BJ (BE)	BK (BD)	CS	AB (CR)	AC (CP)	AD (CM)
BH	BJ (BG)	BK (BF)	BL (BE)				

Bio-Quiz 5

Assume that Joe Namath, born May 31, 1943, opens his 1976 football season on September 12. If a high emotional cycle is indicative of outstanding performance, and if the games occur every 7 days, will Joe be "up" for games: 1, 3, 5 . . ., etc., or for games 2, 4, 6 . . ., etc.?

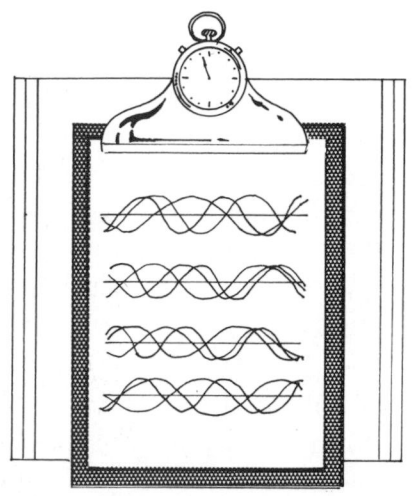

CHAPTER TEN

THE WORLD OF SPORTS
AND BIORHYTHMS

BIORHYTHM CYCLES doubtless affect all areas of achievement. The world of sports, with its great athletic feats and record-breaking performances, is a fertile field for biorhythm application. In this particular area, however, it seems more advantageous to observe *patterns* of behavior rather than isolated results. There are several reasons for this. Some athletes seem to perform best on high days, others on critical days, a few on combinations of lows and criticals. The most favorable combinations will differ from sport to sport as well as individual to individual. Tennis players may lean more heavily on intellect in their game than boxers do, and team players may depend more on emotional and compatibility factors than individual figure skaters would. In addition, many sporting events extend over a period of several consecutive days; for example, golf tournaments, track and field competitions, and Olympic events. In such instances, highs, lows, or criticals for a single day are often less meaningful than a pattern of performance.

It is also difficult to judge the relative importance of physical, emotional, or intellectual attributes in any particular athletic event. Which is more meaningful to Richard Petty if he wins an auto race? Is his victory due more to his physical state than to his mental level? Or is his emotional state the most important ingredient? Emotions play a large role in athletic performance. Football coaches can either tell of the joy or regret of seeing a team emotionally "charged up" for a big game, depending on how their particular team functions while at a high emotional level. Boxers, too, depend on more than physical condition. Professional fighters are usually fairly well matched physically—in size, weight, and training. To a certain extent, their unmatched mental and emotional factors must be important to the outcome of a fight.

Surprisingly, critical and low days are not equivalent to defeat; in some cases they even seem to point the way to victory. An athlete with great natural physical strength may perform better when his emotional or intellectual cycle is at a higher point than his physical curve. A wobbly effect—often observed on critical days—may indicate a relaxation of rigid or concentrated effort, so that the athlete is physically more relaxed, and perhaps his judgment is more finely tuned, so he will not overreach himself. A good example of this is the performance pattern of Jack Nicklaus, described later in this chapter.

Is this apparent lack of definitive interpretation reason for abandoning the use of biorhythm theory in athletics? On the contrary, it is still a very valuable tool both in predicting the athletic performance as well as serving as a training barometer for the athlete himself. This is because a *trend* or *similarity* of experience is often observable, and the athlete can take advantage of this to train for an event and to better control his performance at the time the event occurs. Some of the examples that follow indicate this procedure.

Starting with golf, let's examine the record of the professional circuit for the year 1970. Two well-known golfers of that year were Billy Casper and Jack Nicklaus. During that circuit year, Billy Casper won four tournaments. The tournament dates and his biorhythm record are shown below.

Billy Casper—*1970 Golf Circuit Wins*
Birthdate: June 24, 1931   Birth Codes: P38BL

Date of Win	Place	Physical	Emotional	Intellectual
Jan. 8–11	Los Angeles Open	high 12th	low 7th	low 12th
Apr. 9–13	Augusta, Ga. Masters	critical 8th high 14th	critical 8th high 15th	critical 13th
July 16–19	Phila., Pa. Classic	high 15th critical 20th	critical 15th	high 13th critical 21st
Aug. 20–23	Sutton, Mass. Classic	critical 24th	low 19th	critical 23rd

The reader may wish to check this record or to do a similar analysis of his own golf game. Table look-ups are quite easy. Follow the A, B, C steps below, using the three tables in the back of the book.

	Intellectual	Physical	Emotional
A. *Birth Code Tables*—Look up Birth Codes for Billy Casper, born June 24, 1931.	P	38	BL
B. *Inquiry Year Tables*—For the year 1970, look under the Birth Codes P, 38, BL. This will give you the Inquiry Year Code for each cycle.	P25	S30	M31

C. *Biorhythm Data Tables*—Look under the Inquiry Year Code for each cycle for the months of January, April, July, and August. This will give you the complete biorhythm data for the period covering the 1970 golf tournaments as shown above.

The trend in Billy Casper's biorhythm record during 1970 seems to point to outstanding performance when his physical cycle was high or near a critical, with near-criticals in one or both of the other cycles. With one exception, during the Los Angeles Open, the nearness to critical days in at least two cycles was evident.

During the same pro-circuit year of 1970, Jack Nicklaus had the following biorhythm record:

Jack Nicklaus—*1970 Golf Circuit Wins*
Birthdate: January 21, 1940   Birth Codes: L29BJ

Date of Win	Place	Physical	Emotional	Intellectual
Jan. 22–25	Pebble Beach, Ca. Bing Crosby Open	critical 22nd low 28th	critical 25th	high 27th
Apr. 30–May 3	Dallas, Texas Byron Nelson Classic	low 30th	low 26th critical May 3	critical 28th high May 6
July 8–12	St. Andrews, Scotland British Open	low 8th	critical 12th	high 11th
July 23–26	Ligonier, Pa. PGA Four-Ball	critical 25th	critical 26th	low 27th
Sept. 12–13	Akron, Ohio World Series of Golf	low 15th	low 13th	high 15th

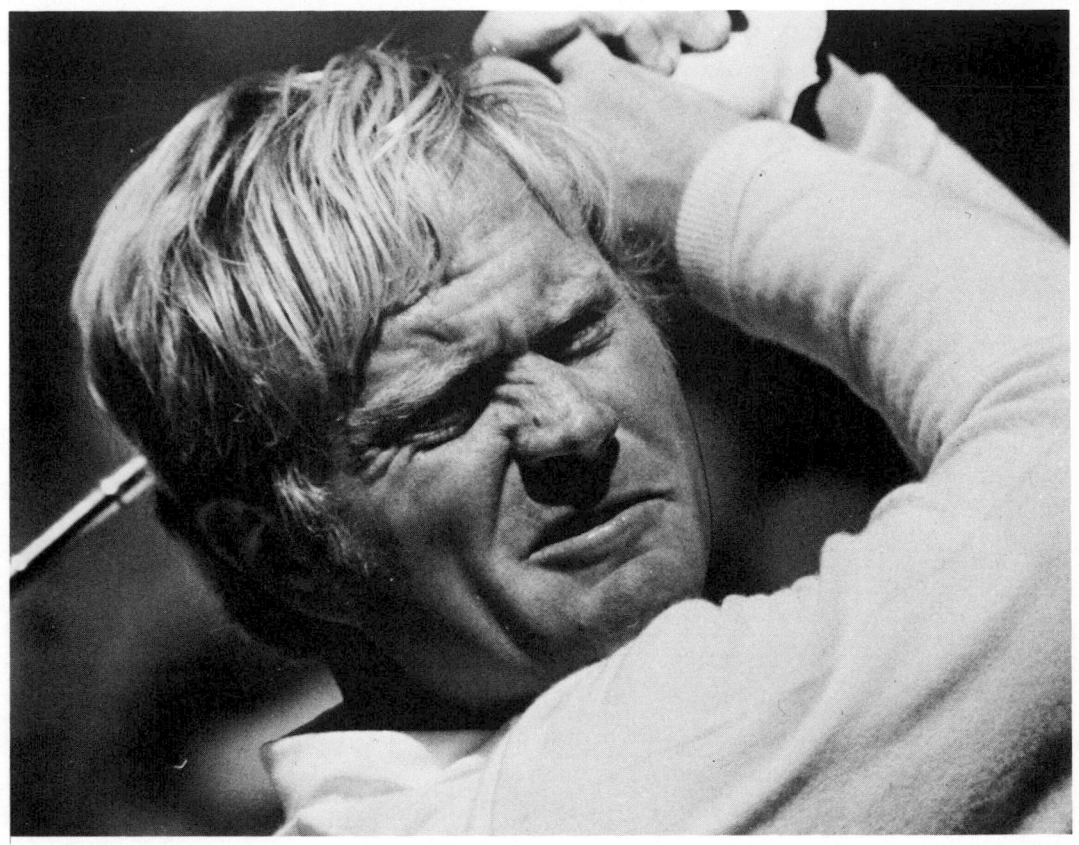

Jack Nicklaus at the Doral Open, 1971

Nearly every time Nicklaus won in 1970, his physical cycle was low or critical, his emotional cycle critical, and his mental or intellectual cycle high. The same pattern seems to exist in other years. Nicklaus is a very strong golfer physically. His biorhythm records suggest that he plays best when his physical strength is somewhat curbed, his mental energies high, and his emotional status on edge.

The Life Cycles Research Foundation in Peoria, Illinois, made a study of golfers who won the PGA co-sponsored tournaments from 1971 to 1975. In each of those years, 61 to 75 percent of the winners were in the high or "ascending" days in their cycles, only 9 to 18 percent in the low or "descending" days in their cycles, and the remainder in combination of mixed or critical days. In 1975 alone, 75 percent of the winners won during high or rising cycles, 9.1 percent during low or falling cycles. It looks as though Nicklaus is the exception—golfers can win championships at any stage in their biorhythms, but a high cycle is usually an enormous boost.

Jack Dempsey, the famous boxer, fought many of his greatest matches on a combination of critical and high days.

Jack Dempsey—*Boxing Fights 1919–1923*
Birthdate: June 24, 1895   Birth Codes: U36AG

Date of Fight	Opponent	Physical	Emotional	Intellectual
July 4, 1919	Jess Willard (KO 4th)	critical 3rd	high 1st	critical 7th
Sept. 6, 1920	Billy Miske (KO 3rd)	high 7th	low 6th	critical 8th
Dec. 14, 1920	Bill Brennan (KO 12th)	critical 13th	high 13th	critical 16th
July 2, 1921	Geo. Carpentier (KO 4th)	high 3rd	critical 4th	critical 2nd
July 4, 1923	Tom Gibbon (on pts. 15th)	critical 3rd	critical 2nd	high 6th
Sept. 14, 1923	Luis Firpo (KO 2nd)	high 16th	high 16th	high 10th

Physically, boxing experts considered Dempsey to be one of the strongest fighters on record. He was also a vicious fighter when he wanted to be. Emotions played a key role in his manner of winning. When he met Willard and Firpo, his emotional state was very high. Willard was the champ and Dempsey was the young upstart challenger at that time. Firpo had the reputation of a great puncher and he had publicly announced his intention to KO Dempsey. After Firpo knocked him clear out of the ring, Dempsey returned to flatten him in the second round. In the fight with Tom Gibbon, it is a matter of record that handlers found it difficult to get Dempsey emotionally up for the fight, which went fifteen rounds.

In 1926, Jack Dempsey lost his title to Gene Tunney. At that time, Tunney was known as the "gentleman" boxer. Most of his fights were won by his boxing skill rather than by knockouts. Dempsey, by this time, had begun to slip. To make matters worse, his biorhythm record indicates that on both dates on which he fought Tunney his emotional state was low or near low.

Gene Tunney's biorhythm record for both of these fight dates indicates that his cycles were either higher or closer to critical points than Dempsey's. The reader may wish to check this.

Jack Dempsey	Birthdate: June 24, 1895	Birth Codes: U36AG
Gene Tunney	Birthdate: May 25, 1898	Birth Codes: K20BG

Jack Dempsey's Biorhythm Record

Date of Fight	Physical	Emotional	Intellectual
Sept. 23, 1926	high 24th	critical 20th	low 19th
Sept. 22, 1927	critical 21st	critical 19th	critical 26th
	high 27th	low 26th	low 17th

Jack Dempsey, 1927

(60)

Turning to track, the case of the famous miler, Jim Ryun, is interesting. Only two events are shown here—not sufficient data to provide a pattern.

Jim Ryun
Birthdate: April 29, 1947   Birth Codes: C16CS

Date	Race	Physical	Emotional	Intellectual
June 23, 1967	1-mile record (3 min., 51.1 sec.)	critical 23rd	critical 27th low 20th	critical 22nd
July 8, 1967	1500 meters (3 min., 33.1 sec.)	critical 4th low 10th	critical 11th high 4th	critical 8th

The most interesting thing about this record is the fact that critical days—or near-criticals—appear in all three curves. During a critical day in any one of the cycles, the individual may be in an unstable state; an athlete may either excel or fail miserably during this time. In Ryun's case, he certainly excelled, and his biorhythm cycles were at a period when they were very closely interrelated.

Jim Ryun at Lawrence, Kansas, April 23, 1966

Many observers of biorhythm cycles have maintained that critical days often signal times when persons go all out in one way or another. For example, one investigator found that suicides are often committed on a person's critical day. A Canadian study found that a large number of those receiving awards for heroism often perform the heroic accomplishment on their critical days, suggesting that on such days caution or self-preservation is often disregarded. Adherents of the biorhythm theory claim this critical day characteristic is often found in sports and is particularly common to those sports that stress individual rather than team competition. In addition to Jim Ryun, other examples of this critical day characteristic may be found in the performances of Olympic contenders Sheila Young, Dorothy Hamill, and Mark Spitz. The cases of Young and Hamill are described in Chapter Eleven.

Mark Spitz won an unprecedented seven gold medals at the Olympics in late August and early September of 1972. Four of these records with his biorhythm data are shown in the following tabulation.

Mark Spitz—1972 Olympic Swimming Competitions
Birthdate: February 10, 1950   Birth Codes: H31CK

Date	Event	Physical	Emotional	Intellectual
Aug. 28, 1972	200-meter butter-fly			
Aug. 31, 1972	100-meter butter-fly	critical 27th	high 25th	critical 27th
Sept. 3, 1972	100-meter free-style			
Sept. 4, 1972	400-meter relay member	high 2nd	high 1st	low 4th

The record indicates his physical cycle was high or critical; his emotional state high; and his mental condition critical on the early days, low near the end of the events.

In baseball, one cannot pass by the greatest team of all time—the 1927 Yankees—and the two immortals: Babe Ruth and Lou Gehrig. Much has been written about the Babe and there has never been his equal. Yet the other great star, who lived in the shadow of Babe Ruth, was Lou Gehrig. On June 3, 1932, against Philadelphia, Lou hit four home runs in four consecutive times at bat—something even the Babe never did. The fifth time at bat should have been a homer too, except for a tremendous catch by Al Simmons. Let's look at Lou's biorhythm record on that day.

Mark Spitz at the Munich Olympics, September, 1972

Lou Gehrig—*June 3, 1932*
Birthdate: June 19, 1903   Birth Codes: T33BJ

Physical	Emotional	Intellectual
critical 6th	low 3rd	critical 2nd

Is the low on the emotional cycle surprising? Hardly. A low emotional state could have been a distinct advantage in this instance, and is particularly appropriate in regard to the legendary Gehrig. Time after time, during his consecutive game streak (2,130 games), he coolly waited for the right pitch and set it on its way. He was one of the calmest batters according to his manager and teammates. The low on his emotional curve during this particular game indicates the biorhythmic pattern in which Gehrig functioned at his best.

There are many factors that contribute to excellence in athletics. It is quite ridiculous to suggest that all contributory influences can be predicted on the ups and downs of three biorhythm cycles. Natural aptitude, including unique physical traits, are often involved. Another big factor is training—not only of one's physical attributes, but training and coordination of one's emotional and mental traits as well.

(63)

Lou Gehrig, 1936

It is within this latter realm—the training grind—that biorhythm cycles can play an important role. Those who have worked in the field of conditioning athletes have recognized "peaks" and "valleys" in effort and output during the training period. The smart coach or athlete can learn to use the biorhythm charts to recognize these varying times. Thus, the low days on the physical cycle can be used for periods of relaxation or "recharge" tactics in the conditioning process, or perhaps to concentrate on strategy more than on physical gains. In a similar vein, high days can be periods of extra concentration or effort.

A second useful technique for biorhythm cycles in sports is careful observance and attention to biorhythm records. Few athletes ever prepare for just one event. Instead, they participate in many competitions, against different adversaries, and in various environments. Yet, for some, a pattern or trend of highs or lows can be discernible in their biorhythm records. If this is true, knowing the facts and coordinating one's efforts accordingly is very valuable. With some sports it is even possible to schedule or arrange competitions to fall in line with favorable periods of the cycles.

Finally, the observance of biorhythm charts of opposing team members or contestants affords opportunities to plan strategies so as to gain competitive advantages. In some cases the game plan may be an open book, merely waiting for someone to turn the pages.

CHAPTER ELEVEN

SOME FAMOUS PEOPLE

IN ADDITION to developing a working knowledge of our own biorhythms, it is often helpful to be aware of how others cope in relation to their rhythmic cycles. You no doubt will want to know and understand the biorhythm data for members of your family, your close friends, and for those with whom you work each day.

Through modern means of communication we also are aware of the daily experiences of many public figures. The famous people in this chapter include writers, actors, television personalities, politicians, rock singers, athletes. If we know their birthdates, and the dates of events in which we are interested, we can learn more about the importance of biorhythms in their lives as well. The events briefly described here may represent times of despair, of triumph or joy, or of just the more usual "up" or "down" days we all experience. In each case we may be reminded of our own unique behavior in response to the rhythms that regulate our lives.

Ernest Hemingway
Born July 21, 1899

Ernest Hemingway put a double-barreled shotgun to his forehead and pulled both triggers on July 2, 1961. He had been ill and depressed for some time and had twice threatened suicide. On the day of Hemingway's death, his physical cycle was at its lowest point. His intellectual cycle was also close to the nadir, and he had experienced an emotional critical just two days earlier, when he had returned home from the hospital.

Virginia Woolf
Born January 25, 1882

On March 28, 1941, the English novelist, Virginia Woolf, wrote a letter to her husband that said, in part, "I have the feeling I shall go mad . . . I have fought against it, but cannot fight any longer." Then she walked to the river bank, put down her hat and walking stick, and waded into the water.

Woolf had experienced earlier bouts of despair. On the day she died, all three of her biorhythm cycles were at or very close to their lowest points.

(67)

George Sanders
Born July 3, 1906

George Sanders, the well-known actor, committed suicide on April 25, 1972. April 25 was a critical day in Sanders' emotional cycle and close to a critical in both his other cycles.

Marilyn Monroe
Born June 1, 1926

Books have been written and rumors spread about the causes of Marilyn Monroe's death. Her loss is still felt by many who knew her only as a technicolor image. But whatever the contributing factors to her death from an overdose of barbiturates on August 6, 1962, criticals in both her emotional and her intellectual cycles occurred within twenty-four hours of that day. If the actress was in the state of depression most sources report, a double-critical day would have been especially hard for her to handle.

Richard M. Nixon
Born January 9, 1913

The "Saturday Night Massacre," which heavily influenced public opinion against President Nixon, occurred on October 20, 1973. Nixon fired Archibald Cox, the Watergate special prosecutor. Attorney General Elliot Richardson and Deputy Attorney General William Ruckelshaus both resigned in protest. Soon after, sentiment in the House of Representatives swung toward an impeachment investigation. Nixon had brought this one on himself.

On October 20, Nixon was one day off the lowest point in his intellectual cycle and very close to the lowest point in his emotional cycle.

Edmund Muskie
Born March 28, 1914

On February 26, 1972, Edmund Muskie was talking to a group of reporters, claiming he had been the brunt of cruel personal news articles written by William Loeb, publisher of the *Manchester Union Leader*. Recalling that Loeb had even made disparaging remarks about his wife, Muskie burst into tears. Although he later stated it would be good for a President to show human emotion, the incident was thought by many to have severely damaged his Presidential campaign. February 26 was a critical day in Muskie's emotional cycle; he was also within twenty-four hours of a critical in his intellectual cycle.

(71)

Bobby Fischer
Born March 9, 1943
Boris Spassky
Born January 30, 1937

While the world looked on, Boris Spassky, the Russian master, beat Bobby Fischer in the 1972 World Chess Championship Tournament on July 11 and 12. On both those days, Fischer's emotional and intellectual curves were at their lowest points, Spassky's emotional and intellectual curves were at their highest points.

Fischer handed the Russian defeats in games played on July 16, 20, 23, and 27. On every one of those dates, Fischer's cycles were higher or rising; whereas Spassky's cycles were lower or falling.

Fischer won again on August 3 and 4 with a double critical in his physical and intellectual curves. He lost to Spassky on the 6th, a day when his emotional curve was low and Spassky's high.

Fischer continued to overwhelm the Russian while his cycles were up and Spassky's down. On August 10 and 11, Fischer had a physical high and Spassky a physical low. On August 31 and September 1, Fischer's intellectual curve was high and Spassky's low. The American won the games, the match, and the money.

Janis Joplin
Born January 19, 1943

Jimi Hendrix
Born November 27, 1942

Janis Joplin, the flamboyant, tormented rock singer, known for her Southern Comfort, wild costumes, and sometimes reckless behavior, died of a drug overdose in the early morning hours of October 4, 1970. Janis was at the critical point in her physical cycle on October 4, within forty-eight hours of an emotional critical, and at the low point in her intellectual cycle.

When rock star Jimi Hendrix died of an overdose of drugs on September 18, 1970, he was at the lowest point in his emotional cycle and close to the bottom of both his physical and intellectual rhythms.

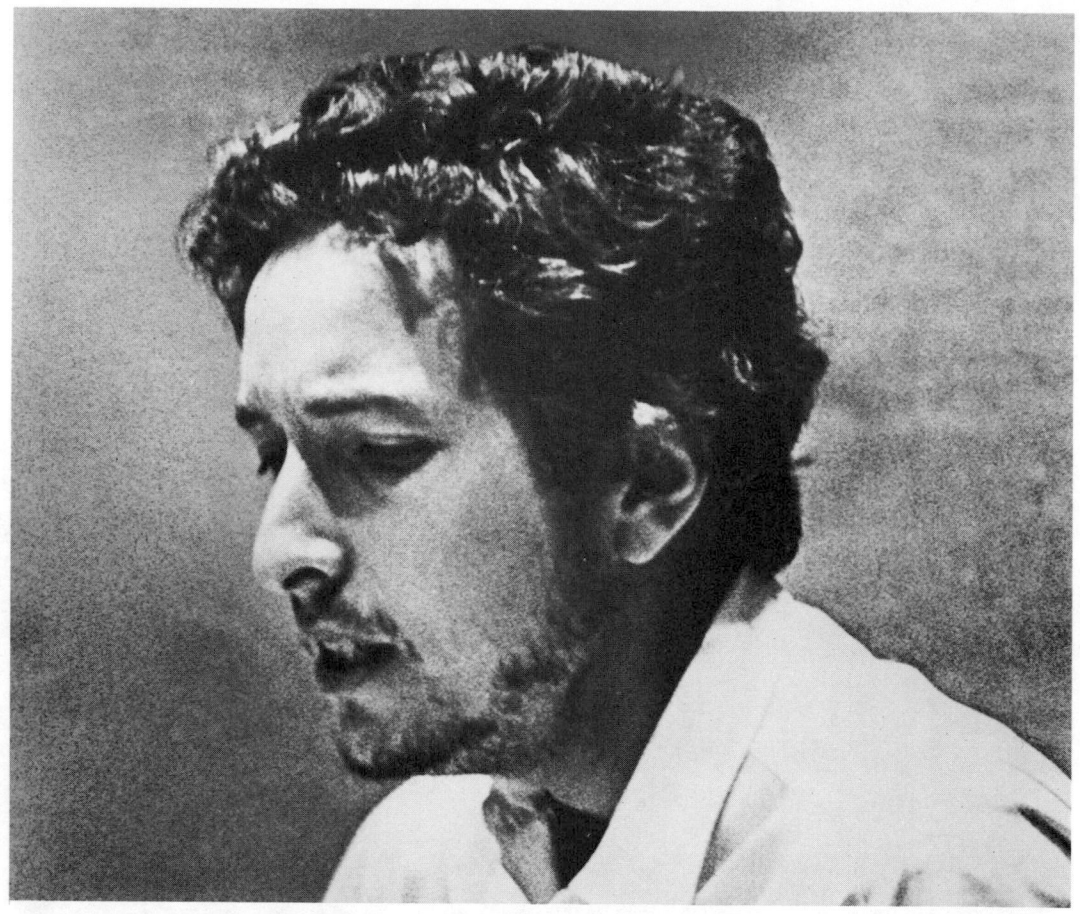

Bob Dylan
Born May 24, 1941

The serious motorcycle accident which nearly cost Bob Dylan his life and profoundly altered his career took place on August 2, 1966. Dylan's neck was broken; a difference of an inch might have spelled his death. On the day of the accident, Dylan was within twenty-four hours of a physical critical, within seventy-two hours of an emotional critical, and very close to the lowest point on his intellectual curve.

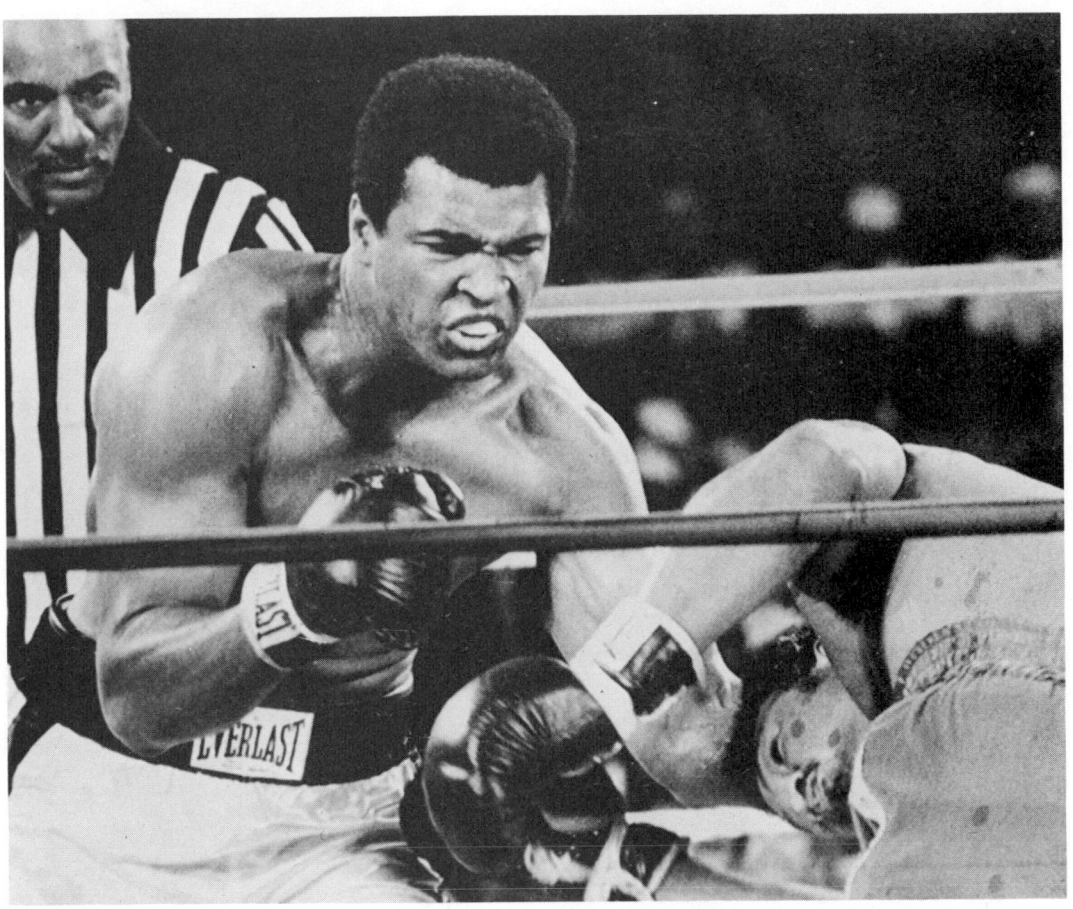

Muhammad Ali
Born January 18, 1942

On March 31, 1973, the boxing world was amazed when champion Muhammad Ali lost his bout against Ken Norton on a split decision. The next day was a double critical for Ali, in both his physical and emotional cycles. As the match was in Africa, the day's difference in dates is quite insignificant.

Sheila Young
Born October 14, 1950

Sheila Young beat out the other Olympic contenders for a gold medal in the Women's 500-Meter Skating Competition on February 6, 1976. She won on her physical critical day—an example of an athlete going all out at a moment when the balance could swing either way. This ability to go all out physically on a critical day has often been noted in situations of high stress and pressure. On the 6th, Young was buoyed by an emotional high and a near-high on her intellectual curve.

Dorothy Hamill
Born July 26, 1956

When nineteen-year-old Dorothy Hamill skated her way to a gold medal in the Women's Figure Skating Competition on February 13, 1976, it was a critical day in her physical cycle and one day after an emotional critical. Once again, an athlete found an extra dose of energy or daring at the critical moment in her rhythms.

(77)

Arthur Godfrey
Born August 31, 1903

Back in the earlier days of television, on October 19, 1953, Julius LaRosa sang as usual on the Arthur Godfrey show. Immediately afterward, Godfrey shocked the TV audience by publicly firing LaRosa—he later said the singer "lacked humility." Television personalities are fired all the time, but rarely on the air. Hardly surprising then, is the fact that October 19 was a critical day in Godfrey's emotional cycle.

You may want to add others to your list—historical figures and dates as well as current celebrities and events. Names and birthdates of many well-known people are listed in Appendix B in the back of the book. All you need is an Inquiry Date—the date of an event for which you seek biorhythmic information.

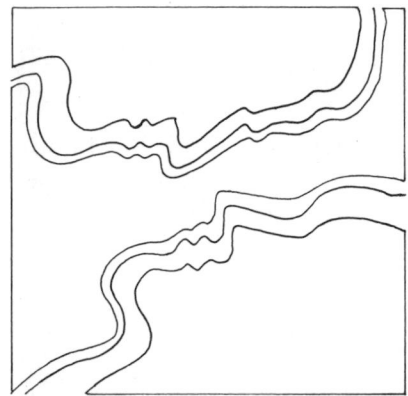

BIORHYTHMIC COMPATIBILITY

ALL OF US get along easily with certain people and find others hard to take, but not all of us feel the same way about the same people. Why? The answer would fill an encyclopedia, several pages of which might be on biorhythmic compatibility.

Fliess, Swoboda, and several more recent researchers have studied people with similar and dissimilar biorhythm cycles. These studies have not yet been numerous or exhaustive, so it is not possible at this date to state definitive conclusions about the fundamental relationship between biorhythmic correlations and compatibility. There is also the problem of trying to measure human values to arrive at psychological interpretations. Most of the studies so far tend to show that people with identical or similar rhythmic cycles are more compatible than those whose cycles are out of step or opposite. It stands to reason that if your moods and aptitudes coincide, you're more likely to enjoy one another. But there are some individuals who will be more compatible with others whose cycles complement rather than coincide with their own. One person may be at an emotional low and need the support of another who is at a high in his sensitivity cycle. In a work situation, for example, it may not always be advantageous to have individuals with compatible cycles together;

(79)

if two individuals are to make important decisions while both are at the lowest points in their intellectual and emotional cycles, it could spell disaster.

Most important, we must remember to recognize the uniqueness of each individual. Age, sex, background, environment, temperament—many factors influence how each of us functions in his daily life. Knowledge of the other person's biorhythms and behavior patterns will give us understanding, and understanding is the true basis of compatibility.

Measuring Compatibility

There are only a limited number of types of people for any one of the biorhythm cycles. In the 23-day physical cycle, for instance, there are only 23 possible types of individuals. Don't worry if you're looking for a mate; it doesn't have to be your twin, born on the same day. Identical types can be born on entirely different birthdays and still have the same biorhythm cycle. If you'd like to check out that statement, compare the physical rhythms for someone born on August 14, 1911, and someone born on October 18, 1918. You'll find they're identical—and 100 percent compatible.

As for the 28-day emotional cycle, the rule is the same: only 28 different types of emotional makeups exist on the biorhythmic scale. The same goes for the 33-day intellectual cycle: the limit is 33 types. Man's mind being what it is, and being his distinctive attribute, it's not surprising that there's more variation among intellectual patterns than among physical.

If anyone should think those numbers add up to a limited view of the human experience, that's worth rethinking. No one cycle exists independently of the others. Everyone is a combination of the three major cycles of life, and once you start combining, the number of types gets out of hand. The number of possible combinations is the product of all three numbers—23 times 28 times 33—which gives 21,252 types of individuals. Multiply this by other non-biorhythmic qualities of human behavior and you soon reach the realm of millions of types. Overall compatibility depends on all three cycles, not one in isolation. With millions of types to choose among, it ought to be difficult to find a perfect match, easy to find a friend.

Comparisons of your biorhythms with your partner's or friend's is interesting, revealing, and fun, but should be done with a dollop of caution and a pinch of salt. If your compatibilities are high, unfortunately that's no guarantee of success, though it's a mighty good beginning. After all, everyone born on the same day in the same year will have 100 percent biorhythmic compatibility, but the odds against all of them doing well together in marriage and business are pretty high. On the other hand, if your percentages turn out to be low, that's not necessarily a signal to give back the ring. You might, indeed, think twice. The first thought might be that opposites attract. The second might be the recognition that your alliance will either be stormier than most or will demand extra effort and understanding.

There are other things that can be checked. When someone says to you, "You're just like your mother," it may mean more than occurs to you.

Measurement of biorhythmic compatibilities at least provides an unbiased perspective on personal relationships. This could be extremely helpful in situations where conflicting compatibilities exist. Instead of mere argument, there are facts to look at, facts that can often be handled with a mixture of tact and good will.

Harold Willis, Director of the Biorhythm Clinic in Joplin, Missouri, believes that family stress can be prevented by measuring the biorhythmic compatibilities of engaged couples. If their compatibilities are low, it's pretty unlikely that they'll call it off, but once aware of problem areas, it is much more likely that they will make satisfactory compensations. Mr. Willis thinks such measurements can be particularly valuable for marriage counselors. He suggests that counselors can "use biorhythm as a functional scapegoat for people." When it's no longer his fault or hers but simply their lack of knowledge of what is happening in their rhythms, the personal bitterness is diminished and the couple has a problem they can tackle together.

Willis also advocates having biorhythms worked out for a year in advance; something that this book enables you to do. In this way, a couple has forewarning about times of low compatibility, about times when one partner is low or critical and could use help from the other, and about times of high compatibility when they could make the most of their relationship.

The technique used in this book is one which enables you to approximate compatibility on a numerical percentage scale ranging from extremes of zero (opposite or no compatibility) to 99 (identical or practically 100 percent compatibility). Table 7, below, divides the areas of compatibility into the five rating categories shown.

Table 7 — Percentages of Compatibility

Biorhythmic Character	Rating Categories				
	Very High	High	Indecisive	Low	Very Low
Physical	82–99	65–73	47–56	22–39	0–13
Sensitivity	85–99	64–78	42–57	21–35	0–14
Intellectual	87–99	63–81	45–57	21–39	0–15
Overall Average	82–99	63–81	42–57	21–39	0–15

As an example: a physical compatibility of 85 percent would be considered very high.

Table 8 is practically self-explanatory. Find the percentage of compatibility by simply reading opposite your Physical Birth Code and under the Physical Birth Code of the other person. Then check the category rating in Table 7 on the preceding page. The example below indicates the method.

Example: Compare the cyclical variations in physical energy, endurance, and drive between former President Lyndon Johnson and his Vice-President, Hubert Humphrey:

	Birthdate	*Physical Birth Code*
Lyndon Baines Johnson	August 27, 1908	A
Hubert H. Humphrey	May 27, 1911	G

Table 8 indicates that the compatibility of their physical cycles (opposite A and under G) is 22 percent. This is categorized as a low percentage of compatibility in Table 7.

NOTE: The reader will appreciate the unique quality of the Birth Codes in this book. Compatibility is really a "built-in" feature of the codes. Thus, in the example above, codes A and G will return exactly the same compatibility regardless of what Inquiry Date is used. Consequently, Inquiry Dates are not required in the procedure.

Table 8 — Percentages of Physical Biorhythmic Compatibility Between Your Physical Code (left Column) and Physical Code of Another Person (Top Row)

CODE	A	B	C	D	E	F	G	H	J	K	L	M	N	P	R	S	T	U	V	W	X	Y	Z
A	99	82	82	39	47	65	22	30	47	73	30	39	90	13	65	22	13	56	56	4	73	4	90
B	82	99	65	56	30	47	39	13	65	56	47	22	90	4	82	4	30	73	39	13	90	22	73
C	82	65	99	22	65	82	4	47	30	90	13	56	73	30	47	39	4	39	73	22	56	13	90
D	39	56	22	99	13	4	82	30	90	13	90	22	47	47	73	39	73	82	4	56	65	65	30
E	47	30	65	13	99	82	30	82	4	73	22	90	39	65	13	73	39	4	90	56	22	47	56
F	65	47	82	4	82	99	13	65	13	90	4	73	56	47	30	56	22	22	90	39	39	30	73
G	22	39	4	82	30	13	99	47	73	4	90	39	30	65	56	56	90	65	22	73	47	82	13
H	30	13	47	30	82	65	47	99	22	56	39	90	22	82	4	90	56	13	73	73	4	65	39
J	47	65	30	90	4	13	73	22	99	22	82	13	56	39	82	30	65	90	4	47	73	56	39
K	73	56	90	13	73	90	4	56	22	99	4	65	65	39	39	47	13	30	82	30	47	22	82
L	30	47	13	90	22	4	90	39	82	4	99	30	39	56	65	47	82	73	13	65	56	73	22
M	39	22	56	22	90	73	39	90	13	65	30	99	30	73	4	82	47	4	82	65	13	56	47
N	90	90	73	47	39	56	30	22	56	65	39	30	99	4	73	13	22	65	47	4	82	13	82
P	13	4	30	47	65	47	65	82	39	39	56	73	4	99	22	90	73	30	56	90	13	82	22
R	65	82	47	73	13	30	56	4	82	39	65	4	73	22	99	13	47	90	22	30	90	39	56
S	22	4	39	39	73	56	56	90	30	47	47	82	13	90	13	99	65	22	65	82	4	73	30
T	13	30	4	73	39	22	90	56	13	82	47	22	73	47	65	90	99	56	30	82	39	90	4
U	56	73	39	82	4	22	65	13	90	30	73	4	65	30	90	22	56	99	13	39	82	47	47
V	56	39	73	4	90	90	22	73	4	82	13	82	47	56	22	65	30	13	99	47	30	39	65
W	4	13	22	56	56	39	73	73	47	30	65	65	4	90	30	82	82	39	47	99	22	90	13
X	73	90	56	65	22	39	47	4	73	47	56	13	82	13	90	4	39	82	30	22	99	30	65
Y	4	22	13	65	47	30	82	65	56	22	73	56	13	82	39	73	90	47	39	90	30	99	4
Z	90	73	90	30	56	73	13	39	39	82	22	47	82	22	56	30	4	47	65	13	65	4	99

Table 9 — Percentages of Emotional Biorhythmic Compatibility Between Your Emotional Code (Left Column) and Another Emotional Code (Top Row)

CODE	11	12	13	14	15	16	17	18	19	20	21	22	23	24	25	26	27	28	29	30	31	32	33	34	35	36	37	38
11	99	78	57	35	14	35	57	85	50	7	50	64	28	42	21	85	64	0	71	42	92	14	7	78	28	21	71	92
12	78	99	78	57	7	14	35	92	71	28	28	42	50	21	0	64	85	21	50	64	85	35	14	57	7	42	92	71
13	57	78	99	78	28	7	14	71	92	50	7	21	71	0	21	42	92	42	28	85	64	57	35	35	14	64	85	50
14	35	57	78	99	50	28	7	50	85	71	14	0	92	21	42	21	71	64	7	92	42	78	57	14	35	85	64	28
15	14	7	28	50	99	78	57	0	35	78	64	50	57	71	92	28	21	85	42	42	7	71	92	35	85	64	14	21
16	35	14	7	28	78	99	78	21	14	57	85	71	35	92	85	50	0	64	64	21	28	50	71	57	92	42	7	42
17	57	35	14	7	57	78	99	42	7	35	92	92	14	85	64	71	21	42	85	0	50	28	50	78	71	21	28	64
18	85	92	71	50	0	21	42	99	64	21	35	50	42	28	7	71	78	14	57	57	92	28	7	64	14	35	85	78
19	50	71	92	85	35	14	7	64	99	57	0	14	78	7	28	35	85	50	21	92	57	64	42	28	21	71	78	42
20	7	28	50	71	78	57	35	21	57	99	42	28	78	50	71	7	42	92	21	64	14	92	85	14	64	85	35	0
21	50	28	7	14	64	85	92	35	0	42	99	85	21	92	71	64	14	50	78	7	42	35	57	71	78	28	21	57
22	64	42	21	0	50	71	92	50	14	28	85	99	7	78	57	78	28	35	92	7	57	21	42	85	64	14	35	71
23	28	50	71	92	57	35	14	42	78	78	21	7	99	28	50	14	64	71	0	85	35	85	64	7	42	92	57	21
24	42	21	0	21	71	92	85	28	7	50	92	78	28	99	78	57	7	57	71	14	35	42	64	64	85	35	14	50
25	21	0	21	42	92	85	64	7	28	71	71	57	50	78	99	50	14	78	50	35	14	64	85	42	92	57	7	28
26	85	64	42	21	28	50	71	71	35	7	64	78	14	57	35	99	50	14	85	28	78	0	21	92	42	7	57	92
27	64	85	92	71	21	0	21	78	85	42	14	28	64	7	14	50	99	35	35	78	71	50	28	42	7	57	92	57
28	0	21	42	64	85	64	42	14	50	92	50	35	71	57	78	14	35	99	28	57	7	85	92	21	71	78	28	7
29	71	50	28	7	42	64	85	57	21	21	78	92	0	71	50	85	35	28	99	14	64	14	35	92	57	7	42	78
30	42	64	85	92	42	21	0	57	92	64	7	7	85	14	35	28	78	57	14	99	50	71	50	21	28	78	71	35
31	92	85	64	42	7	28	50	92	57	14	42	57	35	35	14	78	71	7	64	50	99	21	0	71	21	28	78	85
32	14	35	57	78	71	50	28	28	64	92	35	21	85	42	64	0	50	85	14	71	21	99	78	7	57	92	42	7
33	7	14	35	57	92	71	50	7	42	85	57	42	64	64	85	21	28	92	35	50	0	78	99	28	78	71	21	14
34	78	57	35	14	35	57	78	64	28	14	71	85	7	64	42	92	42	21	92	21	71	7	28	99	50	0	50	85
35	28	7	14	35	85	92	71	14	21	64	78	64	42	85	92	42	7	71	57	28	21	57	78	50	99	50	0	35
36	21	42	64	85	64	42	21	35	71	85	28	14	92	35	57	7	57	78	7	78	28	92	71	0	50	99	50	14
37	71	92	85	64	14	7	28	85	78	35	21	35	57	14	7	57	92	28	42	71	78	42	21	50	0	50	99	64
38	92	71	50	28	21	42	64	78	42	0	57	71	21	50	28	92	57	7	78	35	85	7	14	85	35	14	64	99

Table 10 — Percentages of Intellectual Biorhythmic Compatibility Between Your Intellectual Code (Left Column) and Another Intellectual Code (Top Row)

CODE	AB	AC	AD	AE	AF	AG	AH	AJ	AK	AL	AM	BC	BD	BE	BF	BG	BH	BJ	BK	BL	BM	BP	CD	CE	CF	CH	CJ	CK	CL	CM	CP	CR	CS
AB	99	93	87	81	75	69	63	57	51	45	39	33	27	21	15	9	3	3	9	15	21	27	33	39	45	51	57	63	69	75	81	87	93
AC	93	99	93	87	81	75	69	63	57	51	45	39	33	27	21	15	9	3	3	9	15	21	27	33	39	45	51	57	63	69	75	81	87
AD	87	93	99	93	87	81	75	69	63	57	51	45	39	33	27	21	15	9	3	3	9	15	21	27	33	39	45	51	57	63	69	75	81
AE	81	87	93	99	93	87	81	75	69	63	57	51	45	39	33	27	21	15	9	3	3	9	15	21	27	33	39	45	51	57	63	69	75
AF	75	81	87	93	99	93	87	81	75	69	63	57	51	45	39	33	27	21	15	9	3	3	9	15	21	27	33	39	45	51	57	63	69
AG	69	75	81	87	93	99	93	87	81	75	69	63	57	51	45	39	33	27	21	15	9	3	3	9	15	21	27	33	39	45	51	57	63
AH	63	69	75	81	87	93	99	93	87	81	75	69	63	57	51	45	39	33	27	21	15	9	3	3	9	15	21	27	33	39	45	51	57
AJ	57	63	69	75	81	87	93	99	93	87	81	75	69	63	57	51	45	39	33	27	21	15	9	3	3	9	15	21	27	33	39	45	51
AK	51	57	63	69	75	81	87	93	99	93	87	81	75	69	63	57	51	45	39	33	27	21	15	9	3	3	9	15	21	27	33	39	45
AL	45	51	57	63	69	75	81	87	93	99	93	87	81	75	69	63	57	51	45	39	33	27	21	15	9	3	3	9	15	21	27	33	39
AM	39	45	51	57	63	69	75	81	87	93	99	93	87	81	75	69	63	57	51	45	39	33	27	21	15	9	3	3	9	15	21	27	33
BC	33	39	45	51	57	63	69	75	81	87	93	99	93	87	81	75	69	63	57	51	45	39	33	27	21	15	9	3	3	9	15	21	27
BD	27	33	39	45	51	57	63	69	75	81	87	93	99	93	87	81	75	69	63	57	51	45	39	33	27	21	15	9	3	3	9	15	21
BE	21	27	33	39	45	51	57	63	69	75	81	87	93	99	93	87	81	75	69	63	57	51	45	39	33	27	21	15	9	3	3	9	15
BF	15	21	27	33	39	45	51	57	63	69	75	81	87	93	99	93	87	81	75	69	63	57	51	45	39	33	27	21	15	9	3	3	9
BG	9	15	21	27	33	39	45	51	57	63	69	75	81	87	93	99	93	87	81	75	69	63	57	51	45	39	33	27	21	15	9	3	3
BH	3	9	15	21	27	33	39	45	51	57	63	69	75	81	87	93	99	93	87	81	75	69	63	57	51	45	39	33	27	21	15	9	3
BJ	3	3	9	15	21	27	33	39	45	51	57	63	69	75	81	87	93	99	93	87	81	75	69	63	57	51	45	39	33	27	21	15	9
BK	9	3	3	9	15	21	27	33	39	45	51	57	63	69	75	81	87	93	99	93	87	81	75	69	63	57	51	45	39	33	27	21	15
BL	15	9	3	3	9	15	21	27	33	39	45	51	57	63	69	75	81	87	93	99	93	87	81	75	69	63	57	51	45	39	33	27	21
BM	21	15	9	3	3	9	15	21	27	33	39	45	51	57	63	69	75	81	87	93	99	93	87	81	75	69	63	57	51	45	39	33	27
BP	27	21	15	9	3	3	9	15	21	27	33	39	45	51	57	63	69	75	81	87	93	99	93	87	81	75	69	63	57	51	45	39	33
CD	33	27	21	15	9	3	3	9	15	21	27	33	39	45	51	57	63	69	75	81	87	93	99	93	87	81	75	69	63	57	51	45	39
CE	39	33	27	21	15	9	3	3	9	15	21	27	33	39	45	51	57	63	69	75	81	87	93	99	93	87	81	75	69	63	57	51	45
CF	45	39	33	27	21	15	9	3	3	9	15	21	27	33	39	45	51	57	63	69	75	81	87	93	99	93	87	81	75	69	63	57	51
CH	51	45	39	33	27	21	15	9	3	3	9	15	21	27	33	39	45	51	57	63	69	75	81	87	93	99	93	87	81	75	69	63	57
CJ	57	51	45	39	33	27	21	15	9	3	3	9	15	21	27	33	39	45	51	57	63	69	75	81	87	93	99	93	87	81	75	69	63
CK	63	57	51	45	39	33	27	21	15	9	3	3	9	15	21	27	33	39	45	51	57	63	69	75	81	87	93	99	93	87	81	75	69
CL	69	63	57	51	45	39	33	27	21	15	9	3	3	9	15	21	27	33	39	45	51	57	63	69	75	81	87	93	99	93	87	81	75
CM	75	69	63	57	51	45	39	33	27	21	15	9	3	3	9	15	21	27	33	39	45	51	57	63	69	75	81	87	93	99	93	87	81
CP	81	75	69	63	57	51	45	39	33	27	21	15	9	3	3	9	15	21	27	33	39	45	51	57	63	69	75	81	87	93	99	93	87
CR	87	81	75	69	63	57	51	45	39	33	27	21	15	9	3	3	9	15	21	27	33	39	45	51	57	63	69	75	81	87	93	99	93
CS	93	87	81	75	69	63	57	51	45	39	33	27	21	15	9	3	3	9	15	21	27	33	39	45	51	57	63	69	75	81	87	93	99

Tables 9 and 10 are similar to Table 8. Table 9 provides emotional compatibility percentages; Table 10 measures intellectual compatibility. The reader may want to arrive at a composite compatibility by averaging the three percentages. This is done as follows:

> Example: Continuing the comparison between Lyndon Johnson and Hubert Humphrey:

	Physical Birth Codes and Compatibility	Emotional Birth Codes and Compatibility	Intellectual Birth Codes and Compatibility
Johnson:	A	19	CS
Humphrey:	G 22%	18 64%	BD 21%

> The average percentage of compatibility $= \dfrac{(22\%+64\%+21\%)}{3} = 36\%$; a
>
> low overall compatibility according to Table 7.

Some Compatible and Not So Compatible People

Biorhythmic measurements are only part of the story. All kinds of people live and work together, for good reasons and bad, who don't seem to measure up. Then again, some partnerships which look like they were planned in heaven founder among the difficulties on earth. The biorhythmic percentages are the givens of people's lives, but it's almost more interesting to see what they do with what they're given than to regard the data alone. Following are a few examples; you may make of them what you will.

Consider the well-known case of King Edward VIII and the woman he resigned his throne for—Bessie Wallis Warfield, formerly Mrs. Simpson. King Edward was born on June 23, 1894, Mrs. Simpson on June 19, 1896.

Percentages of Biorhythmic Compatibility

Physical	Emotional	Intellectual	Average Overall Compatibility
22%	92%	93%	69%

Or take another pair of famous lovers: Elizabeth Taylor, born February 27, 1932, and Richard Burton, born November 10, 1925.

Percentages of Biorhythmic Compatibility

Physical	Emotional	Intellectual	Average Overall Compatibility
99%+	71%	39%	70%

Liz and Richard in Montreal, 1964

Jackie Bouvier, born July 28, 1929, was first married to John F. Kennedy, born May 29, 1917.

Percentages of Biorhythmic Compatibility

Physical	Emotional	Intellectual	Average Overall Compatibility
65%	35%	27%	42%

Some time after the death of the President, Jackie Kennedy married Aristotle Onassis, born January 20, 1906.

Percentages of Biorhythmic Compatibility

Physical	Emotional	Intellectual	Average Overall Compatibility
4%	57%	39%	33%

(85)

President and Mrs. John F. Kennedy, 1960

Jackie and Ari on a Nile riverboat, March, 1974

Sonny and Cher

Sonny and Cher, now divorced, are still working and appearing together on the weekly *The Sonny and Cher Show*. Sonny Bono was born on February 16, 1940; Cher (Cherilyn Sarkisian) was born on May 20, 1946.

Percentages of Biorhythmic Compatibility

Physical	Emotional	Intellectual	Average Overall Compatibility
30%	21%	51%	34%

(87)

William H. Masters and Virginia E. Johnson

William H. Masters, born December 27, 1915, and Virginia E. Johnson, born February 11, 1925, first met in 1956 and have been researching, writing, and counseling together ever since. Authors of *Human Sexual Response* and other books, they were married in 1971.

Percentages of Biorhythmic Compatibility

Physical	Emotional	Intellectual	Average Overall Compatibility
90%	85%	93%	89%

Compatibility studies can be done on all kinds of relationships. There have been hundreds of well-known partnerships throughout history, throughout the range of human endeavor—in the theater, music and dance, literature, in science, medicine, politics, business. There are also the relationships and partnerships that are not voluntary, but necessary, results of circumstance and environment—the people we must go to school with, work with, maintain our communities with. The nature of the partnership as well as the characters of the people involved must influence which rhythms predominate in interactions as well as what degrees of difference can be tolerated.

Let's look at some further examples:

Rudolph Nureyev and Dame Margot Fonteyn have thrilled dance audiences with their beautiful executions and interpretations of the classical ballet. Nureyev was born on March 17, 1938; Dame Margot Fonteyn on May 18, 1919.

Percentages of Biorhythmic Compatibility

Physical	Emotional	Intellectual	Average Overall Compatibility
90%	28%	15%	44%

As host and co-host of the *Tonight* show on television, Johnny Carson, born October 23, 1925, and Ed McMahon, born March 6, 1923, have entertained late-night viewers for many years. They first began working together in 1954.

Percentages of Biorhythmic Compatibility

Physical	Emotional	Intellectual	Average Overall Compatibility
65%	28%	69%	54%

Henry Kissinger, born May 27, 1923, was Secretary of State for Richard Nixon, born January 9, 1913.

Percentages of Biorhythmic Compatibility

Physical	Emotional	Intellectual	Average Overall Compatibility
56%	28%	69%	51%

After Nixon's resignation, Kissinger remained as Secretary of State for Gerald Ford, born July 14, 1913.

Percentages of Biorhythmic Compatibility

Physical	Emotional	Intellectual	Average Overall Compatibility
39%	42%	57%	46%

President Gerald Ford and Secretary of State Henry Kissinger in Brussels, 1975

President Nixon and John Erlichman, July, 1972

Another man who worked with and apparently got along well with Nixon, at least until well into the Watergate investigation and the writing of his novel, was John D. Ehrlichman, born March 20, 1925.

Percentages of Biorhythmic Compatibility

Physical	Emotional	Intellectual	Average Overall Compatibility
22%	92%	87%	67%

Senator Walter Mondale and former Governor Jimmy Carter, New York, July, 1976

Jimmy Carter, the 1976 Democratic Presidential nominee, chose Senator Walter F. (Fritz) Mondale as his Vice-Presidential running mate. Carter was born on October 1, 1924; Mondale on January 5, 1928.

Percentages of Biorhythmic Compatibility

Physical	Emotional	Intellectual	Average Overall Compatibility
56%	7%	81%	48%

It will take more skill than the human race possesses to figure out why some people growl at each other while others grin. Biorhythms at least provide a hint.

(91)

Bio-Quiz 6

Is the news you listen to a reflection of your compatibility with . . .

Walter Cronkite, born November 4, 1916;
Eric Sevareid, born November 26, 1912?

John Chancellor, born July 14, 1927;
David Brinkley, born July 10, 1920?

Harry Reasoner, born April 17, 1923;
Barbara Walters, born September 25, 1931?

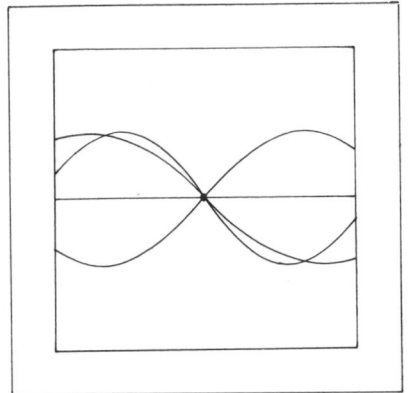

BIORHYTHM'S MULTIPLE CRITICAL DAYS

IN OTHER sections of this book the influence and impact of single critical days have been documented. What about *double*- and *triple*-critical days? Certainly their potential effects must be considerable. How often do double- and triple-critical days occur in the average life span? Is it possible to anticipate when they will occur?

Let's begin by defining each biorhythmic "life span" as one that covers 21,252 days. This number is the product formed by multiplying the three cycle lengths (23 times 28 times 33). The approximate number of years in this span is found by dividing 21,252 days by 365.25. The result is equal to slightly more than 58 years. The distribution of single, double, and triple criticals during this span of years is indicated in Table 11 below:

Table 11

Distribution of Criticals in the Biorhythmic Span of 58 Years

Number of Non-Criticals	Number of Single Criticals	Number of Double Criticals	Number of Triple Criticals	Total Days in Life Span
16,926	4,006	312	8	21,252

The eight triple criticals occur in this first biorhythmic span of years at the times shown below:

Table 12

Triple Critical	Your Age When It Occurs	Exact Day of Your Life
First	1 day	1
Second	19 yrs., 271 days	7,211
Third	21 yrs., 185 days	7,855
Fourth	27 yrs., 121 days	9,983
Fifth	29 yrs., 35 days	10,627
Sixth	48 yrs., 305 days	17,837
Seventh	50 yrs., 218 days	18,481
Eighth	56 yrs., 155 days	20,609

Your 9th triple critical occurs about 68 days after your fifty-eighth birthday. The exact day is 21,253. This is the first day, following birth, when all three biorhythm cycles are not only together but also point in the same direction. On this day the second biorhythmic span begins—a rebirth!

In interpreting the effects of multiple-criticals, it seems reasonable to consider a double critical as more serious than a single critical day. The instability and "wobbly" effects in the system are still present but to a greater degree since more than one cycle is affected. Thus, a woman near childbirth with a double-critical day provides a higher degree of certainty for predicting birth on that day. A double-critical day to a person who is depressed and contemplating suicide would be much more serious than a single critical.

Triple criticals can be interpreted in a similar manner, but with more profound intensities. All three cycles are possibly in unstable states. Depending on circumstances and conditions under which the person finds himself, the results can be most unusual or precipitous. One must keep in mind that although the critical days are often times to be cautious and aware, there is at the same time the potential for unusual or outstanding achievement, the possibility for emotional, physical, or intellectual breakthroughs in one's life. A multiple-critical day compounds this tenuous edge—the possibilities for positive or negative results are that much more intensified. Triple criticals also have the unique characteristic of occurring at certain points in the life cycle. This characteristic could add another dimension to the biorhythm theory. Namely, that an important decision or action taken on or near the

Table 13—Dates When Your Double- and Triple-Critical Days Occur

Find your double- and triple-critical dates by looking opposite your age (left column), and adding the months (top column) and days (right column) shown to the date of your last birthday. Asterisks indicate triple criticals.

Example: Find your first double critical after your 15th birthday. Opposite age 15, first entry (under 2) is 6. This means add 2 months and 6 days to the date of your 15th birthday.

Use the Date of Your Birthday at Age Shown Below	0	1	2	3	4	5	6	7	8	9	10	11	12
					Approximate Number of Months and Days Beyond Birthday at Age Shown in Left Column.								
0	***1			26	7				25	11	23		
1	15		24	8						11			
2	12,29	11			24		14	26		23			2
3		13		19			13			21	19		
4		28				1,23	26			6	22		
5	23		12	17				23			3		
6		3,26		29			9,21					21	
7	21	4		11	26			6				12	
8			13,25		11			21		29	13		
9	14			1				18	16				
10	16	14			23			2	18		24	8	
11		12			21	7,19				25			
12		26			6	22		28	12	17			
13			11,23			3	6,29					5	
14		9				15	21		9		15		1
15			6	9	2			4			18	16	
16			19			13		19		5		14	
17		6	18		8			22	3,20			28	
18						24	8		17		14	26	
19	7				25	6				***1			
20			28	12				18		16			
21	3	6,29					***5				1		
22	15			9	21					11	14	7	
23			4	8			16	3		24	28		
24	13	25							11	23		13	
25	12			3,20			28		2			4,27	
26	24					14	26			21			
27	6			***1					7	1	3		
28	21		18				24	8					
29		***5				1,11				29	15	27	
30		18		27	11						13		
31		16	3,14			28		18		1	27		
32		1	17		23			17			25	23	
33				2			4,27		1		11	26	
34		26		15	21				26			6	
35				7	1	3		13,25		11			
36	19	24	8		14		1			11			
37	11				17,29	15			24			3,17	
38		17		3		12			21	19			
39		21	18		27				6	22		28	
40	7		16		25	11,23					29		
41				1	11	26			2,16	21			
42			14,26			6	6	9	2				
43	3		13				19	24		13		19	
44		1		11	13	6			8			22	
45	15			23			17		23		9		
46	12		9	21		11			25	6,23			
47	26						28	12		21		18	
48	25	11				29	11			***5			
49					2,16				22		21		
50		6	9	2				***8				3	
51		19			13	25					15	18	
52	6			8	12			21	7		28		2
53		17	29							15	27		
54	11	15			6,23				1	5			
55	2,25	28					18		1		24		
56		11				***5				11	4	7	
57		24		22				28	12			2,18	
58			***8				3,14					17	
59	25		22			1,15						17	
60			21	7,18				2	22		4		1
61			5	21		27			21			29	
62	21				5			7	1	3		13	
63	24			1	19	24					1		
64	5				11	4	7	17,29					
65		23	28	12		18		4		14			
66		14				21	2,18			27			
67	1,15		21		7		16			25	23		
68			24	22				1		11	26		
69	27	11		21			29	15,27					3
70					3		13	29			5,19	24	
71					18	1		11	13	6		17	
72		7		17				23	28		17		
73	18		4		14	17	11			12			
74	21	18			26			21		26		12	
75		16		13	25		15			29	11,27		
76			1						2,16		25		
77	17	29	15					3,14				***9	
78						5,19				25		23	
79			11	13	6			***2					

triple-critical day might affect how one enters a new life stage.

At least one factor is clear in relation to multiple criticals: the importance of being able to anticipate when they will occur. A very significant provision of this book is Table 13, which provides the reader with this capability. Table 13 indicates when every double- and triple-critical day occurs in an average life span of seventy-nine years. As each month in the table is assumed to be 30 days in length, dates will be exact to within one to two days. You can doublecheck the given critical dates against your own biorhythm cycles by using your Birth Code and the tables in Appendix A. The following example explains the procedure.

Cher Bono was born on May 20, 1946. Use Table 13 to determine if she had a triple critical in 1975. Her twenty-ninth birthday was May 20, 1975. Look opposite age 29 in the left column, and you will note that a triple critical appears approximately 1 month and 5 days after her twenty-ninth birthday. Adding 35 days (1 month, 5 days) to May 20 gives June 24. By using her Birth Code and the Inquiry Year and Biorhythm Data Tables, it is easy to see that Cher actually had a triple critical on June 23, 1975.

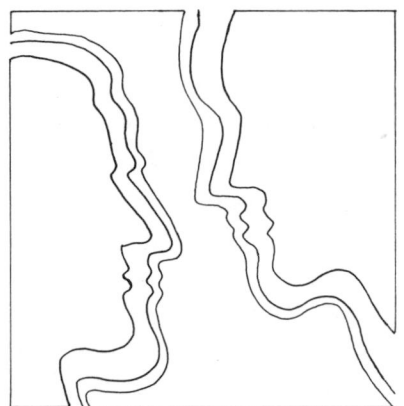

CHAPTER FOURTEEN

BIORHYTHMS IN OUR LIVES

THE BIORHYTHM theory is currently being applied in innumerable ways, and the possibilities for the future are vast. Until the computer took the effort out of calculating individual tables it was not so simple a task to put the biorhythm theory into action. Now it is, and it is easy to see how the theory might be used to advantage in many of the more serious as well as the more leisurely and enjoyable aspects of life.

Certainly enough work has been done in the area of accident prevention to suggest that industry and transport look into the use of biorhythms further. The magazine of the World Health Organization reported last year that about a quarter of a million people are killed on the world's roads every year. Traffic experts estimate that within a few years the number of injured will reach ten million. The figures are staggering, and the efforts to reduce them must be strenuous.

Industrial accidents also claim thousands of maimed and injured every year, and human error costs industry enormous sums. If biorhythm charts, and especially warnings issued on critical days, could help to reduce the high toll that accidents exact, they would be performing an immense service for us all.

Just a few years ago, someone suggested that on holidays of peak travel drivers could turn on their headlights to remind others to be careful. The result was a significant decrease in accidents and fatalities on those days. What's the difference between turning on your headlights and turning on your awareness of precautionary biorhythmic days?

There are other dangers our biorhythms can warn us of, other precautions we can take. Death occurs more often on critical days than the laws of chance would allow. Heart attack victims seem especially prone to the stress of an unstable day in their biorhythms. Heart attack patients and those suffering from other serious illnesses could be given special attention on critical days; an ounce of prevention could save many lives.

Hospitals and doctors in general might make much more extensive use of biorhythms than they do at present. A person will obviously withstand the shock of surgery more easily when his body is in the peak of the physical cycle. He may also respond more patiently and positively to the period of recuperation knowing that his physical cycle is in the recharge phase.

It may be that some day, when biological rhythms are better understood, the course of drug therapy and treatment will be facilitated by synchronizing it with the body's rhythms.

Recent developments in mental health have revealed that there are many individuals who periodically experience days of mental depression. The signals are generally one or more of the following: feelings of despair, loss of self-respect, sleeplessness, loneliness, hypersensitivity, fits of anger, guilt feelings, withdrawal, threats of or attempts at suicide, inability to concentrate. The possible application of the biorhythm cycles, with concentration on the critical days of the intellectual and emotional rhythms, offers a fertile field for investigation.

Treatment of the mentally ill may some day incorporate knowledge of biorhythms. Both drugs and psychotherapeutic techniques might be more effective at different times in the cycles. Since suicides seem to occur suspiciously often on critical or low days, patients with suicidal tendencies could receive special care at such times. Psychologists who treat the less seriously disturbed might find their work made easier if both they and their patients were aware of the normal rhythmic changes in life that influence body and mind.

It is possible that biorhythms could be of help in remedial therapy and special education. With the handicapped and the retarded, it would be particularly helpful to know the periods of maximum receptivity and the times when the individual simply could not be pushed too hard. Both physical therapy and schooling for the child with severe learning difficulties might benefit from an awareness of fluctuations in capacity.

In the sports world, the possibilities of using biorhythms are only beginning to be explored. Individual athletes who follow their biorhythm charts could learn their own best patterns and means of compensation. In some cases they could schedule contests to their own advantage. In every case they could recognize the times to go easy and the times to push hard during training, and would know that if they are in a low between important events it would be more useful to conserve energy than to invite exhaustion through long workouts. This kind of information could be of vital importance to the trainer as well, who must tend to an athlete's psyche as well as his body.

In the area of human relations, knowledge of biorhythms could lead to an increase in compatibility. Knowledge of your own changing rhythms could help you guard against your down times; knowledge of your partner's could go a long way toward making you more tolerant and helpful. Administrators might think about shifting workers who are unlikely to work well together, or creating a work group whose potential promises to be extraordinary because the group members' "high" days coincide.

Marriage and family counselors could probably benefit from learning more about biological rhythms. When there is stress within a family, awareness of the problem is half the battle, and if the family members are able to assign some of the conflict to their inherent rhythms, personal bitterness and antagonisms could be lessened.

In all of these possibilities, knowledge of biorhythms, together with some ability to anticipate their fluctuations, is of prime importance. For most people, the biorhythmic fluctuations or swings may be difficult to anticipate at first because, as humans, we generally have poor feedback mechanisms. A feedback mechanism enables us to know when something is sensed. It is the connection between the sensory object and the brain. One sees because the image travels back to the brain by way of the eye. The nose, and its associated organs, is the feedback mechanism for the sense of smell. But what about the sense of a possible heart attack? Or the sense of approaching mental depression? Or the sense of a sudden loss of emotional control while driving an automobile? Many experts claim that these latter "senses" actually do exist, but that human feedback mechanisms are not developed sufficiently for one to recognize them. However, if one is able to approximate or anticipate the time of these occurrences by careful observation of his biorhythm cycles, he is, in effect, creating a substitute feedback mechanism. Ultimately, after a period of consciously living in tune with his biorhythm cycles, the individual may find that he has developed an unconscious awareness of them, somewhat akin to a "sixth sense."

Awareness of our weak times and our strong times should help us plan, should help us to use and develop a potential that is so often disregarded. Recognition of the cyclic ups and downs in every life should make us a little more tolerant of a friend's bad moments. Knowledge that all of us have similar patterns but that each of us uses his differently should give us an appreciation of the complexity of individual existence.

For all of us, knowledge of our biorhythms opens up the possibility of understanding ourselves and our daily lives a little better. Self-knowledge is a high and difficult goal and every aid to understanding is worth the effort. As the late J. Bronowski said in *The Identity of Man*: "Man is a machine by birth and a self by experience."

Your biorhythms are your inheritance. What you do with them is your choice.

Two Centuries
of
Biorhythmic Projections
1800–2001

TABLES

APPENDIX A:

Birth Code Tables

Inquiry Year Tables

Biorhythm Data Tables

Simple-to-Follow Instructions
for the Use of the Tables
Are on Pages 18 and 19.

FIRST LETTER = YOUR PHYSICAL CODE, MIDDLE NUMBER = SENSITIVITY CODE, LAST TWO LETTERS = INTELLECTUAL CODE

1800

1800	JAN	FEB	MAR	APR	MAY	JUN	JUL	AUG	SEP	OCT	NOV	DEC
1	G38BF	N18BD	F18AJ	Y27AG	R19AD	F23AB	W32CP	R33CL	F35CJ	W24CE	R22BP	K34BK
2	L11BG	A12BE	V12AK	T13AH	X30AE	V36AC	Y20CR	X15CM	V16CK	Y21CF	X29CD	F26BL
3	D31BH	Z37BF	E37AL	G19AJ	B14AF	E32AD	T28CS	B25CP	E24CL	T17CH	B34CE	V38BM
4	J18BJ	C27BG	M27AM	L30AK	N23AG	M20AF	G33AB	N35CR	M21CM	G22CJ	N26CF	E11BP
5	U12BK	K13BH	H13BC	D14AL	A36AH	H28AF	L15AC	A16CS	H17CP	L29CK	A38CH	M31CD
6	R37BL	F19BJ	S19BD	J23AM	Z32AJ	S33AG	D25AD	Z24AB	S22CR	D34CL	Z11CJ	H18CE
7	X27BM	V30BE	P30BE	U36BC	C20AK	P15AH	J35AE	C21AC	P29CS	J26CM	C31CK	S12CF
8	B13BP	E14BL	W14BF	R32BD	K28AL	W25AJ	U16AF	K17AD	W34AB	U38CP	K18CL	P37CH
9	N19CD	M23BM	Y23BG	X20BE	F33AM	Y35AK	R24AG	F22AE	Y25AC	R11CR	F12CM	W27CJ
10	A30CE	H36BP	T36BH	B28BF	V15BG	T16AL	X21AH	V29AF	T38AD	X31CS	V37CP	Y13CK
11	Z14CF	S32CG	G32BJ	N33BG	E25BD	G24AM	B17AJ	E34AG	G11AE	B18AB	E27CR	T19CL
12	C23CH	P20CE	L20BK	A15BH	M35BE	L21BC	N22AK	M26AH	L31AF	N12AC	M13CS	G30CM
13	K36CJ	N26CF	D28BL	Z25BJ	H16BF	D17BD	A29AL	H38AG	D18AG	A37AD	H19AB	L14CP
14	F32CK	Y33CH	J33BM	C35BK	S24BG	J22BE	Z34AM	S11AK	J12AH	Z27AE	S30AC	D23CR
15	V20CL	T15CJ	U15BP	K16BL	P21BH	U29BF	C26BC	P31AL	U37AJ	C13AF	P14AD	J36CS
16	E28CM	G25CK	R25CD	F24BM	W17BJ	R34BG	K38BD	W18AM	R27AK	K19AG	W23AE	U32AB
17	M33CP	L35CL	X35CE	V21BP	Y22BK	X26BH	F11BE	Y12BC	X13AL	F30AH	Y36AF	R20AC
18	H15CR	D16CM	B16CF	E17CD	T29BL	B38BJ	V31BF	T37BD	B19AM	V14AJ	T32AG	X28AD
19	S25CS	J24CP	N24CH	M22CE	G34BM	N11BK	E18BG	G27BE	N30BC	E23AK	G20AH	B33AE
20	P35AB	U21CR	A21CJ	H29CF	L26BP	A31BL	M12BH	L13BF	A14BD	M36AL	L28AJ	N15AF
21	W16AC	R17CS	Z17CK	S34CH	D38CD	Z18BM	H37BJ	D19BG	Z23BE	H32AM	D33AK	A25AG
22	Y24AD	X22AB	C22CL	P26CJ	J11CE	C12BP	S27BK	J30BH	C36BF	S20BC	J15AL	Z35AH
23	T21AE	B29AC	K29CM	W36CK	U31CF	K37CD	P13BL	U14BJ	K32BG	P28BD	U25AM	C16AJ
24	G17AF	N34AD	F34CP	Y11CL	R18CH	F27CE	W19BM	R23BK	F20BH	W33BE	R35BC	K24AK
25	L22AG	A26AE	V26CR	T31CM	X12CJ	V13CF	Y30BP	X36BL	V28BJ	Y15BF	X16BD	F21AL
26	D29AH	Z38AF	E38CS	G18CP	B37CK	E19CH	T14CD	B32BM	E33BK	T25BG	B24BE	V17AM
27	J34AJ	C11AG	M11AB	L12CR	N27CL	M30CJ	G23CE	N20BP	M15BL	G35BH	N21BF	E22BC
28	U26AK	K31AH	H31AC	D37CS	A13CM	H14CK	L36CF	A28CD	H25BM	L16BJ	A17BG	M29BD
29	R38AL		S18AD	J27AB	Z19CP	S23CL	D32CH	Z33CE	S35BP	D24BK	Z22BH	H34BE
30	X11AM		P12AE	U13AC	C30CR	P36CM	J20CJ	C15CF	P16CD	J21BL	C29BJ	S26BF
31	B31BC		W37AF		K14CS		U28CK	K25CH		U17BM		P38BG

1801

1801	JAN	FEB	MAR	APR	MAY	JUN	JUL	AUG	SEP	OCT	NOV	DEC
1	W11BH	R12BF	Z12AL	S13AJ	D30AF	Z36AD	H20CS	D15CP	Z16CL	H21CH	D29CE	A26BM
2	Y31BJ	X37BG	C37AM	P19AK	J14AG	C32AE	S28AB	J25CR	C24CM	S17CJ	J34CF	Z38BP
3	T18BK	B27BH	K37BJ	W30AL	U23AH	K20AF	P33AC	U35CS	K21CP	P22CK	U26CH	C11CD
4	G12BL	N13BJ	F13BD	Y14AM	R36AJ	F28AG	W15AD	R16AB	F17CR	W29CL	R38CJ	K31CE
5	L37BM	A19BK	V19BE	T23BC	X32AK	V33AH	Y25AE	X24AC	V22CS	Y34CM	X11CK	F18CF
6	D27BP	Z30BL	E30BF	G36BD	B20AL	E15AJ	T35AF	B21AD	E29AB	T26CP	B31CL	V12CH
7	J13CD	C14BM	M14BG	L32BE	N28AM	M25AK	G16AG	N17AE	M34AC	G38CR	N18CM	E37CJ
8	U19CE	K23BP	H23BH	D20BF	A33BC	H35AL	L24AH	A22AF	H26AD	L11CS	A12CP	M27CK
9	R30CF	F36CD	S36BJ	J28BG	Z15BD	S16AM	D21AJ	Z29AG	S38AE	D31AB	Z37CM	H13CL
10	X14CH	V32CE	P32BK	U33BH	C25BE	P24BC	J17AK	C34AH	P11AF	J18AC	C27CS	S19CM
11	B23CJ	E20CF	W20BL	R15BJ	K35BF	W21BD	U22AL	K26AJ	W31AG	U12AD	K13AB	P30CP
12	N36CK	M28CH	Y28BM	X25BK	F16BG	Y17BE	R29AM	F38AK	Y18AH	R37AE	F19AC	W14CR
13	A32CL	H33CJ	T33BP	B35BL	V24BH	T22BF	X34BC	V11AL	T12AJ	X27AF	V30AD	Y23CS
14	Z20CM	S15CK	G15CD	N16BM	E21BJ	G29BG	B26BD	E31AM	G37AK	B13AG	E14AE	T36AB
15	C28CP	P25CL	L25CE	A24BP	M17BK	L34BH	N38BE	M18BC	L27AL	N19AH	M23AF	G32AC
16	K33CR	M35CM	D35CF	Z21CD	H22BL	D26BJ	A11BF	H12BD	D13AM	A30AJ	H36AG	L20AD
17	F15CS	Y16CP	J16CH	C17CE	S29BM	J38BK	Z31BG	S37BE	J19BC	Z14AK	S32AH	D28AE
18	V25AB	T24CR	U24CJ	K22CF	P34BP	U11BL	C18BH	P27BF	U30BD	C23AL	P20AJ	J33AF
19	E35AC	G21CS	R21CK	F29CH	W26CD	R31BM	K12BJ	W13BG	R14BE	K36AM	W28AK	U15AG
20	M16AD	L17AB	X17CL	V34CJ	Y38CE	X16BP	F37BK	Y19BH	X23BF	F32BC	Y33AL	R25AH
21	H24AE	D22AC	B22CM	E26CK	T11CF	B12CD	V27BL	T30BJ	B36BG	V20BD	T15AM	X35AJ
22	S21AF	J29AD	N29CP	M38CL	G31CH	N37CE	E13BM	G14BK	N32BH	E28BE	G25BC	B16AK
23	P17AG	U34AE	A34CR	H11CM	L18CJ	A27CF	M19BP	L23BL	A20BJ	M33BF	L35BD	N24AL
24	W22AH	R26AF	Z26CS	S31CP	D12CK	Z13CH	H30CD	D36BM	Z28BK	H15BG	D16BE	A21AM
25	Y29AJ	X38AG	C38AB	P18CR	J37CL	C19CJ	S14CE	J32BP	C33BL	S25BH	J24BF	Z17BC
26	T34AK	B11AH	K11AC	W12CS	U27CM	K30CK	P23CF	U20CD	K15BM	P35BJ	U21BG	C22BD
27	G26AL	N31AJ	F31AD	Y37AB	R13CP	F14CL	W36CH	R28CE	F25BP	W16BK	R17BH	K29BE
28	L38AM	A18AK	V18AE	T27AC	X19CR	V23CH	Y32CJ	X33CF	V35CD	Y24BL	X22BJ	F34BF
29	D11BC		E12AF	G13AD	B30CS	E36CP	T20CK	B15CH	E16CE	T21BM	B29BK	V26BG
30	J31BD		M37AG	L19AE	N14AB	M32CR	G28CL	N25CJ	M24CF	G17BP	U34BM	E38BH
31	U18BE		H27AH		A23AC		L33CH	A35CK		L22CD		M11BJ

1802

1802	JAN	FEB	MAR	APR	MAY	JUN	JUL	AUG	SEP	OCT	NOV	DEC
1	H31BK	D37BH	B37BC	E19AL	T14AH	B32AF	V28AC	T25CS	B24CP	V17CK	T34CH	X38CD
2	S18BL	J27BJ	N27BD	M30AM	G23AJ	N20AG	E33AD	G35AB	N21CR	E22CL	G26CJ	B11CE
3	P12BM	U13BK	A13BE	H14BC	L36AK	A28AH	M15AE	L16AC	A17CS	M29CM	L38CK	N31CF
4	W37BP	R19BL	Z19BF	S23BD	D32AL	Z33AJ	H25AF	D24AD	Z22AB	H34CP	D11CL	A16CH
5	Y27CD	X30BM	C30BG	P36BE	J20AM	C15AK	S35AG	J21AE	C29AC	S26CR	J31CM	Z12CJ
6	T13CE	B14BP	K14BH	W32BF	U28BC	K25AL	P16AH	U17AF	K34AD	P38CS	U16CD	C37CK
7	G19CF	N23CD	F23BJ	Y20BG	R33BD	F35AH	W24AJ	R22AG	F26AE	W11AB	R12CR	K27CL
8	L30CH	A36CE	V36BK	T28BH	X15BE	V16BC	Y21AK	X29AH	V38AF	Y31AC	X37CS	F13CM
9	D14CJ	Z32CF	E32BL	G33BJ	B25BF	E24BD	T17AL	B34AJ	E11AG	T18AD	B27AB	V19CR
10	J23CK	C20CH	M20BM	L15BK	N35BG	M21BE	G22AH	N26AK	M31AH	G12AE	N13AC	E30CR
11	U36CL	K28CJ	H28BP	D25BL	A16BH	H17BF	L29BC	A38AL	H18AJ	L37AF	A19AD	M14CS
12	R32CM	F33CL	S33CD	J35BM	Z24BJ	S22BG	D34BD	Z11AM	S12AK	D27AG	Z30AE	H23AB
13	X20CP	V15CL	P15CE	U16BP	C21BK	P29BH	J26BE	C31BC	P37AL	J13AH	C14AF	S36AC
14	B28CR	E25CM	W25CF	R24CD	K17BL	W34BJ	U38BF	K18BD	W27AM	U19AJ	K23AG	P32AD
15	N33CS	M35CP	Y35CH	X21CE	F22BM	Y26BK	R11BG	F12BE	Y13BC	R30AK	F36AH	W20AE
16	A15AB	H16CR	T16CJ	B17CF	V29BP	T38BL	X31BH	V37BF	T19BD	X14AL	V32AJ	Y28AF
17	Z25AC	S24CS	G24CK	N22CH	E34CD	G11BM	B18BJ	E27BG	G30BE	B23AM	E20AK	T33AG
18	C35AD	P21AB	L21CL	A29CJ	M26CE	L31BP	N12BK	M13BH	L14BF	N36BC	M28AL	G15AM
19	K16AE	W17AC	D17CM	Z34CK	H38CF	D18BD	A37BL	H19BJ	D23BG	A32BD	H33AM	L25AJ
20	F24AF	Y22AD	J22CP	C26CL	S11CH	J12CE	Z27BM	S30BK	J36BH	Z20BE	S15BC	D35AK
21	V21AG	T29AE	U29CR	K38CM	P31CJ	U37CF	C13BP	P14BL	U32BJ	C28BF	P25BD	J16AL
22	E17AH	G34AF	R34CS	F11CP	W18CK	R27CH	K19CD	W23BM	R20BK	K33BG	W35BE	U24AM
23	M22AJ	L26AG	X26AB	V31CR	Y12CL	X13CJ	F30CE	Y36BP	X28BL	F15BH	Y16BF	R21BC
24	H29AK	D38AH	B38AC	E18CS	T37CM	B19CK	V14CF	T32CD	B33BM	V25BJ	T24BG	X17BD
25	S34AL	J11AJ	N11AD	M12AB	G27CP	N30CL	E23CH	G20CE	N15BP	E35BK	G21BH	B22BE
26	P26AM	U31AK	A31AE	H37AC	L13CR	A14CM	M36CJ	L28CF	A25CD	M16BL	L17BJ	N29BF
27	W38BC	R18AL	Z18AF	S27AD	D19CS	Z23CP	H32CK	D33CH	Z35CE	H24BM	D22BK	A34BG
28	Y11BD	X12AM	C12AG	P13AE	J30AB	C36CR	S20CL	J15CJ	C16CF	S21BP	J29BL	Z26BH
29	T31BE		K37AH	W19AF	U14AC	K32CS	P28CM	U25CK	K24CH	P17CD	U34BM	C38BJ
30	G18BF		F27AJ	Y30AG	R23AD	F20AB	W33CP	R35CL	F21CJ	W22CE	R26BP	K11BK
31	L12BG		V13AK		X36AE		Y15CR	X16CH		Y29CF		F31BL

FIRST LETTER = YOUR PHYSICAL CODE, MIDDLE NUMBER = SENSITIVITY CODE, LAST TWO LETTERS = INTELLECTUAL CODE

1803	JAN	FEB	MAR	APR	MAY	JUN	JUL	AUG	SEP	OCT	NOV	DEC
1	V18BM	T27BK	U27BE	K30BC	P23AK	U20AH	C33AE	P35AC	U21CS	C22CM	P26CK	J11CF
2	E12BP	G13BL	R13BF	F14BD	W36AL	R28AJ	K15AF	W16AD	R17AB	K29CP	W38CL	U31CH
3	M37CD	L19BM	X19BG	V23BE	Y32AM	X33AK	F25AG	Y24AE	X22AC	F34CR	Y11CM	R18CJ
4	H27CE	D30BP	B30BH	E36BF	T20BC	B15AL	V35AH	T21AF	B29AD	V26CS	T31CP	X12CK
5	S13CF	J14CD	N14BJ	M32BG	G28BD	N25AM	E16AJ	G17AG	N34AE	E38AB	G18CR	B37CL
6	P19CH	U23CE	A23BK	H20BH	L33BE	A35BC	M24AL	L22AH	A26AF	M11AC	L12CS	N27CH
7	W30CJ	R36CF	Z36BL	S28BJ	D15BF	Z16BD	H21AL	D29AJ	Z38AG	H31AD	D37AB	A13CP
8	Y14CK	X32CM	C32BH	P33BK	J25BG	C24BE	S17AM	J34AK	C11AH	S18AE	J27AC	Z19CR
9	T23CL	B20CJ	K20BP	W15BL	U35BH	K21BF	P22BC	U26AL	K31AJ	P12AF	U13AD	C30CS
10	G36CM	N28CK	F28CD	Y25BM	R16BJ	F17BG	W29BD	R38AM	F18AK	W37AG	R19AE	K14AB
11	L32CP	A33CL	V33CE	T35BP	X24BK	V22BH	Y34BE	X11BC	V12AL	Y27AH	X30AF	F23AC
12	D20CR	Z15CM	E15CF	G16CD	B21BL	E29BJ	T26BF	B31BD	E37AM	T13AJ	B14AG	V36AD
13	J28CS	C25CP	M25CH	L24CE	N17BM	M34BK	G38BG	N18BE	M27BC	G19AK	N23AH	E32AE
14	U33AB	K35CR	H35CJ	D21CF	A22BP	H26BL	L11BH	A12BF	H13BD	L30AL	A36AJ	M20AF
15	R15AC	F16CS	S16CK	J17CH	Z29CD	S38BM	D31BJ	Z37BG	S19BE	D14AM	Z32AK	H28AG
16	X25AD	V24AB	P24CL	U22CJ	C34CE	P11BP	J18BK	C27BH	P30BF	J23BC	C20AL	S33AH
17	B35AE	E21AC	W21CM	R29CK	K26CF	W31CD	U12BL	K13BJ	W14BG	U36BD	K28AM	P15AJ
18	N16AF	M17AD	Y17CP	X34CL	F38CH	Y18CE	R37BM	F19BK	Y23BH	R32BE	F33BC	W25AK
19	A24AG	H22AE	T22CR	B26CM	V11CJ	T12CF	X27BP	V30BL	T36BJ	X20BF	V15BD	Y35AL
20	Z21AH	S29AF	G29CS	N38CP	E31CK	G37CH	B13CD	E14BM	G32BK	B28BG	E25BE	T16AM
21	C17AJ	P34AG	L34AB	A11CR	M18CL	L27CJ	N19CE	M23BP	_20BL	N33BH	M35BF	G24BC
22	K22AK	W26AH	D26AC	Z31CS	H12CM	D13CK	A30CF	H36CD	D28BM	A15BJ	H16BG	L21BD
23	F29AL	Y38AJ	J38AD	C18AB	S37CP	J19CL	Z14CH	S32CE	J33BP	Z25BK	S24BH	D17BE
24	V34AM	T11AK	U11AE	K12AC	P27CR	U30CM	C23CJ	P20CF	U15CD	C35BL	P21BJ	J22BF
25	E26BC	G31AL	R31AF	F37AD	H13CS	R14CP	K36CK	W28CH	K25CE	K16BM	W17BK	U29BG
26	M38BD	L18AM	X18AG	V27AE	Y19AB	X23CR	F32CL	Y33CJ	X35CF	F24BP	Y22BL	R34BH
27	H11BE	D12BC	B12AH	E13AF	T30AC	B36CS	V20CH	T15CK	B16CH	V21CD	T29BM	X26BJ
28	S31BF	J37BD	N37AJ	M19AG	G14AD	N32AB	E28CP	G25CL	N24CJ	E17CE	G34BP	B38BK
29	P18BG		A27AK	H30AH	L23AE	A20AC	M33CR	L35CM	A21CK	M22CF	L26CD	N11BL
30	W12BH		Z13AL	S14AJ	D36AF	Z28AD	H15CS	D16CP	Z17CL	H29CH	D38CE	A31BM
31	Y37BJ		C19AM		J32AG		S25AB	J24CR		S34CJ		Z18BP

1804	JAN	FEB	MAR	APR	MAY	JUN	JUL	AUG	SEP	OCT	NOV	DEC
1	C12CD	P13BM	D19BH	Z23BF	H32BC	D33AL	A25AH	H24AF	D22AD	A34CS	H11CP	L18CK
2	K37CE	W19BP	J30BJ	C36BG	S20BD	J15AM	Z35AJ	S21AG	J29AE	Z26AB	S31CR	D12CL
3	F27CF	Y30CD	U14BK	K32BH	P28BE	U25BC	C16AK	P17AH	U34AF	C38AC	P18CS	J37CH
4	V13CH	T14CE	R23BL	F20BJ	W33BF	R35BD	K24AL	W22AJ	R26AG	K11AD	W12AB	U27CP
5	E19CJ	G23CF	X36BM	V28BK	Y15BG	X16BE	F21AM	Y29AK	X38AH	F31AE	Y37AC	R13CR
6	M30CK	L36CH	B32BP	E33BL	T25BH	B24BF	V17BC	T34AL	B11AJ	V18AF	T27AD	X19CS
7	H14CL	D32CJ	N20CD	M15BM	G35BJ	N21BG	E22BD	G26AM	N31AK	E12AG	G13AE	B30AB
8	S23CM	J20CK	A28CE	H25BP	L16BL	A17BH	M29BE	L38BC	A18AL	M37AH	L19AF	N14AC
9	P36CP	U28CL	Z33CF	S35CD	D24BL	Z22BJ	H34BF	D11BD	Z12AM	H27AJ	D30AG	A23AD
10	W32CR	R33CM	C15CH	P16CE	J21BM	C29BK	S26BG	J31BE	C37BC	S13AK	J14AH	Z36AE
11	Y20CS	X15CP	K25CJ	W24CF	U17BP	K34BL	P38BH	U18BF	K27BD	P19AL	U23AJ	C32AF
12	T28AB	B25CR	F35CK	Y21CH	R22CD	F26BM	W11BJ	R12BG	F13BE	W30AM	R36AK	K20AG
13	G33AC	N35CS	V16CL	T17CJ	X29CE	V38BP	Y31BK	X37BH	V19BF	Y14BC	X32AL	F28AH
14	L15AD	A16AB	E24CM	G22CK	B34CF	E11BD	T18BL	B27BJ	E30BG	T23BD	B20AM	V33AJ
15	D25AE	Z24AC	M21CP	L29CL	N26CH	M31BJ	G12BM	N13BK	M14BH	G36BE	N28BC	E15AK
16	J35AF	C21AD	H17CR	D34CM	A38CJ	H18CF	L37BP	A19BL	H23BJ	L32BF	A33BD	M25AL
17	U16AG	K17AE	S22CS	J26CP	Z11CK	S12CH	D27CD	Z30BM	S36BK	D20BG	Z15BE	H35AM
18	R24AH	F22AF	P29AB	U38CR	C31CL	P37CJ	J13CE	C14BP	P28BH	J28BH	C25BF	S16BC
19	X21AJ	V29AG	W34AM	R11CS	K18CM	W27CK	U19CF	K23CD	W20BM	U33BJ	K35BG	P24BD
20	B17AK	E34AH	Y26AD	X31AB	F12CP	Y13CL	R30CH	F36CE	Y28BP	R15BK	F16BH	W34AM
21	N22AL	M26AJ	T38AE	B18AC	V37CR	T19CM	X14CJ	V32CF	T33CD	X25BL	V24BJ	Y17BF
22	A29AM	H38AK	G11AF	N12AD	E27CS	G30CP	B23CK	E20CH	G15CE	B35BM	E21BK	T22BG
23	Z34BC	S11AL	L31AG	A37AE	M13AB	L14CR	N36CL	M28CJ	L25CF	N16BP	M17BL	G29BH
24	C26BD	P31AM	D18AH	Z27AF	H19AB	D23CS	A32CM	H33CK	D35CH	A24CD	H22BM	L34BJ
25	K38BE	W18BC	J12AJ	C13AG	S30AD	J36AB	Z20CP	S15CL	J16CJ	Z21CE	S29BP	D26BK
26	F11BF	Y12BD	U37AK	K19AH	P14AE	U32AC	C28CR	P25CH	U24CK	C17CF	P34CD	J38BL
27	V31BG	T37BE	R27AL	F30AJ	W23AF	R20AD	K33CS	W35CP	R21CL	K22CH	W26CE	U11BM
28	E18BH	G27BF	X13AM	V14AK	Y36AG	X28AE	F15AB	Y16CR	X17CM	F29CJ	Y38CF	R31BP
29	M12BJ	L13BG	B19BC	E23AL	T32AH	B33AF	V25AC	T24CS	B22CP	V34CK	T11CH	X18CD
30	H37BK		N30BD	M36AM	G20AJ	N15AG	E35AD	G21AB	N29CR	E26CL	G31CJ	B12CE
31	S27BL		A14BE		L28AK		M16AE	L17AC		M38CH		N37CF

1805	JAN	FEB	MAR	APR	MAY	JUN	JUL	AUG	SEP	OCT	NOV	DEC
1	A27CH	H30CE	T30BK	B36BH	V20BE	T15BC	X35AK	V21AH	T29AF	X26AC	V31CS	Y12CM
2	Z13CJ	S14CF	G14BL	N32BJ	E28BF	G25BD	B16AL	E17AJ	G34AG	B38AD	E18AB	T37CP
3	C19CK	P23CH	L23BM	A20BK	M33BG	L35BE	N24AM	M22AK	L26AH	N11AE	M12AC	G27CR
4	K30CL	W36CJ	D36BP	Z28BL	H15BH	D16BF	A21BC	H29AL	D38AJ	A31AF	H37AD	L13CS
5	F14CM	Y32CK	J32CD	C33BM	S25BJ	J24BG	Z17BD	S34AM	J11AK	Z18AG	S27AE	D19AB
6	V23CP	T20CL	U20CE	K15BP	P35BK	U21BH	C22BE	P26BC	U31AL	C12AH	P13AF	J30AC
7	E36CR	G28CM	R28CF	F25CD	W16BL	R17BJ	K29BF	W38BD	R18AM	K37AJ	W19AG	U14AD
8	M32CS	L33CP	X33CH	V35CE	T24BM	X22BK	F34BG	Y11BE	X12BC	F27AK	Y30AH	R23AE
9	H20AB	D15CR	B15CJ	E16CF	T21BP	B29BL	V26BH	T31BF	B37BD	V13AL	T14AJ	X36AF
10	S28AC	J25CS	N25CK	M24CH	G17CD	N34BM	E38BJ	G18BG	N27BE	E19AM	G23AK	B32AG
11	P33AD	U35AB	A35CL	H21CJ	L22CE	A26BP	M11BK	L12BH	A13BF	M30BC	L36AL	N20AH
12	W15AE	R16AC	Z16CM	S17CK	D29CF	Z38CD	H31BD	D37BD	Z19BG	H14BD	D32AM	A28AJ
13	Y25AF	X24AD	C24CP	P22CL	J34CH	C11CE	S18BM	J27BK	C30BH	S23BD	J20BC	Z33AK
14	T35AG	B21AE	K21CR	W29CM	U26CJ	K31CF	P12BP	U13BL	K14BJ	P36BF	U28BD	C15AL
15	G16AH	N17AF	F17CS	Y34CP	R38CK	F18CH	W37CD	R19BM	F23BK	W32BG	R33BE	K25AM
16	L24AJ	A22AG	V22AB	T26CR	X11CL	V12CJ	Y27CE	X30BP	V36BL	Y20BH	X15BF	F35BC
17	D21AK	Z29AH	E29AC	G38CS	B31CM	E37CK	T13CF	B14CD	Z28BM	T28BD	B25BG	V16BD
18	J17AL	C34AJ	M34AD	L11AB	N18CP	M27CL	G19CH	N23CE	M20BP	G33BK	N35BH	E24BE
19	U22AM	K26AK	H26AE	D31AC	A12CR	H13CM	L30CJ	A36CF	H28CD	L15BL	A16BJ	M21BF
20	R29BC	F38AL	S38AF	J18AD	Z37CS	S19CP	D14CK	Z32CH	S33CE	D25BM	Z24BK	H17BG
21	X34BD	V11AM	P11AG	U12AE	C27AB	P30CR	J23CL	C20CJ	P15CF	J35BP	C21BL	S22BH
22	B26BE	E31BC	W31AH	R37AF	K13AC	W14CS	U36CM	K28CK	W25CH	U16CD	K17BM	P29BJ
23	N38BF	M18BD	Y18AJ	X27AG	F19AD	Y23AB	R32CP	F33CL	Y35CJ	R24CE	F22BP	W34BK
24	A11BG	H12BE	T12AK	B13AH	V30AE	T36AC	X20CR	V15CM	T16CK	X21CF	V29CD	Y26BL
25	Z31BH	S37BF	G37AL	N19AJ	E14AF	G32AD	B28CS	E25CP	G24CL	B17CH	E34CE	T38BM
26	C18BJ	P27BG	L27AM	A30AK	M23AG	L20AE	N33AB	M35CR	L21CM	N22CJ	M26CF	G11BP
27	K12BK	W13BH	D13BC	Z14AL	H36AH	D28AF	A15AC	H16CS	D17CP	A29CK	H38CH	L31CD
28	F37BL	Y19BJ	J19BD	C23AM	S32AJ	J33AG	Z25AD	S24AB	J22CR	Z34CL	S11CJ	D18CE
29	V27BM		U30BE	K36BC	P20AK	U15AH	C35AE	P21AC	U29CS	C26CM	P31CK	J12CF
30	E13BP		R14BF	F32BD	W28AL	R25AJ	K16AF	W17AD	R34AB	K38CP	W18CL	U37CH
31	M19CD		X23BG		Y33AM		F24AG	Y22AE		F11CR		R27CJ

FIRST LETTER = YOUR PHYSICAL CODE, MIDDLE NUMBER = SENSITIVITY CODE, LAST TWO LETTERS = INTELLECTUAL CODE

1806

	JAN	FEB	MAR	APR	MAY	JUN	JUL	AUG	SEP	OCT	NOV	DEC
1	X13CK	V14CH	P14BM	U32BK	C28BG	P25BE	J16AM	C17AK	P34AH	J38AE	C18AC	S37CR
2	B19CL	E23CJ	W23BP	R20BL	K33BH	W35BF	U24BC	K22AL	W26AJ	U11AF	K12AD	P27CS
3	N30CM	M36CK	Y36CD	X28BM	F15BJ	Y16BG	R21BD	F29AM	Y38AK	R31AG	F37AE	W13AB
4	A14CP	H32CL	T32CE	B33BP	V25BK	T24BH	X17BE	V34BC	T11AL	X18AH	V27AF	Y19AC
5	Z23CR	S20CM	G20CF	N15CD	E35BL	G21BJ	B22BF	E26BD	G31AM	B12AJ	E13AG	T30AD
6	C36CS	P28CP	L28CH	A25CE	M16BM	L17BK	N29BG	M38BE	L18BC	N37AK	M19AH	G14AE
7	K32AB	M33CR	D33CJ	Z35CF	H24BP	D22BL	A34BH	H11BF	D12BD	A27AL	H30AJ	L23AF
8	F20AC	Y15CS	J15CK	C16CH	S21CD	J29BM	Z26BJ	S31BG	J37BE	Z13AM	S14AK	D36AG
9	V28AD	T25AB	U25CL	K24CJ	P17CE	U34BP	C38BK	P18BH	U27BF	C19BC	P23AL	J32AH
10	E33AE	G35AC	R35CM	F21CK	W22CF	R26CD	K11BL	W12BJ	R13BG	K30BD	W36AM	U20AJ
11	M15AF	L16AD	X16CP	V17CL	Y29CH	X38CE	F31BM	Y37BK	X19BH	F14BE	Y32BC	R28AK
12	H25AG	D24AE	B24CR	E22CM	T34CJ	B11CF	V18BP	T27BL	B30BJ	V23BF	T20BD	X33AL
13	S35AH	J21AF	N21CS	M29CP	G26CK	N31CH	E12CD	G13BM	N14BK	E36BG	G28BE	B15AM
14	P16AJ	U17AG	A17AB	H34CR	L38CL	A18CJ	M37CE	L19BP	A23BL	M32BH	L33BF	N25BC
15	W24AK	R22AH	Z22AC	S26CS	D11CM	Z12CK	H27CF	D30CD	Z36BM	H20BJ	D15BG	A35BD
16	Y21AL	X29AJ	C29AD	P38AB	J31CP	C37CL	S13CH	J14CE	C32BP	S28BK	J25BH	Z16BE
17	T17AM	B34AK	K34AE	W11AC	U18CR	K27CM	P19CJ	U23CF	K20CD	P33BL	U35BJ	C24BF
18	G22BC	N26AL	F26AF	Y31AD	R12CS	F13CP	W30CK	R36CH	F28CE	W15BM	R16BK	K21BG
19	L29BD	A38AM	V38AG	T18AE	X37AB	V19CR	Y14CL	X32CG	V33CF	Y25BP	X24BL	F17BH
20	D34BE	Z11BC	E11AH	G12AF	B27AC	E30CS	T23CM	B20CK	E15CH	T35CD	B21BM	V22BJ
21	J26BF	C31BD	M31AJ	L37AG	N13AD	M14AB	G36CP	N28CL	M25CJ	G16CE	N17BP	E29BK
22	U38BG	K18BE	H18AK	D27AH	A19AE	H23AC	L32CR	A33CM	H35CK	L24CF	A22CD	M34BL
23	R11BH	F12BF	S12AL	J13AJ	Z30AF	S36AD	D20CS	Z15CP	S16CL	D21CH	Z29CE	H26BM
24	X31BJ	V37BG	P37AM	U19AK	C14AG	P32AE	J28AB	C25CR	P24CH	J17CJ	C34CF	S38BP
25	B18BK	E27BH	W27BC	R30AL	K23AH	W20AF	U33AC	K35CS	W21CP	U22CK	K26CH	P11CD
26	N12BL	M13BJ	Y13BD	X14AM	F36AJ	Y28AG	R15AD	F16AB	Y17CR	R29CL	F38CJ	W31CE
27	A37BM	H19BK	T19BE	B23BC	V32AK	T33AH	X25AF	V24AC	T22CS	X34CM	V11CK	Y18CF
28	Z27BP	S30BL	G30BF	N36BD	E20AL	G15AJ	B35AF	E21AD	G29AB	B26CP	E31CL	T12CH
29	C13CD		L14BG	A32BE	M28AM	L25AK	N16AG	M17AE	L34AC	N38CR	M18CM	G37CJ
30	K19CE		D23BH	Z20BF	H33BC	D35AL	A24AH	H22AF	D26AJ	A11CS	H12CP	L27CK
31	F30CF		J36BJ		S15BD		Z21AJ	S29AG		Z31AB		D13CL

1807

	JAN	FEB	MAR	APR	MAY	JUN	JUL	AUG	SEP	OCT	NOV	DEC
1	J19CM	C23CK	M23CD	L20BM	N33BJ	M35BG	G24BD	N22AM	M25AK	G11AG	N12AE	E27AB
2	U30CP	K36CL	H36CE	D28BP	A15BK	H16BH	L21BE	A29BC	H38AL	L31AH	A37AF	M13AC
3	R14CR	F32CM	S32CF	J33CD	Z25BL	S24BJ	D17BF	Z34BD	S11AM	D18AJ	Z27AG	H19AD
4	X23CS	V20CP	P20CH	U15CE	C35BM	P21BK	J22BG	C26BE	P31BC	J12AK	C13AH	S30AE
5	B36AB	E28CR	W28CJ	R25CF	K16BP	W17BL	U29BH	K38BF	W18BD	U37AL	K19AJ	P14AF
6	N32AC	M33CS	Y33CK	X35CH	F24CD	Y22BM	R34BJ	F11BG	Y12BE	R27AM	F30AK	W23AG
7	A20AD	H15AB	T15CL	B16CJ	V21CE	T29BP	X26BK	V31BH	T37BF	X13BC	V14AL	Y36AH
8	Z28AE	S25AC	G25CM	N24CA	E17CF	G34CD	B38BL	E18BJ	G27BG	B19BD	E23AH	T32AJ
9	C33AF	P35AD	L35CP	A21CL	M22CH	L26CE	A31BP	M37BL	-13BH	N30BE	M36BC	G20AK
10	K15AG	W16AE	D16CM	Z17CH	H29CJ	D38CF	A31BP	H37BL	D19BJ	A14BF	H32BD	L28AL
11	F25AH	Y24AF	J24CS	C22CP	S34CK	J11CH	Z18CD	S27BM	J30BK	Z23BG	S20BE	D33AM
12	V35AJ	T21AG	U21AB	K29CR	P26CL	U31CJ	C12CE	P13BP	U14BL	C36BH	P28BF	J15BC
13	E16AK	G17AH	R17AC	F34CS	W28CH	R18CK	K37CF	W19CD	R23BH	K32BJ	W33BG	U25BD
14	M24AL	L22AJ	X22AD	V26AB	Y11CP	X12CL	F27CH	Y30CE	X36BP	F20BK	Y15BH	R35BE
15	H21AM	D29AK	B29AE	E38AC	T31CH	B37CM	V13CL	T14CF	B32CD	V28BL	T25BJ	X16BF
16	S17BC	J34AL	N34AF	M11AD	G18CS	N27CP	E19CK	G23CH	N20CE	E33BM	G35BK	B24BG
17	P22BD	U26AM	A26AG	H31AE	L12AB	A13CR	M30CL	L36CJ	A28CF	M15BP	L16BL	N21BH
18	W29BE	R38BC	Z38AH	S18AF	D37AC	Z19CS	H14CM	D32CK	Z33CH	H25CD	D24BM	A17BJ
19	Y34BF	X11BD	C11AJ	P12AG	J27AD	C30AB	S23CP	J20CL	C15CJ	S35CE	J21BP	Z22BK
20	T26BG	B31BE	K31AK	W37AH	U13AE	K14AC	P36CR	U28CM	K25CK	P16CF	U17CD	C29BL
21	G38BH	N18BF	F18AL	Y27AJ	R19AF	F23AD	W32CS	R33CP	F35CL	W24CH	R22CE	K34BM
22	L11BJ	A12BG	V12AM	T13AK	X30AG	V36AE	Y20AB	X15CR	V16CM	Y21CJ	X29CF	F26BP
23	D31BK	Z37BM	E37BC	G19AL	B14AH	E32AF	T28AC	B25CS	E24CP	T17CK	B34CH	V38CD
24	J18BL	C27BJ	M27BD	L30AM	N23AJ	M20AG	G33AD	N35AB	M21CR	G22CL	N26CJ	E11CE
25	U12BM	K13BK	H13BE	D14BC	A36AK	H28AH	L15AE	A16AC	H17CS	L29CH	A38CK	M31CF
26	R37BP	F19BL	S19BF	J23BD	Z32AL	S33AJ	D25AF	Z24AD	S22AB	D34CP	Z11CL	H18CH
27	X27CD	V30BM	P30BG	U36BE	C20AM	P15AK	J35AG	C21AE	P29AC	J26CR	C31CM	S12CJ
28	B13CE	E14BP	W18BH	R32BF	K28BC	W25AL	U16AH	K17AF	W34AD	U38CS	K18CP	P37CK
29	N19CF		Y23BJ	X20BG	F33BD	Y35AM	R24AJ	F22AG	Y26AE	R11AB	F12CR	W27CL
30	A30CH		T36BK	B28BH	V15BE	T16BC	X21AK	V29AH	T38AF	X31AC	V37CS	Y13CM
31	Z14CJ		G32BL		E25BF		B17AL	E34AJ		B18AD		T19CP

1808

	JAN	FEB	MAR	APR	MAY	JUN	JUL	AUG	SEP	OCT	NOV	DEC
1	G30CR	N36CM	V32CH	T33CE	X25BM	V24BK	T17BG	X34BE	V11BC	Y18AK	X27AH	F19AE
2	L14CS	A32CP	E20CJ	G15CF	B35BP	E21BL	T22BH	B26BF	E31BD	T12AL	B13AJ	V30AF
3	D23AB	Z20CR	M28CK	L25CH	N16CD	M17BM	G29BJ	N38BG	M18BE	G37AM	N19AK	E14AG
4	J36AC	C28CS	H33CL	D35CJ	A24CE	H22BP	L34BK	A11BH	H12BF	L27BC	A30AL	M23AH
5	U32AD	K33AB	S15CM	J16CK	Z21CF	S29CD	D26BL	Z31BJ	S37BG	D13BD	Z14AM	H36AJ
6	R20AE	F15AC	P25CP	U24CL	C17CH	P34CE	J38BH	C18BK	P27BH	J19BE	C23BC	S32AK
7	X28AF	V25AD	W35CR	R21CH	K22CJ	W26CF	U11BP	K12BL	W13BJ	U30BF	K36BD	P20AL
8	B33AG	E35AE	Y16CS	X17CP	F29CK	Y38CH	R31CD	F37BH	Y19BK	R14BG	F32BE	W28AM
9	N15AH	M16AF	T24AB	B22CR	V34CL	T11CJ	X18CC	V27BP	T30BL	X23BH	V20BF	Y33BC
10	A25AJ	H24AG	G21AC	N29CS	E26CM	G31CK	B12CF	E13CD	G14BM	B36BJ	E28BG	T15BD
11	Z35AK	S21AM	L17AD	A34CB	M38CP	L18CL	N37CH	M19CE	L23BP	N32BK	M33BH	G25BE
12	C16AL	P17AJ	D22AE	Z26AC	H11CR	D12CM	A27CJ	H30CF	D35CD	A20BL	H15BJ	L35BF
13	K24AM	W22AK	J29AF	C38AD	S31CS	J37CP	Z13CK	S14CH	J32CE	Z28BM	S25BK	D16BG
14	F21BC	Y29AL	U34AG	K11AE	P18AB	U27CR	C19CL	P23CJ	U20CF	C33BP	P35BL	J24BH
15	V17BD	T34AM	R26AH	F31AE	W12AC	R13CS	K30CM	W36CK	R28CH	K15CD	W16BM	U21BJ
16	E22BE	G26BC	X38AJ	V18AG	Y37AD	X19BM	F14CP	Y32CL	X33CJ	F25CE	Y24BP	R17BK
17	M29BF	L38BD	B11AK	E12AH	T27AE	B30AL	V23CR	T20CM	B15CK	V35CF	T21CD	X22BL
18	H34BG	D11BE	N31AL	M37AJ	G13AF	N14AD	E36CS	G28CP	N25CL	E16CH	G17CD	B29BM
19	S26BH	J31BF	A18AM	H27AK	L19AG	A23AE	M32AB	L33CR	A35CH	M24CJ	L22CF	N34BP
20	P38BJ	U18BG	Z12BC	S13AL	D30AH	J14AJ	C32AG	D15CS	Z16CP	H21CK	D29CH	A26CD
21	W11BK	R12BH	C37BD	P19AM	J14AJ	C32AG	S28AD	J25AB	C24CR	S17CL	J34CJ	Z38CE
22	Y31BL	X37BJ	K27BE	W30BC	U23AK	K20AH	P33AE	U35AC	K21CS	C22CH	U26CK	C11CF
23	T18BM	B27BK	F13BF	Y14BD	R36AL	F28AJ	W15AF	R16AD	F17AB	W29CP	R38CL	K31CH
24	G12BP	N13BL	V19BG	T23BE	G36BF	V33AK	Y25AG	X24AE	V22AC	Y34CR	X11CH	F18CJ
25	L37CD	A19BM	E30BH	G36BF	B20BC	E15AL	T35AH	Y25AG	B21AF	T26CS	X31CM	V12CH
26	D27CE	Z30BP	M14BJ	L32BG	N28BD	M25AH	G16AJ	N17AG	H34AE	G38AB	N18CR	E37CL
27	J13CF	C14CD	H23BK	D20BH	A33BE	H35AG	L24AH	A22AM	H26AF	L11AD	A12CS	M27CM
28	U19CH	K23CE	S36BL	J28BJ	Z15BF	S16BD	D21AL	Z29AJ	S38AG	D31AD	Z37AB	H13CP
29	R30CJ	F36CF	P32BM	U33BK	C25BG	P24BE	J17AM	C34AK	P11AH	J18AE	C27AC	S19CR
30	X14CK		W20BP	R15BL	K35BH	W21BF	U22BC	K26AL	W31AJ	U12AF	K13AD	P30CS
31	B23CL		Y28CD		F16BJ		R29BD	F38AM		R37AG		W14AB

FIRST LETTER = YOUR PHYSICAL CODE, MIDDLE NUMBER = SENSITIVITY CODE, LAST TWO LETTERS = INTELLECTUAL CODE

1809	JAN	FEB	MAR	APR	MAY	JUN	JUL	AUG	SEP	OCT	NOV	DEC
1	Y23AC	X20CS	C20CK	P15CH	J35CD	C21BM	S22BJ	J26BG	C31BE	S12AM	J13AK	Z30AG
2	T36AD	B28AB	K28CL	W25CJ	U16CE	K17BP	P29BK	U38BH	K18BF	P37BC	U19AL	C14AH
3	G32AE	N33AC	F33CM	Y35CK	R24CF	F22CD	W34BL	R11BJ	F12BG	W27BD	R30AM	K23AJ
4	L20AF	A15AD	V15CP	T16CL	X21CH	V29CE	Y26BM	X31BK	V37BH	Y13BE	X14BC	F36AK
5	D28AG	Z25AE	E25CR	G24CM	B17CJ	E34CF	T38BP	B18BL	E27BJ	T19BF	B23BD	V32AL
6	J33AH	C35AF	M35CS	L21CP	N22CK	H26CH	G11CD	N12BM	M13BK	G30BG	N36BE	E20AM
7	U15AJ	K16AG	H16AB	D17CR	A29CL	H38CJ	L31CE	A37BP	H19BL	L14BH	A32BF	M28BC
8	R25AK	F24AH	S24AC	J22CS	Z34CM	S11CK	D18CF	Z27CD	S30BM	D23BJ	Z20BG	H33BD
9	X35AL	V21AJ	P21AD	U29AB	C26CP	P31CL	J12CH	C13CE	P14BP	J36BK	C28BH	S15BE
10	B16AM	E17AK	W17AE	R34AC	K38CR	W18CM	U37CJ	K19CF	W23CD	U32BL	K33BJ	P25BF
11	N24BC	M22AL	Y22AF	X26AD	F11CS	Y12CP	R27CK	F30CH	Y36CE	R20BM	F15BK	W35BG
12	A21BD	H29AM	T29AG	B38AE	V31AB	T37CR	X13CL	V14CJ	T32CF	X28BP	V25BL	Y16BH
13	Z17BE	S34BC	G34AH	N11AF	E18AC	G27CS	B19CM	E23CK	G20CH	B33CD	E35BM	T24BJ
14	C22BF	P26BD	L26AJ	A31AG	M12AD	L13AB	N30CP	M36CL	L28CJ	N15CE	M16BP	G21BK
15	K29BG	W38BE	D38AK	Z18AH	H37AE	D19AC	A14CR	H32CM	D33CK	A25CF	H24CD	L17BL
16	F34BH	Y11BF	J11AL	C12AJ	S27AF	J30AD	Z23CS	S20CP	J15CL	Z35CH	S21CE	D22BM
17	V26BJ	T31BG	U31AM	K37AK	P13AG	U14AE	C36AB	P28CR	U25CM	C16CJ	P17CF	J29BP
18	E38BK	G18BH	R18BC	F27AL	W19AH	R23AF	K32AC	W33CS	R35CP	K24CK	W22CH	U34CD
19	M11BL	L12BJ	X12BD	V13AM	Y30AJ	X36AG	F20AD	Y15AB	X16CR	F21CL	Y29CJ	R26CE
20	H31BM	D37BK	B37BE	E19BC	T14AK	B32AH	V28AE	T25AC	B24CS	V17CM	T34CK	X38CF
21	S18BP	J27BL	N27BF	M30BD	G23AL	N20AJ	E33AF	G35AD	N21AB	E22CP	G26CL	B11CH
22	P12CD	U13BM	H28BG	H14BE	L36AM	A28AK	H15AG	L16AE	A17AC	H29CR	L38CM	N31CJ
23	W37CE	R19BP	Z19BH	S23BF	D32BC	Z33AL	H25AH	D24AF	Z22AD	H34CS	D11CP	A18CK
24	Y27CF	X30CD	C30BJ	P36BG	J20BD	C15AM	S35AJ	J21AG	C29AE	S26AB	J31CR	Z12CL
25	T13CH	B14CE	K14BK	W32BH	U28BE	K25BC	P16AK	U17AH	K34AF	P38AC	U18CS	C37CM
26	G19CJ	N23CF	F23BL	Y20BJ	R33BF	F35BD	W24AL	R22AJ	F26AG	W11AD	R12AB	K27CP
27	L30CK	A35CH	W36BM	T28BK	X15BG	V16BE	Y21AM	X29AK	V38AH	Y31AE	X37AC	F13CR
28	D14CL	Z32CJ	E32BP	G33BL	B25BH	E24BF	T17BC	B34AL	E11AJ	T18AF	B27AD	V19CS
29	J23CM		M20CD	L15BM	N35BJ	M21BG	G22BD	N26AM	M31AK	G12AG	N13AE	E30AB
30	U36CP		H28CE	D25BP	A16BK	H17BH	L29BE	A38BC	H18AL	L37AH	A19AF	M14AC
31	R32CR		S33CF		Z24BL		D34BF	Z11BD		D27AJ		H23AD

1810	JAN	FEB	MAR	APR	MAY	JUN	JUL	AUG	SEP	OCT	NOV	DEC
1	S36AE	J28AC	N28CM	M25CK	G16CF	N17CD	E29BL	G38BJ	N18BG	E37BD	G19AM	B14AJ
2	P32AF	U33AD	A33CP	H35CL	L24CH	A22CE	M34BM	L11BK	A12BH	M27BE	L30BC	N23AK
3	W20AG	R15AE	Z15CR	S16CM	D21CJ	Z29CF	H26BP	D31BL	Z37BJ	H13BF	D14BD	A36AL
4	Y28AH	X25AF	C25CS	P24CP	J17CK	C34CH	S38CD	J18BM	C27BK	S19BG	J23BE	Z32AM
5	T33AJ	B35AG	K35AB	W21CR	U22CL	K26CJ	P11CE	U12BP	K13BL	P30BH	U36BF	C20BC
6	G15AK	N16AH	F16AC	Y17CS	R29CM	F38CK	W31CF	R37CD	F19BM	W14BJ	R32BG	K28BD
7	L25AL	A24AJ	V24AD	T22AB	X34CP	V11CL	Y18CH	X27CE	V30BP	Y23BK	X20BH	F33BE
8	D35AM	Z21AK	E21AE	G29AC	B26CR	E31CH	T12CJ	B13CF	E14CD	T36BL	B28BJ	V15BF
9	J16BC	C17AL	M17AF	L34AD	N38CS	M18CP	G37CK	N19CH	M23CE	G32BM	N33BK	E25BG
10	U24BD	K22AM	H22AG	D26AF	A11AB	H12CR	L27CL	A30CJ	H36CF	L20BP	A15BL	M35BH
11	R21BE	F29BC	S29AH	J38AF	Z31AC	S37CS	D13CM	Z14CK	S32CH	D28CD	Z25BM	H16BJ
12	X17BF	V34BD	P34AJ	U11AG	C18AD	P27AB	J19CP	C23CL	P20CJ	J33CE	C35BP	S24BK
13	B22BG	E26BE	W26AK	R31AH	K12AE	W13AC	U30CR	K36CM	W28CS	U15CF	K16CD	P21BL
14	N29BH	M38BF	Y38AL	X18AJ	F37AF	Y19AD	R14CS	F32CP	Y33CL	R25CH	F24CE	H17BM
15	A34BJ	H11BG	T11AM	B12AK	V27AG	T30AE	X23AB	V20CR	T15CM	X35CJ	V21CF	Y22BP
16	Z26BK	S31BH	G31BC	N37AL	E13AH	G14AF	B36AC	E28CS	G25CP	B16CK	E17CH	T29CD
17	C38BL	P18BJ	L18BD	A27AM	M19AJ	L23AG	N32AD	M33AB	L35CS	N24CL	M22CJ	G34CE
18	K11BM	M12BK	D12BE	Z13BC	H30AK	D36AH	A20AE	H15AC	D16CS	A21CM	H29CK	L26CF
19	F31BP	Y37BL	J37BF	C19BD	S14AL	J32AJ	Z28AF	S25AD	J24AB	Z17CP	S34CL	D38CH
20	V18CD	T27BM	U27BG	K30BE	P23AM	U20AK	C33AG	P35AE	U21AC	C22CR	P26CM	J11CJ
21	E12CE	G13BP	R13BH	F14BF	W36AC	R28AL	K15AM	W16AF	R17AD	K29CS	W38CP	U31CK
22	M37CF	L19CD	X19BJ	V23BG	Y32BD	X33AM	F25AJ	Y24AG	X22AE	F34AB	Y11CR	R18CL
23	H27CH	D30CE	B30BK	E36BH	T20BE	B15BC	V35AK	T21AH	B29AF	V26AC	T31CS	X12CM
24	S13CJ	J14CF	N14BL	M32BJ	G28BF	N25BD	E16AL	G17AJ	N34AG	E38AD	G18AB	B37CP
25	P19CK	U23CH	A23BM	H20BK	L33BG	A35BE	M24AM	L22AK	A26AH	M11AE	L12AC	N27CR
26	W30CL	R36CJ	Z36BP	S28BL	D15BH	Z16BF	H21BC	D29AL	Z38AJ	H31AF	D37AD	A13CS
27	Y14CM	X32CK	C32CD	P33BM	J25BJ	C24BG	S17BD	J34AM	C11AK	S18AG	J27AE	Z19AB
28	T23CP	B20CL	K20CE	W15BP	U35BK	K21BH	P22BE	U26BC	K31AL	P12AH	U13AF	C30AC
29	G36CR		F28CF	Y25CD	R16BL	F17BJ	W29BF	R38BD	F18AM	W37AJ	R19AG	K14AD
30	L32CS		V33CH	T35CE	X24BM	V22BK	Y34BG	X11BE	V12BC	Y27AK	X30AH	F23AE
31	D20AB		E15CJ		B21BP		T26BH	B31BF		T13AL		V36AF

1811	JAN	FEB	MAR	APR	MAY	JUN	JUL	AUG	SEP	OCT	NOV	DEC
1	E32AG	G33AE	R33CR	F35CM	W24CJ	R22CF	K34BP	W11BL	R12BJ	K27BF	W30BD	U23AL
2	M20AH	L15AF	X15CS	V16CP	Y21CK	X29CH	F26CD	Y31BM	X37BK	F13BG	Y14BE	R36AM
3	H28AJ	D25AG	B25AB	E24CR	T17CL	B34CJ	V38CE	T18BP	B27BL	V19BH	T23BF	X32BC
4	S33AK	J35AH	N35AC	M21CS	G22CM	N26CK	E11CF	G12CD	N13BM	E30BJ	G36BG	B20BD
5	P15AL	U16AJ	A16AD	H17AB	L29CP	A38CL	M31CH	L37CE	A19BP	M14BK	L32BH	N28BE
6	W25AM	R24AK	Z24AE	S22AC	D34CR	Z11CM	H18CJ	D27CF	Z30CD	H23BL	D20BJ	A33BF
7	Y35BC	X21AL	C21AF	P29AD	J26CS	C31CP	W27CK	S12CK	J13CH	C14CE	J28BK	Z15BG
8	T16BD	B17AM	K17AG	W34AE	U38AB	K18CR	P37CL	U19CJ	K23CF	P32BP	U33BL	C25BH
9	G24BE	N22BC	F22AH	Y26AF	R11AC	F12CS	W27CH	R30CK	F35CH	W20CD	R15BM	K35BJ
10	L21BF	A29BD	V29AJ	T38AG	X31AD	V37AB	Y13CP	X14CL	V32CJ	Y28CE	X25BP	F16BK
11	D17BG	Z34BE	E34AK	G11AH	B18AB	E27AC	T19CR	B23CM	E20CS	T33CF	B35CD	V24BL
12	J22BH	C26BF	M26AL	L31AJ	N12AF	M13AD	G30CS	N36CP	M28CL	G15CH	N16CE	E21BM
13	U29BJ	K38BG	H38AM	D18AK	A37AG	H19AE	L14AB	A32CR	H33CM	L25CJ	A24CF	M17BP
14	R34BK	F11BH	S11BC	J12AL	Z27AH	S30AF	D23AC	Z20CS	S15CH	D35CK	Z21CH	H22CD
15	X26BL	V31BJ	P31BD	U37AM	C13AJ	P14AG	J36AD	C28AB	P25CR	J16CL	C17CJ	S29CE
16	B38BM	E18BK	W18BK	R27BC	K19AK	W23AH	U32AE	K33AC	W35CS	U24CM	K22CK	P34CF
17	N11BP	M12BL	Y12BF	X13BD	F30AL	Y36AJ	R20AF	F15AD	Y16AB	R21CP	F29CL	W26CH
18	A31CD	H37BM	T37BG	B19BE	V14AM	T32AK	X28AG	V25AE	T24AC	X17CR	V34CM	Y38CJ
19	Z18CE	S27BP	G27BH	N30BF	E23BC	G20AL	B33AH	E35AF	G21AD	B22CS	E26CP	T11CK
20	C12CF	P13CD	L13BJ	A14BG	M36BD	L28AM	N15AJ	M16AG	L17AE	N29AB	M38CR	G31CL
21	K37CH	W19CE	D19BK	Z23BH	H32BE	D33BC	A25AK	H24AH	D22AF	A34AC	H11CS	L18CM
22	F27CJ	Y30CF	J30BL	C36BJ	S20BF	J15BD	Z35AL	S21AJ	J29AG	Z26AD	S31AB	D12CP
23	V13CK	T14CM	U14BM	K32BK	P28BG	U25BD	C16AM	P17AK	U34AH	C38AE	U35AB	J37CR
24	E19CL	G23CJ	R23BP	F20BL	W33BH	R35BF	K24BC	W22AL	R26AJ	K11AF	W12AD	U27CS
25	M30CM	L36CK	X36CD	V28BM	Y15BJ	X16BG	F21BD	Y29AM	X38AK	F31AG	Y37AE	R13AB
26	H14CP	D32CL	B32CE	E33BP	T25BK	B24BH	V17BE	T34CG	B11AL	V18AH	T27AF	X19AC
27	S23CR	J20CM	N20CF	M15CD	G35BL	N21BJ	E22BF	G26BD	N31AM	E12AJ	G13AG	B30AD
28	P36CS	U28CP	A28CH	H25CE	L16BM	A17BK	M29BG	L38BE	A18BC	M37AK	L19AH	N14AE
29	W32AB		Z33CJ	S35CF	D24BP	Z22BL	H34BH	D11BF	Z12BD	H27AL	D30AJ	A23AF
30	Y20AC		C15CK	P16CH	J21CD	C29BM	S26BJ	J31BG	C37BE		P19BC	Z36AG
31	T28AD		K25CL		U17CE		P38BK	U18BH				C32AH

THE COMPLETE BOOK OF BIORHYTHM LIFE CYCLES

FIRST LETTER = YOUR PHYSICAL CODE, MIDDLE NUMBER = SENSITIVITY CODE, LAST TWO LETTERS = INTELLECTUAL CODE

1812	JAN	FEB	MAR	APR	MAY	JUN	JUL	AUG	SEP	OCT	NOV	DEC
1	K20AJ	W15AG	J25AC	C24CS	S17CH	J34CK	Z38CF	S18CD	J27BM	Z19BJ	S23BG	D32BM
2	F28AK	Y25AH	U35AD	K21AB	P22CP	U26CL	C11CH	P12CE	U13BP	C30BK	P36BH	J20BE
3	V33AL	T35AJ	R16AE	F17AC	W29CR	R38CM	K31CJ	W37CF	R19CD	K14BL	W32BJ	U28BF
4	E15AM	G16AK	X24AF	V22AD	Y34CS	X11CP	F18CK	Y27CH	X30CE	F23BH	Y20BK	R33BG
5	M25BC	L24AL	B21AG	E29AE	T26AB	B31CR	V12CL	T13CJ	B14CF	V36BP	T28BL	X15BH
6	H35BD	D21AM	N17AH	M34AF	G38AC	N18CS	E37CM	G19CK	N23CH	E32CD	G33BM	B25BJ
7	S16BE	J17BC	A22AJ	H26AG	L11AD	A12AB	M27CP	L30CL	A36CJ	M20CE	L15BP	N35BK
8	P24BF	U22BD	Z29AK	S38AH	D31AE	Z37AC	H13CN	D14CM	Z32CK	H28CF	D25CD	A16BL
9	W21BG	R29BE	C34AL	P11AJ	J18AF	C27AC	S19CS	J23CP	C20CL	S33CH	J35CE	Z24BM
10	Y17BH	X34BF	K26AM	W31AK	U12AG	K13AE	P30AB	U36CR	K28CM	P15CJ	U16CF	C21BP
11	T22BJ	B26BG	F38BC	Y18AL	R37AH	F19AF	W14AC	R32CS	F33CP	W25CK	R24CH	K17CD
12	G29BK	N38BH	V11BD	T12AM	X27AJ	V30AG	Y23AD	X20AB	V15CR	Y35CL	X21CJ	F22CE
13	L34BL	A11BJ	E31BE	G37BC	B13AK	E14AH	T36AE	B28AC	E25CS	T16CM	B17CK	V29CF
14	D26BM	Z31BK	M18BF	L27BD	N19AL	M23AJ	G32AF	N33AD	M35AB	G24CP	N22CL	E34CH
15	J38BP	C16BL	H12BG	D13BE	A30AM	H36AK	L20AG	A15AE	H16AC	L21CR	A29CM	M26CJ
16	U11CD	K12BM	S37BH	J19BF	Z14BC	S32AL	D28AH	Z25AF	S24AD	D17CS	Z34CP	H38CK
17	R31CE	F37BP	P27BJ	U30BG	C23BD	P20AM	J33AJ	C35AG	P21AE	J22AB	C26CR	S11CL
18	X18CF	V27CD	W13BK	R14BH	K36BE	W28BC	U15AK	K16AH	W17AF	U29AC	K38CS	P31CH
19	B12CH	E13CE	Y19BL	X23BJ	F32BF	Y33BD	R25AL	F24AJ	Y22AG	R34AD	F11AB	W18CP
20	N37CJ	M19CH	T30BM	B36BK	V20BG	T15BE	X35AM	V21AK	T29AH	X26AE	V31AC	Y12CR
21	A27CK	H30CH	G14BP	N32BL	E28BH	G25BF	B16BC	E17AL	G34AJ	B38AF	E18AD	T37CS
22	Z13CL	S14CJ	L23CD	A20BM	M33BJ	L35BG	N24BD	M22AM	L26AK	N11AG	M12AE	G27AB
23	C19CM	P23CK	D36CE	Z28BP	H15BK	D16BH	A21BE	H29BC	D38AL	A31AH	H37AF	L13AC
24	K30CP	W36CL	J32CF	C33CD	S25BL	J24BJ	Z17BF	S34BD	J11AM	Z18AJ	S27AG	D19AD
25	F14CR	Y32CM	U20CH	K15CE	P35BM	U21BK	C22BG	P26BE	U31BC	C12AK	P13AH	J30AE
26	V23CS	T20CP	R28CJ	F25CF	W16BP	R17BL	K29BH	W38BF	R18BD	K37AL	W19AJ	U14AF
27	E36AB	G28CR	X33CK	V35CH	Y24CD	X22BM	F34BJ	Y11BG	X12BE	F27AM	Y30AK	R23AG
28	M32AC	L33CS	B15CL	E16CJ	T21CE	B29BP	V26BK	T31BH	B37BF	V13BC	T14AL	X36AH
29	H20AD	D15AB	N25CM	H24CK	G17CF	N34CD	E38BL	G18BJ	N27BG	E19BD	G23AM	B32AJ
30	S28AE		A35CP	H21CL	L22CH	A26CE	M11BM	L12BK	A13BH	M30BE	L36BC	N20AK
31	P33AF		Z16CR		D29CJ		H31BP	D37BL		H14BF		A28AL

1813	JAN	FEB	MAR	APR	MAY	JUN	JUL	AUG	SEP	OCT	NOV	DEC
1	Z33AM	S35AK	G35AE	N21AC	E22CR	G26CM	B11CJ	E12CF	G13CD	B30BL	E36BJ	T20BF
2	C15BC	P16AL	L16AF	A17AD	M29CS	L38CP	N31CK	M37CH	L19CE	N14BM	M32BK	G28BG
3	K25BD	W24AM	D24AG	Z22AE	H34AB	D11CR	A18CL	H27CJ	D30CF	A23BP	H20BL	L33BH
4	F35BE	Y21BC	J21AH	C29AF	S26AC	J31CS	Z12CM	S13CK	J14CH	Z36CD	S28BM	D15BJ
5	V16BF	T17BD	U17AJ	K34AG	P38AD	U18AB	C37CP	P19CL	U23CJ	C32CE	P33BP	J25BK
6	E24BG	G22BE	R22AK	F26AH	W11AB	R12AC	K27CR	W30CM	R36CK	K20CF	W15CD	U35BL
7	M21BH	L29BF	X29AL	V38AJ	Y31AF	X37AD	F13CS	Y14CP	X32CL	F28CH	Y25CE	R16BM
8	H17BJ	D34BG	B34AM	E11AK	T18AG	B27AE	V19AB	T23CR	B20CM	V33CJ	T35CF	X24BP
9	S22BK	J26BH	N26BC	M31AL	G12AH	N13AF	E30AC	G36CS	N28CP	E15CK	G16CH	B21CD
10	P29BL	U38BJ	A38BD	H18AM	L37AJ	A19AG	M14AD	L32AB	A33CR	M25CL	L24CJ	N17CE
11	W34BM	R11BK	Z11BE	S12BC	D27AK	Z30AH	H23AE	D20AC	Z15CS	H35CM	D21CK	A22CF
12	Y26BP	X31BL	C31BF	P37BD	J13AL	C14AJ	S36AF	J28AD	C25AB	S16CP	J17CL	Z29CH
13	T38CD	B18BM	K18BG	W27BE	U19AM	K23AK	P32AG	U33AE	K35AC	P24CR	U22CM	C34CJ
14	G11CE	N12BP	F12BH	Y13BF	R30BC	F36AL	W20AH	R15AF	F16AD	W21CS	R29CP	K26CK
15	L31CF	A37CD	V37BJ	T19BG	X14BD	V32AM	Y28AJ	X25AG	V24AE	Y17AB	X34CR	F38CL
16	D18CH	Z27CE	E27BK	G30BH	B23BE	E20BC	T33AK	B35AH	E21AF	T22AC	B26CS	V11CM
17	J12CJ	C13CF	M13BL	L14BJ	N36BF	M28BD	G15AL	N16AJ	M17AG	G29AD	N38AB	E31CP
18	U37CK	K19CH	H19BM	D23BK	A32BG	H33BE	L25AM	A24AK	H22AH	L34AE	A11AC	M18CR
19	R27CL	F30CJ	S30BP	J36BL	Z20BH	S15BF	D35BC	Z21AL	S29AJ	D26AF	Z31AD	H12CS
20	X13CM	V14CK	P14CD	U32BM	C28BJ	P25BG	J16BD	C17AM	P34AK	J38AG	C18AE	S37AB
21	B19CP	E23CL	W23CE	R20BP	K33BK	W35BH	U24BE	K22BC	W26AL	U11AH	K12AF	P27AC
22	N30CR	M36CM	Y36CF	X28CD	F15BL	Y16BJ	R21BF	F29BD	Y38AM	R31AJ	F37AG	W13AD
23	A14CS	H32CP	T32CH	B33CE	V25BM	T24BK	X17BG	V34BE	T11BC	X18AK	V27AH	Y19AE
24	Z23AB	S20CR	G20CJ	N15CF	E35BP	G21BL	B22BH	E26BF	G31BD	B12AL	E13AJ	T30AF
25	C36AC	P28CS	L28CK	A25CH	M16CD	L17BM	N29BJ	M38BG	L18BE	N37AM	M19AK	G14AG
26	K32AD	W33AB	D33CL	Z35CJ	H24CE	D22BP	A34BK	H11BH	D12BF	A27BC	H30AL	L23AH
27	F20AE	Y15AC	J15CM	C16CK	S21CF	J29CD	Z26BL	S31BJ	J37BG	Z13BD	S14AM	D36AJ
28	V28AF	T25AD	U25CP	K24CL	P17CH	U34CE	C38BM	P18BK	U27BH	C19BE	P23BC	J32AK
29	E33AG		R35CR	F21CM	W22CJ	R26CF	K11BP	W12BL	R13BJ	K30BF	W36BD	U20AL
30	M15AH		X16CS	V17CP	Y29CK	X38CH	F31CD	Y37BM	X19BK	F14BG	Y32BE	R28AM
31	H25AJ		B24AB		T34CL		V18CE	T27BP		V23BH		X33AL

1814	JAN	FEB	MAR	APR	MAY	JUN	JUL	AUG	SEP	OCT	NOV	DEC
1	B15BD	E16AM	W16AG	R17AE	K29AB	W38CR	U31CL	K37CJ	W19CF	U14BP	K32BL	P28BH
2	N25BE	M24BC	Y24AH	X22AF	F34AC	Y11CS	R18CM	F27CK	Y30CH	R23CD	F20BM	W33BJ
3	A35BF	H21BD	T21AJ	B29AG	V26AD	T31AB	X12CP	V13CL	T14CJ	X36CE	V28BP	Y15BK
4	Z16BG	S17BE	G17AK	N34AH	E38AE	G18AC	B37CR	E19CM	G23CK	B32CF	E33CD	T25BL
5	C24BH	P22BF	L22AL	A26AJ	M11AF	L12AD	N27CS	M30CP	L36CL	N20CH	M15CE	G35BM
6	K21BJ	W29BG	D29AM	Z38AK	H31AG	D37AE	A13AB	H14CR	D32CM	A28CJ	H25CF	L16BP
7	F17BK	Y34BH	J34BC	C11AL	S18AH	J27AF	Z19AC	S23CS	J20CP	Z33CK	S35CH	D24CD
8	V22BL	T26BJ	U26BD	K31AM	P12AJ	U13AG	C30AD	P36AB	U28CR	C15CL	P16CJ	J21CE
9	E29BM	G38BK	R38BE	F18BC	W37AK	R19AF	K14AE	W32AC	R33CS	K25CK	W24CK	U17CF
10	M34BP	L11BL	X11BF	V12BD	Y27AL	X30AJ	F23AF	Y20AD	X15BC	F35CP	Y21CL	R22CD
11	H26CD	D31BM	B31BG	E37BE	T13AM	B14AK	V36AG	T28AE	B25BD	V16CR	T17CM	X29CJ
12	S38CE	J18BP	N18BH	M27BF	G19BC	N23AL	E32AH	G33AF	N35AD	E24CS	G22CP	B34CK
13	P11CF	U12CD	A12BJ	H13BG	L30BD	A36AM	M20AJ	L15AG	A16AE	M21AB	L29CR	N26CL
14	W31CH	R37CF	Z37BK	S19BH	D14BE	Z32AM	H28AK	D25AH	Z24AF	H17AC	D34CS	A38CM
15	Y18CJ	X27CF	C27BL	P30BJ	J23BF	C20AD	S33AL	J35AJ	C21AG	S22AD	J26AB	Z11CP
16	T12CK	B13CH	F19BP	W14BK	U36BG	K28AK	P15AM	U16AK	K17AH	P29AE	U38AC	C31CR
17	G37CL	N19CJ	V19BP	Y23BL	R28BH	F33BF	M25BC	R24AL	F22AJ	M34AF	R11AD	K18CS
18	L27CM	A30CK	V30BL	T36BM	X20BJ	V15BG	Y35BD	X21AM	V29AK	Y26AG	X31AE	F12AB
19	D13CP	Z14CL	E14CE	G32BP	B28BK	E25BH	T16BE	B17BC	E34AL	T38AH	B18AF	V37AC
20	J19CR	C23CM	M23CF	L20CD	N33BL	M35BJ	G24BF	N22BD	M26AM	G11AJ	N12AG	E27AD
21	U30CS	K36CP	H36CH	D28CE	A15BM	H16BK	L21BG	A29BE	H38BC	L31AK	A37AH	M13AE
22	R14AB	F32CR	S32CJ	J33CF	Z25BP	S24BL	D17BH	Z34BF	S11BD	D18AL	Z27AJ	H19AF
23	X23AC	V20CS	P20CK	U15CH	C35CD	P21BM	J22BJ	C22BG	P38BE	J12AM	C13AK	S30AG
24	B36AD	E28AB	W28CL	R25CJ	K16CE	W17BP	U29BK	K38BH	W18BF	U37BC	K19AL	P14AH
25	N32AE	H33AC	T15CP	X35CK	F24CF	U22CD	R34BL	F11BG	Y12BG	R27BD	F30AM	W23AH
26	A20AF	H15AD	T15CP	B16CL	V21CH	T29CE	X26BM	V31BK	T37BH	X13BE	V14BC	Y36AK
27	Z28AG	S25AE	G25CR	N24CM	E17CJ	G34CF	B38BP	E18BL	G27BJ	B19BF	E23BD	T32AL
28	C33AH	P35AF	L35CS	A21CP	M22CK	L26CH	N11CD	M12BM	L13BK	N30BG	M36BF	G20AM
29	K15AJ		D16AB	Z17CR	H29CL	D38CJ	A31CE	H37BP	D19BL	A14BH	N30BG	L28BC
30	F25AK		J24AC	C22CS	S34CM	J11CK	Z16CF	S27CD	J30BM	Z23BJ	S20BG	D33BD
31	V35AL		U21AD		P26CP		C12CH	P13CE		C36BK		J15BE

FIRST LETTER = YOUR PHYSICAL CODE, MIDDLE NUMBER = SENSITIVITY CODE, LAST TWO LETTERS = INTELLECTUAL CODE

1815	JAN	FEB	MAR	APR	MAY	JUN	JUL	AUG	SEP	OCT	NOV	DEC
1	U25BF	K24BD	H24AJ	D22AG	A34AD	H11AB	L18CP	A27CL	H30CJ	L23CE	A20BP	M33BK
2	R35BG	F21BE	S21AK	J29AH	Z26AE	S31AC	D12CR	Z13CM	S14CK	D36CF	Z28CD	H15BL
3	X16BH	V17BF	P17AL	U34AJ	C38AF	P18AD	J37CS	C19CP	P23CL	J32CH	C33CE	S25BM
4	B24BJ	E22BG	W22AM	R26AK	K11AG	W12AE	U27AB	K30CR	W36CM	U20CJ	K15CF	P35BP
5	N21BK	M29BH	Y29BC	X38AL	F31AH	Y37AF	R13AC	F14CS	Y32CP	R28CK	F25CH	W16CD
6	A17BL	H34BJ	T34BD	B11AM	V18AJ	T27AG	X19AD	V23AB	T20CR	X33CL	V35CJ	Y24CE
7	Z22BM	S26BK	G26BE	N31BC	E12AK	G13AH	B30AE	E36AC	S28CS	B15CM	E16CK	T21CF
8	C29BP	P38BL	L38BF	A18BD	M37AL	L19AJ	N14AF	M32AD	L33AB	N25CP	M24CL	G17CH
9	K34CD	W11BM	D11BG	Z12BE	H27AM	D30AK	A23AG	H20AE	D15AC	A35CR	H21CH	L22CJ
10	F26CE	Y31BP	J31BH	C37BF	S13BC	J14AL	Z36AH	S28AF	J25AD	Z16CS	S17CP	D29CK
11	V38CF	T18CD	U18BJ	K27BG	P19BD	U23AM	C32AJ	P33AG	U35AE	C24AB	P22CR	J34CL
12	E11CH	G12CE	R12BK	F13BH	W30BE	R36BC	K20AK	W15AH	R16AF	K21AC	W29CS	U26CM
13	M31CJ	L37CF	X37BL	V19BJ	Y14BF	X32BD	F26AL	Y25AJ	X24AG	F17AD	Y34AB	R38CP
14	H18CK	D27CH	B27BM	E30BK	T23BG	B20BE	V33AM	T35AK	B21AH	V22AE	T26AC	X11CH
15	S12CL	J13CJ	N13BP	M14BL	G36BH	N28BF	E15BC	G16AL	N17AJ	E29AF	G38AD	B31CS
16	P37CM	U19CK	A19CD	H23BM	L32BJ	A33BG	M25BD	L24AM	A22AK	M34AG	L11AE	N18AB
17	W27CP	R30CL	Z30CE	S36BP	D20BK	Z15BH	H35BE	D21BC	Z29AL	H26AH	D31AF	A12AC
18	Y13CR	X14CM	C14CF	P32CD	J28BL	C25BJ	S16BF	J17BD	C34AM	S38AJ	J18AG	Z37AD
19	T19CS	B23CP	K23CH	W20CE	U33BM	K35BK	P24BG	U22BE	K26BC	P11AK	U12AH	C27AE
20	G30AB	N36CR	F36CJ	Y28CF	R15BP	F16BL	W21BH	R29BF	F38BD	W31AL	R37AJ	K13AF
21	L14AC	A32CS	V32CK	T33CH	X25CD	V24BM	Y17BJ	X34BG	V11BE	Y18AM	X27AK	F19AG
22	D23AD	Z20AB	E20CL	G15CJ	B35CE	E21BP	T22BK	B26BH	E31BF	T12BC	B13AL	V30AH
23	J36AE	C28AC	M28CM	L25CK	N16CF	M17CD	G29BL	N38BJ	M18BG	G37BD	N19AM	E14AJ
24	U32AF	K33AD	H33CP	D35CL	A24CH	H22CE	L34BM	A11BK	412BH	L27BE	A30BC	M23AK
25	R20AG	F15AE	S15CR	J16CM	Z21CJ	S29CF	D26BP	Z31BL	S37BJ	D13BF	Z14BD	H36AL
26	X28AH	W25AF	P25CS	U24CP	C17CK	P34CH	J38CD	C18BM	P27BK	J19BG	C23BE	S32AM
27	B33AJ	E35AG	W35AB	R21CR	K22CL	W26CJ	U11CE	K12BP	W13BL	U30BH	K36BF	P20BC
28	N15AK	M16AH	Y16AC	X17CS	F29CM	Y38CK	R31CF	F37CD	Y19BM	R14BJ	F32BG	W28BD
29	A25AL		T24AD	B22AB	V34CP	T11CL	X18CH	V27CE	T30BP	X23BK	V20BH	Y33BE
30	Z35AM		G21AE	N29AC	E26CR	G31CM	B12CJ	E13CF	G14CD	B36BL	E28BJ	T15BF
31	C16BC		L17AF		M38CS		N37CK	M19CH		N32BM		G25BG

1816	JAN	FEB	MAR	APR	MAY	JUN	JUL	AUG	SEP	OCT	NOV	DEC
1	L35BH	A21BF	E17AM	G34AK	B38AG	E18AE	T37AB	B19CR	E23CM	T32CJ	B33CF	V25BP
2	D16BJ	Z17BG	M22BC	L26AL	N11AH	M12AF	G27AC	N30CS	M36CP	G20CK	N15CH	E35CD
3	J24BK	C22BH	H29BD	D38AM	A31AJ	H37AG	L13AD	A14AB	H32CR	L28CL	A25CJ	M16CE
4	U21BL	K29BJ	S34BE	J11BC	Z18AK	S27AH	D19AE	Z23AC	S20CS	D33CM	Z35CK	H24CF
5	R17BM	F34BK	P26BF	U31BD	C12AL	P13AJ	J30AF	C36AD	P28AB	J15CP	C16CL	S21CH
6	X22BP	V26BL	W38BG	R18BE	K37AM	W19AK	U14AG	K32AE	W33AC	U25CR	K24CM	P17CJ
7	B29CD	E38BM	Y11BH	X12BF	F27BC	Y30AL	R23AH	F20AF	Y15AD	R35CS	F21CP	W22CK
8	N34CE	M11BP	T31BJ	B37BG	V13BD	T14AM	X36AJ	V28AG	T25AE	X16AB	V17CR	Y29CL
9	A26CF	H31CD	G18BK	N27BH	E19BE	G23BC	B32AK	E33AH	G35AF	B24AC	E22CS	T34CM
10	Z38CH	S18CE	L12BL	A13BJ	M30BF	L36BD	N20AL	M15AJ	L16AG	N21AD	M29AB	G26CP
11	C11CJ	P12CF	D37BM	Z19BK	H14BG	D32BE	A28AM	H25AK	D24AH	A17AE	H34AC	L38CR
12	K31CK	W37CH	J27BP	C30BL	S23BH	J20BF	Z33BC	S35AL	J21AJ	Z22AF	S26AD	D11CS
13	F18CL	Y27CJ	U13CD	K14BM	P36BJ	U28BG	C15BD	P16AM	U17AK	C29AG	P38AE	J31AB
14	V12CM	T13CK	R19CE	F23BP	W32BK	R33BH	K25BE	W24BC	322AL	K34AH	W11AF	U18AC
15	E37CP	G19CL	X30CF	V36CD	Y20BL	X15BJ	F35BF	Y21BD	X29AM	F26AJ	Y31AG	R12AD
16	M27CR	L30CM	B14CH	E32CE	T28BM	B25BK	V16BG	T17BE	B34BC	V38AK	T18AH	X37AE
17	H13CS	D14CP	N23CJ	M20CF	G33BP	N35BL	E24BH	G22BF	N26BD	E11AL	G12AJ	B27AF
18	S19AB	J23CR	A36CK	H28CH	L15CD	A16BM	M21BJ	L29BG	A33BE	M31AM	L37AK	N13AG
19	P30AC	U36CS	Z32CL	S33CJ	D25CE	Z24BP	H17BK	D34BH	Z11BF	H16BC	D27AL	A19AH
20	W14AD	R32AB	C20CM	P15CK	J35CF	C21CD	S22BL	J26BJ	C31BG	S12BD	J13AM	Z30AJ
21	Y23AE	X20CAL	K28CP	W25CL	U16CH	K17CE	P29BM	U38BK	K18BH	P37BE	U19BC	C14AK
22	T36AF	B28AD	F33CR	Y35CM	R24CJ	F22CF	W34BP	R11BL	F12BJ	W27BF	R30BD	K23AL
23	G32AG	N33AE	V15CS	T16CP	X21CK	V29CH	Y26CD	X31BM	V37BK	Y13BG	X14BE	F36AM
24	L20AH	A15AF	E25AB	G24CR	B17CL	E34CJ	T38CE	B18BP	E27BL	T19BH	B23BF	V32BC
25	D28AJ	Z25AG	M35AC	L21CS	N22CM	M26CK	G11CF	N12CD	M13BM	G30BJ	N36BG	E20BD
26	J33AK	C35AH	H16AD	D17AB	A29CP	H38CL	L31CH	A37CE	H19BP	L14BK	A32BH	M28BE
27	U15AL	K16AJ	S24AE	J22AC	Z34CR	S11CM	D18CJ	Z27CF	S30CD	D23BL	Z20BJ	H33BF
28	R25AM	F24AK	P21AF	U29AD	C26CS	P31CP	J12CK	C13CH	P14CE	J36BM	C28BK	S15BG
29	X35BC	V21AL	W17AG	R34AE	K38AB	W18CR	U37CL	K19CJ	W23CF	U32BP	K33BL	P25BH
30	B16BD		Y22AH	X26AF	F11AC	Y12CS	R27CM	F30CK	Y36CH	R20CD	F15BM	W35BJ
31	N24BE		T29AJ		V31AD		X13CP	V14CL		X28CH		Y16BK

1817	JAN	FEB	MAR	APR	MAY	JUN	JUL	AUG	SEP	OCT	NOV	DEC
1	T24BL	B22BJ	K22BD	W26AM	U11AJ	K12AG	P27AD	U30AB	K36CR	P20CL	U15CJ	C35CE
2	G21BM	N29BK	F29BE	Y38BC	R31AK	F37AH	W13AE	R14AC	F32CS	W28CM	R25CK	K16CF
3	L17BP	A34BL	V34BF	T11BD	X18AL	V27AJ	Y19AF	X23AD	V20AB	Y33CP	X35CL	F24CH
4	D22CD	Z26BM	E26BG	G31BE	B12AM	E13AK	T30AG	B36AE	E28AC	T15CR	B16CM	V21CJ
5	J29CE	C38BP	M38BH	L18BF	N37BC	M19AL	G14AH	N32AF	M33AD	G25CS	N24CP	E17CK
6	U34CF	K11CD	H11BJ	D12BG	A27BD	H30AM	L23AJ	A20AG	H15AE	L35AB	A21CR	M22CL
7	R26CH	F31CE	S31BK	J37BH	Z13BE	S14AM	D36AK	Z28AH	S25AF	D16AC	Z17CS	H29CM
8	X38CJ	V18CF	P18BL	U27BJ	C19BF	P23BD	J32AL	C33AJ	P35AG	J24AD	C22AB	S34CP
9	B11CK	E12CH	W12BM	R13BK	K30BG	W20BC	U20AM	K15AK	W16AH	U21AE	K29AC	P26CR
10	N31CL	M37CJ	Y37BP	X19BL	F14BH	Y32BF	R28BC	F25AL	Y24AJ	R17AF	F34AD	W38CS
11	A18CM	H27CK	T27BD	B30BM	V23BJ	T20BG	X33BD	V35AM	T21AK	X22AG	V26AE	Y11AB
12	Z12CP	S13CL	G13CE	N14BP	E36BK	G28BH	B15BE	E16BC	G17AL	B29AH	E38AF	T31AC
13	C37CR	P19CM	L19CF	A23CD	M32BL	L33BJ	N25BF	M24BD	L22AM	N34AJ	M11AG	G18AD
14	K27CS	W30CP	D30CH	Z36CE	H20BM	D15BL	A35BG	H21BE	D29BC	A26AK	H31AH	L12AE
15	F13AB	Y14CR	J14CJ	C32CF	S28BP	J25BK	Z16BH	S17BF	J34BD	Z38AL	S18AJ	D37AF
16	V19AC	T23CS	U23CK	K20CH	P33CD	U35BL	C24BJ	P22BG	U26BE	C11AM	P12AK	J27AG
17	E30AD	G36AB	R36CL	F28CJ	W15CE	R16BM	K29BK	W29BH	R38BD	K31AL	W37AM	U13AH
18	M14AE	L32AC	X32CM	V33CK	Y25CF	X24BP	F17BL	Y34BJ	X11BG	F18BK	Y27AL	R19AJ
19	H23AF	D20AD	B20CP	E15CL	T35CH	B21BR	V22BM	T26BK	B31BH	V12BJ	T13BC	X30AK
20	S36AG	J28AE	N28CR	M25CH	G16CJ	N17CF	E29BP	G38BL	N18BJ	E37BF	G19BD	B14AL
21	P32AH	U33AF	A33CS	H35CP	L24CK	A22CH	M34CD	L11BM	A12BK	M27BG	L30BE	N23AM
22	W20AJ	R15AG	Z15AB	S16CR	D21CL	Z29CJ	H26CE	D31BP	Z37BL	H13BH	D14BF	A36BC
23	Y28AK	X25AH	C25AC	P24CS	J17CM	C34CK	S38CF	J18CD	C27BM	S19BJ	J23BG	Z32BD
24	T33AL	B35AJ	K35AD	W21AB	U22CP	K26CL	P11CH	U12CD	K13BP	P30BK	U36BH	C20BE
25	G15AM	N16AK	F16AE	Y17AC	R29CR	F38CM	W31CJ	R37CF	F19CD	W14BL	R32BJ	K28BF
26	L25BC	A24AL	V24AF	T22AD	X34CS	V11CP	Y18CK	X27CH	V30CE	Y23BM	X20BK	F33BG
27	D35BD	Z21AM	E21AG	G29AE	B26AB	E31CH	T12CL	B13CJ	E14CF	T36BP	B28BL	V15BH
28	J16BE	C17BC	M17AH	L34AF	N38AC	M18CS	G37CH	N19CK	M23CH	G32CD	N33BM	E25BJ
29	U24BF		H22AJ	D26AG	A11AD	H12AB	L27CP	A30CL	H36CJ	L20CL	A15BP	M35BK
30	R21BG		S29AK	J38AH	Z31AE	S37AC	D13CR	Z14CM	S32CK	D28CF	Z25CD	H16BL
31	X17BH		P34AL		C18AF		J19CS	C23CP		J33CH		S24BM

(107)

FIRST LETTER = YOUR PHYSICAL CODE, MIDDLE NUMBER = SENSITIVITY CODE, LAST TWO LETTERS = INTELLECTUAL CODE

1818

	JAN	FEB	MAR	APR	MAY	JUN	JUL	AUG	SEP	OCT	NOV	DEC
1	P21BP	U29BL	A29BF	H38BD	L31AL	A37AJ	M13AF	L14AD	A32AB	M28CP	L25CL	N16CH
2	W17CD	R34BM	Z34BG	S11BE	D18AM	Z27AK	H19AG	D23AE	Z20AC	H33CR	D35CM	A24CJ
3	Y22CE	X26BP	C26BH	P31BF	J12BG	C13AL	S30AH	J36AF	S28AD	S15CS	J16CP	Z21CK
4	T29CF	B38CD	K38BJ	W18BG	U37BD	K19AM	P14AJ	U32AG	K33AE	P25AB	U24CR	C17CL
5	G34CH	N11CE	F11BK	Y12BH	R27BE	F30BC	W23AK	R20AH	F15AF	W35AC	R21CS	K22CM
6	L26CJ	A31CF	V31BL	T37BJ	X13BF	V14BD	Y36AL	X28AJ	V25AG	Y16AD	X17AB	F29CP
7	D38CK	Z18CH	E18BM	G27BK	B19BG	E23BE	T32AM	B33AK	E35AH	T24AE	B22AC	V34CR
8	J11CL	C12CJ	M12BP	L13BL	N30BH	M36BF	G20BC	N15AL	M16AJ	G21AF	N29AD	E26CS
9	U31CM	K37CK	H37CD	D19BM	A14BJ	H32BG	L28BD	A25AM	H24AK	L17AG	A34AE	M38AB
10	R18CP	F27CL	S27CE	J30BP	Z23BK	S20BH	D33BE	Z35BC	S21AL	D22AH	Z26AH	H11AC
11	X12CR	V13CM	P13CF	U14CD	C36BL	P28BJ	J15BF	C16BD	P17AM	J29AJ	C38AG	S31AD
12	B37CS	E19CP	W19CH	R23CE	K32BM	W33BK	U25BG	K24BE	W22BC	U34AK	K11AH	P18AE
13	N27AB	M30CR	Y30CJ	X36CF	F20BP	Y15BL	R35BH	F21BF	Y29BD	R26AL	F31AJ	W12AF
14	A13AC	H14CS	T14CK	B32CH	V28CD	T25BM	X16BJ	V17BG	T34BE	X38AM	V18AK	Y37AG
15	Z19AD	S23AB	G23CL	N20CJ	E33CE	G35BP	B24BK	E22BH	G26BF	B11BC	E12AL	T27AH
16	C30AE	P36AC	L36CM	A28CK	M15CF	L16CD	N21BL	M29BJ	L38BG	N31BD	M37AM	G13AJ
17	K14AF	W32AD	D32CP	Z33CL	H25CH	D24CE	A17BM	H34BK	D11BH	A18BE	H27BC	L19AK
18	F23AG	Y20AE	J20CR	C15CM	S35CJ	J21CF	Z22BP	S26BL	J31BJ	Z12BF	S13BD	D30AL
19	V36AH	T28AF	U28CS	K25CP	P16CK	U17CH	C29CD	P38BM	U18BK	C37BG	P19BE	J14AM
20	E32AJ	G33AG	R33AB	F35CR	W24CL	R22CJ	K34CE	W11BP	R12BL	K27BH	W30BF	U23BC
21	M20AK	L15AH	X15AC	V16CS	Y21CM	X29CK	F26CF	Y31CD	X37BM	F13BJ	Y14BG	R36BD
22	H28AL	D25AJ	B25AD	E24AB	T17CP	B34CL	V38CH	T18CE	B27BP	V19BK	T23BH	X32BE
23	S33AM	J35AK	N35AE	M21AC	G22CR	N26CM	E11CJ	G12CF	N13CD	E30BL	G36BJ	B20BF
24	P15BC	U16AL	A16AF	H17AD	L29CS	A38CP	M31CK	L37CH	A19CE	M14BM	L32BK	N28BG
25	W25BD	R24AM	Z24AG	S22AE	D34AB	Z11CR	H18CL	D27CJ	Z30CF	H23BP	D20BL	A33BH
26	Y35BE	X21BC	C21AH	P29AF	J26AC	C31CS	S12CM	J13CK	C14CH	S36CD	J28BM	Z15BJ
27	T16BF	B17BD	K17AJ	W34AG	U38AD	K18AB	P37CP	U19CL	K23CJ	P32CE	U33BP	C25BK
28	G24BG	N22BE	F22AK	Y26AH	R11AE	F12AC	W27CR	R30CM	F36CK	W20CF	R15CD	K35BL
29	L21BH		V29AL	T38AJ	X31AF	V37AD	Y13CS	X14CP	V32CL	Y28CH	X25CE	F16BM
30	D17BJ		E34AM	G11AK	B18AG	E27AE	T19AB	B23CR	E20CM	T33CJ	B35CF	V24BP
31	J22BK		M26BC		N12AH		G30AC	N36CS		G15CK		E21CD

1819

	JAN	FEB	MAR	APR	MAY	JUN	JUL	AUG	SEP	OCT	NOV	DEC
1	M17CE	L34BP	X34BH	V11BF	Y18BC	X27AL	F19AH	Y23AF	X20AD	F33CS	Y35CP	R24CK
2	H22CF	D26CD	B26BJ	E31BG	T12BD	B13AM	V30AJ	T36AG	B28AE	V15AB	T16CR	X21CL
3	S29CH	J38CE	N38BK	M18BH	G37BE	N19BC	E14AK	G32AH	N33AF	E25AC	G24CS	B17CM
4	P34CJ	U11CF	A11BL	H12BJ	L27BF	A30BD	M23AL	L20AJ	A15AG	M35AD	L21AB	N22CP
5	W26CK	R31CH	Z31BM	S37BK	D13BG	Z14BE	H36AM	D28AK	Z25AH	H16AF	D17AC	A29CR
6	Y38CL	X18CJ	C18BP	P27BL	J19BH	C23BF	S32BC	J33AL	C35AJ	S24AF	J22AD	Z34CS
7	T11CM	B12CK	K12CD	W13BM	U30BJ	K36BG	P20BD	U15AM	K16AK	P21AG	U29AE	C26AB
8	G31CP	N37CL	F37CE	Y19BP	R14BK	F32BH	W28BE	R25BC	F24AL	W17AH	R34AF	K38AC
9	L18CM	A27CM	V27CF	T30CD	X23BL	V20BJ	Y33BF	X35BD	V21AM	Y22AJ	X26AG	F11AD
10	D12CS	Z13CP	E13CH	G14CE	B36BM	E28BH	T15BG	B16BE	E17BC	T29AK	B38AH	V31AE
11	J37AB	C19CR	M19CJ	L23CF	N32BP	M33BL	G25BH	N24BF	M22BD	G34AL	N11AJ	E18AF
12	U27AC	K30CS	H30CK	D36CH	A20CD	H15BM	L35BJ	A21BG	H29BE	L26AM	A31AK	M12AG
13	R13AD	F14AB	S14CL	J32CJ	Z28CE	S25BP	D16BK	Z17BH	S34BF	D38BC	Z18AL	H37AH
14	X19AE	V23AC	P23CM	U20CK	C33CF	P35CD	J24BL	C22BJ	P26BG	J11BD	C12AM	S27AJ
15	B30AF	E36AD	W36CP	R28CL	K15CH	W16CE	U21BM	K29BK	W38BH	U31BE	K37BC	P13AK
16	N14AG	M32AE	Y32CR	X33CM	F25CJ	Y24CF	R17BP	F34BL	Y11BJ	R18BF	F27BD	W19AL
17	A23AH	H20AF	T20CS	B15CP	V35CK	T21CH	X22CD	V26BM	T31BK	X12BG	V13BE	Y30AM
18	Z36AJ	S28AG	G28AB	N25CR	E16CL	G17CJ	B29CE	E38BP	G18BL	B37BH	E19BF	T14BC
19	C32AK	P33AH	L33AC	A35CS	M24CM	L22CK	N34CF	M11CD	L28BM	N27BJ	M30BG	G23BD
20	K20AL	W15AJ	D15AD	Z16AB	H21CP	D29CL	A26CH	H31CE	D37BP	A13BK	H14BH	L36BE
21	F28AM	Y25AK	J25AE	C24AC	S17CR	J34CM	Z38CJ	S16CF	J27CD	Z19BL	S23BJ	D32BF
22	V33BC	T35AL	U35AF	K21AD	P22CS	U26CP	C11CK	P12CH	U13CE	C30BM	P36BK	J20BG
23	E15BD	G16AM	R16AG	F17AE	W29AB	R38CR	K31CL	W37CJ	R19CF	K14BP	W32BL	U28BH
24	M25BE	L24BC	X24AH	V22AF	Y34AC	X11CS	F18CM	Y27CK	X30CH	F23CD	Y20BM	R33BJ
25	H35BF	D21BD	B21AJ	E29AG	T26AD	B31AB	V12CP	T13CL	B14CJ	V36CE	T28BP	X15BK
26	S16BG	J17BE	N17AK	M34AH	G38AE	N18AC	E37CR	G19CM	N23CK	E32CF	G33CD	B25BL
27	P24BH	U22BF	A22AL	H26AJ	L11AF	A12AD	M27CS	L30CP	A36CL	M20CH	L15CE	N35BM
28	W21BJ	R29BG	Z29AM	S38AK	D31AG	Z37AE	H13AB	D14CR	Z32CM	H28CJ	D25CF	A16BP
29	Y17BK		C34BC	P11AL	J18AH	C27AF	S19AC	J23CS	C20CP	S33CK	J35CH	Z24CD
30	T22BL		K26BD	W31AM	U12AJ	K13AG	P30AD	U36AB	K28CR	P15CL	U16CJ	C21CE
31	G29BM		F38BE		R37AK		W14AE	R32AC		W25CM		K17CF

1820

	JAN	FEB	MAR	APR	MAY	JUN	JUL	AUG	SEP	OCT	NOV	DEC
1	F22CH	Y26CE	U38BL	K18BJ	P37BF	U19BD	C14AL	P32AJ	U33AG	C25AD	P24AB	J17CP
2	V29CJ	T38CF	R11BM	F12BK	R30BE	K23AM	W20AK	R15AH	K35AE		W21AC	U22CR
3	E34CK	G11CH	X31BP	V37BL	Y13BH	X14BF	F36BC	Y28AL	X25AJ	F16AF	Y17AD	R29CS
4	M26CL	L31CJ	B18CD	E27BM	T19BJ	B23BG	V32BD	T33AM	B35AK	V24AG	T22AE	X34AB
5	H38CM	D18CK	N12CE	M13BP	G30BK	N36BH	E20BE	G15BN	N16AL	E21AH	G29AF	B26AC
6	S11CP	J12CL	A37CF	H19CD	L14BL	A32BJ	M28BF	L25BD	A24AM	M17AJ	L34AG	N38AD
7	P31CR	U37CM	Z27CH	P14CF	J36BP	C28BL	S15BH	J16BF	C17BD	S29AL	J38AJ	A11AE
8	W16CS	R27CP	C13CJ	P14CF	J36BP	C28BL	S15BH	J16BF	C17BD	S29AL	J38AJ	Z31AF
9	Y12AB	X13CR	K19CK	W23CH	U32CD	K33BM	P25BJ	U24BG	K22BE	P34AM	U11AK	C18AG
10	T37AC	B19CS	F30CL	Y36CJ	R20CE	F15BP	W35BK	R21BH	F29BF	W26BC	R31AL	K12AH
11	G27AD	N30AB	V14CM	T32CK	X28CF	V25CD	Y16BL	X17BF	V34BG	Y38BD	X18AM	F37AJ
12	L13AE	A14AC	E23CP	G20CL	B33CH	E35CE	T24BM	B22BK	E26BH	T11BE	B22...	V27AK
13	D19AF	Z23AD	M36CR	L28CM	N15CJ	M16CF	G21BP	N29BL	M38BJ	G31BF	B12BC	E13AL
14	J30AG	C36AE	H32CS	D33CP	H24CH	H24CH	L17CD	A34BK	H11BK	L18BG	A27BD	M19AM
15	U14AH	K32AF	S20AB	J15CR	Z35CK	S21CJ	D22CE	Z26BP	S31BL	D12BH	Z13BF	H30BC
16	R23AJ	F20AG	P28AC	U25CS	C16CM	P17CK	W22CL	U34CH	K11CE	P18BM	J37BG	S14BD
17	X36AK	V28AH	W33AD	R35AB	K24CP	W22CL	U34CH	K11CE	M12BP	U27BK	K30BH	P23BE
18	B32AL	E33AJ	Y15AE	X16AC	F21CR	Y29CM	R26CJ	F31CF	Y37CD	R13BL	F14BJ	W36BF
19	N20AM	M15AK	T25AF	B24AD	V17CS	T34CP	X38CH	V18CH	T27CE	X19BM	V23BK	Y32BG
20	A28BC	H25AL	G35AG	N21AE	E22AB	G26CR	B11CL	E12CJ	G13CF	B30BP	E36BL	T20BH
21	Z33BD	S35AM	L16AH	A17AF	M29AC	L38CS	N31CM	M37CL	L19CH	A23CE	H20BP	G28BJ
22	C15BE	P16BC	D24AJ	Z22AG	H34AD	D11AB	A18CP	H27CL	D30CJ	A23CE	H20BP	L33BK
23	K25BF	W24BD	J21AK	C29AH	S26AE	J31AL	Z12CR	P12CH	R36CM	Z36CF	H20BP	D15BL
24	F35BG	Y21BE	U17AL	K34AJ	P38AF	U18AD	C37CS	P19CP	U23CL	C32CH	P33CE	J25BM
25	V16BH	T17BF	R22AM	F26AK	W11AG	R12AE	K27AB	H30CM	R36CM	K20CJ	W15CF	U35BP
26	E24BJ	G22BG	X29BC	V38AL	Y31AH	X37AF	F13AC	Y14CS	X32CP	F28CK	Y25CH	R16CD
27	M21BK	L29BH	B34BD	E11AM	T18AJ	B27AG	V19AD	T23AB	B20CR	V33CL	T35CJ	X24CE
28	H17BL	D34BJ	N26BE	M31BC	G12AK	N13AH	E30AE	G36AC	N28CS	E15CM	G16CK	B21CF
29	S22BM	J26BK	A38BF	H18BD	L37AL	A19AJ	M14AF	L32AD	A33AB	M25CP	L24CL	N17CH
30	P29BP		Z11BG	S12BE	D27AM	Z30AK	H23AG	D20AE	Z15AC	H35CR	D21CM	A22CJ
31	W34CD		C31BH		J13BC		S36AH	J28AF		S16CS		Z29CK

FIRST LETTER = YOUR PHYSICAL CODE, MIDDLE NUMBER = SENSITIVITY CODE, LAST TWO LETTERS = INTELLECTUAL CODE

1821	JAN	FEB	MAR	APR	MAY	JUN	JUL	AUG	SEP	OCT	NOV	DEC
1	C34CL	P11CJ	L11BP	A12BL	M27BH	L30BF	N23BC	M20AL	L15AJ	N35AF	M21AD	G22CS
2	K26CM	W31CK	D31CD	Z37BM	H13BJ	D14BG	A36BD	H28AM	D25AK	A16AG	H17AE	L29AB
3	F38CP	Y18CL	J18CE	C27BP	S19BK	J23BH	Z32BE	S33BC	J35AL	Z24AH	S22AF	D34AC
4	V11CR	T12CM	U12CF	K13CD	P30BL	U36BJ	C20BF	P15BD	U16AM	C21AJ	P29AG	J26AD
5	E31CS	G37CP	R37CH	F19CE	W14BH	R32BK	K28BG	W25BE	R24BC	K17AK	W34AH	U38AE
6	M18AB	L27CM	X27CJ	V30CF	Y23BP	X20BL	F33BH	Y35BF	X21BD	F22AL	Y26AJ	R11AF
7	H12AC	D13CS	B13CK	E14CH	T36CD	B28BM	V15BJ	T16BG	B17BE	V29AM	T38AK	X31AG
8	S37AD	J19AB	N19CL	M23CJ	G32CE	N33BP	E25BK	G24BH	N22BF	E34BC	G11AL	B18AH
9	P27AE	U30AC	A30CM	H36CK	L20CF	A15CD	M35BL	L21BJ	A29BG	M26BD	L31AM	N12AJ
10	W13AF	R14AD	Z14CP	S32CL	D28CH	Z25CE	H16BM	D17BK	Z34BH	H38BE	D18BC	A37AK
11	Y19AG	X23AE	C23CR	P20CM	J33CJ	C35CF	S24BP	J22BL	C26BJ	S11BF	J12BD	Z27AL
12	T30AH	B36AF	K36CS	W28CP	U15CK	K16CM	P21CD	U29BM	K38BK	P31BG	U37BE	C13AM
13	G14AJ	N32AG	F32AB	Y33CR	R25CL	F24CJ	W17CE	R34BP	F11BL	W18BH	R27BF	K19BC
14	L23AK	A20AM	V20AC	T15CS	X35CM	V21CK	Y22CF	X26CD	V31BM	Y12BJ	X13BG	F30BD
15	D36AL	Z28AJ	E28AD	G25AB	B16CP	E17CL	T29CH	B38CE	E18BP	T37BK	B19BH	V14BE
16	J32AM	C33AK	M33AE	L35AC	N24CP	M22CM	G34CJ	N11CF	M12CD	G27BL	N30BJ	E23BF
17	U20BC	K15AL	H15AF	D16AD	A21CS	H29CP	L26CK	A31CH	H37CE	L13BM	A14BK	M36BG
18	R28BD	F25AM	S25AG	J24AE	Z17AB	S34CR	D38CL	Z16CJ	S27CF	D19BP	Z23BL	H32BH
19	X33BE	V35BC	P35AH	U21AF	C22AC	P26CS	J11CM	C12CK	P13CH	J30CD	C36BM	S20BJ
20	B15BF	E16BD	W16AJ	R17AG	K29AD	W38AB	U31CP	K37CL	W19CJ	U14CE	K32BP	P28BK
21	N25BG	H24BF	Y24AK	X22AH	F34AE	Y11AC	R18CR	F27CH	Y30CK	R23CF	F20CD	W33BL
22	A35BH	H21BF	T21AL	B29AJ	V26AF	T31AD	X12CS	V13CP	T14CL	X36CH	V28CE	Y15BM
23	Z16BJ	S17BG	G27BG	N34AK	E38AG	G18AE	B37AB	E19CR	G23CM	B32CJ	E33CF	T25BP
24	C24BK	P22BH	L22BC	A26AL	M11AH	L12AF	N27AC	M30CS	L36CP	N20CK	M15CH	G35CD
25	K21BL	W29BJ	D29BD	Z38AM	H31AJ	D37AG	A13AD	H14AB	D32CR	A28CL	H25CJ	L16CE
26	F17BM	Y34BK	J34BE	C11BC	S18AK	J27AH	Z19AE	S23AC	J20CS	Z33CM	S35CK	D24CF
27	V22BP	T26BL	U26BF	K31BD	P12AL	U13AJ	C30AF	P36AD	U28AB	C15CP	P16CL	J21CH
28	E29CD	G38BM	R38BG	F18BE	W37AM	R19AK	K14AG	W32AE	R33AC	K25CR	W24CH	U17CJ
29	M34CE		X11BH	V12BF	Y27BC	X30AL	F23AH	Y20AF	X15AD	F35CS	Y21CP	R22CK
30	H26CF		B31BJ	E37BG	T13BD	B14AM	V36AJ	T28AG	B25AE	V16AB	T17CR	X29CL
31	S38CH		N18BK		G19BE		E32AK	G33AH		E24AC		B34CM

1822	JAN	FEB	MAR	APR	MAY	JUN	JUL	AUG	SEP	OCT	NOV	DEC
1	N26CP	H31CL	Y31CE	X37BP	F13BK	Y14BH	R36BE	F28BC	Y25AL	R16AH	F17AF	W29AC
2	A38CR	H18CM	T18CF	B27CD	V19BL	T23BJ	X32BF	V33BD	T35AM	X24AJ	V22AG	Y34AD
3	Z11CS	S12CP	G12CH	N13CE	E30BM	G36BK	B20BG	E15BE	G15BC	B21AK	E29AH	T26AE
4	C31AB	P37CR	L37CJ	A19CF	M14BP	L32BL	N28BH	M25BF	L24BD	N17AL	M34AJ	G38AF
5	K18AC	W27CS	D27CK	Z30CH	H23CD	D20BM	A33BJ	H35BG	D21BE	A22AM	H26AK	L11AG
6	F12AD	Y13AB	J13CL	C14CJ	S36CE	J28BP	Z15BK	S16BH	J17BF	Z29AM	S38AL	D31AH
7	V37AE	T19AC	U19CM	K23CK	P32CF	U33CD	C25BL	P24BJ	U22BG	C34BD	P11AM	J18AJ
8	E27AF	G30AD	R30CP	F36CL	W20CH	R15CE	K35BM	W21BK	R29BH	K26BE	W31BC	U12AK
9	M13AG	L14AE	X14CR	V32CM	Y28CJ	X25CF	F16BP	Y17BL	X34BJ	F38BF	Y18BD	R37AL
10	H19AH	D23AF	B23CS	E20CP	T33CK	B35CH	V24CD	T22BM	B26BK	V11BG	T12BE	X27AM
11	S30AJ	J36AG	N36AB	M28CR	G15CL	N16CJ	E21CE	G29BP	N38BL	E31BH	G37BF	B13BC
12	P14AK	U32AH	A32AC	H33CS	L25CM	A24CK	M17CF	L34CD	A11BM	M18BJ	L27BG	N19BD
13	W23AL	R20AJ	Z20AD	S15AB	D35CP	Z21CL	H22CH	D26CE	Z31BP	H12BK	D13BH	A30BE
14	Y36AM	X28AK	C28AE	P25AC	J16CR	C17CH	S29CJ	J38CF	C18CD	S37BL	J19BJ	Z14BF
15	T32BC	B33AL	K33AF	W35AD	U24CS	K22CP	P34CK	U11CH	K12CE	P27BM	U30BK	C23BG
16	G20BD	N15AM	F15AG	Y16AE	R21AB	F29CR	W26CL	R31CJ	F37CF	W13BP	R14BL	K36BH
17	L28BE	A25BC	V25AH	T24AF	X17AC	V34CS	Y38CM	X16CK	V27CH	Y19CD	X23BM	F32BJ
18	D33BF	Z35BD	E35AJ	G21AG	B22AD	E26AB	T11CP	B12CL	E13CJ	T30CE	B36BP	V20BK
19	J15BG	C16BE	M16AK	L17AH	N29AE	M38AC	G31CR	N37CM	M19CK	G14CF	N32CD	E28BL
20	U25BH	K24BD	H24AL	D22AJ	A34AF	H11AD	L18CS	A27CP	H30CL	L23CH	A20CE	M33BM
21	R35BJ	F21BG	S21AM	J29AK	Z26AG	S31AE	D12AB	Z13CR	S14CM	D36CJ	Z28CF	H15BP
22	X16BK	V17BH	P17BC	U34AL	C38AH	P18AF	J37AC	C19CS	P23CP	J32CK	C33CH	S25CD
23	B24BL	E22BJ	W22BD	R26AM	K11AJ	W12AG	U27AD	K30AB	W36CR	U20CL	K15CJ	P35CE
24	N21BM	H29BK	Y29BE	X38BC	F31AK	Y37AH	R13AE	F14AC	Y32CS	R28CM	F25CK	W16CF
25	A17BP	H34BL	T34BF	B11BD	V18AL	T27AJ	X19AF	V23AD	T20AB	X33CP	V35CL	Y24CH
26	Z22CD	S26BM	G26BG	N31BE	E12AM	G13AK	B30AG	E36AE	G28AC	B15CR	E16CM	T21CJ
27	C29CE	P38BP	L38BH	H27BD	M37BC	L19AL	N14AH	M32AF	L33AD	N25CS	M24CP	G17CK
28	K34CF	W11CD	D11BJ	Z12BG	H27BD	D30AM	A23AJ	H20AG	D15AE	A35AB	H21CR	L22CL
29	F26CH		J31BK	C37BM	S13BE	J14BC	Z36AK	S28AH	J25AF	Z16AC	S17CS	D29CM
30	V38CJ		U18BL	K27BJ	P19BF	U23BD	C32AL	P33AJ	U35AG	C24AD	P22AB	J34CP
31	E11CK		R12BM		W30BG		K20AM	W15AK		K21AE		U26CR

1823	JAN	FEB	MAR	APR	MAY	JUN	JUL	AUG	SEP	OCT	NOV	DEC
1	R38CS	F18CP	S18CH	J27CE	Z19BM	S23BK	D32BG	Z33BE	S35BC	D24AK	Z22AH	H34AE
2	X11AB	V12CR	P12CJ	U13CF	C30BP	P36BL	J20BH	C15BF	P16BD	J21AL	C29AJ	S26AF
3	B31AC	E37CS	W37CK	R19CH	K14CD	W32BM	U28BJ	K25BG	W24BE	U17AM	K34AK	P38AG
4	N18AD	M27AB	Y27CL	X30CJ	F23CE	Y20BP	R33BK	F35BH	Y21BF	R22BC	F26AL	W11AH
5	A12AE	H13AC	T13CM	B14CK	V36CF	T28CD	X15BL	V16BJ	T17BG	X29BD	V38AM	Y31AJ
6	Z37AF	S19AD	G19CP	N23CL	C33CG	G33CE	B25BM	C24BK	G22BH	B34BE	E11BC	T18AK
7	C27AG	P30AE	L30CP	A36CM	M20CJ	L15CF	N35BP	M21BL	L29BJ	N26BF	M31BD	G12AL
8	K13AH	W14AF	D14CS	Z32CP	H28CK	D25CH	A15CD	H17BM	D34BK	A38BG	H18BE	L37AH
9	F19AJ	Y23AG	J23AB	C20CR	S33CL	J35CJ	Z24CE	S22BP	J26BL	Z11BH	S12BF	D27BC
10	V30AK	T36AH	U36AC	K28CS	P15CM	U16CK	C21CF	P29CD	U38BM	C31BF	P13BD	J13BC
11	E14AL	G32AJ	R32AD	F33AB	W25CP	R24CL	K17CH	W34CE	R11BP	K18BK	W27BH	U19BE
12	M23AM	L20AK	X20AE	V15AC	Y35CR	X21CM	F22CJ	Y26CF	X31CD	F12BL	Y13BJ	R30BF
13	H36BC	D28AL	B28AF	E25AD	T16CS	B17CP	V29CK	T38CH	B18CE	V37BM	T19BK	X14BG
14	S32BD	J33AM	N33AE	M35AE	G24AB	N22CR	E34CL	G11CJ	N12CF	E27BP	G30BL	B23BH
15	P20BE	U15BC	A15AM	H16AF	L21AH	A29CS	M26CH	L31CK	A37CH	M13BD	L14BM	N36BJ
16	W28BF	R25BD	Z25AJ	S24AG	D17AD	Z34AB	H38CP	D16CL	Z27CJ	H19CE	D23BP	A32BK
17	Y33BG	X35BE	G35AK	P21AH	J22AE	G26AC	S11CR	J12CH	C13CK	S30CF	J36CD	Z20BL
18	T15BH	B16BF	K16AL	W17AJ	U29AF	K38AD	P31CS	U37CP	K19CL	P14CH	U32CE	C28BM
19	G25BJ	N24BG	F24AB	Y22AK	R34AG	F11AE	W18AB	R27CR	F30CM	W23CJ	R20CF	K33BP
20	L35BK	A21BH	V21BC	T29AL	X26AH	V31AF	Y12AC	X13CS	V14CP	Y36CK	X28CH	F15CD
21	D16BL	Z17BJ	E17BD	G34AM	B38AJ	E18AG	T37AD	B19AB	E23CR	T32CL	B33CJ	V25CE
22	J24BM	C22BK	H22BE	L26BC	N11AK	H12AH	G27AE	N30AC	H32AB	M36CS	N15CK	E35CF
23	U21BP	K29BL	H29BF	D38BD	A31AL	H37AJ	L13AF	A14AD	M32AB	L28CP	A25CL	M16CH
24	R17CD	F34BM	S34BG	J11BE	Z16AM	S27AK	D19AG	Z23AE	S20AC	D33CR	Z35CH	H24CJ
25	X22CE	V26BP	P26BH	U31BF	C12BC	P13AL	J30AH	C36AF	P28AD	J15CS	C16CP	S21CK
26	B29CF	E38CD	W38BJ	R18BG	K37BD	W19AM	U14AJ	K32AG	W33AE	U25AB	K24CR	P17CL
27	N34CH	M11CE	Y11BK	X12BH	F27BE	Y30BC	R23AK	F20AH	Y15AF	R35AC	F21CS	W22CM
28	A26CJ	H31CF	T31BL	B37BJ	V13BF	T14BD	X36AL	V28AJ	T25AG	X16AD	V17AB	Y29CP
29	Z38CK		G18BM	N27BK	E19BG	G23BE	B32AM	E33AK	G35AH	B24AE	E22AC	T34CR
30	C11CL		L12BP	A13BL	M30BH	L36BF	N20BC	M15AL	L16AJ	N21AF	M29AD	G26CS
31	K31CM		D37CD		H14BJ		A28BD	H25AM		A17AG		L38AB

FIRST LETTER = YOUR PHYSICAL CODE, MIDDLE NUMBER = SENSITIVITY CODE, LAST TWO LETTERS = INTELLECTUAL CODE

1824	JAN	FEB	MAR	APR	MAY	JUN	JUL	AUG	SEP	OCT	NOV	DEC
1	D11AC	Z12CS	M37CL	L19CJ	N14CE	M32BP	G28BK	N25BH	M24BF	G17BC	N34AL	E38AH
2	J31AD	C37AB	H27CM	D30CK	A23CH	H20CD	L33BM	A35BJ	H21BG	L22BD	A26AM	M11AJ
3	U18AE	K27AC	S13CP	J14CL	Z36CH	S28CE	D15BM	Z16BK	S17BH	D29BE	Z38BC	H31AK
4	R12AF	F13AD	P19CR	U23CH	C32CJ	P33CF	J25BP	C24BL	P22BJ	J34BF	C11BD	S18AL
5	X37AG	V19AE	W30CS	R36CP	K20CK	W15CH	U35CD	K21BM	W29BK	U26BG	K31BE	P12AM
6	B27AH	E30AF	Y14AB	X32CR	F28CL	Y25CJ	R16CE	F17BP	Y34BL	R38BH	F18BF	W37BC
7	N13AJ	M14AG	T23AC	B20CS	V33CM	T35CK	X24CF	V22CD	T26BM	X11BJ	V12BG	Y27BD
8	A19AK	H23AH	G36AD	N28AB	E15CP	G16CL	B21CH	E29CE	G38BP	B31BK	E37BH	T13BE
9	Z30AL	S36AJ	L32AE	A33AC	M25CR	L24CM	N17CJ	M34CF	L11CD	N18BL	M27BJ	G19BF
10	C14AM	P32AK	D20AF	Z15AD	H35CS	D21CP	A22CK	H26CH	D31CE	A12BM	H13BK	L30BG
11	K23BC	W20AL	J28AG	C25AE	S16AB	J17CR	Z29CL	S38CJ	J18CF	Z37BP	S19BL	D14BH
12	F36BD	Y28AM	U33AH	K35AF	P24AC	U22CS	C34CM	P11CK	U12CH	C27CD	P30BM	J23BJ
13	V32BE	T33BC	R15AJ	F16AG	W21AD	R29AB	K26CP	M31CL	R37CJ	K13CE	W14BP	U36BK
14	E20BF	G15BD	X25AK	V24AH	Y17AE	X34AC	F38CR	Y18CM	X27CK	F19CF	Y23CD	R32BL
15	M28BG	L25BE	B35AL	E21AJ	T22AF	B26AD	V11CS	T12CP	B13CL	V30CH	T36CE	X20BM
16	H33BH	D35BF	N16AM	M17AK	G29AG	N38AE	E31AB	G37CR	N19CM	E14CJ	G32CF	B28BP
17	S15BJ	J16BG	A24BC	H22AL	L34AH	A11AF	M18AC	L27CS	A30CP	M23CK	L20CH	N33CD
18	P25BK	U24BH	Z21BD	S29AM	D26AJ	Z31AG	H12AD	D13AB	Z14CR	H36CL	D28CJ	A15CE
19	W35BL	R21BJ	C17BE	P34BC	J38AK	C18AH	S37AE	J19AC	C23CS	S32CM	J33CK	Z25CF
20	Y16BM	X17BK	K22BF	W26BD	U11AL	K12AJ	P27AF	U30AD	K36AB	P20CP	U15CL	C35CH
21	T24BP	B22BL	F29BG	Y38BE	R31AM	F37AK	W13AG	R14AE	F32AC	W28CR	R25CM	K16CJ
22	G21CD	N29BM	V34BH	T11BF	X18BC	V27AL	Y19AH	X23AF	V20AD	Y33CS	X35CP	F24CK
23	L17CE	A34BP	E26BJ	G31BG	B12BD	E13AM	T30AJ	B36AG	E28AE	T15AB	B16CR	V21CL
24	D22CF	Z26CD	M38BK	L18BH	N37BE	M19AC	G14AK	N32AH	M33AF	G25AC	N24CS	E17CM
25	J29CH	C38CE	H11BL	D12BJ	A27BF	H30BD	L23AL	A20AJ	415AG	L35AD	A21AB	M22CP
26	U34CJ	K11CF	S31BM	J37BK	Z13BG	S14BE	D36AM	Z28AK	S25AH	D16AE	Z17AC	H29CR
27	R26CK	F31CH	P18BP	U27BL	C19BH	P23BF	J32BC	C33AL	P35AJ	J24AF	C22AD	S34CS
28	X38CL	V18CJ	W12CD	R13BM	K30BJ	W36BG	U20BD	K15AM	W16AK	U21AG	K29AE	P26AB
29	B11CM	E12CK	Y37CE	X19BP	F14BK	Y32BH	R28BE	F25BC	Y24AL	R17AH	F34AF	W38AC
30	N31CP		T27CF	B30CD	V23BL	T20BJ	X33BF	V35BD	T21AM	X22AJ	V26AG	Y11AD
31	A18CR		G13CH		E36BM		B15BG	E16BE		B29AK		T31AE

1825	JAN	FEB	MAR	APR	MAY	JUN	JUL	AUG	SEP	OCT	NOV	DEC
1	G18AF	N27AD	F27CP	Y30CL	R23CH	F20CE	W33BM	R35BK	F21BH	W22BE	R26BC	K11AK
2	L12AG	A13AE	V13CR	T14CM	X36CJ	V28CF	Y15BP	X16BL	V17BJ	Y29BF	X38BD	F31AL
3	D37AH	Z19AF	E19CS	G23CP	B32CK	E33CH	T25CD	B24BH	E22BK	T34BG	B11BE	V18AM
4	J27AJ	C30AG	M30AB	L36CR	N20CL	M15CJ	G35CE	N21BP	M29BL	G26BH	N31BF	E12BC
5	U13AK	K14AH	H14AC	D32CS	A28CM	H25CK	L16CF	A17CD	H34BM	L38BJ	A18BG	M37BD
6	R19AL	F23AJ	S23AD	J20AB	Z33CP	S35CL	D24CH	Z22CE	S26BP	D11BK	Z12BH	H27BE
7	X30AM	V36AK	P36AE	U28AC	C15CR	P16CM	J21CJ	C29CF	P38CD	J31BL	C37BJ	S13BF
8	B14BC	E32AL	W32AF	R33AD	K25CS	W24CP	U17CK	K34CH	W11CE	U18BM	K27BK	P19BG
9	N23BD	M20AM	Y20AG	X15AE	F35AB	Y21CR	R22CL	F26CJ	Y31CF	R12BP	F13BL	W30BH
10	A36BE	H28BC	T28AH	B25AF	V16AC	T17CS	X29CM	V38CK	T18CH	X37CD	V19BM	Y14BJ
11	Z32BF	S33BD	G33AJ	N35AG	E24AD	G22AB	B34CP	E11CL	G12CJ	B27CE	E30BP	T23BK
12	C20BG	P15BE	L15AK	A16AH	M21AE	L29AC	N26CR	M31CL	L37CK	N13CF	M14CD	G36BL
13	K28BH	W25BF	D25AL	Z24AJ	H17AF	D34AD	A38CS	H18CP	D27CL	A19CH	H23CE	L32BM
14	F33BJ	Y35BG	J35AM	C21AK	S22AG	J26AE	Z11AB	S12CR	J13CM	Z30CJ	S36CF	D20BP
15	V15BK	T16BH	U16BC	K17AL	P29AH	U38AF	C31AC	P37CS	U19CP	C14CK	P32CH	J28CD
16	E25BL	G24BJ	R24BD	F22AM	W34AJ	R11AG	K18AD	W27AB	R30CR	K23CL	W20CJ	U33CE
17	M35BM	L21BK	X21BE	V29BC	Y26AK	X31AH	F12AE	Y13AC	X14CS	F36CM	Y28CK	R15CF
18	H16BP	D17BL	B17BF	E34BD	T38AL	B18AJ	V37AF	T19AD	B23AB	V32CP	T33CL	X25CH
19	S24CD	J22BM	N22BG	M26BE	G11AM	N12AK	E27AG	G30AE	N36AC	E20CR	G15CM	B35CJ
20	P21CE	U29BP	A29BH	H38BF	L31BC	A37AL	M13AH	L14AF	A32AD	M28CS	L25CP	N16CK
21	W17CF	R34CD	Z34BJ	S11BG	D18BD	Z27AM	H19AJ	D23AG	Z20AE	H33AB	D35CM	A24CL
22	Y22CH	X26CE	C26BK	P31BH	J12BE	C13BC	S30AK	J36AH	C28AF	S15AC	J16CS	Z21CM
23	T29CJ	B38CF	K38BL	W18BJ	U37BF	K19BD	P14AL	U32AJ	K33AG	P25AD	U24AB	C17CP
24	G34CK	N11CH	F11BM	Y12BK	R27BG	F30BE	W23AM	R20AK	F15AH	W35AE	R21AC	K22CR
25	L26CL	A31CJ	V31BP	T37BL	X13BH	V14BF	Y36BC	X28AL	V25AJ	Y16AF	X17AD	F29CS
26	D38CM	Z18CK	E18CD	G27BM	B19BJ	E23BG	T32BD	B33AM	E35AK	T24AG	B22AE	V34AB
27	J11CP	C12CL	M12CE	L13BP	N30BK	M36BH	G20BE	N15BC	M16AL	G21AH	N29AF	E26AC
28	U31CR	K37CM	H37CF	D19CD	A14BL	H32BJ	L28BF	A25BD	H24AM	L17AJ	A34AG	M38AD
29	R18CS		S27CH	J30CE	Z23BM	S20BK	D33BG	Z35BE	S21BC	D22AK	Z26AH	H11AE
30	X12AB		P13CJ	U14CF	C36BP	P28BL	J15BM	C16BF	P17BD	J29AL	C38AJ	S31AF
31	B37AC		W19CK		K32CD		U25BJ	K24BG		U34AM		P18AG

1826	JAN	FEB	MAR	APR	MAY	JUN	JUL	AUG	SEP	OCT	NOV	DEC
1	W12AH	R13AF	Z13CS	S14CP	D36CK	Z28CH	H15CD	D16BM	Z17BK	H29BG	D38BE	A31AM
2	Y37AJ	X19AG	C19AB	P23CR	J32CL	C33CJ	S25CE	J24BP	C22BL	S34BH	J11BF	Z18BC
3	T27AK	B30AH	K30AC	W36CS	U20CM	K15CK	P35CF	U21CD	K29BM	P26BJ	U31BG	C12BD
4	G13AL	N14AJ	F14AD	Y32AB	R28CP	F25CL	W16CH	R17CE	F34BP	W38BK	R18BH	K37BE
5	L19AM	A23AK	V23AE	T20AC	X33CR	V35CM	Y24CJ	X22CF	V25CD	Y11BL	X12BJ	F27BF
6	D30BC	Z36AL	E36AF	G28AD	B15CS	E16CP	T21CK	B29CH	E38CE	T31BM	B37BK	V13BG
7	J14BD	C32AM	M32AG	L33AE	N25AB	M24CR	G17CL	N34CJ	M11CF	G18BP	N27BL	E19BH
8	U23BE	K20BC	H20AH	D15AF	A35AC	H21CH	L22CM	A26CK	H31CH	L12CD	A13BM	M30BJ
9	R36BF	F28BD	S28AJ	J25AG	Z16AD	S17AB	D29CP	Z38CL	S18CJ	D37CE	Z19BP	H14BK
10	X32BG	V33BE	P33AK	U35AH	C24AE	P22AC	J34AB	C11CP	P12CK	J27CF	C30CD	S23BL
11	B20BH	E15BF	W15AL	R16AJ	K21AF	W29AD	U26CS	K31CL	W37CJ	U13CH	K14CE	P36BM
12	N28BJ	M25BG	Y25AM	X24AK	F17AG	Y34AE	R38AB	F18CR	Y27CL	R19CJ	F23CF	W32BP
13	A33BK	H35BH	T35BC	B21AL	V22AH	T26AF	X11AC	V12CS	T13CP	X30CK	V36CH	Y20CD
14	Z15BL	S16BJ	G16BD	N17AM	E29AJ	G38AG	B31AD	E37AB	G19CR	B14CL	E32CJ	T28CE
15	C25BM	P24BK	L24BE	A22BC	M34AK	L11AH	N18AE	M27AC	L30CS	N23CM	M20CK	G33CF
16	K35BP	W21BL	D21BF	Z29BD	H26AL	D31AJ	A12AF	H13AD	D14BB	A36CP	H28CL	L15CH
17	F16CD	Y17BM	J17BG	C34BE	S38AM	J18AK	Z37AG	S19AE	J23AC	Z32CR	S33CM	D25CJ
18	V24CE	T22BP	U22BH	K26BF	P11BC	U12AL	C27AH	P30AF	U36AD	C20CS	P15CP	J35CK
19	E21CF	G29CD	R29BJ	F38BG	W31BD	R37AM	K13AJ	W14AG	R32AE	K28AB	W25CL	U16CL
20	M17CH	L34CE	X34BK	V11BH	Y18BE	X27BC	F19AK	Y23AH	X20AF	F33AC	Y35CS	R24CH
21	H22CJ	D26CF	B26BL	E31BJ	T12BF	B13BD	V30AL	T36AJ	B28AG	V15AD	T16AB	X21CP
22	S29CK	J38CH	N38BM	M18BK	G37BG	N19BE	E14AM	G32AK	N33AH	E25AE	G24AC	B17CR
23	P34CL	U11CJ	A11BP	H12BL	L27BH	A30BF	M23BC	L20AL	A15AJ	M35AF	L21AD	N22CS
24	W26CM	R31CK	Z31CD	S37BM	D13BJ	Z14BG	H36BD	D28AM	Z25AK	H16AG	D17AE	A29AB
25	Y38CP	X18CL	C18CE	P27BP	J19BK	C23BH	S32BE	J33BC	C35AL	S24AH	J22AF	Z34AC
26	T11CR	B12CM	K12CF	W13CD	U30BL	K36BJ	P28BF	U15BD	K16AM	P21AJ	U29AG	C26AD
27	G31CS	N37CP	F37CH	Y19CE	R14BM	F32BK	W28BG	R25BE	F24BC	W17AH	R34AH	K38AC
28	L18AB	A27CR	V27CJ	T30CF	X23BP	V20BL	Y33BD	X35BF	V21BD	Y22AL	X26AJ	F11AD
29	D12AC		E13CK	G14CH	B36CD	E28BM	T15BJ	B16BG	E17BE	T29AH	B38AK	V31AG
30	J37AD		M19CL	L23CJ	N32CE	M33BP	G25BK	N24BH	M22BF	G34BC	N11AL	E18AH
31	U27AE		H30CH		A20CF		L35BL	A21BJ		L26BD		M12AJ

FIRST LETTER = YOUR PHYSICAL CODE, MIDDLE NUMBER = SENSITIVITY CODE, LAST TWO LETTERS = INTELLECTUAL CODE

1827	JAN	FEB	MAR	APR	MAY	JUN	JUL	AUG	SEP	OCT	NOV	DEC
1	H37AK	D19AH	B19AC	E23CS	T32CM	B33CK	V25CF	T24CD	B22BM	V34BJ	T118G	X18BD
2	S27AL	J30AJ	N30AD	M36AB	G20CP	N15CL	E35CH	G21CE	N29BP	E26BK	G31BH	B12BE
3	P13AM	U14AK	A14AE	H32AC	L28CR	A25CM	M16CJ	L17CF	A34CD	M38BL	L18BJ	N37BF
4	W19BC	R23AL	Z23AF	S20AD	O33CS	Z35CP	H24CK	D22CH	Z26CE	H11BM	O12BK	A27BG
5	Y30BD	X36AM	C36AG	P28AE	J15AB	C16CR	S21CL	J29CJ	C38CF	S31BP	J37BL	Z13BH
6	T14BE	B32BC	K32AH	W33AF	U25AC	K24CS	P17CM	U34CK	K11CH	P18CD	U27BM	C19BJ
7	G23BF	N20BD	F20AJ	Y15AG	R35AD	F21AB	W22CP	R26CL	F31CJ	M12CE	R13BP	K30BK
8	L36BG	A28BE	V28AK	T25AH	X16AE	V17AC	Y29CR	X38CM	V18CK	Y37CF	X19CD	F14BL
9	D32BH	Z33BF	E33AL	G35AJ	B24AF	E22AD	T34CS	B11CP	E12CL	T27CH	B30CE	V23BM
10	J20BJ	C15BG	M15AM	L16AK	N21AG	M29AE	G26AB	N31CR	M37CH	G13CJ	N14CF	E36BP
11	U28BK	K25BH	H25BC	D24AL	A17AH	H34AF	L38AC	A18CS	H27CP	L19CK	A23CH	M32CD
12	R33BL	F35BJ	S35BD	J21AM	Z22AJ	S26AG	D11AD	Z12AB	S13CR	O30CL	Z36CJ	H20CE
13	X15BM	V16BK	P16BE	U17BC	C29AK	P38AH	J31AE	C37AC	P19CS	J14CM	C32CK	S28CF
14	B25BP	E24BL	W24BF	R22BD	K34AL	W11AJ	U18AF	K27AD	W30AB	U23CP	K20CL	P33CH
15	N35CD	M21BM	Y21BG	X29BE	F26AM	Y31AK	R12AG	F13AE	Y14AC	R36CR	F28CM	W15CJ
16	A16CE	H17BP	T17BH	B34BF	V38BC	T18AL	X37AH	V19AF	T23AD	X32CS	V33CP	Y25CK
17	Z24CF	S22CD	G22BJ	N26BG	E11BD	G12AM	B27AJ	E30AG	G36AE	B20AB	E15CR	T35CL
18	C21CH	P29CE	L29BK	A38BH	M31BE	L37BC	N13AK	M14AH	L32AF	N28AC	M25CS	G16CM
19	K17CJ	H34CF	D34BL	Z11BJ	H18BF	D27BD	A19AL	H23AJ	D20AG	A33AD	H35AB	L24CP
20	F22CK	Y26CH	J26BM	C31BK	S12BG	J13BE	Z30AM	S36AK	J28AH	Z15AE	S16AC	D21CR
21	V29CL	T38CJ	U38BP	K18BL	P37BH	U19BF	C14BC	P32AL	U33AJ	C25AF	P24AD	J17CS
22	E34CM	G11CK	R11CD	F12BM	W27BJ	R30BG	K23BD	W20AM	R15AK	K35AB	W21AE	U22AB
23	M26CP	L31CL	X31CF	V37BP	Y13BK	X14BH	F36BE	Y28BC	X25AL	F16AH	Y17AF	R29AC
24	H38CR	D18CM	B18CF	E27CD	T19BL	B23BJ	V32BF	T33BD	B35AM	V24AJ	T22AG	X34AD
25	S11CS	J12CP	N12CH	M13CE	G30BM	N36BK	E20BG	G15BE	N16BC	E21AK	G29AH	B26AE
26	P31AB	U37CR	A37CJ	H19CF	L14BP	A32BL	M28BH	L25BF	A24BD	M17AL	L34AJ	N38AF
27	W18AC	R27CS	Z27CK	S30CH	O23CD	Z20BM	H33BJ	O35BG	Z21BE	H22AM	O26AK	A11AG
28	Y12AD	X13AB	C13CL	P14CJ	J36CE	C28BP	S15BK	J16BH	O17BF	S29BC	J38AL	Z31AH
29	T37AE		K19CM	W23CK	U32CF	K33CD	P25BL	U24BJ	K22BG	P34BD	U11AM	C18AJ
30	G27AF		F30CP	Y36CL	R20CH	F15CE	W35BM	R21BK	F29BH	W26BE	R31BC	K12AK
31	L13AG		V14CR		X28CJ		Y16BP	X17BL		Y38BF		F37AL

1828	JAN	FEB	MAR	APR	MAY	JUN	JUL	AUG	SEP	OCT	NOV	DEC
1	V27AM	T30AK	R14AF	F32AD	W28CS	R25CP	K16CK	W17CH	R34CE	K38BM	W18BK	U37BG
2	E13BC	G14AL	X23AG	V20AE	Y33AB	X35CR	F24CL	Y22CJ	X26CF	F11BP	Y12BL	R27BH
3	M19BD	L23AM	B36AH	E28AF	T15AC	B16CS	V21CM	T29CK	B38CH	V31CD	T37BM	X13BJ
4	H30BE	D36BC	N32AJ	M33AG	G25AD	N24AB	E17CP	G34CL	N11CJ	E18CE	G27BP	B19BK
5	S14BF	J32BD	A20AK	H15AH	L35AE	A21AC	M22CS	L26CM	A31CK	M12CF	L13CD	N30BL
6	P23BG	U20BE	Z28AL	S25AJ	D16AF	Z17AD	H29CS	D38CP	Z18CL	H37CH	D19CE	A14BM
7	W36BH	R28BF	C33AM	P35AK	J24AG	C22AE	S34AB	J11CR	C12CM	S27CJ	J30CF	Z23BP
8	Y32BJ	X33BG	K15BC	W16AL	U21AH	K34AF	Y26AC	U31CS	K37CP	P13CK	U14CH	C36CD
9	T20BK	B15BH	F25BD	Y24AM	R17AJ	F34AG	W38AD	R18AB	F27CR	W19CL	R23CJ	K32CE
10	G28BL	N25BJ	V35BE	T21BC	X22AK	V26AH	H11AE	X12AC	V13CS	Y30CM	X36CK	F20CE
11	L33BM	A35BK	E16BF	G17BD	B29AL	E38AJ	T31AF	B37AD	E19AB	T14CP	B32CL	V28CH
12	D15BP	Z16BL	M24BG	L22BE	N34AM	M11AK	G18AG	N27AE	M30AC	G23CR	N20CM	E33CJ
13	J25CD	C24BM	H21BH	D29BF	A26BC	H31AL	L12AH	A13AF	H14AD	L36CS	A28CP	M15CK
14	U35CE	K21BP	S17BJ	J34BG	Z38BD	S18AM	D37AJ	Z19AG	S23AE	D32AB	Z33CR	H25CL
15	R16CF	F17CD	P22BK	U26BH	C11BE	P12BC	J27AK	C30AH	P36AF	J20AC	C15CS	S35CM
16	X24CH	V22CE	N29BL	R38BJ	K31BF	N37BD	W37BD	U13AL	K14AJ	W32AG	U28AD	P16CP
17	B21CJ	E29CF	Y34BM	X11BK	F18BG	Y27BE	R19AM	F23AK	Y20AH	R33AE	F35AC	W24CR
18	N17CK	M34CH	T26BP	B31BL	V12BH	T13BF	X30BC	V36AL	T28AJ	X15AF	V16AD	Y21CS
19	A22CL	H26CJ	G38CD	N18BM	E37BJ	G19BG	B14BD	E32AM	G33AK	B25AG	E24AE	T17AB
20	Z29CM	S38CK	L11CE	A12BP	M27BK	L30BH	N23BE	M20BC	L15AL	N35AH	M21AF	G22AC
21	C34CP	P11CL	D31CF	Z37CD	H13BL	D14BJ	A36BF	H28BD	D25AM	A16AJ	H17AG	L29AD
22	K26CR	W31CM	J18CH	C27CE	S19BM	J23BK	Y32BG	S33BE	J35BC	Y24AK	S22AH	D34AE
23	F38CS	Y18CP	U12CJ	K13CF	P30BP	U36BL	C20BH	P15BF	U16BD	C21AL	P29AJ	J26AF
24	V11AB	T12CR	R37CK	F19CH	W14CD	R32BM	K28BJ	W25BG	R24BE	K17AM	W34AK	U38AG
25	E31AC	G37CS	X27CL	V30CJ	Y23CE	X20BP	F33BK	Y35BH	X21BF	F22BC	Y26AL	R11AH
26	M18AD	L27AB	B13CM	E14CK	T36CF	B28CD	V15BL	T16BJ	B17BG	V29BD	T38AM	X31AJ
27	H12AE	D13AC	N19CP	M23CL	G32CH	N33CE	E25BM	G24BK	N22BH	E34BE	G11BC	B18AK
28	S37AF	J19AD	A30CR	H36CM	L20CJ	A15CF	M35BP	L21BL	A29BJ	M26BF	L31BD	N12AL
29	P27AG	U30AE	Z14CS	S32CP	D28CK	Z25CH	H16CD	D17BM	Z34BK	H38BG	D18BE	A37AM
30	W13AH		K23AB	P20CR	J33CL	C35CJ	S24CE	J22BP	C26BL	S11BH	J12BF	Z27BC
31	Y19AJ		K36AC		U15CM		P21CF	U29CD		P31BJ		C13BD

1829	JAN	FEB	MAR	APR	MAY	JUN	JUL	AUG	SEP	OCT	NOV	DEC
1	K19BE	W23BC	D23AH	Z20AF	H33AC	D35CS	A24CM	H22CK	D26CH	A11CD	H12BM	L27BJ
2	F30BF	Y36BD	J36AJ	C28AG	S15AD	J16AB	Z21CP	S29CL	J38CJ	Z31CE	S37BP	D13BK
3	V14BG	T32BE	U32AK	K33AH	P25AE	U24AC	C17CR	P34CM	U11CK	C18CF	P27CD	J19BL
4	E23BH	G20BF	R20AL	F15AJ	W35AF	R21AD	K22CS	W26CP	R31CL	K12CH	W13CE	U30BM
5	M36BJ	L28BG	X28AM	V25AK	Y16AG	X17AE	F29AB	Y38CR	X18CM	F37CJ	Y19CF	R14BP
6	H32BK	D33BH	B33BC	E35AL	T24AH	B22AF	V34AC	T11CS	B12CP	V27CK	T30CH	X23CD
7	S20BL	J15BJ	N15BD	M16AM	G21AJ	N29AG	E26AD	G31AB	N37CR	E13CL	G14CJ	B36CE
8	P28BM	U25BK	A25BE	H24BC	L17AK	A34AH	M38AE	L18AC	A27CS	M19CM	L23CH	N32CF
9	W33BP	R35BL	Z35BF	S21BD	D22AL	Z26AJ	H11AF	D12AD	Z13AB	H30CP	D36CL	A20CH
10	Y15CD	X16BM	C16BG	P17BE	J29AM	C38AK	S31AG	J37AE	C19AC	S14CR	J32CM	Z28CJ
11	T25CE	B24BP	K24BH	W22BF	U34BC	K11AL	P18AH	U27AF	K30AD	P23CS	U20CM	C33CK
12	G35CF	N21CD	F21BJ	Y29BG	R26BD	F31AM	W12AJ	R13AG	F14AE	W36AB	R28CR	K15CL
13	L16CH	A17CE	V17BK	T34BH	X38BE	V18BC	Y37AK	X19AH	V23AF	Y32AC	X33CS	F25CM
14	D24CJ	Z22CF	E22BL	G26BJ	B11BD	E12BD	T27AL	B30AD	E36AC	T20AD	B15AB	V35CP
15	J21CK	C29CH	M29BM	L38BK	N31BG	M37BE	G13AM	N14AK	M32AH	G28AE	N25AC	E16CR
16	U17CL	K34CK	H34BP	D11BL	A18BH	H27BD	L19BC	A23AL	H20AJ	L33AF	A35AD	M24CS
17	R22CM	F26CK	S26CD	J31BM	Z12BK	S13BG	D30BD	Z36AM	S28AK	D15AG	Z16AE	H21AB
18	X29CP	V38CL	P38CE	U18BP	C37BK	P19BE	W30BJ	U23BF	P33AL	J25AH	C24AF	S17AC
19	B34CR	E11CM	E11CH	R12CD	K27BJ	W30BJ	U23BF	K20BD	W15AM	U35AJ	K21AG	P22AD
20	N26CS	M31CP	Y31CH	X37CE	F13BM	Y14BK	R36BG	F28BE	Y25BC	R16AK	F17AH	W29AE
21	A38AB	H18CR	T18CJ	B27CF	V19BP	T23BL	X32BH	V33BF	T35BD	X24AL	V22AJ	Y24AF
22	Z11AC	S12CS	G12CK	N13CH	E30CD	G36BM	B20BJ	E15BG	G16BE	B21AM	E29AK	T26AG
23	C31AD	P37AB	L37CL	A19CJ	M14CE	L32BP	N28BK	M25BH	L24BF	N17BC	M34AL	G38AH
24	K18AE	W27AC	D27CM	Z30CK	H23CF	D20CD	A33BL	H35BJ	D21BG	A22BD	H26AM	L11AJ
25	F12AF	Y13AD	J13CP	C14CL	S36CH	J28CE	Z15BM	S16BK	J17BH	Z29BE	S38BC	D31AK
26	V37AG	T19AE	U19CR	K23CH	P32CJ	U33CF	C25BP	P24BL	U22BJ	C34BF	P11BD	J18AL
27	E27AH	G30AF	R30CS	F36CP	W20CK	R15CH	K35CD	W29BM	R29BK	K26BG	W31BE	U12AM
28	M13AJ	L14AG	X14AB	V32CR	Y28CM	B35CK	V24CF	T22CD	Y17AH	X34BL	F38BH	R37BC
29	H19AK		B23AC	E20CS	T33CM	B35CK	V24CF	T22CD	B26BK	V11BJ	T12BG	X27BD
30	S30AL		N36AD	M28AB	G15CP	N16CL	E21CH	G29CE	N38BP	E31BK	G37BH	B13BE
31	P14AM		A32AE		L25CR		M17CJ	L34CF		M18BL		N19BF

FIRST LETTER = YOUR PHYSICAL CODE, MIDDLE NUMBER = SENSITIVITY CODE, LAST TWO LETTERS = INTELLECTUAL CODE

1830	JAN	FEB	MAR	APR	MAY	JUN	JUL	AUG	SEP	OCT	NOV	DEC
1	A30BG	H36BE	T36AK	B28AH	V15AE	T16AC	X21CR	V29CM	T38CK	X31CF	V37CD	Y13BL
2	Z14BH	S32BF	G32AL	N33AJ	E25AF	G24AD	B17CS	E34CP	G11CL	B18CH	E27CE	T19BM
3	C23BJ	P20BG	L20AM	A15AK	M35AG	L21AE	N22AB	M26CR	L31CM	N12CJ	M13CF	G30BP
4	K36BH	N28BH	D28BC	Z25AL	H16AH	D17AF	A29AC	H38CS	D18CP	A37CK	H19CH	L14CD
5	F32BL	Y33BJ	J33BD	C35AM	S24AJ	J22AG	Z34AD	S11AB	J12CR	Z27CL	S30CJ	D23CE
6	V20BM	T15BK	U15BE	K16BC	P21AK	U29AH	C26AE	P31AC	U37CS	C13CM	P14CK	J36CF
7	E28BP	G25BL	R25BF	F24BD	M17AL	R34AJ	K38AF	W18AD	R27AB	K19CP	W23CL	U32CH
8	M33CD	L35BM	X35BG	V21BE	Y22AM	X26AK	F11AG	Y12AE	X13AC	F30CR	Y36CH	R20CJ
9	H15CE	D16BP	B16BH	E17BF	T29BL	B38AL	V31AH	T37AF	B19AD	V14CS	T32CP	X28CH
10	S25CF	J24CD	N24BJ	M22BG	G34BD	N11AM	E18AJ	G27AG	N30AE	E23AB	G20CR	B33CL
11	P35CH	U21CE	A21BK	H29BH	L26BE	A31BC	M12AK	L13AH	A14AF	M36AC	L28CS	N15CH
12	W16CJ	R17CF	Z17BL	S34BJ	D38BF	Z18BD	H37AL	D19AJ	Z23AG	H32AD	D33AB	A25CP
13	Y24CK	X22CH	C22BM	P26BK	J11BG	C12BE	S27AM	J30AK	C36AH	S20AE	J15AC	Z35CR
14	T21CL	B29CJ	K29BP	W38BL	U31BH	K37BF	P13BC	U14AL	K32AJ	P28AF	U25AD	C16CS
15	G17CM	N34CK	F34CD	Y11BM	R18BJ	F27BG	W19BD	R23AM	F20AK	W33AG	R35AE	K24AB
16	L22CP	A26CL	V26CE	T31BP	X12BK	V13BH	Y30BE	X36BC	V28AL	Y15AH	X16AF	F21AC
17	D29CR	Z38CM	E38CF	G18CD	B37BL	E19BJ	T14BF	B32BD	E33AM	T25AJ	B24AG	V17AD
18	J34CS	C11CP	M11CH	L12CE	N27BM	M30BK	G23BG	N20BE	M15BC	G35AK	N21AH	E22AE
19	U26AB	K31CR	H31CJ	D37CF	A13BP	H14BL	L36BH	A28BF	H25BD	L16AL	A17AJ	M29AF
20	R38AC	F18CS	S18CK	J27CH	Z19CD	S23BH	D32BJ	Z33BG	S35BE	D24AH	Z22AK	H34AG
21	X11AD	V12AB	P12CL	U13CJ	C30CE	P36BP	J20BK	C15BH	P16BF	J21BC	C29AL	S26AH
22	B31AE	E37AC	W37CM	R19CK	K14CF	W32CD	U28BL	K25BJ	W24BG	U17BD	K34AM	P38AJ
23	N18AF	M27AD	Y27CP	X30CL	F23CH	Y20CE	R33BM	F35BK	Y21BH	R22BE	F26BC	W11AK
24	A12AG	H13AE	T13CR	B14CM	V36CJ	T28CF	X15BP	V16BL	T17BJ	X29BF	V38BD	Y31AL
25	Z37AH	S19AF	G19CS	N23CP	E32CK	G33CH	B25CD	E24BM	G22BK	B34BG	E11BE	T18AM
26	C27AJ	P30AG	L30AB	A36CR	M20CL	L15CJ	N35CE	M21BP	L29BL	N26BH	M31BF	G12BC
27	K13AK	W14AH	D14AC	Z32CS	H28CM	D25CK	A16CF	H17CD	D34BM	A38BJ	H18BG	L37BD
28	F19AL	Y23AJ	J23AD	C20AB	S33CP	J35CL	Z24CH	S22CE	J26BP	Z11BK	S12BH	D27BE
29	V30AM		U36AE	K28AC	P15CR	U16CH	C21CJ	P29CF	U38CD	C31BL	P37BJ	J13BF
30	E14BC		R32AF	F33AD	W25CS	R24CP	K17CK	W34CH	R11CE	K18BP	W27BK	U19BG
31	M23BD		X20AG		Y35AB		F22CL	Y26CJ		F12BP		R30BH

1831	JAN	FEB	MAR	APR	MAY	JUN	JUL	AUG	SEP	OCT	NOV	DEC
1	X14BJ	V32BG	P32AM	U33AK	C25AG	P24AE	J17AB	C34CR	P11CM	J18CJ	C27CF	S19BP
2	B23BK	E20BH	H20BC	R15AL	K35AH	W21AF	U22AC	K26CS	W31CP	U12CK	K13CH	P30CD
3	N36BL	M28BJ	Y28BD	X25AM	F16AJ	Y17AG	R29AD	F38AB	Y18CR	R37CL	F19CJ	W14CE
4	A32BM	H33BK	I33BE	B35BC	V24AK	I22AH	X34AE	V11AC	I12CS	X27CM	V30CK	Y23CF
5	Z20BP	S15BL	G15BF	N16BD	E21AL	G29AJ	B26AF	E31AD	G37AB	B13CP	E14CL	T36CH
6	C28CD	P25BM	L25BG	A24BE	M17AM	L34AK	N38AG	M18AE	L27AC	N19CR	M23CH	G32CJ
7	K33CE	W35BP	D35BH	Z21BF	H22BC	D26AL	A11AH	H12AF	D13AD	A30CS	H36CP	L20CK
8	F15CF	Y16CD	J16BJ	C17BG	S29BD	J38AM	Z31AJ	S37AG	J19AE	Z14AB	S32CR	D28CL
9	V25CH	T24CE	U24BK	K22BH	P34BE	U11BC	C18AK	P27AH	U30AF	C23AC	P20CS	J33CM
10	E35CJ	G21CF	R21BL	F29BJ	N26BF	R31BD	K12AL	M13AJ	R14AG	K36AD	M28AB	U15CP
11	M16CK	L17CH	X17BM	V34BK	Y38BG	X18BE	F37AM	Y19AK	X23AH	F32AE	Y33AC	R25CR
12	H24CL	D22CJ	B22BP	E26BL	T11BH	B12BF	V27BC	T30AL	B36AJ	V20AF	T15AD	X35CS
13	S21CM	J29CK	N29CD	M38BM	G31BK	N37BG	E13BD	G14AM	N32AK	E28AG	G25AE	B16AB
14	P17CP	U34CL	A34CE	H11BP	L18BK	A27BH	H30BF	M19BD	A20AL	M33AH	L35AF	N24AC
15	W22CR	R26CM	Z26CF	S31CD	D12BL	Z13BJ	H30BD	D36BD	Z28AM	H15AJ	D16AG	A21AD
16	Y29CS	X38CP	C38CH	P18CE	J37BM	C19BK	S14BG	J32BE	C33BC	S25AK	J24AH	Z17AE
17	T34AB	B11CR	K11CJ	W12CF	U27BP	K30BL	P23BH	U20BF	K15BD	P35AL	U21AJ	C22AF
18	G26AC	N31CS	F31CK	Y37CH	R13CD	F14BH	W36BJ	R28BG	F25BE	W16AM	R17AK	K29AG
19	L38AD	A18AB	V18CL	T27CJ	X19CE	V23BP	Y32BK	X33BH	V35BF	Y24BC	X22AL	F34AH
20	D11AE	Z12AC	E12CM	G13CK	B30CF	E36CD	T20BL	B15BJ	E16BG	T21BD	B29AM	V26AJ
21	J31AF	C37AD	M37CP	L19CL	N14CH	M32CE	G28BM	N25BK	M24BH	G17BE	N34BC	E38AK
22	U18AG	K27AE	H27CR	D30CM	A23CJ	H20CF	L33BP	A35BL	H21BJ	L22BF	A26BD	M11AL
23	R12AH	F13AF	S13CS	J14CP	Z36CK	S28CH	D15CD	Z16BM	S17BK	D29BG	Z38BE	H31AM
24	X37AJ	V19AG	P19AB	U23CR	C32CL	P33CJ	J25CE	C24BP	P22BL	J34BH	C11BF	S18BC
25	B27AK	E30AH	W30AC	R36CS	K20CM	W15CK	U35CF	K21CD	W29BM	U26BJ	K31BG	P12BD
26	N13AL	M14AJ	Y14AD	X32AB	F28CP	Y25CL	R16CH	F17CF	Y34BP	R38BK	F18BH	W37BE
27	A19AM	H23AK	T23AE	B20AC	V33CR	T35CM	X24CJ	V22CF	T26CD	X11BL	V12BJ	Y27BF
28	Z30BC	S36AL	G36AF	N28AD	E15CS	G16CP	B21CK	E29CH	G38CE	B31BM	E37BL	T13BG
29	C14BD		L32AG	A33AE	M25AB	L24CR	N17CL	M34CJ	L11CF	N18BP	M27BL	G19BH
30	K23BE		D20AH	Z15AF	H35AC	D21CS	A22CM	H26CK	D31CH	A12CD	H13BM	L30BJ
31	F36BF		J28AJ		S16AD		Z29CP	S38CL		Z37CE		D14BK

1832	JAN	FEB	MAR	APR	MAY	JUN	JUL	AUG	SEP	OCT	NOV	DEC
1	J23BL	C20BJ	H28BE	D25BC	A16AK	H17AH	L29AE	A38AC	H18CS	L37CM	A19CK	M14CF
2	U36BM	K28BK	S33BF	J35BD	Z24AL	S22AJ	D34AF	Z11AD	S12AB	D27CP	Z30CL	H23CH
3	R32BP	F33BL	P15BG	U16BE	C21AM	P29AK	J26AG	C31AE	P37AC	J13CR	C14CM	S36CJ
4	X20CD	V15BM	W25BH	R24BF	K17BC	W34AL	U38AH	K18AF	W27AD	U19CS	K23CP	P32CK
5	B28CE	E25BP	Y35BJ	X21BG	F22BD	Y26AM	R11AJ	F12AG	Y13AE	R30AB	F36CR	W20CL
6	N33CF	M35CD	T16BK	B17BH	V29BE	T38BC	X31AK	V37AH	T19AF	X14AC	V32CS	Y28CM
7	A15CH	H16CE	G24BL	N22BJ	E34BF	G11BD	B18AL	E27AF	G30AD	B23AB	E20CR	T33CP
8	Z25CJ	S24CF	L21BM	A29BK	M26BG	L31BE	N12AM	M13AK	L14AH	N36AE	M28AC	G15CR
9	C35CK	P21CH	D17BP	Z34BL	H38BJ	D18BF	A37BC	H19AL	D23AJ	A32AF	H33AD	L25CS
10	K16CL	W17CJ	J22CD	C26BM	S11BJ	J12BG	Z27BD	S30AM	J36AK	Z20AG	S15AE	D35AB
11	F24CM	Y22CH	U29CE	K38BP	P31BH	U37BF	C13BE	P14BC	U32AL	C28AH	P25AF	J16AC
12	V21CP	T29CL	R34CF	W11CD	W18BL	R27BJ	K19BF	W23BD	R20AM	K33AJ	W35AG	U24AD
13	E17CR	G34CH	X26CH	V31CE	Y12BM	X13BK	F30BG	Y36BE	X28BC	F15AK	Y16AH	R21AE
14	M22CS	L26CP	G26CP	E18CF	T37BP	G27BD	N30BM	T32BD	G20BK	N15BE	T24AJ	X17AF
15	H29AB	D38CR	N11CK	M12CH	G27CD	N30BM	E23BJ	G20BG	N15BD	E35AM	G21AK	B22AG
16	S34AC	J11CS	A31CL	M37CJ	L13CE	A14BP	M38BK	L28BH	A25BF	M16BC	L17AL	N29AH
17	P26AD	U31AB	Z18CM	S27CK	D19CF	Z23CD	H32BL	D33BJ	Z35BG	H24BD	D22AM	A34AJ
18	W38AE	R18AC	C12CP	P13CL	J30CH	C36CE	S20BM	J15BK	C16BH	S21BE	J29BC	C38AL
19	Y11AF	X12AD	K37CR	W19CM	U14CJ	K32CF	P28BP	U25BL	K24BJ	P17BF	U34BD	U38BD
20	T31CG	G18AG	F27CS	Y38CP	R23CK	F20CH	W33CD	R35BM	F21BK	W22BG	R26BE	K11AM
21	G18AH	N27AF	V13AB	T14CR	X36CL	V28CJ	Y15CE	X16BP	V17BL	Y29BH	X38BF	F31BG
22	L12AJ	A13AG	E19AC	G23CS	B32CK	E33CK	T25CF	B24CD	E22BM	T34BJ	B11BG	V18BD
23	D37AK	Z19AH	H30AB	M30AD	N20CP	M15CL	G35CM	N21CE	M29BP	G26BK	N31BH	E12BE
24	J27AL	C30AJ	H14AE	D32AC	A28CR	H25CH	L16CJ	A17CF	H34CD	L38BL	A18BJ	M37BF
25	U13AM	K14AK	S23AF	J20AD	Z33CS	S35CP	D24CK	Z22CH	S26CE	D11BM	Z12BK	H27BG
26	R19BC	F23AL	P36AG	U28AE	C15AB	P16CR	J21CL	C29CJ	P38CF	J31BP	C37BL	S13BH
27	X30BD	V36AM	W32AH	R33AF	K25AC	W24CS	U17CM	K34CK	W11CH	R12CE	K27BM	P19BJ
28	B14BE	E32BC	Y20AJ	X15AG	F35AD	Y21AB	R22CP	F26CL	Y31CJ	R12CF	N30BP	W30BK
29	N23BF	M20BD	T28AK	B25AH	V16AE	T17AC	X29CR	V30CM	T18CK	X37CF	V19CD	Y14BL
30	A36BG		G33AL	N35AJ	E24AF	G22AD	B34CS	E11CP	G12CL	B27CH	E30CE	T23BM
31	Z32BH		L15AM		M21AG		N26AB	M31CR		N13CJ		G36BP

FIRST LETTER = YOUR PHYSICAL CODE, MIDDLE NUMBER = SENSITIVITY CODE, LAST TWO LETTERS = INTELLECTUAL CODE

1833	JAN	FEB	MAR	APR	MAY	JUN	JUL	AUG	SEP	OCT	NOV	DEC
1	L32CD	A33BM	V33BG	T35BE	X24AM	V22AK	Y34AG	X11AE	V12AC	Y27CR	X30CM	F23CJ
2	O20CE	Z15BP	E15BH	G16BF	B21BC	E29AL	T26AH	B31AF	E37AD	T13CS	B14CP	V36CK
3	J28CF	C25CD	M25BJ	L24BG	N17BD	M34AM	G38AJ	N18AG	M27AE	G19AB	N23CR	E32CL
4	U33CH	K35CE	H35BK	D21BH	A22BE	H26BC	L11AK	A12AH	H13AF	L30AC	A36CS	M20CM
5	R15CJ	F16CF	S16BL	J17BJ	Z29BF	S38BD	D31AL	Z37AJ	S19AG	D14AD	Z32AB	H28CP
6	X25CK	V24CH	P24BM	U22BK	C34BG	P11BE	J18AM	C27AK	P30AH	J23AE	C20AC	S33CR
7	B35CL	E21CJ	W21BP	R29BL	K26BH	W31BF	U12BC	K13AL	M14AJ	U36AF	K28AD	P15CS
8	N16CM	M17CK	Y17CD	X34BM	F38BJ	Y18BG	R37BD	F19AM	Y23AK	R32AG	F33AE	W25AB
9	A24CP	H22CL	T22CE	B26BP	V11BK	T12BH	X27BE	V30BC	T36AL	X20AH	V15AF	Y35AC
10	Z21CR	S29CM	G29CF	N38CD	E31BL	G37BJ	B13BF	E14BD	G32AM	B28AJ	E25AG	T16AD
11	C17CS	P34CP	L34CH	A11CE	M18BM	L27BK	N19BG	M23BE	L20BC	N33AK	M35AH	G24AE
12	K22AB	W26CR	D26CJ	Z31CF	H12BP	D13BL	A30BH	H36BF	D28BD	A15AL	H16AJ	L21AF
13	F29AC	Y38CS	J38CK	C18CH	S37CD	J19BM	Z14BJ	S32BG	J33BE	Z25AM	S24AK	D17AG
14	V34AD	T11AB	U11CL	K12CJ	P27CE	U30BP	C23BK	P20BH	U15BF	C35BC	P21AL	J22AH
15	E26AE	G31AC	R31CM	F37CK	W13CF	R14CD	K36BL	W28BJ	R25BG	K16BD	W17AM	U29AJ
16	M38AF	L18AD	X18CP	V27CL	Y19CM	X23CE	F32BH	Y33BK	X35BH	F24BE	Y22BC	R34AK
17	H11AG	D12AE	B12CR	E13CM	T30CJ	B36CF	V20BP	T15BL	B16BJ	V21BF	T29BD	X26AL
18	S31AH	J37AF	N37CS	M19CP	G14CK	N32CH	E28CD	G25BH	N24BK	E17BG	G34BE	B38AM
19	P18AJ	U27AG	A27AB	H30CR	L23CL	A20CJ	M33CE	L35BP	A21BL	M22BH	L26BF	N11BC
20	W12AK	R13AH	Z13AC	S14CS	D36CM	Z28CK	H15CF	D16CD	Z17BM	H29BJ	D38BG	A31BD
21	Y37AL	X19AJ	C19AD	P23AB	J32CP	C33CL	S25CH	J24CE	C22BP	S34BK	J11BH	Z18BE
22	T27AM	B30AK	K30AE	W36AC	U20CR	K15CM	P35CJ	U21CF	K29CD	P26BL	U31BJ	C12BF
23	G13BC	N14AL	F14AF	Y32AD	R28CS	F25CP	W16CK	R17CH	F34CE	W38BH	R18BK	K37BG
24	L19BD	A23AM	V23AG	T20AE	X33AB	V35CR	Y24CL	X22CJ	V26CF	Y11BP	X12BL	F27BH
25	D30BE	Z36BC	E36AH	G28AF	B15AC	E16CS	T21CM	B29CK	E38CH	T31CD	B37BM	V13BJ
26	J14BF	C32BD	M32AJ	L33AG	N25AD	M24AB	G17CP	N34CL	M11CJ	G18CE	N27BP	E19BK
27	U23BG	K20BE	H20AK	D15AH	A35AE	H21AC	L22CR	A26CM	H31CK	L12CF	A13CD	M30BL
28	R36BH	F28BF	S28AL	J25AJ	Z16AF	S17AD	D29CS	Z38CP	S18CL	D37CH	Z19CE	H14BM
29	X32BJ		P33AM	U35AK	C24AG	P22AE	J34AB	C11CR	P12CM	J27CJ	C30CF	S23BP
30	B20BK		W15BC	R16AL	K21AH	W29AF	U26AC	K31CS	W37CP	U13CK	K14CH	P36CD
31	N28BL		Y25BD		F17AJ		R38AD	F18AB		R19CL		W32CE

1834	JAN	FEB	MAR	APR	MAY	JUN	JUL	AUG	SEP	OCT	NOV	DEC
1	Y20CF	X15CD	C15BJ	P16BG	J21BD	C29AM	S26AJ	J31AG	C37AE	S13AB	J14CR	Z36CL
2	T28CH	B25CE	K25BK	W24BH	U17BE	K34BC	P38AK	U18AH	K27AF	P19AC	U23CS	C32CH
3	G33CJ	N35CF	F35BL	Y21BJ	R22BF	F26BD	W11AL	R12AJ	F13AG	W30AD	R36AB	K20CP
4	L15CK	A16CH	V16BH	T17BK	X29BG	V38BE	Y31AM	X37AK	V19AH	Y14AE	X32AC	F28CR
5	D25CL	Z24CJ	E24BP	G22BL	B34BH	E11BF	T18BC	B27AL	E30AJ	T23AF	B20AD	V33CS
6	J35CM	C21CK	M21CD	L29BM	N26BJ	M31BG	G12BD	N13AM	M14AK	G36AG	N28AE	E15AB
7	U16CP	K17CL	H17CE	D34BP	A38BK	H18BH	L37BE	A19BC	H23AL	L32AH	A33AF	M25AC
8	R24CR	F22CM	S22CF	J26CD	Z11BL	S12BJ	D27BF	Z30BD	S36AM	D20AJ	Z15AG	H35AD
9	X21CS	V29CP	P29CH	U38CE	C31BM	P37BK	J13BG	C14BE	P32BC	J28AK	C25AH	S16AE
10	B17AB	E34CR	W34CJ	R11CF	K18BP	W27BL	U19BH	K23BF	W20BD	U33AL	K35AJ	P24AF
11	N22AC	M26CS	Y26CK	X31CH	F12CD	Y13BM	R30BJ	F36BG	Y28BE	R15AM	F16AK	W21AG
12	A29AD	H38AB	T38CL	B18CJ	V37CE	T19BP	X14BK	V32BH	T33BF	X25BC	V24AL	Y17AH
13	Z34AE	S11AC	G11CM	N12CK	E27CF	G30CD	B23BL	E20BJ	G15BG	B35BD	E21AM	T22AJ
14	C26AF	P31AD	L31CP	A37CL	M13CH	L14CE	N36BM	M28BK	L25BH	N16BE	M17BC	G29AK
15	K38AG	W18AE	D18CR	Z27CM	H19CJ	D23CF	A32BP	H33BL	D35BJ	A24BF	H22BD	L34AL
16	F11AH	Y12AF	J12CS	C13CP	S30CK	J36CH	Z20CC	S15BM	J16BK	Z21BG	S29BE	D26AM
17	V31AJ	T37AG	U37AB	K19CR	P14CL	U32CJ	C28CE	P25BP	U24BL	C17BH	P34BF	J38BC
18	E18AK	G27AH	R27AC	F30CS	W23CM	R20CK	K33CF	W35CD	R21BM	K22BJ	W26BG	U11BD
19	M12AL	L13AJ	X13AD	V14AB	Y36CP	X28CL	F15CH	Y16CE	X17BP	F29BK	Y38BH	R31BE
20	H37AM	D19AM	B19AE	E23AC	T32CP	B33CM	V25CJ	T24CF	B22CD	V34BL	T11BJ	X18BF
21	S27BC	J30AL	N30AF	M36AD	G20CS	N15CP	E35CK	G21CH	N29CE	E26BM	G31BK	B12BG
22	P13BD	U14AM	A14AG	H32AE	L28AB	A25CR	M16CL	L17CJ	A34CF	M38BP	L18BL	N37BH
23	W19BE	R23BC	Z23AH	S20AF	D33AC	Z35CS	H24CM	D22CK	Z26CH	H11CD	D12BM	A27BJ
24	Y30BF	X36BD	C36AJ	P28AG	J15AD	C16AB	S21CP	J29CL	C38CJ	S31CE	J37BP	Z13BK
25	T14BG	B32BE	K32AK	W33AH	U25AE	K24AC	P17CR	U34CH	K11CL	P18CF	U27CD	C19BL
26	G23BH	N20BF	F20AL	Y15AJ	R35AF	F21AD	W22CS	R26CP	F31CL	W12CH	R13CE	K30BM
27	L36BJ	A28BG	V28AM	T25AK	X16AG	V17AE	Y29AB	X38CR	V18CM	Y37CJ	X19CF	F14BP
28	D32BK	Z33BH	W28AM	E33BC	G35AL	B24AH	E22AF	T34AC	B11CS	E12CP	T27CK	V23CD
29	J20BL		M15BD	L16AM	N21AJ	M29AG	G26AD	W31AB	M37CR	G13CL	N14CJ	E36CE
30	U28BM		H25BE	D24BC	A17AH	H34AH	L38AE	A18AC	H27CS	L19CM	A23CK	M32CF
31	R33BP		S35BF		Z22AL		D11AF	Z12AD		D30CP		H20CH

1835	JAN	FEB	MAR	APR	MAY	JUN	JUL	AUG	SEP	OCT	NOV	DEC
1	S28CJ	J25CF	N25BL	M24BJ	G17BF	N34BD	E38AL	G18AJ	M27AG	E19AD	G23AB	B32CP
2	P33CK	U35CH	A35BM	H21BK	L22BG	A26BE	M11AM	L12AK	A13AH	M30AE	L36AC	N20CR
3	W15CL	R16CJ	Z16BP	S17BL	D29BH	Z38BF	H31BC	D37AL	Z19AJ	H14AF	D32AD	A28CS
4	Y25CM	X24CK	C24CD	P22BM	J34BJ	C11BG	S18BD	J27AM	C30AK	S23AG	J20AE	Z33AB
5	T35CP	B21CL	K21CE	W29BP	U26BK	K31BH	P12BE	U13BC	K14AL	P36AH	U28AF	C15AC
6	G16CR	N17CM	F17CF	Y34CD	R38BL	F18BJ	W37BF	R19BD	F23AM	W32AJ	R33AG	K25AD
7	L24CS	A22CP	V22CH	T26CE	X11BM	V12BK	Y27BG	X30BE	V36BC	Y20AK	X15AH	F35AE
8	D21AB	Z29CR	E29CJ	G38CF	B31BP	E37BJ	T13BH	B14BF	E32BD	T28AL	B25AJ	V16AF
9	J17AC	C34CS	M34CK	L11CH	N18CD	M27BM	G19BJ	N23BG	M20BE	G33AM	N35AK	E24AG
10	U22AD	K26AB	H26CL	D31CJ	A12CE	H13BP	L30BK	A36BH	H28BF	L15BC	A16AL	M21AH
11	R29AE	F38AC	S38CM	J16CK	Z37CF	S19CD	D14BL	Z32BJ	S33BG	D25BD	Z24AH	M17AJ
12	X34AF	V11AD	P11CP	U12CL	C27CH	P30CE	J23BM	C20BK	P15BH	J35BE	C21BC	S22AK
13	B26AG	E31AE	W31CR	R37CM	K13CJ	W14CF	U36BP	K28BL	W25BJ	U16BF	K17BD	P29AL
14	N38AH	M18AF	Y18CS	X27CP	F19CK	Y23CH	R32CD	F33BM	Y35BK	R24BG	F22BE	W34AM
15	A11AJ	H12AG	T12AB	B13CR	V30CL	T36CJ	X20CC	V15BP	T16BL	X21BG	V29BF	Y26BC
16	Z31AK	S37AH	G37AC	N19CS	E14CM	G32CK	B28CF	E25CD	G24BM	B17BJ	E34BG	T38BD
17	C18AL	P27AH	L27AC	A30AB	M23CP	L20CL	N33CH	M35CE	L21BP	N22BK	M26BH	G11BE
18	K12AM	W13AK	D13AE	Z14AC	H36CM	D28CH	A15CJ	H16CF	D17CD	A29BL	H38BJ	L31BF
19	F37BC	Y19AL	J19AF	C23AD	S32CS	J33CP	Z25CK	S24CH	J22CE	Z34BM	S11BK	D18BG
20	V27BD	T30AM	U30AG	K36AE	P20AB	U15CR	C35CL	P21CJ	U29CF	C26BP	P31BL	J12BH
21	E13BE	G15BC	R14AH	F32AF	W28AC	R25CS	K16CM	W17CK	R34CH	K38CD	W18BM	U37BJ
22	M19BF	L23BD	X23AJ	V20AG	Y33AD	X35AB	V24CR	Y22CM	X26CJ	V11CF	Y12BP	R27BK
23	H30BG	D36BE	B36AK	E28AH	T15AE	B16AC	V21CR	T29CM	B38CJ	V31CF	T37CD	X13BL
24	S14BH	J32BF	N32AM	M33AJ	G25AF	N24AD	M22AB	G34CL	N11CJ	M12CJ	G27CE	B19BM
25	P23BJ	U20BG	A20AM	H15AK	L35AG	A21AE	H22AB	L26CR	A31CM	H12CJ	L13CF	N30BP
26	W36BK	R28BH	Z28BC	S25AL	D16AH	Z17AF	H29AC	D38CS	Z18CP	H37CK	D19CH	A14CD
27	Y32BL	X33BJ	C33BD	P35AM	Z22AG	C22AB	S34AD	J11AB	C12CR	S27CL	J30CJ	Z23CE
28	T20BM	B15BK	K15BE	W16BC	U21AK	K29AH	P26AE	U31AC	K37CS	P13CM	U14CK	C36CF
29	G28BP		F25BF	Y24BD	R17AL	F34AJ	W38AF	R18AD	F27AB	W19CP	R23CL	K32CH
30	L33CD		V35BG	T21BE	X22AM	V26AK	Y11AG	X12AE	V13AC	Y30CR	X36CM	F20CJ
31	D15CE		E16BH		B29BC		T31AH	B37AF		T14CS		V28CK

FIRST LETTER = YOUR PHYSICAL CODE, MIDDLE NUMBER = SENSITIVITY CODE, LAST TWO LETTERS = INTELLECTUAL CODE

1836	JAN	FEB	MAR	APR	MAY	JUN	JUL	AUG	SEP	OCT	NOV	DEC
1	E33CL	G35CJ	X16CD	V17BM	Y29BJ	X38BG	F31BD	Y37AM	X19AK	F14AG	Y32AE	R28AB
2	M15CM	L16CK	B24CE	E22BP	T34BK	B11BH	V18BE	T27BC	B30AL	V23AH	T20AF	X33AC
3	H25CP	D24CL	N21CF	M29CD	G26BL	N31BJ	E12BF	G13BD	N14AM	E36AJ	G28AG	B15AD
4	S35CR	J21CM	A17CH	H34CE	L38BM	A18BK	M37BG	L19BE	A23BC	M32AK	L33AH	N25AE
5	P16CS	U17CP	Z22CJ	S26CF	D11BP	Z12BL	H27BH	D30BF	Z36BD	H20AL	D15AJ	A35AF
6	W24AB	R22CR	C29CK	P38CH	J31CD	C37BM	S13BJ	J14BG	C32BE	S28AM	J25AK	Z16AG
7	Y21AC	X29CS	K34CL	W11CJ	U18CE	K27BP	P19BK	U23BH	K20BF	P33BC	U35AL	C24AH
8	T17AD	B34AB	F26CM	Y31CK	R12CF	F13CD	W30BL	R36BJ	F28BG	W15BD	R16AM	K21AJ
9	G22AE	N26AC	V38CP	T18CL	X37CH	V19CE	Y14BM	X32BK	V33BH	Y25BE	X24BC	F17AK
10	L29AF	A38AD	E11CR	G12CM	B27CJ	E30CF	T23BP	B20BL	E15BJ	T35BF	B21BD	V22AL
11	D34AG	Z11AE	M31CS	L37CP	N13CK	M14CH	G36CD	N28BM	M25BK	G16BG	N17BE	E29AM
12	J26AH	C31AF	H18AB	D27CR	A19CL	H23CJ	L32CE	A33BP	H35BL	L24BH	A22BF	M34BC
13	U38AJ	K18AG	S12AC	J13CS	Z30CM	S36CK	D20CF	Z15BD	S16BM	D21BJ	Z29BG	H26BD
14	R11AK	F12AH	P37AD	U19AB	C14CP	P32CL	J28CH	C25CE	P24BP	J17BK	C34BH	S38BE
15	X31AL	V37AJ	W27AE	R30AC	K23CR	W20CH	U33CJ	K35CF	W21CD	U22BL	K26BJ	P11BF
16	B18AM	E27AK	Y13AF	X14AD	F36CS	Y28CP	R15CK	F16CH	Y17CE	R29BM	F38BK	W31BG
17	N12BC	M13AL	T19AG	B23AE	V32AB	T33CR	X25CL	V24CJ	T22CF	X34BP	V11BL	Y18BH
18	A37BD	H19AM	G30AH	N36AF	E20AC	G15CS	B35CM	E21CK	G29CH	B26CD	E31BM	T12BJ
19	Z27BE	S30BC	L14AJ	A32AG	M28AD	L25AB	N16CP	M17CL	L34CJ	N38CE	M18BP	G37BK
20	C13BF	P14BD	D23AK	Z20AH	H33AE	D35AC	A24CR	H22CM	D26CK	A11CF	H12CD	L27BL
21	K19BG	W23BE	J36AL	C28AJ	S15AF	J16AD	Z21CS	S29CP	J38CL	Z31CH	S37CE	D13BM
22	F30BH	Y36BF	U32AM	K33AK	P25AG	U24AE	C17AB	P34CR	U11CH	C18CJ	P27CF	J19BP
23	V14BJ	T32BG	R20BC	F15AL	W35AH	R21AF	K22AC	W26CS	R31CP	K12CK	W13CH	U30CD
24	E23BK	G20BM	X28BD	V25AM	Y16AJ	X17AG	F29AD	Y38AB	X18CR	F37CL	Y19CJ	R14CE
25	M36BL	L28BJ	B33BE	E35BC	T24AK	B22AH	V34AE	T11AC	B12CS	V27CM	T30CK	X23CF
26	H32BM	D33BK	N15BF	M16BD	G21AL	N29AJ	E26AF	G31AD	N37AB	E13CP	G14CL	B36CH
27	S20BP	J15BL	A25BG	H24BE	L17AM	A34AK	M38AG	L18AE	A27AC	M19CR	L23CH	N32CJ
28	P28CD	U25BM	Z35BH	S21BF	D22BC	Z26AL	H11AH	D12AF	Z13AD	H30CS	D36CP	A20CK
29	W33CE	R35BP	C16BJ	P17BG	J29BD	C38AM	S31AJ	J37AG	C19AE	S14AB	J32CR	Z28CL
30	Y15CF		K24BK	W22BH	U34BE	K11BC	P18AK	U27AH	K30AF	P23AC	U20CS	C33CM
31	T25CH		F21BL		R26BF		W12AL	R13AJ		W36AD		K15CP

1837	JAN	FEB	MAR	APR	MAY	JUN	JUL	AUG	SEP	OCT	NOV	DEC
1	F25CR	Y24CM	J24CF	C22CD	S34BL	J11BJ	Z18BF	S27BD	J30AM	Z23AJ	S20AG	D33AD
2	V35CS	T21CP	U21CM	X29CE	P26BM	U31BH	C12BG	P13BE	U14BC	C36AK	P28AH	J15AE
3	E16AB	G17CR	R17CJ	F34CF	W38BP	R18BL	K37BH	W19BF	R23BD	K32AL	W33AJ	U25AF
4	M24AC	L22CS	X22CK	V26CH	Y11CD	X12BM	F27BJ	Y30BG	X36BE	F20AM	Y15AK	R35AG
5	H21AD	D29AB	B29CL	E38CJ	T31CE	B37BP	V13BK	T14BH	B32BF	V28BC	T25AL	X16AH
6	S17AE	J34AC	N34CM	M11CK	G18CF	N27CD	E19BL	G23BJ	N20BG	E33BD	G35AM	B24AJ
7	P22AF	U26AD	A26CP	H31CL	L12CH	A13CE	M30BH	L36BK	A28BH	M15BE	L16BC	N21AK
8	W29AG	R38AE	Z38CR	S16CM	D37CJ	Z19CF	H14BP	D32BL	Z33BJ	H25BF	D24BD	A17AL
9	Y34AH	X11AF	C11CS	P12CP	J27CK	C30CH	S23CD	J20BM	C15BK	S35BG	J21BE	Z22AM
10	T26AJ	B31AG	K31AB	W37CR	U13CL	K14CJ	P36CE	U28BP	K25BL	P16BH	U17BF	C29BC
11	G38AK	N18AH	F18AC	Y27CS	R19CM	F23CK	W32CF	R33CD	F35BM	W24BJ	R22BG	K34BD
12	L11AL	A12AJ	V12AD	T13AB	X30CP	V36CL	Y20CH	X15CE	V16BP	Y21BK	X29BH	F26BE
13	D31AM	Z37AK	E37AE	G19AC	B14CR	E32CH	T28CJ	B25CF	E24CD	T17BL	B34BJ	V38BF
14	J18BC	C27AL	M27AF	L30AD	N23CS	M20CP	G33CK	N35CH	M21CE	G22BM	N26BK	E11BG
15	U12BD	K13AM	H13AG	D14AE	A36AB	H28CR	L15CL	A16CJ	H17CF	L29BP	A38BL	M31BH
16	R37BE	F19BC	S19AH	J23AF	Z32AC	S33CS	D25CM	Z24CK	S22CH	D34CD	Z11BM	H18BJ
17	X27BF	V30BD	P30AJ	U36AG	C20AD	P15AB	J35CP	C21CL	P29CJ	J26CE	C31BP	S12BK
18	B13BG	E14BE	W14AK	R32AH	K28AE	W25AC	U16CR	K17CH	W34CK	U38CF	K18CD	P37BL
19	N19BH	M23BF	Y23AL	X20AJ	F33AF	Y35AD	R24CS	F22CP	Y26CL	R11CH	F12CE	W27BM
20	A30BJ	H36BG	T36AM	B28AK	V15AG	T16AE	X21AB	V29CR	T38CE	X31CJ	V37CF	Y13BP
21	Z14BK	S32BH	G32BC	N33AL	E25AH	G24AF	B17AC	E34CS	G11CP	B18CK	E27CH	T19CD
22	C23BL	P20BJ	L20BD	A15AM	M35AJ	L21AG	N22AD	M26AB	L31CR	N12CL	M13CJ	G30CE
23	K36BM	W28BK	D28BE	Z25BD	H16AK	D17AH	A29AE	H38AC	D18CS	A37CH	H19CK	L14CF
24	F32BP	Y33BL	J33BF	C35BD	S24AL	J22AJ	Z34AF	S11AD	J12AB	Z27CP	S30CL	D23CH
25	V20CD	T15BM	U15BG	K16BE	P21AM	U29AG	C26AD	P31AE	U37AC	C13CR	P14CM	J36CJ
26	E28CE	G25BP	R25BH	F24BF	W17BC	R34AL	K38AH	W18AF	R27AD	K19CS	W23CP	U32CK
27	M33CF	L35CD	X35BJ	V21BG	Y22BD	X26AM	F11AJ	Y12AG	X13AB	F30AB	Y36CR	R20CL
28	H15CH	D16CE	B16BK	E17BH	T29BE	B38BC	V31AK	T37AH	B19AF	V14AC	T32CS	X28CM
29	S25CJ		N24BL	M22BJ	G34BF	N11BD	E18AL	G27AJ	N30AG	E23AD	G20AB	B33CP
30	P35CK		A21BM	H29BK	L26BG	A31BE	M12AM	L13AK	A14AH	M36AE	L28AC	N15CP
31	W16CL		Z17BP		D38BH		H37BC	D19AL		H32AF		A25CS

1838	JAN	FEB	MAR	APR	MAY	JUN	JUL	AUG	SEP	OCT	NOV	DEC
1	Z35AB	S21CR	G21CJ	N29CF	E26BP	G31BL	B12BH	E13BF	G14BD	B36AL	E28AJ	T15AF
2	C16AC	P17CS	L17CK	A34CH	M38CD	L18BM	N37BJ	M19BG	L23BE	N32AM	M33AK	G25AG
3	K24AD	W22AB	D22CL	Z26CJ	H11CE	D12BP	A27BK	H30BH	D36BF	A20BC	H15AL	L35AH
4	F21AE	Y29AC	J29CM	C36CK	S31CF	J37CD	Z13BL	S14BJ	J32BG	Z28BD	S25AM	D16AJ
5	V17AF	T34AD	U34CP	K11CL	P18CH	U27CE	C19BM	P23BK	U20BH	C33BE	P35BC	J24AK
6	E22AG	G26AE	R26CR	F31CH	W12CJ	R13CD	K30BP	W36BL	R28BJ	K15BF	W16BD	U21AL
7	M29AH	L38AF	X38CS	V18CP	Y37CK	X19CH	F14CD	Y32BM	X33BK	F25BG	Y24BE	R17AM
8	H34AJ	D11AG	B11AB	E12CR	T27CL	B30CJ	V23CE	T20BP	B15BL	V35BH	T21BF	X22BC
9	S26AK	J31AH	N31AC	M37CS	G24CM	N14CK	E36CF	G28CD	N25BM	E16BJ	G17BG	B29BD
10	P38AL	U18AJ	A18AD	H27AB	L19CP	A23CL	M32CH	L33CE	A35BP	M24BK	L22BH	N34BE
11	W11AM	R12AK	Z12AE	S13AC	D30CR	Z36CH	H20CJ	D15CF	Z16CD	H21BL	D29BJ	A26BF
12	Y31BC	X37AL	C37AF	P19AD	J14CS	C32CP	S28CK	J25CH	C24CE	S17BM	J34BK	Z38BG
13	T18BD	B27AM	K27AG	W30AE	U23AB	K20CR	P33CL	U35CJ	K21CF	P22BP	U26BL	C11BH
14	G12BE	N13BC	F13AH	Y14AF	R36AC	F28CS	W15CM	R16CK	F17CH	W29CD	R38BM	K31BJ
15	L37BF	A19BD	V19AJ	T23AG	X32AD	V33AB	Y25CP	X24CL	V22CD	Y34CE	X11BP	F18BK
16	D27BG	Z30BE	E30AK	G36AH	B20AE	E15AC	T35CR	B21CM	E29CK	T26CF	B31CD	V12BL
17	J13BH	C14BF	M14AL	L32AJ	N28AF	M25AD	G16CS	N17CP	M34CL	G38CH	N18CE	E37BM
18	U19BJ	K23BG	H23AM	D20AK	A33AG	H35AE	L24AB	A22CR	H26CH	L11CJ	A12CF	M27BP
19	R30BK	F36BH	S36BC	J28AL	Z15AH	S16AF	D21AC	Z29CS	S38CP	D31CK	Z37CH	H13CD
20	X14BL	V32BJ	P32BD	U33AM	C25AJ	P24AG	J17AD	C34AB	P11CR	J18CL	C27CJ	S19CE
21	B23BM	E20BK	W20BE	R15BC	K35AK	W21AH	U22AE	K26AC	W31CS	U12CL	K13CK	P30CF
22	N36BP	M28BM	Y28BF	X25BD	F16AL	Y17AJ	R29AF	F38AD	Y18AB	R37CR	F19CL	W14CH
23	A32CD	H33BM	T33BG	B35BE	V24AM	T22AK	X34AG	V11AE	T12AC	X27CR	V30CH	Y23CJ
24	Z20CE	S15BP	G15BH	N16BF	E21BC	G29AL	B26AH	E31AF	G37AD	B13CS	E14CP	T36CK
25	C28CF	P25CD	L25BJ	A24BG	M17BD	L34AM	N38AJ	M18AG	L27AE	N19AB	M23CR	G32CL
26	K33CH	W35CE	D35BK	Z21BH	H22BE	D26BC	A11AK	H12AH	D13AF	A30AC	H36CS	L20CH
27	F15CJ	Y16CF	J16BL	C17BJ	S29BF	J38BD	Z31AL	S37AJ	J19AG	Z14AD	S32AB	D28CP
28	V25CK	T24CM	U24BM	K22BK	P34BG	U11BE	C18AM	P27AK	U30AH	C23AE	P20AC	J33CR
29	E35CL		R21BP	F29BL	W26BH	R31BF	K12AJ	W13AK	R14AJ	K36AF	W28AD	U15CS
30	M16CM		X17CD	V34BM	Y38BJ	X18BG	F37BD	Y19AM	X23AK	F32AG	Y33AE	R25AB
31	H24CP		B22CE		T11BK		V27BE	T30BC		V20AH		X35AC

FIRST LETTER = YOUR PHYSICAL CODE, MIDDLE NUMBER = SENSITIVITY CODE, LAST TWO LETTERS = INTELLECTUAL CODE

1839

	JAN	FEB	MAR	APR	MAY	JUN	JUL	AUG	SEP	OCT	NOV	DEC
1	B16AD	E17AB	W17CL	R34CJ	K38CE	W18BP	U37BK	K19BH	W23BF	U32BC	K33AL	P25AH
2	N24AE	M22AC	Y22CM	X26CK	F11CF	Y12CD	R27BL	F30BJ	Y36BG	R20BD	F15AM	W35AJ
3	A21AF	H29AD	T29CP	B38CL	V31CH	T37CE	X13BM	V14BK	T32BH	X28BE	V25BC	Y16AK
4	Z17AG	S34AE	G34CR	N11CM	E18CJ	G27CF	B19BP	E23BL	G20BJ	B33BF	E35BD	T24AL
5	C22AH	P26AF	L26CS	A31CP	M12CK	L13CH	N30CD	M36BM	L28BK	N15BG	M16BE	G21AM
6	K29AJ	W38AG	D38AB	Z18CR	H37CL	D19CJ	A14CE	H32BP	D33BL	A25BH	H24BF	L17BC
7	F34AK	Y11AH	J11AC	C12CS	S27CM	J30CK	Z23CF	S20CD	J15BM	Z35BJ	S21BG	D22BD
8	V26AL	T31AJ	U31AD	K37AB	P13CP	U14CL	C36CH	P28CE	U25BP	C16BK	P17BH	J29BE
9	E38AM	G18AK	R18AE	F27AC	W19CR	R23CM	K32CJ	W33CF	R35CD	K24BL	W22BJ	U34BF
10	M11BC	L12AL	X12AF	V13AD	Y30CS	X36CP	F20CK	Y15CH	X16CE	F21BH	Y29BK	R26BG
11	H31BD	D37AM	B37AG	E19AE	T14AB	B32CR	V28CL	T25CJ	B24CF	V17BP	T34BL	X38BH
12	S18BE	J27BC	N27AH	M30AF	G23AC	N20CS	E33CM	G35CK	N21CH	E22CD	G26BM	B11BJ
13	P12BF	U13BD	A13AJ	H14AG	L36AD	A28AB	M15CP	L16CL	A17CJ	M29CE	L38BP	N31BK
14	W37BG	R19BE	Z19AK	S23AH	D32AE	Z33AC	H25CR	D24CM	Z22CK	H34CF	D11CD	A18BL
15	Y27BH	X30BF	C30AL	P36AJ	J20AF	C15AD	S35CS	J21CP	C29CL	S26CH	J31CE	Z12BM
16	T13BJ	B14BG	K14AM	W32AK	U28AG	K25AE	P16AB	U17CR	K34CM	P38CJ	U18CF	C37BP
17	G19BK	N23BH	F23BC	Y20AL	R33AH	F35AF	M24AC	R22CS	F26CP	M11CK	R12CH	K27CD
18	L30BL	A36BJ	V36BD	T28AM	X15AJ	V16AG	Y21AD	X29AB	V38CR	Y31CL	X37CJ	F13CE
19	D14BM	Z32BK	E32BE	G33BC	B25AK	E24AH	T17AE	B34AC	E11CS	T18CM	B27CK	V19CF
20	J23BP	C20BL	M20BF	L15BD	N35AL	M21AJ	G22AF	N26AD	M31AB	G12CP	N13CL	E30CH
21	U36CD	K28BM	H28BG	D25BE	A16AM	H17AK	L29AG	A38AE	H18AC	L37CR	A19CM	M14CJ
22	R32CE	F33BP	S33BH	J35BF	Z24BC	S22AL	D34AH	Z11AF	S12AD	D27CS	Z30CP	H23CK
23	X20CF	V15CD	P15BJ	U16BG	C21BD	P29AM	J26AJ	C31AG	P37AE	J13AB	C14CR	S36CL
24	B28CH	E25CE	W25BK	R24BH	K17BE	W34BF	U38AK	K18AH	W27AF	U19AC	K23CS	P32CM
25	N33CJ	M35CF	Y35BL	X21BJ	F22BF	Y26BD	R11AL	F12AJ	Y13AG	R30AD	F36AB	W20CP
26	A15CK	H16CH	T16BM	B17BK	V29BG	T38BE	X31AM	V37AK	T19AH	X14AE	V32AC	Y28CR
27	Z25CL	S24CJ	G24BP	N22BL	G11BF	L31BF	B18BC	E27AL	G30AJ	B23AF	E20AD	T33CS
28	C35CM	P21CK	L21CE	A29BM	H26BJ	L31BG	N12BD	M13AM	L14AK	N36AG	M28AE	G15AB
29	K16CP		D17CE	Z34BP	H38BK	D18BH	A37BE	H19BC	D23AL	A32AH	H33AF	L25AC
30	F24CR		J22CF	C26CD	S11BL	J12BJ	Z27BF	S30BD	J36AM	Z20AJ	S15AG	D35AD
31	V21CS		U29CH		P31BH		C13BG	P14BE		C28AK		J16AE

1840

	JAN	FEB	MAR	APR	MAY	JUN	JUL	AUG	SEP	OCT	NOV	DEC
1	U24AF	K22AD	S29CR	J38CM	Z31CJ	S37CF	D13BP	Z14BL	S32BJ	D28BF	Z25BD	H16AL
2	R21AG	F29AE	P34CS	U11CP	E18CK	P27CH	J19CD	C23BM	P20BK	J33BG	C35BE	S24AM
3	X17AH	V34AF	W26AB	R31CR	K12CL	W13CJ	U30CE	K36BP	W28BL	U15BH	K16BF	P21BC
4	B22AJ	E26AG	Y38AC	X18CS	F37CM	Y19CK	R14CF	F32CD	Y33BM	R25BJ	F24BG	M17BD
5	N29AK	M38AH	T11AD	B12AB	V27CP	T30CL	X23CH	V20CE	T15BP	X35BK	V21BH	Y22BE
6	A34AL	H11AJ	G31AE	N37AC	E13CR	G14CM	B36CJ	E28CF	G25CD	B16BL	E17BJ	T29BF
7	Z26AM	S31AK	L18AF	A27AD	M19CS	L23CP	N32CK	M33CH	L35CE	N24BH	M22BK	G34BG
8	C38BC	P18AL	D12AG	Z13AF	H30AB	D36CR	A20CL	H15CJ	D16CF	A21BP	H29BL	L26BH
9	K11BD	W12AM	J37AH	C19AF	S14AC	J32CS	Z28CM	S25CK	J24CH	Z17CD	S34BM	D38BJ
10	F31BE	Y37BC	U27AJ	K30AG	P23AD	U20AB	G33CP	P35CL	U21CJ	G22CE	P26BP	J11BK
11	V18BF	T27BD	R13AK	F14AH	W36AE	R28AC	K15CR	W16CM	R17CK	K29CF	W38CD	U31BL
12	E12BG	G13BE	X19AL	V23AJ	Y32AF	X33AD	F25CS	Y24CP	X22CL	F34CH	Y11CE	R18BM
13	M37BH	L19BF	B30AM	E36AK	T20AG	B15AE	V35AB	T21CR	B29CM	V26CJ	T31CF	X12BP
14	H27BJ	D33BG	N14BC	M32AL	G28AH	N25AF	E16AC	G17CS	N34CP	E38CK	G18CH	B37CD
15	S13BK	J14BH	A23BD	H20AM	L33AJ	A35AG	M24AD	L22AB	A26CR	M11CL	L12CJ	N27CE
16	P19BL	U23BJ	Z36BE	S28BC	D15AK	Z16AH	H21AE	D29AC	Z38CS	H31CM	D37CK	A13CF
17	W30BM	R36BK	C32BF	P33BD	J25AL	C24AJ	S17AF	J34AD	C11AB	S18CP	J27CL	Z19CH
18	Y14BP	X32BL	K20BG	W15BE	U35AM	K21AK	P22AG	U26AE	K31AC	P12CR	U13CM	C30CJ
19	T23CD	B20BM	F28BH	Y25BF	R16BC	F17AL	W29AH	R38AF	F18AD	W37CS	R19CP	K14CK
20	G36CE	N28BP	V33BJ	T35BG	X24BD	V22AM	Y34AJ	X11AG	V12AE	Y27AD	X30CR	F23CL
21	L32CF	A33CD	E15BK	G16BH	B21BE	E29BC	T26AK	B31AH	E37AF	T13AC	B14CS	V36CM
22	D20CH	Z15CE	M25BL	L24BJ	N17BF	M34BD	G38AL	N18AJ	M27AG	G19AD	N23AB	E32CP
23	J28CJ	C25CF	H35BM	D21BK	A22BG	H26BE	L11AM	A12AK	H13AH	L30AE	A36AC	M20CR
24	U33CK	K35CM	S16BP	J17BL	Z29BH	S38BF	D31BC	Z37AL	S19AJ	D14AF	Z32AD	H28CS
25	R15CL	F16CJ	P24CD	U22BM	C34BJ	P11BG	J18BD	C27AM	P30AK	J23AG	C20AE	S33AB
26	X25CM	V24CK	W21CE	R29BP	K26BK	W31BH	U12BE	K13BC	W14AL	U36AH	K28AF	P15AC
27	B35CP	E21CL	Y17CF	X34CD	F38BL	Y18BJ	R37BF	F19BD	Y23AM	R32AJ	F33AG	W25AD
28	N16CR	M17CM	T22CH	B26CE	V11BM	T12BK	X27BG	V30BE	T36BC	X20AK	V15AH	Y35AE
29	A24CS	H22CP	G29CJ	N38CF	E31BP	G37BL	B13BH	E14BF	G32BD	B28AL	E25AJ	T16AF
30	Z21AB		L34CK	A11CM	M18CD	L27BM	N19BJ	M23BG	L20BE	N33AM	M35AK	G24AG
31	C17AC		D26CL		H12CE		A30BK	H36BH		A15BC		L21AH

1841

	JAN	FEB	MAR	APR	MAY	JUN	JUL	AUG	SEP	OCT	NOV	DEC
1	D17AJ	Z34AG	E34AB	G11CR	B18CL	E27CJ	T19CE	B23BP	E20BL	T33BH	B35BF	V24BC
2	J22AK	C26AM	H26AC	L31CS	N12CK	M13CK	G30CF	N36CD	M28BM	G15BJ	N16BG	E21BD
3	U29AL	K38AJ	H38AD	D18AB	A37CP	H19CL	L14CH	A32CE	H33BP	L25BK	A24BH	M17BE
4	R34AM	F11AK	S11AE	J12AC	Z27CR	S30CM	D23CJ	Z20CF	S15CD	D35BL	Z21BJ	H22BF
5	X26BC	V31AL	P31AF	U37AD	C13CS	P14CP	J36CK	C28CH	P25CE	J16BM	C17BK	S29BG
6	B38BD	E18AM	W18AG	R27AE	K19AB	W23CR	U32CL	K33CJ	W35CF	U24BP	K22BL	P34BF
7	N11BE	M12BC	Y12AH	X13AF	F30AC	Y36CS	R20CM	F15CK	Y16CH	R21CD	F29BM	W26BJ
8	A31BF	H37BD	T37AJ	B19AG	V14AD	T32AB	X28CR	V25CM	T24CJ	X17CE	V34BP	Y38BK
9	Z18BG	S27BE	G27AK	N30AH	E23AE	G20AC	B33CR	E35CM	G21CK	B22CF	E26CD	T11BL
10	C12BH	P13BF	L13AL	A14AJ	M36AF	L28AD	D33AB	M15AB	L17CL	D22CJ	M34CE	G31BM
11	K37BJ	W19BG	D19AM	Z23AK	H32AG	D33AE	A25AB	H24CR	D22CR	A34CJ	H11CF	L18BP
12	F27BK	Y30BH	J30BC	C36AL	S20AH	J15AF	Z35AC	S21CS	J29CP	Z26CS	S31CH	D12CD
13	V13BL	T14BJ	U14BD	K32AG	P28AJ	U25AG	C16AB	P17AB	U34CL	C38CL	P18CJ	J37CE
14	E19BM	G23BK	R23BE	F20BC	W33AK	R35AH	K24AE	W22AC	R26CS	K11CM	W12CK	U27CF
15	M30BP	L36BL	X36BF	Y28BD	Y15AH	X16AJ	F21AF	Y29AD	X38AB	F31CP	Y37CL	R13CH
16	H14CD	D32BM	B32BG	E33BE	T25AM	B24AK	V17AH	T34AE	B11AC	V18CR	T27CM	X19CJ
17	S23CE	J20BP	N20BH	M15BF	G35BC	N21AL	E22AH	G26AF	N31AD	E12CS	G13CP	B30CK
18	P36CF	U28CD	A28BP	H25BG	L16BD	A17AM	M29AJ	L38AG	A18AE	M37AB	L19CR	N14CL
19	W32CH	R33CE	Z33BK	S35BH	D24BE	Z22BH	H34AK	D11AH	Z12AF	H27AC	D30CS	A23CM
20	Y20CJ	X15CF	C15BL	P16BJ	J21BF	C29BD	S26AL	J31AJ	C37AG	S13AD	J14AB	Z36CP
21	T28CK	B25CH	K25BM	W24BK	U17BG	K34BE	P38AM	U18AL	K27AH	P19AF	U23AC	C32CR
22	G33CL	N35CM	F35BP	Y21BL	R22BH	F26BF	M11BC	R12AL	F13AJ	M30AF	R36AD	K20CS
23	L15CM	A16CK	V16CD	T17BM	X29BJ	V38BF	Y31BD	X37AM	V19AK	Y14AG	X32AE	F28AB
24	D25CP	Z24CL	E24CE	G22BP	B34BK	E11BH	T18BF	B27BC	E30AL	T23AH	B20AF	V33AC
25	J35CR	C21CM	M21CF	L29CD	N26BL	M31BJ	G12BE	N13BD	M14AM	G36AJ	N28AG	E15AD
26	U16CS	K17CP	H17CH	D34CE	A38BM	H18BK	L37BG	A19BE	H23BD	L32AK	A33AH	M25AE
27	R24AB	F22CR	S22CJ	J26CF	Z11BH	S12BL	D27BH	Z30BF	S35BD	D20AL	Z15AJ	H35AF
28	X21AC	V29CS	P29CK	U38CH	C31BD	P37BH	J13BJ	C14BD	P32BE	J28AM	C25AK	S16AG
29	B17AD		W34CL	R11CJ	K18CK	W27BH	U19BJ	K23BH	W20BF	U33BC	K35AL	P24AH
30	N22AE		Y26CM	X31CK	F12CF	Y13CD	R30BL	F36BJ	Y28BG	R15BD	F16AM	W21AJ
31	A29AF		T38CP		V37CH		X14BM	V32BK		X25BE		Y17AK

FIRST LETTER = YOUR PHYSICAL CODE, MIDDLE NUMBER = SENSITIVITY CODE, LAST TWO LETTERS = INTELLECTUAL CODE

1842	JAN	FEB	MAR	APR	MAY	JUN	JUL	AUG	SEP	OCT	NOV	DEC
1	T22AL	B26AJ	K26AD	W31AB	U12CP	K13CL	P30CH	U36CE	K28BP	P15BK	U16BH	C21BE
2	G29AM	N38AK	F38AE	Y18AC	R37CP	F19CH	W14CJ	R32CF	F33CD	W25BL	R24BJ	K17BF
3	L34BC	A11AL	V11AF	T12AD	X27CS	V30CP	Y23CK	X20CH	V15CE	Y35BM	X21BK	F22BG
4	D26BD	Z31AM	E31AG	G37AE	B13AB	E14CR	T36CL	B28CJ	E25CF	T16BP	B17BL	V29BH
5	J38BE	C16BC	M18AH	L27AF	N19AC	M23CS	G32CM	N33CK	M35CH	G24CD	N22BM	E34BJ
6	U11BF	K12BD	H12AJ	D13AG	A30AD	H36AB	L20CP	A15CL	H16CJ	L21CE	A29BP	M26BK
7	R31BG	F37BE	S37AK	J19AH	Z14AE	S32AC	D28CR	Z25CM	S24CK	D17CF	Z34CD	H38BL
8	X18BH	V27BF	P27AL	U30AJ	C23AF	P20AD	J33CS	C35CP	P21CL	J22CH	C26CE	S11BM
9	B12BJ	E13BG	W13AM	R14AK	K36AG	W28AE	U15AB	K16CR	M17CM	U29CJ	K38CF	P31BP
10	N37BK	M19BH	Y19BC	X23AL	F32AH	Y33AF	R25AC	F24CS	Y22CP	R34CK	F11CH	W18CD
11	A27BL	H30BJ	T30BD	B36AM	V20AJ	T15AG	X35AD	V21AB	T29CR	X26CL	V31CJ	Y12CE
12	Z13BM	S14BK	G14BE	N32BC	E28AK	G25AH	B16AE	E17AC	G34CS	B38CM	E18CK	T37CF
13	C19BP	P23BL	L23BF	A20BD	M33AL	L35AJ	N24AF	M22AD	L26AB	N11CP	M12CL	G27CH
14	K30CD	W36BM	D36BG	Z28BE	H15AM	D16AK	A21AG	H29AE	D38AC	A31CR	H37CM	L13CJ
15	F14CE	Y32BP	J32BH	C33BF	S25BC	J24AL	Z17AH	S34AF	J11AD	Z18CS	S27CP	D19CK
16	V23CF	T20CD	U20BJ	K15BG	P35BD	U21AM	C22AJ	P26AG	U31AE	C12AB	P13CR	J30CL
17	E36CH	G28CE	R28BK	F25BH	W16BE	R17BC	K29AK	W38AH	R18AF	K37AC	W19CS	U14CM
18	M32CJ	L33CF	X33BL	V35BJ	Y24BF	X22BF	F34AL	Y11AJ	X12AG	F27AD	Y30AB	R23CP
19	H20CK	D15CH	B15BM	E16BK	T21BG	B29BE	V26AM	T31AK	B37AH	V13AE	T14AC	X36CR
20	S28CL	J25CJ	N25BP	M24BL	G17BH	N34BF	E38BC	G18AL	N27AJ	E19AF	G23AD	B32CS
21	P33CM	U35CK	A35CD	H21BM	L22BJ	A26BG	M11BD	L12AM	A13AK	M30AG	L36AE	N20AB
22	W15CP	R16CL	Z16CE	S17BP	D29BK	Z38BH	H31BE	D37BC	Z19AL	H14AM	D32AF	A28AC
23	Y25CR	X24CM	C24CF	P22CD	J34BL	C11BJ	S18BF	J27BD	C30AM	S23AJ	J20AG	Z33AD
24	T35CS	B21CP	K21CH	W29CE	U26BM	K31BK	P12BG	U13BE	K14BC	P36AK	U28AH	C15AE
25	G16AB	N17CR	F17CJ	Y34CF	R38BP	F18BL	W37BJ	R19BF	F23BD	W32AL	R33AJ	K25AF
26	L24AC	A22CS	V22CK	T26CH	X11CD	V12BM	Y27BJ	X30BG	V36BE	Y20AM	X15AK	F35AG
27	D21AD	Z29AB	E29CL	G38CJ	B31CE	E37BP	T13BK	B14BH	E32BF	T28BC	B25AL	V16AH
28	J17AE	C34AC	M34CM	L11CK	N18CF	M27CD	G19BL	N23BJ	M20BG	G33BD	N35AM	E24AJ
29	U22AF		H26CP	D31CL	A12CH	H13CE	L30BM	A36BK	H28BH	L15BE	A16BC	M21AK
30	R29AG		S38CR	J18CM	Z37CJ	S19CF	D14BP	Z32BL	S33BJ	D25BF	Z24BD	H17AL
31	X34AH		P11CS		C27CK		J23CD	C20BH		J35BG		S22AM

1843	JAN	FEB	MAR	APR	MAY	JUN	JUL	AUG	SEP	OCT	NOV	DEC
1	P29BC	U38AL	A38AF	H18AD	L37CS	A19CP	M14CH	L32CH	A33CE	M25BM	L24BK	N17BG
2	W34BD	R11AM	Z11AG	S12AE	D27AB	Z30CR	H23CL	D20CJ	Z15CF	H35BP	D21BL	A22BH
3	Y26BE	X31BC	C31AH	P37AF	J13AC	C14CS	S36CM	J28CK	C25CH	S16CD	J17BM	Z29BJ
4	T38BF	B18BD	K18AJ	W27AG	U19AD	K23AB	P32CP	U33CL	K35CJ	P24CE	U22BP	C34BK
5	G11BG	N12BE	F12AK	Y13AH	R30AE	F36AC	W20CR	R15CM	F16CK	W21CF	R29CD	K26BL
6	L31BH	A37BF	V37AL	T19AJ	X14AF	V32AD	Y28CS	X25CP	V24CL	Y17CH	X34CE	F38BM
7	D18BJ	Z27BG	E27AM	G30AK	B23AG	E20AE	T33AB	B35CR	E21CM	T22CJ	B26CF	V11BP
8	J12BK	C13BH	M13BC	L14AL	N36AH	M28AF	G15AC	N16CS	M17CP	G29CK	N38CH	E31CD
9	U37BL	K19BJ	H19BD	D23AM	A32AJ	H33AG	L25AD	A24AB	H22CR	L34CL	A11CJ	M18CE
10	R27BM	F30BK	S30BE	J36BC	Z20AK	S15AH	D35AE	Z21AC	S29CS	D26CM	Z31CK	H12CF
11	X13BP	V14BL	P14BF	U32BD	C28AL	P25AJ	J16AF	C17AD	P34AB	J38CP	C18CL	S37CH
12	B19CD	E23BM	W23BG	R20BE	K33AM	W35AK	U24AG	K22AE	M26AC	U11CR	K12CM	P27CJ
13	N30CE	M36BP	Y36BH	X28BF	F15BC	Y16AL	R21AH	F29AF	Y38AD	R31CS	F37CP	W13CK
14	A14CF	H32CD	T32BJ	B33BG	V25BD	T24AM	X17AJ	V34AG	T11AE	X18AB	V27CR	Y19CL
15	Z23CH	S20CE	G20BK	N15BH	E35BE	G21BC	B22AK	E26AH	G31AF	B12AC	E13CS	T30CM
16	C36CJ	P28CF	L28BL	A25BJ	M16BF	L17BD	N29AL	M38AJ	L18AG	N37AD	M19AB	G14CP
17	K32CK	W33CH	D33BM	Z35BK	H24BG	D22BE	A34AM	H11AK	D12AH	A27AE	H30AC	L23CR
18	F20CL	Y15CJ	J15BP	C16BL	S21BH	J29BF	Z26BC	S31AL	J37AJ	Z13AF	S14AD	D36CS
19	V28CM	T25CK	U25CD	K24BM	P17BJ	U34BG	C38BD	P18AM	U27AK	C19AG	P23AE	J32AB
20	E33CP	G35CL	R35CE	F21BP	W22BK	R26BH	K11BE	W12BC	R13AL	K30AH	W36AF	U20AC
21	M15CR	L16CM	X16CF	V17CD	Y29BL	X38BJ	F31BF	Y37BD	X19AM	F14AJ	Y32AG	R28AD
22	H25CS	D24CP	B24CH	E22CE	T34BM	B11BK	V18BG	T27BE	B30BC	V23AK	T20AH	X33AE
23	S35AB	J21CR	N21CJ	M29CF	G26BP	N31BL	E12BH	G13BF	N14BD	E36AL	G28AJ	B15AF
24	P16AC	U17CS	A17CK	H34CH	L38CD	A18BM	M37BJ	L19BG	A23BE	M32AH	L33AK	N25AG
25	W24AD	R22AB	Z22CL	S26CJ	D11CE	Z12BP	H27BK	D30BH	Z36BF	H20BC	D15AL	A35AH
26	Y21AE	X29AC	C29CM	P38CK	J31CF	C37CD	S13BL	J14BJ	C32BG	S28BD	J25AM	Z16AJ
27	T17AF	B34AD	K34CP	W11CL	U18CH	K27CE	P19BM	U23BK	K20BH	P33BE	U35BC	C24AK
28	G22AG	N26AE	F26CR	Y31CM	R12CJ	F13CF	W30BP	R36BL	F28BJ	W15BF	R16BD	K21AL
29	L29AH		V38CS	T18CP	X37CK	V19CH	Y14BG	X32BM	V33BK	Y25BG	X24BE	F17AM
30	D34AJ		E11AB	G12CR	B27CL	E30CJ	T23CE	B20BP	E15BL	T35BH	B21BD	V22BC
31	J26AK		M31AC		N13CM		G36CF	N28CD		G16BJ		E29BD

1844	JAN	FEB	MAR	APR	MAY	JUN	JUL	AUG	SEP	OCT	NOV	DEC
1	M34BE	L11BC	B31AJ	E37AG	T13AD	B14AB	V36CP	T28CL	B25CJ	V16CE	T17BP	X29BK
2	H26BF	D31BD	N18AK	M27AH	G19AE	N23AC	E32CR	G33CH	N35CK	E24CF	G22CD	B34BL
3	S38BG	J18BE	A12AL	H13AJ	L30AF	A36AD	M20CS	L15CP	A16CL	M21CH	L29CE	N26BM
4	P11BH	U12BF	Z37AM	S19AK	D14AG	Z32AE	H28AB	D25CR	Z24CM	H17CJ	D34CF	A38BP
5	W31BJ	R37BG	C27BC	P30AL	J23AH	C20AF	S33AC	J35CS	C21CH	S22CL	J26CH	Z11CD
6	Y18BK	X27BH	K13BD	W14AM	U36AJ	K28AG	P15AD	U16AB	K17CR	P29CL	U38CJ	C31CE
7	T12BL	B13BJ	F19BE	Y23BC	R32AK	F33AH	W25AE	R24AC	F22CS	W34CM	R11CK	K18CF
8	G37BM	N19BK	V38BF	T36BD	X20AL	V15AK	Y35AF	X21AB	V29AB	Y26CP	X31CL	F12CH
9	L27BP	A38BL	E14BG	G32BE	B28AM	E25AK	T16AG	B17AE	E34AC	T38CR	B18CK	V37CJ
10	D13CD	Z14BM	M23BH	L28BG	N33AL	M35AH	G24AF	N22AD	M26AB	G11CS	N12CP	E27CK
11	J19CE	C23BP	H36BJ	D28BG	A15BD	H16AM	L21AJ	A29AG	H38AE	L31AB	A37CR	M13CL
12	U30CF	K36CD	S32BK	J33BH	Z25BE	S24BC	D17AM	Z34AK	S11AF	D18AC	Z27CS	H19CM
13	R14CH	F32CE	P20BL	U15BJ	C35BF	P21BD	J22AL	C26AJ	P31AG	J12AC	C13AB	S30CP
14	X23CJ	V20CF	W28BM	R25BK	K16BG	W17BE	U29AH	K38AK	W18AH	U37AE	K19AC	P14CR
15	B36CK	E28CH	Y33BP	X35BL	F24BH	Y22BF	T29BG	R34BC	F11AL	Y12AJ	R27AF	N23CS
16	N32CL	M33CJ	T15CD	B16BM	V21BJ	T29BG	X26BD	V31AM	T37AK	X13AG	V14AE	Y36AB
17	A20CM	H15CK	G25CE	M24BP	E17BK	G34BH	B38BE	E18BC	G27AL	B19AH	E23AF	T32AC
18	Z28CP	S25CL	L35CF	A21CD	M22BL	L26BJ	N11BF	M12BD	L13AM	N30AJ	M36AG	G20AD
19	C33CR	P35CM	D16CH	Z17CE	H29BM	D38BK	A31BG	H37BE	D19BC	A14AM	H32AH	L28AE
20	K15CS	W16CP	J24CJ	C22CF	S34BP	J11BL	Z18BH	S27BF	J30BD	Z23AL	S20AJ	D33AF
21	F25AB	Y24CR	U21CK	K29CP	P13BG	U31BM	C12BH	P18BD	U14BE	C36AM	P28AK	J15AG
22	V35AC	T21CS	R17CL	F34CJ	W38CE	R18BP	K37BK	W19BH	R23BF	K32BC	W33AL	U25AH
23	E16AD	G17AB	X22CM	V26CK	Y11CF	X12CD	F27BL	Y30BJ	X36BG	F20BD	Y15AM	R35AJ
24	M24AE	L22AC	B29CP	E38CL	T31CH	B37CE	V13BM	T14BK	B32BH	V28BE	T25BC	X16AK
25	H21AF	D29AD	N34CR	M11CM	G18CJ	N27CF	E19BP	G23BL	N20BJ	E33BF	G35BD	B24AL
26	S17AG	J34AE	A26CS	H31CP	L12CK	A13CH	M30CD	L36BM	A28BK	M15BG	L16BE	N21AM
27	P22AH	U26AF	Z38AB	S18CR	D37CL	Z19CJ	H14CD	D32BP	Z33BL	H25BH	D24BF	A17BC
28	W29AJ	R38AG	C11AD	P12CS	J27CM	C30CK	S23CF	J20CD	C15BM	S35BJ	J21BG	Z22BD
29	Y34AK	X11AH	K31AD	W37AB	U13CP	K14CL	P36CH	U28CE	K25BP	P16BK	U17BH	C29BE
30	T26AL		F18AE	Y27AC	R19CR	F23CH	W32CJ	R33CF	F35CD	W24BL	R22BJ	K34BF
31	G38AM		V12AF		X30CS		Y20CK	X15CH		Y21BM		F26BG

FIRST LETTER = YOUR PHYSICAL CODE, MIDDLE NUMBER = SENSITIVITY CODE, LAST TWO LETTERS = INTELLECTUAL CODE

1845	JAN	FEB	MAR	APR	MAY	JUN	JUL	AUG	SEP	OCT	NOV	DEC
1	V38BH	T18BF	U18AL	K27AJ	P19AF	U23AD	C32CS	P33CP	U35CL	C24CH	P22CE	J34BM
2	E11BJ	G12BG	R12AM	F13AK	W30AG	R36AE	K20AB	W15CR	R16CM	K21CJ	W29CF	U26BP
3	M31BK	L37BH	X37BC	V19AL	Y14AH	X32AF	F28AC	Y25CS	X24CP	F17CK	Y34CH	R38CD
4	H18BL	D27BJ	B27BD	E30AM	T23AJ	B20AG	V33AD	T35AB	B21CR	V22CL	T26CJ	X11CE
5	S12BM	J13BK	N13BE	M14BC	G36AK	N28AH	E15AE	G16AC	N17CS	E29CM	G38CK	B31CF
6	P37BP	U19BL	A19BF	H23BD	L32AL	A33AJ	M25AF	L24AD	A22AB	M34CP	L11CL	N18CH
7	W27CD	R30BM	Z30BG	S36BE	D20AM	Z15AK	W35AG	D21AE	Z29AC	H26CR	D31CM	A12CJ
8	Y13CE	X14BP	C14BH	P32BF	J28BC	C25AL	S16AH	J17AF	C34AD	S38CS	J18CP	Z37CK
9	T19CF	B23CD	K23BJ	W20BG	U33BD	K35AM	P24AJ	U22AG	K26AE	P11AB	U12CR	C27CL
10	G30CH	N36CE	F36BK	Y28BH	R15BE	F16BC	W21AK	R29AH	F38AF	W31AC	R37CS	K13CH
11	L14CJ	A32CF	V32BL	T33BJ	X25BF	V24BD	Y17AL	X34AJ	V11AG	Y18AD	X27AB	F19CP
12	D23CK	Z20CH	E20BM	G15BK	B35BG	E21BE	T22AM	B26AK	E31AH	T12AE	B13AC	V30CR
13	J36CL	C28CJ	M28BP	L25BL	N16BH	M17BF	G29BC	N38AL	M18AJ	G37AF	N19AD	E14CS
14	U32CM	K33CK	H33CD	D35BM	A24BJ	H22BG	L34BD	A11AM	H12AK	L27AG	A30AE	M23AB
15	R20CP	F15CL	S15CE	J16BP	Z21BK	S29BH	D26BE	Z31BC	S37AL	D13AH	Z14AF	H36AC
16	X28CR	V25CM	P25CF	U24CD	C17BL	P34BJ	J38BF	C18BD	P27AM	J19AJ	C23AG	S32AD
17	B33CS	E35CP	W35CH	R21CE	K22BM	W26BK	U11BG	K12BE	W13BC	U30AK	K36AH	P20AE
18	N15AB	M16CR	Y16CJ	X17CF	F29BP	Y38BL	R31BH	F37BF	Y19BD	R14AL	F32AJ	W28AF
19	A25AC	H24CS	T24CK	B22CH	V34CD	T11BM	X18BJ	V27BG	T30BE	X23AM	V20AK	Y33AG
20	Z35AD	S21AB	G21CL	N29CJ	E26CE	G31BP	B12BK	E13BH	G14BF	B36BC	E28AL	T15AH
21	C16AE	P17AC	L17CM	A34CK	M38CF	L18CD	N37BL	M19BJ	L23BG	N32BD	M33AM	G25AJ
22	K24AF	W22AD	D22CP	Z26CL	H11CH	D12CE	A27BM	H30BK	D36BH	A20BE	H15BC	L35AK
23	F21AG	Y29AE	J29CR	C38CM	S31CJ	J37CF	Z13BP	S14BL	J32BJ	Z28BF	S25BD	D16AL
24	V17AH	T34AF	U34CS	K11CP	P18CK	U27CH	C19CD	P23BM	U20BK	C33BG	P35BE	J24AH
25	E22AJ	G26AG	R26AB	F31CR	W12CL	R13CJ	K30CE	W36BP	R28BL	K15BH	W16BF	U21BC
26	M29AK	L38AH	X38AC	V18CS	Y37CM	X19CK	F14CF	Y32CD	X33BM	F25BJ	Y24BG	R17BD
27	H34AL	D11AJ	B11AD	E12AB	T27CP	B30CL	V23CH	T20CE	B15BP	V35BK	T21BH	X22BF
28	S26AM	J31AK	N31AE	M37AC	G13CR	N14CM	E36CJ	G28CF	N25CD	E16BL	G17BJ	B29BF
29	P38BC		A18AF	H27AD	L19CS	A23CP	M32CK	L33CH	A35CE	M24BM	L22BK	N34BG
30	W11BD		Z12AG	S13AE	D30AB	Z36CR	H20CL	D15CJ	Z16CF	H21BP	D29BL	A26BH
31	Y31BE		C37AH		J14AC		S28CH	J25CK		S17CD		Z38BJ

1846	JAN	FEB	MAR	APR	MAY	JUN	JUL	AUG	SEP	OCT	NOV	DEC
1	C11BK	P12BH	L12BC	M13AL	M30AH	L36AF	N20AC	M15CS	L16CP	N21CK	M29CH	G26CD
2	K31BL	H37BJ	D37BD	Z19AM	H14AJ	D32AG	A28AD	H25AB	D24CR	A17CL	H34CJ	L38CE
3	F18BM	Y27BK	J27BE	C30BC	S23AK	J20AH	Z33AE	S35AC	J21CS	Z22CM	S26CK	D11CF
4	V12BP	T13BL	U13BF	K14BD	P36AL	U28AJ	C15AF	P16AD	U17AB	C29CP	P38CL	J31CH
5	E37CD	G19BM	R19BG	F23BE	W32AM	R33AK	K25AG	W24AE	R22AC	K34CR	W11CH	U18CJ
6	M27CE	L30BP	X30BH	V36BF	Y20BC	X15AL	F35AH	Y21AF	X29AD	F26CS	Y31CP	R12CK
7	H13CF	D14CD	B14BJ	E32BG	T28BD	B25AM	V16AJ	T17AG	B34AE	V38AB	T18CR	X37CL
8	S19CH	J23CE	N23BK	M20BH	G33BE	N35BC	E24AK	G22AH	N26AF	E11AC	G12CS	B27CM
9	P30CJ	U36CF	A36BL	H28BJ	L15BF	A16BD	M21AL	L29AJ	A38AG	M31AD	L37AB	N13CP
10	W14CK	R32CH	Z32BM	S33BK	D25BG	Z24BE	H17AM	D34AK	Z11AH	H18AE	D27AC	A19CR
11	Y23CL	X20CJ	C20BP	P15BL	J35BH	C21BF	S22BC	J26AL	C31AJ	S12AF	J13AD	Z30CS
12	T36CM	B28CK	K28CD	W25BM	U16BJ	K17BG	P29BD	U38AM	K18AK	P37AG	U19AE	C14AB
13	G32CP	N33CL	F33CE	Y35BP	R24BK	F22BH	W34BE	R11BC	F12AL	W27AH	R30AF	K23AC
14	L20CR	A15CM	V15CF	T16CD	X21BL	V29BJ	Y26BF	X31BD	V37AM	Y13AJ	X14AG	F36AD
15	D28CS	Z25CP	E25CH	G24CE	B17BM	E34BK	T38BG	B18BE	E27BC	T19AK	B23AH	V32AE
16	J33AB	C35CR	M35CJ	L21CF	N22BP	M26BL	G11BH	N12BF	M13BD	G30AL	N36AJ	E20AF
17	U15AC	K16CS	H16CK	D17CH	A29CD	H38BM	L31BJ	A37BG	H19BE	L14AM	A32AK	M28AG
18	R25AD	F24AB	S24CL	J22CJ	Z34CE	S11BP	D18BK	Z27BH	S30BF	D23BC	Z20AL	H33AH
19	X35AE	V21AC	P21CM	U29CK	C26CF	P31CD	J12BL	C13BJ	P14BG	J36BD	C28AM	S15AJ
20	B16AF	E17AD	W17CP	R34CL	K38CH	W18CE	U37BM	K19BK	W23BH	U32BE	K33BC	P25AK
21	N24AG	M22AE	Y22CR	X26CM	F11CJ	Y12CF	R27BP	F30BL	Y35BJ	R20BF	F15BD	W35AL
22	A21AH	H29AF	T29CS	B38CP	V31CK	T37CH	X13CD	V14BM	T32BK	X28BG	V25BE	Y16AM
23	Z17AJ	S34AG	G34AB	N11CR	E18CL	G27CJ	B19CE	E23BP	G20BL	B33BH	E35BF	T24BC
24	C22AK	P26AH	L26AC	A31CS	M12CM	L13CK	N30CF	M36CD	L28BM	N15BJ	M16BG	G21BD
25	K29AL	W38AJ	D38AD	Z18AB	H37CP	D19CL	A14CH	H32CE	D33BP	A25BK	H24BH	L17BE
26	F34AM	Y11AK	J11AE	C12AC	S27CR	J30CM	Z23CJ	S20CF	J15CD	Z35BL	S21BJ	D22BF
27	V26BC	T31AL	U31AF	K37AD	P13CS	U14CP	C36CK	P28CH	U25CE	C16BM	P17BK	J29BG
28	E38BD	G18AM		F27AE	W19AB	R23CR	K32CL	W33CJ	R35CF	K24BP	W22BL	U34BH
29	M11BE		X12AH	V13AF	Y30AC	X36CS	F20CM	Y15CK	X16CH	F21CD	Y29BM	R26BJ
30	H31BF		B37AJ	E19AG	T14AD	B32AB	V28CP	T25CL	B24CJ	V17CE	T34BP	X38BK
31	S18BG		N27AK		G23AE		E33CR	G35CM		E22CF		B11BL

1847	JAN	FEB	MAR	APR	MAY	JUN	JUL	AUG	SEP	OCT	NOV	DEC
1	N31BM	M37BK	Y37BE	X19BC	F14AK	Y32AH	R28AE	F25AC	Y24CS	R17CH	F34CK	W38CF
2	A18BP	H27BL	T27BF	B30BD	V23AL	T20AJ	X33AF	V35AD	T21AB	X22CP	V26CL	Y11CH
3	Z12CD	S13BM	G13BG	N14BE	E36AM	G28AK	B15AG	E16AE	G17AC	B29CR	E38CM	T31CJ
4	C37CE	P19BP	L19BH	A23BF	M32BC	L33AL	N25AH	M24AF	L22AD	N34CS	M11CP	G18CK
5	K27CF	W30CD	D30BJ	Z36BG	H20BD	D15AM	A35AJ	H21AG	D29AE	A26AB	H31CR	L12CL
6	F13CH	Y14CE	J14BK	C32BH	S28BE	J25BC	Z16AK	S17AH	J34AF	Z38AC	S18CS	D37CH
7	V19CJ	T23CF	U23BL	K20BJ	P33BF	U35BD	C24AL	P22AJ	U26AG	C11AD	P12AB	J27CP
8	E30CK	G36CH	R36BM	F28BK	W15BG	R16BE	K21AM	W29AK	R38AH	K31AE	W37AC	U13CR
9	M14CL	L32CJ	X32BP	V33BL	Y25BH	X24BF	F17BC	Y34AL	X11AJ	F18AF	Y27AD	R19CS
10	H23CM	D20CK	B20CD	E15BM	T35BJ	B21BG	V22BD	T26AM	B31AK	V12AG	T13AE	X30AB
11	S36CP	J28CL	N28CE	M25BP	G16BK	N17BH	E29BE	G38BC	N18AL	E37AH	G19AF	B14AC
12	P32CR	U33CM	A33CF	H35CD	L24BL	A22BJ	M34BF	L11BD	A12AM	M27AJ	L30AG	N23AD
13	W20CS	R15CP	Z15CH	G25CE	D21BM	Z29BK	H26BG	D31BE	Z37BC	H13AK	D14AH	A36AE
14	Y28AB	X25CR	C25CJ	P24CF	J17BP	C34BL	S38BH	J18BF	C27BD	S19AL	J23AJ	Z32AF
15	T33AC	B35CS	K35CK	W21CH	U22CD	K26BM	P11BF	U12BG	K13BE	P30AM	U36AK	C20AG
16	G15AD	N16AB	F16CL	Y17CJ	R29CE	F38BP	M31BH	R37BH	F19BF	M14BC	R32AL	K28AH
17	L25AE	A24AC	V24CM	T22CK	X34CF	V11CD	Y18BL	X27BJ	V30BG	Y23BD	X20AH	F33AJ
18	D35AF	Z21AD	E21CP	G29CL	B26CH	E31CE	T12BM	B13BK	E14BH	T36BE	B28BC	V15AK
19	J16AG	C17AE	M17CR	L34CM	N38CJ	M18CF	G37BP	N19BL	M23BJ	G32BF	N33BD	E25AL
20	U24AH	K22AF	H22CS	D26CP	A11CM	H12CH	L27CD	A30BM	H36BK	L20BG	A15BE	M35AM
21	R21AJ	F29AG	S29AB	J38CR	Z31CL	S37CJ	D13CE	Z14BP	S32BL	D28BH	Z25BF	H16BC
22	X17AK	V34AH	P34AC	U11CS	C18CM	P27CK	J19CF	C23BD	P20BH	J33BJ	C35BG	S24BD
23	B22AL	E26AJ	W26AD	R31AB	K12CP	W13CL	N30CH	K36CE	W28BP	N15BK	K16BH	P21BE
24	N29AM	H38AK	Y38AE	X18AC	F37CR	Y19CM	R14CJ	F32CF	Y33CD	R25BL	F24BJ	W17BF
25	A34BC	H11AL	T11AF	B12AD	V27CS	T30CP	X23CK	V20CH	T15CE	X35BM	V21BK	Y22BG
26	Z26BD	S31AM	G31AG	N37AE	E13AB	G14CR	B36CL	E28CJ	G25CF	B16BP	E17BL	T29BH
27	C38BE	P18BC	L18AH	A27AF	M19AC	L23CS	N32CJ	M33CH	L35CD	N24CD	M22BM	G34BJ
28	K11BF	W12BD	D12AJ	Z13AG	H30AD	D36AB	A20CP	H15CL	D16CJ	A21CE	H29BP	L26BK
29	F31BG		J37AK	C19AH	S14AH	J32AC	Z28CR	S25CM	J24CK	Z17CF	S34CD	D38BL
30	V18BH		U27AL	K30AJ	P23AF	U20AD	C33CS	P35CP	U21CL	C22CH	P26CE	J11BM
31	E12BJ		R13AM		W36AG		K15AB	W16CR		K29CJ		U31BP

THE COMPLETE BOOK OF BIORHYTHM LIFE CYCLES

FIRST LETTER = YOUR PHYSICAL CODE, MIDDLE NUMBER = SENSITIVITY CODE, LAST TWO LETTERS = INTELLECTUAL CODE

1848

	JAN	FEB	MAR	APR	MAY	JUN	JUL	AUG	SEP	OCT	NOV	DEC
1	R18CD	F27BM	P13BH	U14BF	C36BC	P28AL	J15AH	C16AF	P17AD	J29CS	C38CP	S31CK
2	X12CE	V13BP	W19BJ	R23BG	K32BD	W33AM	U25AJ	K24AG	W22AE	U34AB	K11CR	P18CL
3	B37CF	E19CD	Y30BK	X36BH	F20BE	Y15BC	R35AK	F21AH	Y29AF	R26AC	F31CS	W12CM
4	N27CH	M30CE	T14BL	B32BJ	V28BF	T25BD	X16AL	V17AJ	T34AG	X38AD	V18AB	Y37CP
5	A13CJ	H14CF	G23BM	N20BK	E33BG	G35BE	B24AM	E22AK	G26AH	B11AE	E12AC	T27CR
6	Z19CK	S23CH	L36BP	A28BL	M15BH	L16BF	N21BC	M29AL	L38AJ	N31AF	M37AD	G13CS
7	C30CL	P36CJ	D32CD	Z33BM	H25BJ	D24BG	A17BD	H34AM	D11AK	A18AG	H27AE	L19AB
8	K14CM	W32CK	J20CE	C15BP	S35BN	J21BH	Z22BE	S26BC	J31AL	Z12AH	S13AF	D30AC
9	F23CP	Y20CL	U28CF	K25CD	P16BL	U17BJ	C29BF	P38BD	U18AM	C37AJ	P19AG	J14AD
10	V36CR	T28CM	R33CH	F35CE	W24BM	R22BK	K34BG	W11BE	R12BC	K27AK	W30AH	U23AE
11	E32CS	G33CP	X15CJ	V16CF	Y21BP	X29BL	F26BH	Y31BF	X37BD	F13AL	Y14AJ	R36AF
12	M20AB	L15CR	B25CK	E24CH	T17CD	B34BM	V38BJ	T18BG	B27BE	V19AM	T23AK	X32AG
13	H28AC	D25CS	N35CL	M21CJ	G22CE	N26BP	E11BK	G12BH	N13BF	E30BC	G36AL	B20AH
14	S33AD	J35AB	A16CM	H17CK	L29CF	A38CD	M31BL	L37BJ	A19BG	M14BD	L32AM	N28AJ
15	P15AE	U16AC	Z24CP	S22CL	D34CH	Z11CE	H18BM	D27BK	Z30BH	H23BE	D20BC	A33AK
16	W25AF	R24AD	C21CR	P29CM	J26CJ	C31CF	S12BP	J13BL	C14BJ	S36BF	J28BD	Z15AL
17	Y35AG	X21AE	K17CS	W34CP	U38CK	K18CH	P37CD	U19BM	K23BK	P32BG	U33BE	C25AM
18	T16AH	B17AF	F22AB	Y26CR	R11CL	F12CJ	W27CE	R30BP	F36BL	W20BH	R15BF	K35BC
19	G24AJ	N22AG	V29AC	T38CS	X31CM	V37CK	Y13CF	X14CD	V32BM	Y28BJ	X25BG	F16BD
20	L21AK	A29AH	E34AD	G11AB	B18CP	E27CL	T19CH	B23CE	E20BP	T33BK	B35BH	V24BE
21	D17AL	Z34AJ	M26AE	L31AC	N12CR	M13CH	G30CJ	N36CF	M28CD	G15BL	N16BJ	E21BF
22	J22AM	C26AK	H38AF	D18AD	A37CS	H19CP	L14CK	A32CH	H33CE	L25BM	A24BK	M17BG
23	U29BC	K38AL	S11AG	J12AE	Z27AB	S30CR	D23CL	Z20CJ	S15CF	D35BP	Z21BL	H22BH
24	R34BD	F11AM	P31AH	U37AF	C13AC	P14CS	J36CH	C28CK	P25CH	J16CD	C17BM	S29BJ
25	X26BE	V31BC	W18AJ	R27AG	K19AD	W23AB	U32CP	K33CL	W35CJ	U24CE	K22BP	P34BK
26	B38BF	E18BD	Y12AK	X13AH	F30AE	Y36AC	R20CR	F15CM	Y16CK	R21CF	F29CD	W26BL
27	N11BH	M12BE	I37AL	B19AJ	V14AF	T32AD	X28CS	V25CP	T24CL	X17CH	V34CE	Y38BM
28	A31BH	H37BF	G27AM	N30AK	E23AG	G20AE	B33AB	E35CR	G21CM	B22CJ	E26CF	T11BP
29	Z18BJ	S27BG	L13BC	A14AL	M36AH	L28AF	N15AC	M16CS	L17CP	N29CK	M38CH	G31CD
30	C12BK		D19BD	Z23AM	H32AJ	D33AG	A25AD	H24AB	D22CR	A34CL	H11CJ	L18CE
31	K37BL		J30BE		S20AK		Z35AE	S21AC		Z26CM		D12CF

1849

	JAN	FEB	MAR	APR	MAY	JUN	JUL	AUG	SEP	OCT	NOV	DEC
1	J37CH	C19CE	M19BK	L23BH	N32BE	M33BC	G25AK	N24AH	M22AF	G34AC	N11CS	E18CM
2	U27CJ	K30CF	H30BL	D36BJ	A20BF	H15BD	L35AL	A21AJ	H29AG	L26AD	A31AB	M12CP
3	R13CK	F14CM	S14BM	J32BK	Z28BG	S25BE	D16AM	Z17AK	S34AH	D38AE	Z18AC	H37CR
4	X19CL	V23CJ	P23BP	U20BL	C33BH	P35BF	J24BC	C22AL	P26AJ	J11AF	C12AD	S27CS
5	B30CM	E36CK	W36CO	R28BM	K15BJ	W16BG	U21BD	K29AM	W38AK	U31AG	K37AE	P13AB
6	N14CP	M32CL	Y32CE	X33BP	F25BK	Y24BH	R17BE	F34BC	Y11AL	R18AH	F27AF	W19AC
7	A23CR	H20CM	T20CF	B15CD	V35BL	T21BJ	X22BF	V26BD	T31AM	X12AJ	V13AG	Y30AD
8	Z36CS	S28CP	G28CH	N25CE	E16BM	G17BK	B29BG	E38BC	G18BC	B37AK	E19AH	T14AE
9	C32AB	P33CR	L33CJ	A35CF	M24BP	L22BL	N34BH	M11BF	L12BD	N27AL	M30AJ	G23AF
10	K20AC	W15CS	D15CK	Z16CH	H21CD	D29BM	A26BJ	H31BG	D37BE	A13AM	H14AK	L36AG
11	F28AD	Y25AB	J25CL	C24CJ	S17CE	J34BP	Z38BK	S18BH	J27BF	Z19BC	S23AL	D32AH
12	V33AE	T35AC	U35CM	K21CK	P22CF	U26CD	C11BL	P12BJ	U13BG	C30BD	P36AM	J20AJ
13	E15AF	G16AD	R16CP	F17CL	W29CH	R38CE	K31BM	W37BK	R19BH	K14BE	W32BC	U28AK
14	M25AG	L24AE	X24CR	V22CM	Y34CJ	X11CF	F18BP	Y27BL	X30BJ	F23BF	Y20BD	R33AL
15	H35AH	D21AF	B21CS	E29CP	T26CK	B31CH	V12CD	T13BM	B14BK	V36BG	T28BE	X15AM
16	S16AJ	J17AG	N17AB	M34CR	G38CL	N18CJ	E37CE	G19BP	N23BL	E32BH	G33BF	B25BC
17	P24AK	U22AM	A22AC	H26CS	L11CM	A12CK	M27CF	L30CD	A36BM	M20BJ	L15BG	N35BD
18	W21AL	R29AJ	Z29AD	S38AB	D31CP	Z37CL	H13CH	D14CE	Z32BP	H28BK	D25BH	A16BE
19	Y17AM	X34AK	C34AE	P11AC	J18CR	C27CM	S19CJ	J23CF	C20CD	S33BL	J35BJ	Z24BF
20	T22BC	B26AL	K26AF	W31AD	U12CS	K13CP	P30CK	U36CH	K28CE	P15BM	U16BK	C21BG
21	G29BD	N38AM	F38AG	Y18AE	R37AB	F19CR	W14CL	R32CJ	F33CF	W25BP	R24BL	K17BH
22	L34BE	A11BC	V11AH	T12AF	X27AC	V30CS	Y23CH	X20CK	V15CH	Y35CD	X21BM	F22BJ
23	D26BF	Z31BD	E31AJ	G37AG	B13AD	E14AB	T36CP	B28CL	E25CJ	T16CE	B17BP	V29BK
24	J38BG	C18BE	M18AK	L27AH	N19AE	M23AC	G32CR	N33CH	M35CK	G24CF	N22CD	E34BL
25	U11BH	K12BF	H12AL	D13AJ	A30AF	H36AD	L20CS	A15CP	H16CL	L21CH	A29CE	M26BM
26	R31BJ	F37BG	S37AM	J19AK	Z14AG	S32AE	D28AB	Z25CR	S24CM	D17CJ	Z34CF	H38BP
27	X18BK	V27BH	P27BC	U30AL	C23AH	P20AF	J33AC	C35CS	P21CP	J22CK	C26CH	S11CD
28	B12BL	E13BJ	W13BD	R14AM	K36AJ	W28AG	U15AD	K16AB	W17CR	U29CL	K38CJ	P31CE
29	N37BM		Y19BE	X23BC	F32AK	Y33AH	R25AE	F24AC	Y22CS	R34CM	F11CK	W18CF
30	A27BP		T30BF	B36BD	V20AL	T15AJ	X35AF	V21AD	T29AB	X26CP	V31CL	Y12CH
31	Z13CD		G14BG		E28AM		B16AG	E17AE		B38CR		T37CJ

1850

	JAN	FEB	MAR	APR	MAY	JUN	JUL	AUG	SEP	OCT	NOV	DEC
1	G27CK	N30CM	F30BM	Y36BK	R20BG	F15BE	W35AM	R21AK	F29AH	W26AE	R31AC	K12CR
2	L13CL	A14CJ	V14BP	T32BL	X28BH	V25BF	Y16BC	X17AL	V34AJ	Y38AF	X18AD	F37CS
3	D19CM	Z23CK	E23CD	G20BM	B33BJ	E35BG	T24BD	B22AM	E26AK	T11AG	B12AE	V27AB
4	·J30CP	C36CL	M36CE	L28BP	N15BK	M16BH	G21BE	N29BC	M38AL	G31AH	N37AF	E13AC
5	U14CR	K32CM	H32CF	D33CD	A25BL	H24BJ	L17BF	A34BD	H11AM	L18AJ	A27AK	M19AD
6	R23CS	F20CP	S20CH	J15CE	Z35BM	S21BK	D22BG	Z26BE	S31BC	D12AK	Z13AH	H30AE
7	X36AB	V28CR	P28CJ	U25CF	C16BP	P17BL	J29BH	C38BF	P18BD	J37AL	C19AJ	S14AF
8	B32AC	E33CS	W33CK	R35CH	K24CD	W22BM	U34BJ	K11BG	W12BE	U27AH	K30AK	P23AG
9	N20AD	M15AB	Y15CL	X16CJ	F21CE	Y29BP	R26BK	F31BH	Y37BF	R13BC	F14AL	W36AH
10	A28AE	H25AC	T25CM	B24CK	V17CF	T34CD	X38BL	V18BJ	T27BG	X19BD	V23AM	Y32AJ
11	Z33AF	S35AD	G35CP	N21CL	E22CH	G26CE	B11BM	E12BK	G13BH	B30BE	E36BC	T20AK
12	C15AG	P16AE	L16CR	A17CH	M29CJ	L38CF	N31BP	M37BL	L19BJ	N14BF	M32BD	G28AL
13	K25AH	N24AC	D24CS	Z22CJ	H34CK	D11CH	A18CD	H27BM	D30BH	A23BG	H20BE	L33AM
14	F35AJ	Y21AJ	J21AB	C29CR	S26CL	J31CJ	Z12CE	S13BP	J14BL	Z36BH	S28BF	D15BC
15	V16AK	T17AH	U17AC	K34CS	P38CM	U18CK	C37CF	P19CD	U23BM	C32BJ	P33BG	J25BD
16	E24AL	G22AJ	R22AD	F26AB	W11CP	R12CL	K27CH	W30CE	R36BP	K20BK	W15BH	U35BE
17	M21AM	L29AK	X29AE	V38AC	Y31CR	X37CM	F13CJ	Y14CF	X32CD	F28BL	Y25BJ	R16BF
18	H17BC	D34AL	B34AF	E11AD	T18CS	B27CP	V19CK	T23CH	B20CE	V33BM	T35BH	X24BG
19	S22BD	J26AM	N26AG	M31AE	G12AB	N13CR	E30CL	G36CJ	N28CF	E15BP	G16BL	B21BH
20	P29BE	U38BC	A38AH	H18AF	L37AC	A19CS	M14CR	L32CK	A33CH	M25CD	L24BP	N17BJ
21	W34BF	R11BD	Z11AJ	S12AG	D27AD	Z30AB	H23CP	D20CL	Z15CJ	H35CE	D21BP	A22BK
22	Y26BG	X31BE	C31AK	P37AH	J13AE	C14AC	S36CR	J28CM	U33CP	S16CF	J17CD	Z29BL
23	T38BH	B18BF	K18AL	W27AJ	U19AF	K23AD	P32CS	U33CP	K35CL	P24CH	U22CE	C34BM
24	G11BJ	N12BG	F12AM	Y13AK	R30AG	F36AE	W20AB	R15CR	F16CM	W21CJ	R29CF	K26BP
25	L31BK	A37BH	V37BC	T19AL	X14AH	V32AF	Y28AC	X25CS	V24CP	Y17CK	X34CH	F38CD
26	D18BL	Z27BJ	E27BD	G30AM	B23AJ	E20AG	T33AD	B35AB	E21CR	T22CL	B26CJ	V11CE
27	J12BM	C13BK	M13BE	L14BC	N36AK	M28AH	G15AE	N16AC	M17CS	G29CH	N38CK	E31CF
28	U37BP	K19BL	H19BF	D23BD	A32AL	H33AJ	L25AF	A24AD	H22AB	L34CP	A11CL	M18CH
29	R27CD		S30BG	J36BE	Z20AM	S15AK	D35AG	Z21AE	S29AC	D26CR	Z31CH	H12CJ
30	X13CE		P14BH	U32BF	C28BC	P25AL	J16AH	C17AF	P34AD	J38CS	C18CP	S37CK
31	B19CF		W23BJ		K33BD		U24AJ	K22AG		U11AB		P27CL

(118)

FIRST LETTER = YOUR PHYSICAL CODE, MIDDLE NUMBER = SENSITIVITY CODE, LAST TWO LETTERS = INTELLECTUAL CODE

1851

	JAN	FEB	MAR	APR	MAY	JUN	JUL	AUG	SEP	OCT	NOV	DEC
1	W13CM	R14CK	Z14CD	S32BM	D28BJ	Z25BG	H16BD	D17AM	Z34AK	H38AG	D18AE	A37AB
2	Y19CP	X23CL	C23CE	P20BP	J33BK	C35BH	S24BE	J22BC	C26AL	S11AH	J12AF	Z27AC
3	T30CR	B36CM	K36CF	W28CD	U15BL	K16BJ	P21BF	U29BD	K38AM	P31AJ	U37AG	C13AD
4	G14CS	N32CP	F32CH	Y33CE	R25BM	F24BK	W17BG	R34BE	F11BC	W18AK	R27AH	K19AE
5	L23AB	A20CR	V20CJ	T15CF	X35BP	V21BL	Y22BH	X26BF	V31BD	Y12AL	X13AJ	F30AF
6	D36AC	Z28CS	E28CK	G25CH	B16CD	E17BM	T29BJ	B38BG	E18BE	T37AM	B19AK	V14AG
7	J32AD	C33AB	M33CL	L35CJ	N24CE	M22BP	G34BK	N11BH	M12BF	G27BC	N30AL	E23AH
8	U20AE	K15AC	H15CM	D16CK	A21CF	H29CD	L26BL	A31BJ	H37BG	L13BD	A14AM	M36AJ
9	R28AF	F25AD	S25CP	J24CL	Z17CH	S34CE	D38BM	Z18BK	S27BH	D19BE	Z23BC	H32AK
10	X33AG	V35AE	P35CR	U21CM	C22CJ	P26CF	J11BP	C12BL	P13BJ	J30BF	C36BD	S20AL
11	B15AH	E16AF	W16CS	R17CP	K29CK	W38CH	U31CD	K37BH	W19BK	U14BG	K32BE	P28AM
12	N25AJ	M24AG	Y24AB	X22CR	F34CL	Y11CJ	R18CE	F27BP	Y30BL	R23BH	F20BF	W33BC
13	A35AK	H21AH	T21AC	B29CS	V26CH	T31CK	X12CF	V13CD	T14BM	X36BJ	V28BG	Y15BD
14	Z16AL	S17AJ	G17AD	N34AB	E38CP	G18CL	B37CH	E19CE	G23BP	B32BK	E33BH	T25BE
15	C24AM	P22AK	L22AE	A26AC	M11CR	L12CH	N27CJ	M30CF	L36CD	N20BL	M15BJ	G35BF
16	K21BC	W29AL	D29AF	Z38AD	H31CS	D37CP	A13CK	H14CH	D32CE	A28BM	H25BK	L16BG
17	F17BD	Y34AM	J34AG	C11AE	S18AB	J27CR	Z19CL	S23CJ	J20CF	Z33BP	S35BL	D24BH
18	V22BE	T26BC	U26AH	K31AF	P12AC	U13CS	C30CM	P36CK	U28CH	C15CD	P16BM	J21BJ
19	E29BF	G38BD	R38AJ	F18AG	W37AD	R19AB	K14CP	W32CL	R33CJ	K25CE	W24BP	U17BK
20	M34BG	L11BE	X11AK	V12AH	Y27AE	X30AC	F23CR	Y20CM	X15CK	F35CF	Y21CD	R22BL
21	H26BH	D31BF	B31AL	E37AJ	T13AF	B14AD	V36CS	T28CP	B25CL	V16CH	T17CE	X29BM
22	S38BJ	J18BG	N18AM	M27AK	G19AG	N23AE	E32AB	G33CR	N35CM	E24CJ	G22CF	B34BP
23	P11BK	U12BH	A12BC	H13AL	L30AH	A36AF	M20AC	L15CS	A16CP	M21CK	L29CH	N26CD
24	W31BL	R37BJ	Z37BD	S19AM	D14AJ	Z32AG	H28AD	D25AB	Z24CR	H17CL	D34CJ	A38CE
25	Y18BM	X27BK	C27BE	P30BC	J23AK	C20AH	S33AE	J35AC	C21CS	S22CM	J26CK	Z11CF
26	T12BP	B13BL	K13BF	W14BD	U36AL	K28AJ	P15AF	U16AD	K17AB	P29CP	U38CL	C31CH
27	G37CD	N19BM	F19BG	Y23BE	R32AM	F33AK	W25AG	R24AE	F22AC	W34CR	R11CM	K18CJ
28	L27CE	A30BP	V30BH	T36BF	X20BC	V15AL	Y35AM	X21AF	V29AD	Y26CS	X31CP	F12CK
29	D13CF		E14BJ	G32BG	B28BD	E25AM	T16AJ	B17AG	E34AE	T38AB	B18CR	V37CL
30	J19CH		M23BK	L20BH	N33BE	M35BC	G24AK	N22AH	M26AF	G11AC	N12CS	E27CM
31	U30CJ		H36BL		A15BF		L21AL	A29AJ		L31AD		M13CP

1852

	JAN	FEB	MAR	APR	MAY	JUN	JUL	AUG	SEP	OCT	NOV	DEC
1	H19CR	D23CM	N36CH	M28CE	G15BM	N16BK	E21BG	G29BE	N38BC	E31AK	G37AH	B13AE
2	S30CS	J36CP	A32CJ	H33CF	L25BP	A24BL	M17BH	L34BF	A11BD	M18AL	L27AJ	N19AF
3	P14AB	U32CR	Z20CK	S15CH	D35CD	Z21BM	H22BJ	D26BG	Z31BE	H12AM	D13AK	A30AG
4	W23AC	R20CS	C28CL	P25CJ	J16CE	C17BP	S29BK	J38BH	C18BF	S37BC	J19AL	Z14AH
5	Y36AD	X28AB	K33CH	W35CK	U24CF	K22CD	P34BL	U11BJ	K12BG	P27BD	U30AM	C23AJ
6	T32AE	B33AC	F15CP	Y16CL	R21CH	F29CE	W26BH	R31BK	F37BH	W13BE	R14BC	K36AK
7	G20AF	N15AD	V25CR	T24CM	X17CJ	V34CF	Y38BP	X18BL	V27BJ	Y19BF	X23BD	F32AL
8	L28AG	A25AE	E35CS	G21CP	B22CK	E26CH	T11CD	B12BM	E13BK	T30BG	B36BE	V20AM
9	D33AH	Z35AF	M16AB	L17CR	N29CL	M38CJ	G31CE	N37BP	M19BL	G14BH	N32BF	E28BC
10	J15AJ	C16AG	H24AC	D22CS	A34CM	H11CK	L18CF	A27CD	H30BM	L23BJ	A20BG	M33BD
11	U25AK	K24AH	S21AD	J29AB	Z26CP	S31CL	D12CH	Z13CE	S14BP	D36BK	Z28BH	H15BE
12	R35AL	F21AJ	P17AE	U34AC	C38CR	P18CH	J37CJ	C19CF	P23CD	J32BL	C33BJ	S25BF
13	X16AM	V17AK	W22AF	R26AD	K11CS	W12CP	U27CK	K30CH	W36CE	U20BM	K15BK	P35BG
14	B24BC	E22AL	Y29AG	X38AE	F31AB	Y37CR	R13CL	F14CJ	Y32CF	R28BP	F25BL	W16BH
15	N21BD	M29AM	T34AH	B11AF	V18AC	T27CS	X19CM	V23CK	T20CH	X33CD	V35BM	Y24BJ
16	A17BE	H34BC	G26AJ	N31AG	E12AD	G13AB	B30CP	E36CL	G28CJ	B15CE	E16BP	T21BK
17	Z22BF	S26BD	L38AK	A18AH	M37AE	L19AC	N14CR	M32CH	L33CK	N25CF	M24CD	G17BL
18	C29BG	P38BE	D11AL	Z12AJ	H27AF	D30AD	A23CS	H20CP	D15CL	A35CH	H21CE	L22BM
19	K34BH	W11BF	J31AM	C37AK	S13AG	J14AE	Z36AB	S28CR	J25CM	Z16CJ	S17CF	D29BP
20	F26BJ	Y31BG	U18BC	K27AL	P19AH	U23AF	C32AC	P33CS	U35CP	C24CK	P22CH	J34CD
21	V38BK	T18BH	R12BD	F13AM	W30AJ	R36AG	K20AD	W15AB	R16CR	K21CL	W29CJ	U26CE
22	E11BL	G12BJ	X37BE	V19BC	Y14AK	X32AH	F28AE	Y25AC	X24CS	F17CM	Y34CK	R38CF
23	M31BM	L37BK	B27BF	E38BD	T23AL	B20AJ	V33AF	T35AD	B21AB	V22CP	T26CL	X11CH
24	H18BP	D27BL	N13BG	M14BE	G36AM	N28AK	E15AG	G16AE	N17AC	E29CR	G38CM	B31CJ
25	S12CD	J13BM	Z30BJ	H23BF	L32BC	A33AL	H25AH	L24AF	A22AD	M34CS	L11CP	N18CK
26	P37CE	U19BP	C14BH	S36BG	D20BD	Z15AM	H35AJ	D21AG	Z29AE	H26AB	D31CR	A12CL
27	W27CF	R30CD	C14BH	P32BH	J28BE	C25BC	S16AK	J17AH	C34AF	S38AC	J18CS	Z37CH
28	Y13CH	X14CE	K23BL	W20BJ	U33BF	K35BD	P24AL	U22AJ	K26AG	P11AD	U12AB	C27CP
29	T19CJ	B23CF	F36BM	Y28BK	R15BG	F16BE	W21AM	R29AK	F38AH	W31AE	R37AC	K13CR
30	G30CK		V32BP	T33BL	X25BH	V24BF	Y17BC	X34AL	V11AJ	Y18AF	X27AD	F19CS
31	L14CL		E20CD		B35BJ		T22BD	B26AM		T12AG		V30AB

1853

	JAN	FEB	MAR	APR	MAY	JUN	JUL	AUG	SEP	OCT	NOV	DEC
1	E14AC	G32CS	R32CK	F33CH	W25CD	R24BM	K17BJ	W34BG	R11BE	K18AM	W27AK	U19AG
2	M23AD	L20AB	X20CL	V15CJ	Y35CE	X21BP	F22BK	Y26BH	X31BF	F12BC	Y13AL	R30AH
3	H36AE	D28AC	B28CM	E25CK	T16CF	B17CD	V29BL	T38BJ	B18BG	V37BD	T19AM	X14AJ
4	S32AF	J33AD	N33CP	M35CL	G24CH	N22CE	E34BH	G11BK	N12BH	E27BE	G30BC	B23AK
5	P20AG	U15AE	A15CR	H16CH	L21CJ	A29CF	M26BP	L31BL	A37BJ	M13BF	L14BD	N36AL
6	W28AH	R25AF	Z25CS	S24CP	D17CK	Z34CH	H38CD	D18BM	Z27BK	H19BG	D23BE	A32AM
7	Y33AJ	X35AG	C35AB	P21CR	J22CL	C26CJ	S11CE	J12BP	C13BL	S30BH	J36BF	Z20BC
8	T15AK	B16AH	K16AC	W17CS	U29CH	K38CK	P31CF	U37CD	K19BM	P14BJ	U32BG	C28BD
9	G25AL	N24AJ	F24AD	Y22AB	R34CP	F11CL	W18CH	R27CE	F30BP	W23BK	R20BH	K33BE
10	L35AM	A21AK	V21AE	T29AC	X26CS	V31CM	Y12CJ	X13CF	V14CD	Y36BL	X28BJ	F15BF
11	D16BC	Z17AL	E17AF	G34AD	B38CS	E18CP	T37CK	B19CH	E23CE	T32BM	B33BK	V25BG
12	J24BD	C22AM	M22AG	L26AE	N11AB	M12CR	G27CL	N30CJ	M36CF	G20BP	N15BL	E35BH
13	U21BE	K29BC	H29AH	D38AF	A31AC	H37CS	L13CH	A14CK	H32CH	L28CD	A25BM	M16BJ
14	R17BF	F34BD	S34AJ	J11AG	Z18AD	S27AB	D19CP	Z23CL	S20CJ	D33CE	Z35BP	H24BK
15	X22BG	V26BE	P26AK	U31AH	C12AE	P13AC	J30CR	C36CM	P28CK	J15CF	C16CD	S21BL
16	B29BH	E38BF	W38AL	R18AJ	K37AF	W19AD	U14CS	K32CP	W33CL	U25CH	K24CE	P17BM
17	N34BJ	M11BG	Y11AM	X12AK	F27AG	Y30AE	R23AB	F20CR	Y15CM	R35CJ	F21CF	W22BP
18	A26BK	H31BH	T31BC	B37AL	V13AH	T14AF	X36AC	V28CS	T25CP	X16CK	V17CH	Y29CD
19	Z38BL	S18BJ	G18BD	N27AM	E19AJ	G23AG	B32AD	E33AB	G35CR	B24CL	E22CJ	T34CE
20	C11BM	P12BK	L12BE	A13BC	M30AK	L36AH	N20AE	M15AC	L16CS	N21CH	M29CK	G26CF
21	K31BP	W37BL	D37BF	Z19BD	H14AL	D32AJ	A28AF	H25AD	D24AB	A17CP	H34CL	L38CH
22	F18CD	Y27BM	J27BG	C30BE	S23AM	J20AK	Z33AG	S35AE	J21AC	Z22CS	S26CH	D11CJ
23	V12CE	T13BP	U13BH	K14BF	P36BC	U28AL	C15AH	P16AF	U17AD	C29CS	P38CP	J31CK
24	E37CF	G19CD	R19BJ	F23BG	W32BD	R33AM	K25AJ	W24AG	R22AE	K34AB	W11CR	U18CL
25	M27CH	L30CE	X30BK	V36BH	Y20BE	X15BC	F35AK	Y21AH	X29AF	F26AC	Y31CS	R12CM
26	H13CJ	D14CF	B14BL	E32BJ	T28BF	B25BD	V16AL	T17AJ	B34AG	V38AD	T18AB	X37CP
27	S19CK	J23CH	N23BH	M20BK	G33BG	N35BE	E24AM	G22AK	N26AH	E11AE	G12AC	B27CR
28	P30CL	U36CJ	A36BH	H28BL	L15BH	A16BF	H21AB	L29AL	A38AJ	H31AF	L37AD	N13CS
29	W14CM		Z32CD	S33BM	D25BJ	Z24BG	H17BD	D34AM	Z11AK	H18AG	D27AE	A19AB
30	Y23CP		C20CE	P15BP	J35BK	C21BH	S22BE	J26BC	C31AL	S12AH	J13AF	Z30AC
31	T36CR		K28CF		U16BL		P29BF	U38BD		P37AJ		C14AD

FIRST LETTER = YOUR PHYSICAL CODE, MIDDLE NUMBER = SENSITIVITY CODE, LAST TWO LETTERS = INTELLECTUAL CODE

1854	JAN	FEB	MAR	APR	MAY	JUN	JUL	AUG	SEP	OCT	NOV	DEC
1	K23AE	W20AC	D20CM	Z15CK	H35CF	D21CD	A22BL	H26BJ	D31BG	A12BD	H13AM	L30AJ
2	F36AF	Y28AD	J28CP	C25CL	S16CH	J17CE	Z29BM	S38BK	J18BH	Z37BE	S19BC	D14AK
3	V32AG	T33AE	U33CR	K35CM	P24CJ	U22CF	C34BP	P11BL	U12BJ	C27BF	P30BD	J23AL
4	E20AH	G15AF	R15CS	F16CP	W21CK	R29CH	K26CD	W31BM	R37BK	K13BG	W14BE	U36AM
5	M28AJ	L25AG	X25AB	V24CR	Y17CL	X34CJ	F38CE	Y18BP	X27BL	F19BH	Y23BF	R32BC
6	H33AK	D35AH	B35AC	E21CS	T22CM	B26CK	V11CF	T12CD	B13BM	V30BJ	T36BG	X20BD
7	S15AL	J16AJ	N16AD	M17AB	G29CP	N38CL	E31CH	G37CE	N19BP	E14BK	G32BH	B28BE
8	P25AM	U24AK	A24AE	H22AC	L34CR	A11CM	M18CJ	L27CF	A30CD	M23BL	L20BJ	N33BF
9	W35BC	R21AL	Z21AF	S29AD	D26CS	Z31CP	H12CK	D13CH	Z14CE	H36BM	D28BK	A15BG
10	Y16BD	X17AM	C17AG	P34AE	J38AB	C18CR	S37CL	J19CJ	C23CF	S32BP	J33BL	Z25BH
11	T24BE	B22BC	K22AH	W26AF	U11AC	K12CS	P27CM	U30CK	K36CH	P20CD	U15BM	C35BJ
12	G21BF	N29BD	F29AJ	Y38AG	R31AD	F37AB	W13CP	R14CL	F32CJ	W28CE	R25BP	K16BK
13	L17BG	A34BE	V34AK	T11AH	X18AE	V27AC	Y19CR	X23CH	V20CK	Y33CF	X35CD	F24BL
14	D22BH	Z26BF	E26AL	G31AJ	B12AF	E13AD	T30CS	B36CP	E28CL	T15CH	B16CE	V21BM
15	J29BJ	C38BG	M38AM	L18AK	N37AG	M19AE	G14AB	N32CR	M33CM	G25CJ	N24CF	E17BP
16	U34BK	K11BH	H11BC	D12AL	A27AH	H30AF	L23AC	A20CS	H15CP	L35CK	A21CH	M22CD
17	R26BL	F31BJ	S31BD	J37AM	Z13AJ	S14AG	D36AD	Z28AB	S25CR	D16CL	Z17CJ	H29CE
18	X38BM	V18BK	P18BE	U27BC	C19AK	P23AH	J32AE	C33AC	P35CS	J24CM	C22CK	S34CF
19	B11BP	E12BL	M12BF	R13BD	K30AL	W36AJ	U20AF	K15AD	W16AB	U21CP	K29CL	P26CH
20	N31CD	M37BM	Y37BG	X19BE	F14AM	Y32AK	R28AG	F25AE	Y24AC	R17CR	F34CM	W38CJ
21	A18CE	H27BP	T27BH	B30BF	V23BC	T20AL	X33AH	V35AF	T21AD	X22CS	V26CP	Y11CK
22	Z12CF	S13CD	G13BJ	N14BG	E36BD	G28AM	B15AJ	E16AG	G17AE	B29AB	E38CR	T31CL
23	C37CH	P19CE	L19BK	A23BH	M32BE	L33BC	N25AK	M24AH	L22AF	N34AC	M11CS	G18CM
24	K27CJ	W30CF	D30BL	Z36BJ	H20BF	D15BD	A35AL	H21AJ	D29AG	A26AD	H31AB	L12CP
25	F13CK	Y14CH	J14BM	C32BK	S28BG	J25BE	Z16AM	S17AK	J34AH	Z38AE	S18AC	D37CR
26	V19CL	T23CJ	U23BP	K20BL	P33BH	U35BF	C24BC	P22AL	U26AJ	C11AF	P12AD	J27CS
27	E30CM	G36CK	R36CD	F26BM	W15BJ	R16BG	K21BD	W29AM	R38AK	K31AG	W37AE	U13AB
28	M14CP	L32CL	X32CE	V33BP	Y25BK	X24BH	F17BE	Y34BC	X11AL	F18AH	Y27AF	R19AC
29	H23CR		B20CF	E15CD	T35BL	B21BJ	V22BF	T26BD	B31AM	V12AJ	T13AG	X30AD
30	S36CS		N28CH	M25CE	G16BM	N17BK	E29BG	G38BE	N18BC	E37AK	G19AH	B14AE
31	P32AB		A33CJ		L24BP		M34BH	L11BF		M27AL		N23AF

1855	JAN	FEB	MAR	APR	MAY	JUN	JUL	AUG	SEP	OCT	NOV	DEC
1	A36AG	H28AE	T28CR	B25CM	V16CJ	T17CF	X29BP	V38BL	T18BJ	X37BF	V19BD	Y14AL
2	Z32AH	S33AF	G33CS	N35CP	E24CK	G22CH	B34CD	F11BM	G12BK	B27BG	E30BE	T23AM
3	C20AJ	P15AG	L15AB	A16CR	M21CL	L29CJ	N26CE	M31BP	L37BL	N13BH	M14BF	G36BC
4	K28AK	W25AH	D25AC	Z24CS	H17CM	D34CK	A38CE	H18CD	D27BM	A19BJ	H23BG	L32BD
5	F33AL	Y35AJ	J35AD	C21AB	S22CP	J26CL	Z11CH	S12CE	J13BP	Z30BK	S36BH	D20BE
6	V15AM	T16AK	U16AE	K17AC	P29CR	U38CM	C31CJ	P37CF	U19CD	C14BL	P32BJ	J28BF
7	E25BC	G24AL	R24AF	F22AD	W34CS	R11CP	K18CK	W27CH	R30CE	K23BM	W20BK	U33BG
8	M35BD	L21AM	X21AG	V29AE	Y26AB	X31CR	F12CL	Y13CJ	X14CF	F36BP	Y28BL	R15BH
9	H16BE	D17BC	B17AH	E34AF	T38AC	B18CS	V37CM	T19CK	B23CH	V32CD	T33BM	X25BJ
10	S24BF	J22BD	N22AJ	M26AG	G11AD	N12AB	E27CP	G30CL	N36CJ	E20CE	G15BP	B35BK
11	P21BG	U29BE	A29AK	H38AH	L31AE	A37AC	M13CR	L14CM	A32CK	M28CF	L25CD	N16BL
12	W17BH	R34BF	Z34AL	S11AJ	D18AF	Z27AD	H19CS	D23CP	Z20CL	H33CH	D35CE	A24BM
13	Y22BJ	X26BG	C26AM	P31AK	J12AG	C13AE	S30AB	J36CR	C28CM	S15CJ	J16CF	Z21BP
14	T29BK	B38BH	K38BC	W18AL	U37AH	K19AF	P14AC	U32CS	K33CP	P25CK	U24CH	C17CD
15	G34BL	N11BJ	F11BD	Y12AM	R27AJ	F30AG	W23AD	R20AB	F15CR	W35CL	R21CJ	K22CE
16	L26BM	A31BK	V31BE	T37BC	X13AK	V14AH	Y36AE	X28AC	V25CS	Y16CM	X17CK	F29CF
17	D38BP	Z18BL	E18BF	G27BD	B19AL	E23AJ	T32AF	B33AD	E35AB	T24CP	B22CL	V34CH
18	J11CD	C12BM	M12BG	L13BE	N30AM	M36AK	G20AG	N15AE	M16AC	G21CR	N29CM	E26CJ
19	U31CE	K37BP	H37BH	D19BF	A14BC	H32AL	L28AH	A25AF	H24AD	L17CS	A34CP	H38CK
20	R18CF	F27CD	S27BJ	J30BG	Z23BD	S20AM	D33AJ	Z35AG	S21AE	D22AB	Z26CR	H11CL
21	X12CH	V13CE	P13BK	U14BH	C36BE	P28BC	J15AK	C16AH	P17AF	J29AC	C38CS	S31CM
22	B37CJ	E19CF	W19BL	R23BJ	K32BF	W33BD	U25AL	K24AJ	W22AG	U34AD	K11AB	P18CP
23	N27CK	M30CH	Y30BM	X36BK	F20BG	Y15BE	R35AM	F21AK	Y29AH	R26AE	F31AC	W12CR
24	A13CL	H14CJ	T14BP	B32BL	V28BH	T25BF	X16BC	V17AL	T34AJ	X38AF	V18AD	Y37CS
25	Z19CM	S23CM	G23CD	N20BM	E33BJ	G35BG	B24BD	E22AM	G26AK	B11AG	E12AE	T27AB
26	C30CP	P36CL	L36CE	A28BP	M15BK	L16BH	N21BE	M29BC	L38AL	N31AH	M37AF	G13AC
27	K14CR	W32CM	D32CF	Z33CD	H25BL	D24BJ	A17BF	H34BD	D11AM	A18AJ	H27AG	L19AD
28	F23CS	Y20CP	J20CH	C15CE	S35BM	J21BK	Z22BG	S26BE	J31BC	Z12AK	S13AH	D30AE
29	V36AB		U28CJ	K25CF	P16BP	U17BL	C29BM	P38BF	U18BD	C37AL	P19AJ	J14AF
30	E32AC		R33CK	F35CH	W24CD	R22BM	K34BJ	W11BG	R12BE	K27AM	W30AK	U23AG
31	M20AD		X15CL		Y21CE		F26BK	Y31BH		F13BC		R36AH

1856	JAN	FEB	MAR	APR	MAY	JUN	JUL	AUG	SEP	OCT	NOV	DEC
1	X32AJ	V33AG	W15AC	R16CS	K21CM	W29CK	U26CF	K31CD	W37BM	U13BJ	K14BG	P36BD
2	B20AK	E15AH	Y25AD	X24AB	F17CP	Y34CL	R38CH	F18CE	Y27BP	R19BK	F23BH	W32BE
3	N28AL	M25AJ	T35AE	B21AC	V22CR	T26CM	X11CJ	V12CF	T13CD	X30BL	V36BJ	Y20BF
4	A33AM	H35AK	G16AF	N17AD	E29CS	G38CP	B31CK	E37CH	G19CE	B14BM	E32BK	T28BG
5	Z15BC	S16AL	L24AG	A22AE	M34AB	L11CR	N18CL	M27CJ	L30CF	N23BP	M20BL	G33BH
6	C25BD	P24AM	D21AH	Z29AF	H26AC	D31CS	A12CM	H13CK	D14CH	A36CD	H28BM	L15BJ
7	K35BE	W21BC	J17AJ	C34AG	S38AD	J18AB	Z37CP	S19CL	J23CJ	Z32CE	S33BP	D25BK
8	F16BF	Y17BD	U22AK	K26AH	P11AE	U12AC	C27CR	P30CM	U36CK	C20CF	P15CD	J35BL
9	V24BG	T22BE	R29AL	F38AJ	W31AF	R37AD	K13CS	W14CP	R32CL	K28CH	W25CE	U16BM
10	E21BH	G29BF	X34AM	V11AK	Y18AG	X27AE	F19AB	Y23CR	X20CM	F33CJ	Y35CF	R24BP
11	M17BJ	L34BG	E31AL	B26BC	T12AH	B13AF	V30AC	T36CS	B28CP	V15CK	T16CH	X21CD
12	H22BK	D26BH	N38BD	M18AM	G37AJ	N19AG	E14AD	G32AB	N33CR	E25CL	G24CJ	B17CE
13	S29BL	J38BJ	A11BE	L27AK	A30AH	M23AE	L20AC	A15CS	M35CH	L21CK	N22CC	N23CF
14	P34BM	U11BK	Z31BF	S37BD	D13AL	Z14AJ	H36AF	D28AD	Z25AB	H16CP	D17CL	A29CH
15	W26BP	R31BL	C18BG	P27BE	J19AM	C23AK	S32AG	J33AE	C35AC	S24CR	J22CM	Z34CJ
16	Y38CD	X18BM	K12BH	W13BF	U30BC	K36AL	P20AH	U15AF	K16AD	P21CS	U29CP	C26CK
17	T11CE	B12BP	F37BJ	Y19BG	R14BD	F32AM	W28AK	R25AG	F24AE	W17AB	R34CR	K38CL
18	G31CF	N37CD	V27BK	T30BH	X23BE	V20BC	Y33AK	X35AH	V21AF	Y22AC	X26CS	F11CM
19	L18CH	A27CE	E13BL	G14BJ	B36BF	E28BD	T15AL	B16AJ	E17AG	T29AD	B38AB	V31CP
20	D12CJ	Z13CF	M19BM	L23BK	N32BG	M33BE	G25AM	N24AK	M22AH	G34AE	N11AC	E18CR
21	J37CK	C19CH	H30BP	D36BL	A20BH	H15BF	L35BC	A21AL	H29AJ	L26AF	A31AD	M12CS
22	U27CL	K30CJ	S14CD	J32BM	Z28BJ	S25BG	D16BD	Z17AM	S34AK	D38AG	Z18AE	H37AB
23	R13CM	F14CK	P23CE	U20BP	C33BK	P35BH	J24BE	C22BC	P26AL	J11AH	C12AF	S27AC
24	X19CP	V23CL	W36CF	R28CD	K15BL	W16BJ	U21BF	K29BD	W38AM	U31AJ	K37AG	P13AD
25	B30CR	E36CM	Y32CH	X33CE	F25BM	Y24BK	R17BG	F34BE	Y11BC	R18AK	F27AH	W19AE
26	N14CS	M32CP	T20CJ	B15CF	V35BP	T21BL	X22BH	V26BF	T31BD	X12AL	V13AJ	Y30AF
27	A23AB	H20CR	G28CK	N25CH	E16CD	G17BM	B29BJ	E38BG	G18BE	B37AM	E19AK	T14AG
28	Z36AC	S28CS	L33CL	A35CJ	M24CE	L22BP	N34BK	M11BH	L28AD	N27BC	M30AL	G23AH
29	C32AD	P33AB	D15CM	Z16CK	H21CF	D29CD	A26BL	H31BJ	D37BG	A13BD	H14AM	L36AJ
30	K20AE		J25CP	C24CL	S17CH	J34CE	Z38BM	S18BK	J27BH	Z19BE	S23BC	D32AK
31	F28AF		U35CR		P22CJ		C11BP	P12BL		C30BF		J20AL

FIRST LETTER = YOUR PHYSICAL CODE, MIDDLE NUMBER = SENSITIVITY CODE, LAST TWO LETTERS = INTELLECTUAL CODE

1857	JAN	FEB	MAR	APR	MAY	JUN	JUL	AUG	SEP	OCT	NOV	DEC
1	U28AM	K25AK	H25AE	D24AC	A17CR	H34CM	L38CJ	A18CF	H27CD	L19BL	A23BJ	M32BF
2	R33BC	F35AL	S35AF	J21AD	Z22CS	S26CP	D11CK	Z12CH	S13CE	D30BM	Z36BK	H20BG
3	X15BD	V16AM	P16AG	U17AE	C29AB	P38CR	J31CL	C37CJ	P19CF	J14BP	C32BL	S28BH
4	B25BE	E24BC	W24AH	R22AF	K34AC	W11CS	U18CM	K27CK	W30CH	U23CD	K20BM	P33BJ
5	N35BF	M21BD	Y21AJ	X29AG	F26AD	Y31AB	R12CP	F13CL	Y14CJ	R36CE	F28BP	W15BK
6	A16BG	H17BE	T17AK	B34AH	V38AE	T18AC	X37CR	V19CM	T23CK	X32CF	V33CD	Y25BL
7	Z24BH	S22BF	G22AL	N26AJ	E11AF	G12AD	B27CS	E30CP	G36CL	B20CH	E15CE	T35BM
8	C21BJ	P29BG	L29AM	A38AK	M31AG	L37AE	N13AB	M14CR	L32CM	N28CJ	M25CF	G16BP
9	K17BK	W34BH	D34BC	Z11AL	H18AH	D27AF	A19AC	H23CS	D20CP	A33CK	H35CH	L24CD
10	F22BL	Y26BJ	J26BD	C31AM	S12AJ	J13AG	Z30AD	S36AB	J28CR	Z15CL	S16CJ	D21CE
11	V29BM	T38BK	U38BE	K18BD	P37AL	U19AH	C14AE	P32AC	U33CS	C25CM	P24CK	J17CF
12	E34BP	G11BL	R11BF	F12BD	W27AL	R30AJ	K23AF	W20AD	R15AB	K35CP	W21CL	U22CH
13	M26CD	L31BM	X31BG	V37BE	Y13AM	X14AK	F36AG	Y28AE	X25AC	F16CR	Y17CM	R29CJ
14	H38CE	D18BP	B18BH	E27BF	T19BC	B23AL	V32AH	T33AF	B35AD	V24CS	T22CP	X34CK
15	S11CF	J12CD	N12BJ	M13BG	G30BD	N36AM	E20AJ	G15AG	N16AE	E21AB	G29CR	B26CL
16	P31CH	U37CE	A37BK	H19BH	L14BE	A32BC	M28AK	L25AH	A24AF	M17AC	L34CS	N38CM
17	W18CJ	R27CF	Z27BL	S30BJ	D23BF	Z20BD	H33AL	D35AJ	Z21AG	H22AD	D26AB	A11CP
18	Y12CK	X13CH	C13BM	P14BK	J36BG	C28BE	S15AM	J16AK	C17AH	S29AE	J38AC	Z31CH
19	T37CL	B19CJ	K19BP	W23BL	U32BH	K33BF	P25BC	U24AL	K22AJ	P34AF	U11AD	C18CS
20	G27CM	N30CK	F30CD	Y36BM	R20BJ	F15BG	W35BD	R21AM	F29AK	W26AG	R31AE	K12AB
21	L13CP	A14CL	V14CE	T32BP	X28BK	V25BH	Y16BE	X17BC	V34AL	Y38AH	X18AF	F37AC
22	D19CR	Z23CM	E23CF	G20CD	B33BL	E35BJ	T24BF	B22BD	E26AM	T11AJ	B12AG	V27AD
23	J30CS	C36CP	M36CH	L28CE	N15BM	M16BK	G21BG	N29BE	M38BC	G31AK	N37AH	E13AE
24	U14AB	K32CR	H32CJ	D33CF	A25BP	H24BL	L17BH	A34BF	H11BD	L18AL	A27AJ	M19AF
25	R23AC	F28CS	S28CR	J15CH	Z35CD	S21BM	D22BJ	Z26BG	S31BE	D12AM	Z13AK	H30AG
26	X36AD	V28AB	P28CL	U25CJ	C16CE	P17BP	J29BK	C38BH	P18BF	J37BC	C19AL	S14AH
27	B32AE	E33AC	W33CM	R35CK	K24CF	W22CQ	U34BL	K11BJ	W12BG	U27BD	K30AM	P23AJ
28	N20AF	M15AD	Y15CP	X16CL	F21CH	Y29CE	R26BM	F31BK	Y37BH	R13BE	F14BC	W36AK
29	A28AG		T25CR	B24CH	V17CJ	T34CF	X38BP	V18BL	T27BJ	X19BF	V23BD	Y32AL
30	Z33AH		G35CS	N21CP	E22CK	G26CH	B11CD	E12BM	G13BK	B30BG	E36BE	T20AM
31	C15AJ		L16AB		M29CL		N31CE	M37BP		N14BH		G28BC

1858	JAN	FEB	MAR	APR	MAY	JUN	JUL	AUG	SEP	OCT	NOV	DEC
1	L33BD	A35AM	V35AG	T21AE	X22AB	V26CR	Y11CL	X12CJ	V13CF	Y30BP	X36BL	F20BH
2	D15BE	Z16BC	E16AH	G17AF	J23AC	E38CS	T31CM	B37CK	E19CH	T14CD	B32BM	V28BJ
3	J25BF	C24BD	M24AJ	L22AG	N34AD	M11AB	G18CP	N27CL	M30CJ	G23CE	N20BP	E33BK
4	U35BG	K21BE	H21AK	D29AH	A26AE	H31AC	L12CR	A13CM	H14CK	L36CF	A28CD	M15BL
5	R16BH	F17BF	S17AL	J34AJ	Z38AF	S18AD	D37CS	Z19CP	S23CL	D32CH	Z33CE	H25BM
6	X24BJ	V22BG	P22AM	U26AK	C11AG	P12AE	J27AB	C30CR	P36CM	J20CJ	C15CF	S35BP
7	B21BK	E29BH	W29BC	R38AL	K31AH	W37AF	U13AC	K14CS	W32CP	U28CK	K25CH	P16CD
8	N17BL	M34BJ	Y34BD	X11AM	F18AJ	Y27AG	R19AD	F23AB	Y20CR	R33CL	F35CJ	W24CE
9	A22BM	H26BK	T26BE	B31BC	V12AK	T13AH	X30AE	V36AC	T28CS	X15CM	V16CK	Y21CF
10	Z29BP	S38BL	G38BF	N18BD	E37AL	G19AJ	B14AF	E32AD	G33AB	B25CP	E24CL	T17CH
11	C34CD	P11BM	L11BG	A12BE	H27AM	L30AK	N23AG	H28AF	L25AD	N35CR	M21CM	G22CJ
12	K26CE	W31BP	D31BH	Z37BF	H13BC	D14AL	A36AH	H28AF	D25AD	A16CS	H17CP	L29CK
13	F38CF	Y18CD	J18BJ	C27BG	S19BD	J23AM	Z32AJ	S33AG	J35AE	Z24AB	S22CR	D34CL
14	V11CH	T12CE	U12BK	K13BH	P30BE	U36BC	C20AK	P15AH	U16AF	C21AC	P29CS	J26CM
15	E31CJ	G37CF	R37BL	F19BJ	W14BF	R32BD	K28AL	W25AJ	R24AG	K17AD	W34AB	U38CP
16	M18CK	L21CH	K27BM	V30BK	Y23BG	K20BE	F33AM	Y35AK	K21AH	F22AE	Y26AC	R11CR
17	H12CL	D13CJ	B13BP	E14BL	T36BH	B28BF	V15BC	T16AL	B17AJ	V29AF	T38AD	X31CS
18	S37CM	J19CK	N19CD	M23BM	G32BJ	N33BG	E25BD	G24AM	N22AK	E34AG	G11AE	B18AB
19	P27CP	U30CL	A30CE	H36BP	L20BK	A15BH	M35BE	L21BC	A29AL	M26AH	L31AF	N12AC
20	W13CR	R14CM	Z14CF	S32CD	D28BL	Z25BJ	H16BF	D17BD	Z34AM	H38AJ	D18AG	A37AD
21	Y19CS	X23CP	C23CH	P20CE	J33BM	C35BK	S24BG	J22BE	C26BC	S11AK	J12AH	Z27AE
22	T30AB	B36CR	K36CJ	W28CF	U15BP	K16BL	P21BH	U29BF	K38BD	P31AL	U37AJ	C13AF
23	G14AC	N32CS	F32CK	Y33CH	R25CD	F24BM	W17BJ	R34BG	F11BE	W18AM	R27AK	K19AG
24	L23AD	A20AB	V20CL	T15CJ	X35CE	V21BP	Y22BK	X26BH	V31BF	Y12BC	X13AL	F30AH
25	D36AE	Z28AC	E28CH	G25CK	B16CF	E17CD	T29BL	B38BJ	E18BG	T37BD	B19AM	V14AJ
26	J32AF	C33AD	M33CP	L35CL	N24CH	M22CE	G34BM	N11BK	M12BH	G27BE	N30BC	E23AK
27	U20AG	K15AE	H15CR	D16CM	A21CJ	H29CF	L26BP	A31BL	H37BJ	L13BF	A14BD	M36AL
28	R28AH	F25AF	S25CS	J24CP	Z17CK	S34CH	D38CD	Z18BM	S27BK	D19BG	Z23BE	H32AM
29	X33AJ		P35AB	U21CR	C22CL	P26CJ	J11CE	C12BP	P13BL	J30BH	C36BF	S20BC
30	B15AK		W16AC	R17CS	K29CM	W38CK	U31CF	K37CD	W19BM	U14BJ	K32BG	P28BD
31	N25AL		Y24AD		F34CP		R18CH	F27CE		R23BK		W33BE

1859	JAN	FEB	MAR	APR	MAY	JUN	JUL	AUG	SEP	OCT	NOV	DEC
1	Y15BF	X16BD	C16AJ	P17AG	J29AD	C38AB	S31CP	J37CL	C19CJ	S14CE	J32BP	Z28BK
2	T25BG	B24BE	K24AK	W22AH	U34AE	K11AC	P18CR	U27CM	K30CK	P23CF	U20CD	C33BL
3	G35BH	N21BF	F21AL	Y29AJ	R26AF	F31AD	W12CS	R13CP	F14CL	W36CH	R28CE	K15BM
4	L16BJ	A17BG	V17AM	T34AK	X38AG	V18AE	Y37AB	X19CR	V23CM	Y32CJ	X33CF	F25BP
5	D24BK	Z22BH	E22BC	G26AL	B11AH	E12AF	T27AC	B30CS	E36CP	T20CK	B15CH	V35CD
6	J21BL	C29BJ	M29BD	L38AM	N31AG	M37AG	G13AD	N14AB	M32CR	G28CL	N25CJ	E16CE
7	U17BM	K34BK	H34BE	D11BC	A18AK	H27AH	L19AE	A23AC	H20CS	L33CM	A35CK	M24CF
8	R22BP	F26BL	S26BF	J31BD	Z12AL	S13AJ	D30AF	Z36AD	S28AB	D15CP	Z16CL	H21CH
9	X29CD	V38BM	P38BG	U18BE	C37AM	P19AK	J14AG	C32AE	P33AC	J25CR	C24CM	S17CJ
10	B34CE	E18BP	W11BH	R12BF	K27BC	W30AL	U23AH	K20AF	W15AD	U35CS	K21CP	P22CK
11	N26CF	M31CD	Y31BJ	X37BG	F13BD	Y14AM	R36AJ	F28AG	Y25AE	R16AB	F17CR	W29CL
12	A38CH	H18CE	T18BK	B27BH	V19BE	T23BC	X32AK	V33AH	T35AF	X24AC	V22CS	Y34CM
13	Z11CJ	S12CF	G12BL	N13BJ	E30BF	G20BD	B20AL	E15AH	G16AG	B21AD	E29AB	T26CP
14	C31CK	P37CH	L37BM	A19BK	M14BG	L32BE	N28AM	M25AK	L24AH	N17AE	M34AC	G38CR
15	K18CL	W27CJ	D27BP	Z30BL	H23BH	D20BF	A33AM	H35AK	D21AJ	A22AF	H26AD	L11CS
16	F12CM	Y13CK	J13CD	C14BM	S36BJ	J28BG	Z15BD	S16AM	J17AK	Z29AG	S38AE	D31AB
17	V37CP	T19CL	U19CE	K23BP	P32BK	U33BH	C25BE	P24BC	U22AL	C34AH	P11AF	J18AC
18	E27CR	G30CM	R30CF	F36CD	W20BL	R35BJ	K35BF	W21BD	R29AM	K26AJ	W31AG	U12AD
19	M13CS	L14CP	X14CH	V32CF	Y28BM	X25BH	F16BG	Y17BE	X34BC	F38AK	Y18AH	R37AE
20	H19AB	D23CR	B23CJ	E20CF	T33BP	B35BL	V24BH	T22BF	B26BD	V11AL	T12AJ	X27AF
21	S30AC	J36CS	N36CK	M28CH	G15CD	N16BM	E21BJ	G29BG	N38BE	E31AM	G37AK	B13AG
22	P14AD	U32AB	A32CL	H33CJ	L25CE	A24BP	M17BK	L34BH	A11BF	M18CS	L27AL	N19AH
23	W23AE	R20AC	Z20CM	S15CK	D35CF	Z21CD	H22BL	D26BJ	Z31BG	H12BD	D13AM	A30AJ
24	Y36AF	X28AD	C28CP	P25CL	J16CH	C17CE	S29BH	J38BK	C18BH	S37BF	J19BC	Z14AK
25	T32AG	B33AE	K33CR	W35CH	U24CJ	K22CF	P34BP	U11BL	K12BJ	P27BF	U30BD	C23AL
26	G20AH	N15AF	F15CS	Y16CP	R21CK	F29CH	W26CD	R31BM	F37BK	W13BG	R14BE	K36AM
27	L28AJ	A25AG	V25AB	T24CR	X17CL	V34CJ	Y38CE	X18BP	V27BL	Y19BH	X23BF	F32BC
28	D33AK	Z35AH	E35AC	G21CS	B22CM	E26CP	T11CF	B12CD	E13BM	T30BJ	B36BG	V20BD
29	J15AL		M16AD	L17AB	N29CP	M38CL	G31CH	N37CE	M19BP	G14BK	N32BH	E28BE
30	U25AM		H24AE	D22AC	A34CR	H11CM	L18CJ	A27CF	H30CD	L23BL	A20BJ	M33BF
31	R35BC		S21AF		Z26CS		D12CK	Z13CH		D36BH		H15BG

(121)

FIRST LETTER = YOUR PHYSICAL CODE, MIDDLE NUMBER = SENSITIVITY CODE, LAST TWO LETTERS = INTELLECTUAL CODE

1860	JAN	FEB	MAR	APR	MAY	JUN	JUL	AUG	SEP	OCT	NOV	DEC
1	S25BH	J24BF	A21AM	H29AK	L26AG	A31AE	M12AB	L13CR	A14CM	M36CJ	L28CF	N15BP
2	P35BJ	U21BG	Z17BC	S34AL	D38AH	Z18AF	H37AC	D19CS	Z23CP	H32CK	D33CH	A25CD
3	W16BK	R17BH	C22BD	P26AM	J11AJ	C12AG	S27AD	J30AB	C36CR	S20CL	J15CJ	Z35CE
4	Y24BL	X22BJ	K29BE	W38BC	U31AK	K37AH	P13AE	U14AC	K32CS	P28CM	U25CK	C16CF
5	T21BM	B29BK	F34BF	Y11BD	R18AL	F27AJ	W19AF	R23AD	F20AB	W33CP	R35CL	K24CH
6	G17BP	N34BL	V26BG	T31BE	X12AM	V13AK	Y30AG	X36AE	V28AC	Y15CR	X16CM	F21CJ
7	L22CD	A26BM	E38BH	G18BF	B37BC	E19AL	T14AH	B32AF	E33AD	T25CS	B24CP	V17CK
8	D29CE	Z38BP	M11BJ	L12BG	N27BD	M30AM	G23AJ	N20AG	M15AE	G35AB	N21CR	E22CL
9	J34CF	C11CD	H31BK	D37BH	A13BE	H14BC	L36AK	A28AH	H25AF	L16AC	A17CS	M29CM
10	U26CH	K31CE	S18BL	J27BJ	Z19BF	S23BD	D32AL	Z33AJ	S35AG	D24AD	Z22AB	H34CP
11	R38CJ	F18CF	P12BM	U13BK	C30BG	P36BE	J20AM	C15AK	P16AH	J21AE	C29AC	S26CR
12	X11CK	V12CH	W37BP	R19BL	K14BH	W32BF	U28BC	K25AL	W24AJ	U17AF	K34AD	P38CS
13	B31CL	E37CJ	Y27CD	X30BM	F23BJ	Y20BG	R33BD	F35AM	Y21AK	R22AD	F26AE	M11AB
14	N18CM	M27CK	T13CE	B14BP	V36BK	T28BH	X15BE	V16BC	T17AL	X29AH	V38AF	Y31AC
15	A12CP	H13CL	G19CF	N23CD	E32BL	G33BJ	B25BF	E24BD	G22AM	B34AJ	E11AG	T18AD
16	Z37CR	S19CM	L30CH	A36CE	M20BM	L15BK	N35BG	M21BE	L29BC	N26AK	M31AH	G12AE
17	C27CS	P30CP	D14CJ	Z32CF	H28BP	D25BL	A16BH	H17BF	D34BD	A38AL	H18AJ	L37AF
18	K13AB	W14CR	J23CK	C20CH	S33CD	J35BM	Z24BJ	S22BG	J26BE	Z11AM	S12AK	D27AG
19	F19AC	Y23CS	U36CL	K28CJ	P15CE	U16BP	C21BK	P29BH	U38BF	C31BC	P37AL	J13AH
20	V30AD	T36AB	R32CM	F33CK	W25CF	R24CD	K17BL	W34BJ	R11BG	K18BD	W27AM	U19AJ
21	E14AE	G32AC	X20CP	V15CL	Y35CH	X21CE	F22BM	Y26BK	X31BH	F12BE	Y13BC	R30AK
22	M23AF	L20AD	B28CR	E25CM	T16CJ	B17CF	V29BP	T38BL	B18BJ	V37BF	T19BD	X14AL
23	H36AG	D28AE	N33CS	M35CP	G24CK	N22CH	E34CD	G11BM	N12BK	E27BG	G30BE	B23AM
24	S32AH	J33AF	A15AB	H16CR	L21CL	A29CJ	M26CE	L31BP	A37BL	M13BH	L14BF	N36BC
25	P20AJ	U15AG	Z25AC	S24CS	D17CM	Z34CK	H38CF	D18CD	Z27BM	H19BJ	D23BG	A32BD
26	W28AK	R25AH	C35AD	P21AB	J22CP	C26CL	S11CH	J12CK	C13BP	S30BK	J36BM	Z20BE
27	Y33AL	X35AJ	K16AE	H17AC	U29CR	K38CM	P31CJ	U37CF	K19CD	P14BL	U32BJ	C28BF
28	T15AM	B16AK	F24AF	Y22AD	R34CS	F11CP	W18CK	R27CH	F30CE	W23BM	R20BK	K33BG
29	G25BC	N24AL	V21AG	T29AE	X26AB	V31CR	Y12CL	X13CJ	V14CF	Y36BP	X28BL	F15BH
30	L35BD		E17AH	G34AF	B38AC	E18CS	T37CM	B19CK	E23CH	T32CD	B33BM	V25BJ
31	D16BE		M22AJ		N11AD		G27CP	N30CL		G20CE		E35BK

1861	JAN	FEB	MAR	APR	MAY	JUN	JUL	AUG	SEP	OCT	NOV	DEC
1	M16BL	L17BJ	X17BD	V34AM	Y38AJ	X18AG	F37AD	Y19AB	X23CR	F32CL	Y33CJ	R25CE
2	H24BM	D22BK	B22BF	E26BC	T11AK	B12AH	V27AE	T30AC	B36CS	V20CH	T15CK	X35CF
3	S21BP	J29BL	N29BF	M38BD	G31AL	N37AJ	E13AF	G14AD	N32AB	E28CP	G25CL	B16CH
4	P17CD	U34BM	A34BG	H11BE	L18AM	A27AK	M19AG	L23AF	A20AC	M33CR	L35CM	N24CJ
5	W22CE	R26BP	Z26BH	S31BF	D12BC	Z13AL	H30AH	D36AF	Z28AD	H15CS	D16CP	A21CK
6	Y29CF	X38CD	C30BJ	P18BG	J37BD	C19AM	S14AJ	J32AG	C33AE	S25AB	J24CR	Z17CL
7	T34CH	B11CE	K11BK	W12BH	U27BE	K30BC	P23AK	U20AH	K15AF	P35AC	U21CS	C22CM
8	G26CJ	N31CF	F31BL	Y37BJ	R13BF	F14BD	W36AL	R28AJ	F25AG	W16AD	R17AB	K29CP
9	L38CK	A18CH	V18BM	T27BK	X19BG	V23BE	Y32AM	X33AK	V35AH	Y24AE	X22AC	F34CR
10	D11CL	Z12CJ	E12BP	G13BL	B30BH	E36BF	T20BC	B15AL	E16AJ	T21AF	B29AD	V26CS
11	J31CM	C37CK	M37CD	L19BM	N14BJ	M32BG	G28BD	N25AM	M24AK	G17AG	N34AE	E38AB
12	U18CP	K27CL	H27CE	D30BP	A23BK	H20BH	L33BE	A35BC	H21AL	L22AH	A26AF	M11AC
13	R12CR	F13CM	S13CF	J14CD	Z36BM	S28BJ	D15BF	Z16BD	S17AM	D29AJ	Z38AG	H31AD
14	X37CS	V19CP	P19CH	U23CE	C32BM	P33BK	J25BG	C24BF	P22BC	J34AK	C11AH	S18AE
15	B27AB	E30CR	W30CJ	R36CF	K20BP	W15BL	U35BH	K21BF	W29BD	U26AL	K31AJ	P12AF
16	N13AC	M14CS	Y14CK	X32CH	F28CD	Y25BM	R16BJ	F17BG	Y34BE	R38AM	F18AK	W37AG
17	A19AD	H23AB	T23CL	B20CJ	V33CE	T35BP	X24BK	V22BH	T26BF	X11BC	V12AL	Y27AH
18	Z30AE	S36AC	G36CM	N28CK	E15CF	G16CD	B21BL	E29BJ	G38BG	B31BD	E37AM	T13AJ
19	C14AF	P32AB	L32CP	A33CL	M25CH	L24CE	N17BM	M34BK	L11BH	N18BE	M27BC	G19AK
20	K23AG	N20AE	D20CR	Z15CM	H35CJ	D21CF	A22BP	H26BL	D31BJ	A12BF	H13BD	L30AL
21	F36AH	Y28AF	J28CS	C25CP	S16CK	J17CH	Z29CD	S38BM	J18BK	Z37BG	S19BE	D14AM
22	V32AJ	T33AG	U33AB	K35CR	P24CL	U22CJ	C34CE	P11BP	U12BL	C27BH	P30BF	J23BC
23	E20AK	G15AH	R15AC	F16CS	W21CH	R29CK	K26CF	W31CD	R37BH	K13BJ	W14BG	U36BD
24	M28AL	L25AJ	X25AD	V24AB	Y17CP	X34CL	F38CH	Y18CE	X27BP	F19BK	Y23BH	R32BE
25	H33AM	D35AK	B35AE	E21AC	T22CR	B26CM	V11CJ	T12CF	B13CD	V30BL	T36BJ	X20BF
26	S15BC	J16AL	N16AF	M17AD	G29CS	N38CP	E31CK	G37CH	N19CE	E14BM	G32BK	B28BG
27	P25BD	U24AM	A24AG	H22AF	L34AB	A11CR	M18CL	L27CJ	A30CF	M23BP	L20BL	N33BH
28	W35BE	R21BC	Z21AH	S29AF	D26AC	Z31CS	H12CM	D13CK	Z14CH	H36CD	D28BM	A15BJ
29	Y16BF		C17AJ	P34AG	J38AD	C18AB	S37CP	J19CL	C23CJ	S32CE	J33BP	Z25BK
30	T24BG		K22AK	W26AH	U11AE	K12AG	P27CR	U30CH	K36CK	P20CF	U15CD	C35BL
31	G21BH		F29AL		R31AF		W13CS	R14CP		W28CH		K16BM

1862	JAN	FEB	MAR	APR	MAY	JUN	JUL	AUG	SEP	OCT	NOV	DEC
1	F24BF	Y22BL	J22BF	C26BD	S11AL	J12AJ	Z27AF	S30AD	J36AB	Z20CP	S15CL	D35CH
2	V21CD	T29BM	U29BG	K38BE	P31AM	U37AK	C13AG	P14AE	U32AC	C28CR	P25CM	J16CJ
3	E17CE	G34BP	R34BH	F11BF	W18BG	R27AL	K19AH	W23AF	R20AD	K33CS	W35CP	U24CK
4	M22CF	L26CD	X26BJ	V31BG	Y12BD	X13AM	F30AJ	Y36AG	X28AE	F15AB	Y16CR	R21CL
5	H29CH	D38CE	B38BK	E18BH	T37BE	B19BC	V14AK	T32AH	B33AF	V25AC	T24CS	X17CM
6	S34CJ	J11CF	N11BL	M12BJ	G27BF	N30BD	E23AL	G20AJ	N15AG	E35AD	G21AB	B22CP
7	P26CK	U31CH	A31BM	H37BK	L13BG	A14BE	M36AM	L28AK	A25AH	M16AE	L17AC	N29CR
8	W38CL	R18CJ	Z18BP	S27BL	D19BH	Z23BF	H32AM	D33AJ	Z35AJ	H24AF	D22AD	A34CS
9	Y11CM	X12CK	C12CD	P13BM	J30BJ	C36BG	S20BD	J15AM	C16AK	S21AG	J29AE	Z26AB
10	T31CP	B37CL	K37CE	W19BP	U14BK	K32BH	P28BE	U25BC	K24AL	P17AH	U34AF	C38AC
11	G18CR	N27CM	F27CF	Y30CD	R23BL	F20BJ	W33BF	R35BD	F21AM	W22AJ	R26AG	K11AD
12	L12CS	A13CP	V13CH	T14CE	X36BH	V28BK	Y15BG	X16BE	V17BC	Y29AK	X38AH	F31AE
13	D37AB	Z19CR	E19CJ	G23CF	B32BP	E33BL	T25BH	B24BF	E22BD	T34AL	B11AJ	V18AF
14	J27AC	C30CS	M30CK	L36CH	N20CD	M15BM	G35BJ	N21BG	M29BE	G26AM	N31AK	E12AG
15	U13AD	K14AB	H14CL	D32CJ	A28CE	H25BP	L16BK	A17BH	H34BF	L38BC	A18AL	M37AH
16	R19AE	F23AD	S23CM	J20CK	Z33CF	S35CD	D24BL	Z22BJ	S26BG	D11BD	Z12AM	H27AJ
17	X30AF	V36AD	P36CP	U28CL	C15CH	P16CE	J21BM	C29BK	P38BH	J31BE	C37BC	S13AK
18	B14AG	E32AE	W32CR	R33CH	K25CJ	W24CF	U17BP	K34BL	W11BJ	U18BF	K27BD	P19AL
19	N23AH	M20AF	Y20CS	X15CP	F35CK	Y21CH	R22CD	F26BM	Y31BK	R12BG	F13BE	W30AM
20	A36AJ	H28AG	T28AB	B25CR	V16CL	T17CJ	X29CE	V38BP	T18BL	X37BH	V19BF	Y14BC
21	Z32AK	S33AH	G33AC	N35CS	E24CM	G22CK	B34CF	E11CD	G12BM	B27BJ	E30BG	T23BD
22	C20AL	P15AJ	L15AD	A16AB	M21CP	L29CL	N26CH	M31CJ	L37BP	N13BK	M14BH	G36BE
23	K28AM	W25AK	D25AE	Z24AC	H17CR	D34CM	A38CJ	H18CF	D27CD	A19BL	H23BJ	L32BF
24	F33BC	Y35AL	J35AF	C21AD	S22CS	J26CP	Z11CK	S12CH	J13CE	Z30BM	S36BK	D20BG
25	V15BD	T16AM	U16AG	K17AE	P29AB	U38CR	C31CL	P37CJ	U19CF	C14BP	P32BL	J28BH
26	E25BE	G24BC	R24AH	F22AF	W34AC	R11CS	K18CH	W27CK	R30CH	K23CD	W20BM	U33BJ
27	M35BF	L21BD	X21AJ	V29AG	Y26AD	X31AB	F12CP	Y13CL	X14CJ	F36CE	Y28BP	R15BK
28	H16BG	D17BE	B17AK	E34AH	T38AE	B18AC	V37CR	T19CM	B23CK	V32CF	T33CD	X25BL
29	S24BH		N22AL	M26AJ	G11AF	N12AD	E27CS	G30CP	N36CL	E20CH	G15CE	B35BM
30	P21BJ		A29AM	H38AK	L31AG	A37AE	M13AB	L14CR	A32CM	M28CJ	L25CF	N16BP
31	W17BK		Z34BC		D18AH		H19AC	D23CS		H33CK		A24CD

FIRST LETTER = YOUR PHYSICAL CODE, MIDDLE NUMBER = SENSITIVITY CODE, LAST TWO LETTERS = INTELLECTUAL CODE

1863	JAN	FEB	MAR	APR	MAY	JUN	JUL	AUG	SEP	OCT	NOV	DEC
1	Z21CE	S29BP	G29BH	N38BF	E31BC	G37AL	B13AH	E14AF	G32AD	B28CS	E25CP	T16CK
2	C17CF	P34CD	L34BJ	A11BG	M18BD	L27AM	N19AJ	M23AG	L20AE	N33AB	M35CR	G24CL
3	K22CH	W26CE	D26BK	Z31BH	H12BE	D13BC	A30AK	H36AH	D28AF	A15AC	H16CS	L21CH
4	F29CJ	Y38CF	J38BL	C18BJ	S37BF	J19BD	Z14AL	S32AJ	J33AG	Z25AD	S24AB	D17CP
5	V34CK	T11CH	U11BM	K12BK	P27BG	U30BE	C23AM	P20AK	U15AH	C35AE	P21AC	J22CR
6	E26CL	G31CJ	R31BP	F37BL	W13BH	R14BF	K36BC	W28AL	R25AJ	K16AF	W17AD	U29CS
7	M38CM	L18CK	X18CD	V27BM	Y19BJ	X23BG	F32BD	Y33AM	X35AK	F24AG	Y22AE	R34AB
8	H11CP	D12CL	B12CE	E13BP	T30BK	B36BH	V20BE	T15BC	B16AL	V21AH	T29AF	X26AC
9	S31CR	J37CM	N37CF	M19CD	G14BL	N32BJ	E28BF	G25BD	N24AM	E17AJ	G34AG	B38AD
10	P18CS	U27CP	A27CH	H30CE	L23BH	A20BK	M33BG	L35BE	A21BC	M22AK	L26AH	N11AE
11	H12AB	R13CR	Z13CJ	S14CF	D36BP	Z28BL	H15BH	D16BF	Z17BD	H29AL	D38AJ	A31AF
12	Y37AC	X19CS	C19CK	P23CH	J32CD	C33BM	S28BJ	J24BG	C22BE	S34AM	J11AK	Z18AG
13	T27AD	B30AB	K30CL	W36CJ	U20CE	K15BP	P35BK	U21BH	K29BF	P26BC	U31AL	C12AH
14	G13AE	N14AC	F14CM	Y32CK	R28CF	F25CD	W16BL	R17BJ	F34BG	W38BD	R18AM	K37AJ
15	L19AF	A23AD	V23CP	T20CL	X33CH	V35CE	Y24BH	X22BK	V26BH	Y11BE	X12BC	F27AK
16	D30AG	Z36AE	E36CR	G28CM	B15CJ	E16CF	T21BP	B29BL	E38BJ	T31BF	B37BD	V13AL
17	J14AH	C32AF	M32CS	L33CP	N25CK	M24CH	G17CD	N34BM	M18BK	G18BG	N27BE	E19AM
18	U23AJ	K20AG	H20AB	D15CR	A35CL	H21CJ	L22CE	A26BP	H31BL	L12BH	A13BF	M30BC
19	R36AK	F28AH	S28AC	J25CS	Z16CM	S17CK	D29CF	Z38CD	S18BM	D37BJ	Z19BG	H14BD
20	X32AL	V33AJ	P33AD	U35AB	C24CP	P22CL	J34CH	C11CE	P12BP	J27BK	C30BH	S23BE
21	B20AM	E15AK	M15AE	R16AC	K21CR	H29CM	U26CJ	K31CF	H37CD	U13BL	K14BJ	P36BF
22	N28BC	M25AL	Y25AF	X24AD	F17CS	Y34CP	R38CK	F18CH	Y27CE	R19BM	F23BK	W32BG
23	A33BD	H35AM	T35AG	B21AE	V22AB	T26CR	X11CL	V12CJ	T13CF	X30BP	V36BL	Y20BH
24	Z15BE	S16BC	G16AH	N17AF	E29AC	G38CS	B31CM	E37CK	G19CH	B14CD	E32BM	T28BJ
25	C25BF	P24BD	L24AJ	A22AG	M34AD	L11AB	N18CP	M27CL	L30CJ	N23CE	M20BP	G33BK
26	K35BG	W21BE	D21AK	Z29AH	H26AE	D31AC	A12CR	H13CM	D14CK	A36CF	H28CD	L15BL
27	F16BH	Y17BF	J17AL	C34AJ	S38AF	J18AD	Z37CS	S19CP	J23CL	Z32CH	S33CE	D25BM
28	V24BJ	T22BG	U22AM	K26AK	P11AG	U12AE	C27AB	P30CR	U36CM	C20CJ	P15CF	J35BP
29	E21BK		R29BC	F38AL	W31AH	R37AF	K13AC	W14CS	R32CP	K28CK	W25CH	U16CD
30	M17BL		X34BD	V11AM	Y18AJ	X27AG	F19AD	Y23AB	X20CR	F33CL	Y35CJ	R24CE
31	H22BM		B26BE		T12AK		V30AE	T36AC		V15CM		X21CF

1864	JAN	FEB	MAR	APR	MAY	JUN	JUL	AUG	SEP	OCT	NOV	DEC
1	B17CH	E34CE	Y26BL	X31BJ	F12BF	Y13BD	R30AL	F36AJ	Y28AG	R15AD	F16AB	W21CP
2	N22CJ	M26CF	T38BM	B18BK	V37BG	T19BE	X14AM	V32AK	T33AH	X25AE	V24AC	Y17CR
3	A29CK	H38CH	G11BP	N12BL	E27BH	G30BF	B23BC	E20AL	G15AJ	B35AF	E21AD	T22CS
4	Z34CL	S11CJ	L31CD	A37BM	M13BJ	L14BG	N36BD	M28AM	L25AK	N16AG	M17AE	G29AB
5	C26CM	P31CK	D18CE	Z27BP	H19BK	D23BH	A32BE	H33BC	D35AL	A24AH	H22AF	L34AC
6	K38CP	W18CL	J12CF	O13CD	S30BL	J36BJ	Z20BF	S15BD	J16AM	Z21AJ	S29AG	D26AD
7	F11CR	Y12CM	U37CH	K19CE	P14BM	U32BK	C28BG	P25BE	U24BC	C17AK	P34AH	J38AE
8	V31CS	T37CP	R27CJ	F30CF	W23BP	R20BL	K33BH	W35BF	R21BD	K22AL	H26AJ	U11AF
9	E18AB	G27CR	X13CK	V14CH	Y36CD	X28BM	F15BJ	Y16BG	X17BE	F29AM	Y38AK	R31AG
10	M12AC	L13CS	B19CL	E23CJ	T32CE	B33BP	V25BK	T24BH	B22BF	V34BC	T11AL	X18AH
11	H37AD	D19AB	N30CM	M36CK	G20CF	N15CD	E35BL	G21BJ	N29BG	E26BD	G31AM	B12AJ
12	S27AE	J30AC	A14CP	H32CL	L28CH	A25CE	M16BM	L17BK	A34BH	M38BE	L18BC	N37AK
13	P13AF	U14AD	Z23CR	S20CM	D33CJ	Z35CF	H24BP	D22BL	Z26BJ	H11BF	D12BD	A27AL
14	W19AG	R23AE	C36CS	P28CP	J15CK	C16CH	S21CD	J29BM	C38BK	S31BG	J37BE	Z13AM
15	Y30AH	X36AF	K32AB	W33CR	U25CL	K24CJ	P17CE	U34BP	K11BL	P18BH	U27BF	C19BC
16	T14AJ	B32AG	F20AC	Y15CS	R35CM	F21CK	W22CF	R26CD	F31BM	W12BJ	R13BG	K30BD
17	G23AK	N20AH	V28AD	T25AB	X16CP	V17CL	Y29CH	X38CE	V18BP	Y37BK	X19BH	F14BE
18	L36AL	A28AJ	E33AE	G35AC	B24CR	E22CM	T34CJ	B11CF	E12CD	T27BL	B30BJ	V23BF
19	D32AM	Z33AM	M15AF	L16AD	N21CS	M29CP	G26CK	N31CH	M37CE	G13BM	N14BK	E36BG
20	J20BC	C15AL	H25AG	D24AE	A17AB	H34CR	L38CL	A16CJ	H27CF	L19BP	A23BL	M32BH
21	U28BD	K25AM	S35AH	J21AF	Z22AC	S26CS	D11CM	Z12CK	S13CH	D30CD	Z36BM	H20BJ
22	R33BE	F35BC	P16AJ	U17AG	C29AD	P38AB	J31CP	C37CL	P19CJ	J14CE	C32BP	S28BK
23	X15BF	V16BD	W24AK	R22AH	K34AE	W11AC	U18CR	K27CM	W30CK	U23CF	K20CD	P33BL
24	B25BG	E24BE	Y21AL	X29AJ	F26AF	Y31AD	R12CS	F13CP	Y14CL	R36CH	F28CE	W15BM
25	N35BH	M21BF	T17AM	B34AK	V38AG	T18AE	X37AB	V19CR	T23CM	X32CJ	V33CF	Y25BP
26	A16BJ	H17BG	G22BC	N26AL	E11AH	G12AF	B27AC	E30CS	G36CP	B20CK	E15CH	T35CD
27	Z24BK	S22BH	L29BD	A38AM	M31AJ	L37AG	N13AD	M14AB	L32CR	N28CL	M25CJ	G16CE
28	C21BL	P29BJ	D34BE	Z11BG	H18AK	D27AH	A19AE	H23AC	D20CS	A33CM	H35CK	L24CF
29	K17BM	W34BK	J26BF	C31BD	S12AL	J13AJ	Z30AF	S36AD	J28AB	Z15CP	S16CL	D21CH
30	F22BP		U38BG	K18BE	P37AM	U19AK	C14AG	P32AE	U33AC	C25CR	P24CM	J17CJ
31	V29CD		R11BH		W27BC		K23AH	W20AF		K35CS		U22CK

1865	JAN	FEB	MAR	APR	MAY	JUN	JUL	AUG	SEP	OCT	NOV	DEC
1	R29CL	F38CJ	S38BP	J18BL	Z37BH	S19BF	D14BC	Z32AL	S33AJ	D25AF	Z24AD	H17CS
2	X34CM	V11CK	P11CD	U12BM	C27BJ	P30BG	J23BD	C20AM	P15AK	J35AG	C21AE	S22AB
3	B26CP	E31CL	W31CE	R37BP	K13BK	W14BH	U36BE	K28BC	W25AL	U16AH	K17AF	P29AC
4	N38CR	M18CM	Y18CF	X27CD	F19BL	Y23BJ	R32BF	F33BD	Y35AM	R24AJ	F22AG	W34AD
5	A11CS	H12CP	T12CH	B13CE	V30BM	T36BK	X20BG	V15BE	T16BC	X21AK	V29AH	Y26AE
6	Z31AB	S37CR	G37CJ	N19CF	E14BP	G32BL	B28BH	E25BF	G24BD	B17AL	E34AJ	T38AF
7	C18AC	P27CS	L27CK	A30CH	M23CD	L20BM	N33BJ	M35BG	L21BE	N22AM	M26AK	G11AG
8	K12AD	W13AB	D13CL	Z14CJ	H36CE	D28BP	A15BK	H16BH	D17BF	A29BC	H38AL	L31AH
9	F37AE	Y19AC	J19CM	C23CK	S32CF	J33CD	Z25BL	S24BJ	J22BG	Z34BD	S11AM	D18AJ
10	V27AF	T30AD	U30CP	K36CL	P20CH	U15CE	C35BM	P21BK	U29BH	C26BE	P31BC	J12AK
11	E13AG	G14AE	R14CR	Z32CH	W28CJ	R25CF	K16BP	W17BL	R34BJ	K38BF	W18BD	U37AL
12	M19AH	L23AF	X23CS	V20CP	Y33CK	X35CH	F24CD	Y22BM	X26BK	F11BG	Y12BE	R27AM
13	H30AJ	D36AG	B36AB	E28CR	T15CL	B16CJ	V21CE	T29BP	B38BL	V31BH	T37BF	X13BC
14	S14AK	J32AH	N32AC	M33CS	G25CM	N24CK	E17CF	G34BD	N11BM	E18BJ	G27BG	B19BE
15	P23AL	U20AJ	A20AD	H15AB	L35CP	A21CL	M22CH	L26CE	A31BP	M12BK	L13BH	N30BE
16	W36AM	R28AK	Z28AE	S25AC	D16CR	Z17CM	H29CJ	D38CF	Z18CD	H37BL	D19BJ	A14BF
17	Y32BC	X33AL	C33AF	P35AD	J24CS	C22CP	S34CK	J11CH	C12CE	S27BM	J30BK	Z23BG
18	T20BD	B15AM	K15AG	W16AE	U21AB	K29CR	P26CL	U31CJ	K37CF	P13BP	U14BL	C36BH
19	G28BE	N25BC	F25AH	Y24AF	R17AC	F34CS	W38CM	R18CK	F27CD	W19CD	R23BM	K32BJ
20	L33BF	A35BD	V35AJ	T21AG	X22AD	V26AB	Y11CP	X12CL	V13CJ	Y30CE	X36BP	F20BK
21	D15BG	Z16BE	E16AK	G17AH	B29AE	E38AC	T31CR	B37CM	E19CK	T14CF	B32CD	V28BL
22	J25BH	C24BF	M24AL	L22AJ	N34AF	M11AD	G18CS	N27CP	M30CL	G23CH	N20CE	E33BP
23	U35BJ	K21BG	H21AM	D29AK	A26AG	H31AE	L12AD	A13CR	H14CM	L36CJ	A28CF	M15BP
24	R16BK	F17BM	S17BC	J34AL	Z38AH	S18AF	D37AC	Z19CS	S23CP	D32CK	Z33CH	H25CD
25	X24BL	V22BJ	P22BD	U26AM	C11AJ	P12AG	J27AD	C30AB	P36CR	J20CL	C15CJ	S35CE
26	B21BM	E29BK	W29BE	R38BC	K31AK	W37AH	U13AE	K14AC	W32CS	U28CH	K25CK	P16CF
27	N17BP	H34BJ	Y34BF	X11BD	F18AL	Y27AJ	R19AF	F23AD	Y20AB	R33CP	F35CL	H24CH
28	A22CD	H26BM	T26BG	B31BE	V12AM	T13AK	X30AG	V36AE	T28AC	X15CR	V16CM	Y21CJ
29	Z29CE		G38BH	N18BF	E37BC	G19AL	B14AH	E32AF	G33AD	B25CS	E24CP	T17CK
30	C34CF		L11BJ	A12BG	M27BD	L30AH	N23AJ	M20AG	L15AE	N35AB	M21CR	G22CL
31	K26CH		D31BK		H13BE		A36AK	H28AH		A16AC		L29CM

FIRST LETTER = YOUR PHYSICAL CODE, MIDDLE NUMBER = SENSITIVITY CODE, LAST TWO LETTERS = INTELLECTUAL CODE

1866	JAN	FEB	MAR	APR	MAY	JUN	JUL	AUG	SEP	OCT	NOV	DEC
1	D34CP	Z11CL	E11CE	G12BP	B27BK	E30BH	T23BE	B20BC	E15AL	T35AH	B21AF	V22AC
2	J26CR	C31CM	M31CF	L37CD	N13BL	M14BJ	G36BF	N28BD	M25AM	G16AJ	N17AG	E29AD
3	U38CS	K18CP	H18CH	D27CE	A19BM	H23BK	L32BG	A33BE	H35BC	L24AK	A22AH	M34AE
4	R11AB	F12CR	S12CJ	J13CF	Z30BP	S36BL	D20BH	Z15BF	S16BD	D21AL	Z29AJ	H26AF
5	X31AC	V37CS	P37CK	U19CH	C14CD	P32BM	J28BJ	C25BG	P24BE	J17AM	C34AK	S38AG
6	B18AD	E27AB	W27CL	R30CJ	K23CE	W20BP	U33BK	K35BH	W21BF	U22BC	K26AL	P11AH
7	N12AE	M13AC	Y13CH	X14CK	F36CF	Y28CD	R15BL	F16BJ	Y17BG	R29BD	F38AM	W31AJ
8	A37AF	H19AD	T19CP	B23CL	V32CH	T33CE	X25BM	V24BK	T22BH	X34BE	V11BC	Y18AK
9	Z27AG	S30AE	G30CR	N36CM	E20CJ	G15CF	B35BP	E21BL	G29BJ	B26BF	E31BD	T12AL
10	C13AH	P14AF	L14CS	A32CP	M28CK	L25CH	N16CD	M17BM	L34BK	N38BG	M18BE	G37AM
11	K19AJ	W23AG	D23AB	Z20CR	H33CL	D35CJ	A24CE	H22BP	D26BL	A11BH	H12BF	L27BC
12	F30AK	Y36AH	J36AC	C28CS	S15CM	J16CK	Z21CF	S29CD	J38BM	Z31BJ	S37BG	D13BD
13	V14AL	T32AJ	U32AD	K33AB	P25CP	U24CL	C17CH	P34CE	J11BP	C18BK	P27BH	J19BE
14	E23AM	G20AK	R20AE	F15AC	W35CR	R21CM	K22CJ	W26CF	R31CD	K12BL	W13BJ	U30BF
15	M36BC	L28AL	X28AF	V25AD	Y16CS	X17CP	F29CK	Y38CH	X18CE	F37BM	Y19BK	R14BG
16	H32BD	D33AM	B33AG	E35AE	T24AB	B22CR	V34CL	T11CJ	B12CF	V27BP	T30BL	X23BH
17	S20BE	J15BC	N15AH	M16AF	G21AC	N29CS	E26CM	G31CK	N37CH	E13CD	G14BM	B36BJ
18	P28BF	U25BD	A25AJ	H24AG	L17AD	A34AB	M38CP	L18CL	A27CJ	M19CE	L23BP	N32BK
19	W33BG	R35BE	Z35AK	S21AH	D22AE	Z26AC	H11CN	D12CM	Z13CK	H30CF	D36CD	A20BL
20	Y15BH	X16BF	C16AL	P17AJ	J29AF	C38AD	S31CS	J37CP	C19CL	S14CH	J32CE	Z28BM
21	T25BJ	B24BG	K24AM	W22AK	U34AG	K11AE	P18AB	U27CR	K30CM	P23CJ	U20CF	C33BP
22	G35BK	N21BH	F21BC	Y29AL	R26AH	F31AF	W12AC	R13CS	F14CP	W36CK	R28CH	K15CD
23	L16BL	A17BJ	V17BD	T34AM	X38AJ	V18AG	Y37AD	X19AB	V23CR	Y32CL	X33CJ	F25CE
24	D24BM	Z22BK	E22BE	G26BC	B11AK	E12AH	T27AE	B30AC	E36CS	T20CM	B15CK	V35CF
25	J21BP	C29BL	M29BF	L38BD	N31AL	M37AJ	G13AF	N14AD	M32AB	G28CP	N25CL	E16CH
26	U17CD	K34BM	H34BG	D11BE	A18AM	H27AK	L19AG	A23AE	H20AC	L33CJ	A35CM	M24CJ
27	R22CE	F26BP	S26BH	J31BF	Z12BC	S13AL	D30AH	Z36AF	S28AD	D15CS	Z16CP	H21CK
28	X29CF	V38CD	P38BJ	U18BG	C37BD	P19AM	J14AJ	C32AG	P33AE	J25AB	C24CR	S17CL
29	B34CH		W11BK	R12BH	K27BE	W30BC	U23AK	K20AH	W15AF	U35AC	K21CS	P22CM
30	N26CJ		Y31BL	X37BJ	F13BF	Y14BD	R36AL	F28AJ	Y25AG	R16AD	F17AB	W29CP
31	A38CK		T18BM		V19BG		X32AM	V33AK		X24AE		Y34CR

1867	JAN	FEB	MAR	APR	MAY	JUN	JUL	AUG	SEP	OCT	NOV	DEC
1	T26CS	B31CP	K31CH	W37CE	U13BM	K14BK	P36BG	U28BE	K25BC	P16AK	U17AH	C29AE
2	G38AB	N18CR	F18CJ	Y27CF	R19BP	F23BL	W32BH	R33BF	F35BD	W24AL	R22AJ	K34AF
3	L11AC	A12CS	V12CK	T13CH	X30CD	V36BM	Y20BJ	X15BG	V16BE	Y21AM	X29AK	F26AG
4	D31AD	Z37AB	E37CL	G19CJ	B14CE	E32BP	T28BK	B25BH	E24BF	T17BC	B34AL	V38AH
5	J18AE	C27AC	M27CH	L30CK	N23CF	M20CD	G33BL	N35BJ	M21BG	G22BD	N26AM	E11AJ
6	U12AF	K13AD	H13CP	D14CL	A36CH	H28CE	L15BM	A16BK	H17BH	L29BE	A38BC	M31AK
7	R37AG	F19AE	S19CR	J23CM	Z32CJ	S33CF	D25BP	Z24BL	S22BJ	D34BF	Z11BD	H18AL
8	X27AH	V30AF	P30CS	U36CP	C20CK	P15CH	J35CD	C21BM	P29BK	J26BG	C31BE	S12AM
9	B13AJ	E14AG	W14AB	R32CR	K28CL	W25CJ	U16CE	K17BP	W34BL	U38BH	K18BF	P37BC
10	N19AK	M23AH	Y23AC	X20CS	F33CM	Y35CK	R24CF	F22CD	Y26BM	R11BJ	F12BG	W27BD
11	A30AL	H36AJ	T36AD	B28AB	V15CP	T16CL	X21CH	V29CE	T38BP	X31BK	V37BH	Y13BE
12	Z14AM	S32AK	G32AE	N33AC	E25CR	G24CM	B17CJ	E34CF	G11CD	B18BL	E27BJ	T19BF
13	C23BC	P20AL	L20AF	A15AD	M35CS	L21CP	N22CK	M26CH	L31CE	N12BM	M13BK	G30BG
14	K36BD	W28AM	D28AG	Z25AE	H16AB	D17CR	A29CL	H38CJ	D18CF	A37BP	H19BL	L14BH
15	F32BE	Y33BC	J33AH	C35AF	S24AC	J22CS	Z34CM	S11CK	J12CH	Z27CD	S30BM	D23BJ
16	V20BF	T15BD	U15AJ	K16AG	P21AD	U29AB	C26CP	P31CL	U37CJ	C13CE	P14BP	J36BK
17	E28BG	G25BE	R25AK	F24AH	W17AE	R34AC	K38CR	W18CM	R27CK	K19CF	W23CD	U32BL
18	M33BH	L35BF	X35AL	V21AJ	Y22AF	X26AD	F11CS	Y12CP	X13CL	F30CH	Y36CE	R20BM
19	H15BJ	D16BG	B16AM	E17AK	T29AG	B38AE	V31AB	T37CR	B19CM	V14CJ	T32CF	X28BP
20	S25BK	J24BH	N24BC	M22AL	G34AH	N11AF	E18AC	G27CS	N30CP	E23CK	G20CH	B33CD
21	P35BL	U21BJ	A21BD	H29AM	L26AJ	A31AG	M12AD	L13AB	A14CR	M36CL	L28CJ	N15CE
22	W16BM	R17BK	Z17BE	S34BC	D38AK	Z18AH	H37AE	D19AC	Z23CS	H32CM	D33CK	A25CF
23	Y24BP	X22BL	C22BF	P26BD	J11AL	C12AJ	S27AF	J30AD	C36AB	S20CP	J15CL	Z35CH
24	T21CD	B29BM	K29BG	W38BD	U31AM	K37AK	P13AG	U14AE	K32AC	P28CR	U25CM	C16CJ
25	G17CE	N34BP	F34BH	Y11BF	R18BD	F27AL	W19AH	R23AF	F20AD	W33CS	R35CP	K24CK
26	L22CF	A26CD	V26BJ	T31BG	X12BD	V13AM	Y30AJ	X36AG	V28AE	Y15AB	X16CR	F21CL
27	D29CH	Z38CF	E38BK	G18BH	B37BE	E19BC	T14AK	B32AH	E33AF	T25AC	B24CS	V17CH
28	J34CJ	C11CF	M11BL	L12BJ	N27BF	M30BD	G23AL	N20AJ	M15AG	G35AD	N21AB	E22CP
29	U26CK		H31BM	D37BK	A13BG	H14BE	L36AM	A28AK	H25AH	L16AE	A17AC	M29CR
30	R38CL		S18BP	J27BL	Z19BH	S23BF	D32BC	Z33AL	S35AJ	D24AF	Z22AD	H34CS
31	X11CM		P12CD		C30BJ		J20BD	C15AM		J21AG		S26AB

1868	JAN	FEB	MAR	APR	MAY	JUN	JUL	AUG	SEP	OCT	NOV	DEC
1	P38AC	U18CS	Z12CL	S13CJ	D30CE	Z36BP	H20BK	D15BH	Z16BF	H21BC	D29AL	A26AH
2	W11AD	R12AB	C37CM	P19CK	J14CF	C32CD	S28BL	J25BJ	C24BG	S17BD	J34AM	Z38AJ
3	Y31AE	X37AC	K27CP	W30CL	U23CH	K20CE	P33BM	U35BK	K21BH	P22BE	U26BC	C11AK
4	T18AF	B27AD	F13CR	Y14CM	R36CJ	F28CF	W15BP	R16BL	F17BJ	W29BF	R38BD	K31AL
5	G12AG	N13AE	V19CS	T23CP	X32CK	V33CH	Y25CD	X24BM	V22BK	Y34BG	X11BE	F18AH
6	L37AH	A19AF	E30AB	G36CR	B20CL	E15CJ	T35CE	B21BP	E29BL	T26BH	B31BF	V12BC
7	D27AJ	Z30AG	M14AC	L32CS	N28CM	M25CK	U17CD	N34BM	M33BG	U13BD	N18BC	E37BD
8	J13AK	C14AH	H23AD	D20AB	A33CP	H35CL	L24CH	A22CE	H26BP	L11BK	A12BH	M27BE
9	U19AL	K23AJ	S36AE	J28AC	Z15CR	S16CM	D21CJ	Z29CF	S38CD	D31BL	Z37BJ	H13BF
10	R30AM	F36AK	P32AF	U33AD	C25CS	P24CP	J17CK	C34CH	P11CE	J18BM	C27BK	S19BG
11	X14BC	V32AL	N20AG	R15AE	K35AB	W19AC	K26CJ	W31CF	K13BL	P30BH		
12	B23BD	E20AM	Y28AG	X25AF	F16AC	Y17CS	R29CM	F38CK	Y18CH	R37CD	F19BM	W14BJ
13	N36BE	M28BE	T33AJ	B35AG	V24AD	T22BE	X34AF	V11CL	T12CL	X27CE	V30BP	Y23BK
14	A32BF	H33BD	G15AK	N16AH	E21AE	G29AC	B26CR	E31CM	G37CK	B13CF	E14CD	T36BL
15	Z20BG	S15BE	L25AL	A24AJ	M17AF	L34AD	N38CS	M18CP	L27CL	N19CH	M23CE	G32BM
16	C28BH	P25BF	D35AM	Z21AK	H22AG	D26AE	A11AB	H12CR	D13CM	A30CJ	H36CF	L20BP
17	K33BJ	W35BG	J16BC	C17AL	S29AH	J38AF	Z31AC	S37CS	J19CP	Z14CK	S32CH	D28CD
18	F15BK	Y16BH	U24BD	K22AM	P34AJ	U11AG	C18AD	P27AB	U30CR	C23CL	P20CJ	J33CE
19	V25BL	T24BJ	R21BE	F29BC	W26AK	R31AH	K12AF	W13AD	R14CS	K36CM	W28CK	U15CF
20	E35BM	G21BK	X17BF	V34BD	Y38AL	X18AJ	F37AG	Y19AD	X23AB	F32CP	Y33CL	R25CH
21	M16BP	L17BL	B22BG	E26BE	T11AM	B12AK	V27AG	T30AE	B36AC	V20CR	T15CM	X35CJ
22	H24CD	D22BM	N29BP	M38BF	G31BC	N37AL	E13AH	G14AF	N32AD	E28CS	G25CP	B16CK
23	S21CE	J29BP	A34BJ	H11BD	L18AD	A27AB	M19AJ	L23AG	A20AE	M33AB	L35CR	N24CL
24	P17CF	U34CD	Z26BK	S31BH	D12BE	Z13BC	H30AK	D36AH	Z28AF	H15AC	D16CS	A21CM
25	W22CH	R26CE	C38BL	P18BJ	J37BF	C19BD	S14AL	J32AJ	C33AG	S25AD	J24AB	Z17CP
26	Y29CJ	X38CF	K11BM	W12BK	U27BG	K30BE	P23AM	U20AH	K15AH	P35AE	U21AC	C22CR
27	T34CK	B11CH	F31BP	Y37BL	R13BH	F14BF	W36AG	R28AF	F25AJ	W16AF	R17AD	K29CS
28	G26CL	N31CJ	V18CD	T27BM	X19BJ	V23BG	Y32BD	X33AK	V35AK	Y24AG	X22AE	F34AE
29	L38CM	A18CK	E12CE	G13BF	B30BK	E36BH	T20BE	B15BC	E16AL	T21AH	B29AF	V26AC
30	D11CP		M37CF	L19CD	N14BL	M32BJ	G28BF	N25BD	M24AM	G17AJ	N34AG	E38AD
31	J31CR		H27CH		A23BM		L33BG	A35BE		L22AK		M11AE

FIRST LETTER = YOUR PHYSICAL CODE, MIDDLE NUMBER = SENSITIVITY CODE, LAST TWO LETTERS = INTELLECTUAL CODE

1869	JAN	FEB	MAR	APR	MAY	JUN	JUL	AUG	SEP	OCT	NOV	DEC
1	H31AF	D37AD	B37CP	E19CL	T14CH	B32CE	V28BM	T25BK	B24BH	V17BE	T34BC	X38AK
2	S18AG	J27AE	N27CR	M30CM	G23CJ	N20CF	E33BP	G35BL	N21BJ	E22BF	G26BD	B11AL
3	P12AH	U13AF	A13CS	H14CP	L36CK	A28CH	M15CD	L16BM	A17BK	M29BG	L38BE	N31AM
4	W37AJ	R19AG	Z19AB	S23CR	D32CL	Z33CJ	H25CE	D24BP	Z22BL	H34BH	D11BF	A18BC
5	Y27AK	X38AH	C30AC	P36CS	J20CM	C15CK	S35CF	J21CD	C29BM	S26BJ	J31BG	Z12BD
6	T13AL	B14AJ	K14AD	W32AB	U28CP	K25CL	P16CH	U17CE	K34BP	P38BK	U18BH	C37BE
7	G19AM	N23AK	F23AE	Y20AC	R33CR	F35CM	N24CJ	R22CF	F26CD	M11BL	R12BJ	K27BF
8	L30BC	A36AL	V36AF	T28AD	X15CS	V16CP	Y21CK	X29CH	V38CE	Y31BM	X37BK	F13BG
9	D14BD	Z32AM	E32AG	G33AE	B25AB	E24CR	T17CL	B34CJ	E11CF	T18BP	B27BL	V19BH
10	J23BE	C20BC	M20AH	L15AP	N35AC	M21CS	G22CM	N26CK	M31CH	G12CD	N13BM	E30BJ
11	U36BF	K28BD	H28AJ	D25AG	A16AD	H17AB	L29CP	A38CL	H18CJ	L37CE	A19BP	M14BK
12	R32BG	F33BE	S33AK	J35AH	Z24AE	S22AC	D34CR	Z11CM	S12CK	D27CF	Z30CD	H23BL
13	X20BH	V15BF	P15AL	U16AJ	C21AF	P29AD	J26CS	C31CP	P37CL	J13CH	C14CE	S36BM
14	B28BJ	E25BG	W25AM	R24AK	K17AG	W34AE	U38AB	K18CR	W27CH	U19CJ	K23CF	P32BP
15	N33BK	M35BH	Y35BC	X21AL	F22AH	Y26AF	R11AC	F12CS	Y13CP	R30CK	F36CH	W20CD
16	A15BL	H16BJ	T16BD	B17AM	V29AJ	T38AG	X31AD	V37AB	T19CR	X14CL	V32CJ	Y28CE
17	Z25BM	S24BK	G24BE	N22BC	E34AK	G11AH	B18AE	E27AC	G30CS	B23CH	E20CK	T33CF
18	C35BP	P21BL	L21BF	A29BD	M26AL	L31AJ	N12AF	M13AD	L14AB	N36CP	M28CL	G15CH
19	K16CD	W17BM	D17BG	Z34BE	H38AM	D18AK	A37AG	H19AE	D23AC	A32CR	H33CM	L25CJ
20	F24CE	Y22BP	J22BH	C26BF	S11BC	J12AL	Z27AH	S30AF	J36AD	Z20CS	S15CP	D35CK
21	V21CF	T29CD	U29BJ	K38BG	P31BD	U37AM	C13AJ	P14AG	U32AE	C28AB	P25CR	J16CL
22	E17CH	G34CE	R34BK	F11BH	W18BE	R27BC	K19AK	W23AH	R20AF	K33AC	W35CS	U24CH
23	M22CJ	L26CF	X26BL	V31BJ	Y12BF	X13BD	F30AL	Y36AJ	X28AG	F15AD	Y16AB	R21CP
24	H29CK	D38CH	B38BM	E18BK	T37BG	B19BE	V14AM	T32AK	B33AH	V25AE	T24AC	X17CR
25	S34CL	J11CJ	N11BP	M12BL	G27BH	N30BF	E23BC	G20AL	N15AJ	E35AF	G21AD	B22CS
26	P26CM	U31CK	A31CD	H37BM	L13BJ	A14BG	M36BD	L28AM	A25AK	M16AG	L17AE	N29AB
27	W38CP	R18CL	Z18CE	S27BP	D19BK	Z23BH	H32BE	D33BC	Z35AL	H24AH	D22AF	A34AC
28	Y11CR	X12CM	C12CF	P13CD	J30BL	C36BJ	S20BF	J15BD	C16AM	S21AJ	J29AG	Z26AD
29	T31CS		K37CH	W19CE	U14BM	K32BK	P28BG	U25BE	K24BC	P17AK	U34AH	C38AE
30	G18AB		F27CJ	Y30CF	R23BP	F20BL	W33BH	R35BF	F21BD	W22AL	R26AJ	K11AF
31	L12AC		V13CK		X36CD		Y15BJ	X16BG		Y29AM		F31AG

1870	JAN	FEB	MAR	APR	MAY	JUN	JUL	AUG	SEP	OCT	NOV	DEC
1	V18AH	T27AF	U27CS	K30CP	P23CK	U20CH	C33CD	P35BM	U21BK	C22BG	P26BE	J11AM
2	E12AJ	G13AG	R13AB	F14CR	H36CL	R28CJ	K15CE	W16BP	R17BL	K29BH	W38BF	U31BC
3	M37AK	L19AH	X19AC	V23CS	Y32CH	X33CK	F25CF	Y24CD	X22BH	F34BJ	Y11BG	R18BD
4	H27AL	D30AJ	B30AD	E36AB	T20CP	B15CL	V35CH	T21CE	B29BP	V26BK	T31BH	X12BE
5	S13AM	J14AK	N14AE	M32AC	G28CR	N25CH	E16CJ	G17CF	N34CD	E38BL	G18BJ	B37BF
6	P19BC	U23AL	A23AF	H20AD	L33CS	A35CP	M24CK	L22CH	A26CE	M11BM	L12BK	N27BG
7	W30BD	R36AM	Z36AG	S28AE	D15AB	Z16CR	H21CL	D29CJ	Z38CF	H31BP	D37BL	A13BH
8	Y14BE	X32BC	C32AH	P33AF	J25AC	C24CS	S17CM	J34CK	C11CH	S18CD	J27BM	Z19BJ
9	T23BF	B20BD	K20AJ	W15AG	U35AD	K21AB	P22CP	U26CL	K31CJ	P12CE	U13BP	C30BK
10	G36BG	N28BE	F28AK	Y25AH	R16AE	F17AC	W29CR	R38CM	F18CK	W37CF	R19CD	K14BL
11	L32BH	A33BF	V33AL	T35AJ	X24AF	V22AD	Y34CS	X11CP	V12CL	Y27CH	X30CE	F23BM
12	D20BJ	Z15BG	E15AM	G16AK	B21AG	E29AE	T26AB	B31CR	E37CM	T13CJ	B14CF	V36BP
13	J28BK	C25BH	M25BC	L24AL	N17AH	M34AF	G38AC	N16CS	M27CP	G19CK	N23CH	E32CD
14	U33BL	K35BJ	H35BD	D21AM	A22AJ	H26AG	L11AD	A12AB	H13CR	L30CL	A36CJ	M20CE
15	R15BM	F16BK	S16BE	J17BC	Z29AK	S38AH	D31AE	Z37AC	S19CS	D14CH	Z32CK	H28CF
16	X25BP	V24BL	P24BF	U22BD	C34AL	P11AJ	J18AF	C27AD	P30AB	J23CP	C20CL	S33CH
17	B35CD	E21BM	W21BG	R29BE	K26AM	W31AK	U12GA	K13AE	W14AC	U36CR	K28CH	P15CJ
18	N16CE	M17BP	Y17BH	X34BF	F38BC	Y18AL	R37AH	F19AF	Y23AD	R32CS	F33CP	W25CK
19	A24CF	H22CD	T22BJ	B26BG	V11BD	T12AM	X27AJ	V30AG	T36AE	X20AB	V15CR	Y35CL
20	Z21CH	S29CE	G29BK	N38BH	E31BE	G37BC	B13AK	E14AH	G32AF	B28AC	E25CS	T16CM
21	C17CJ	P34CF	L34BL	A11BJ	M18BF	L27BD	N19AL	M23AJ	L20AG	N33AD	M35AB	G24CP
22	K22CK	W26CH	D26BM	Z31BK	H12BG	D13BE	A30AM	H36AK	D28AH	A15AE	H16AC	L21CR
23	F29CL	Y38CJ	J38BP	C18BL	S37BH	J19BF	Z14BC	S32AL	J33AJ	Z25AF	S24AD	D17CS
24	V34CM	T11CK	U11CD	K12BM	P27BJ	U30BG	C23BD	P20AM	U15AK	C35AG	P21AE	J22AB
25	E26CP	G31CL	R31CE	F37BP	W13BK	R14BH	K36BE	W28BC	R25AL	K16AH	W17AF	U29AC
26	M38CR	L18CM	X18CF	V27CD	Y19BL	X23BJ	F32BD	Y33BD	X35AM	F24AJ	Y22AG	R34AD
27	H11CS	D12CP	B12CH	E13CE	T30BM	B36BK	V20BG	T15BE	B16BC	V21AK	T29AH	X26AE
28	S31AM	J37CR	N37CJ	M19CF	G14BP	N32BL	E28BH	G25BF	N24BD	E17AL	G34AJ	B38AF
29	P18AC		A27CK	H30CH	L23CD	A20BM	H33BJ	L35BG	A21BE	H22AM	L26AK	N11AG
30	W12AD		Z13CL	S14CJ	D36CE	Z28BP	H15BK	D16BH	Z17BF	H29BC	D38AL	A31AH
31	Y37AE		C19CH		J32CF		S25BL	J24BJ		S34BD		Z18AJ

1871	JAN	FEB	MAR	APR	MAY	JUN	JUL	AUG	SEP	OCT	NOV	DEC
1	C12AK	P13AH	L13AC	A14CS	M36CM	L28CK	N15CF	M16CD	L17BM	N29BJ	M38BG	G31BD
2	K37AL	W19AJ	D19AD	Z23AB	H32CP	D33CL	A25CH	H24CE	D22BP	A34BK	H11BH	L18BE
3	F27AM	Y30AK	J30AE	C36AC	S20CR	J15CM	Z35CJ	S21CF	J29CD	Z26BL	S31BJ	D12BF
4	V13BC	T14AL	U14AF	K32AD	P28CS	U25CP	C16CK	P17CH	U34CE	C38BM	P18BK	J37BG
5	E19BD	G23AM	R23AG	F20AE	W33AB	R35CR	K24CL	W22CH	R26CF	K11BP	W12BL	U27BH
6	W30BE	L36BC	X36AH	V28AF	Y15AC	X16CS	F21CH	Y29CK	X38CH	F31CD	Y37BM	R13BJ
7	H14BF	D32BD	B32AJ	E33AG	T25AD	B24AB	V17CP	T34CL	B11CJ	V18CE	T27BP	X19BK
8	S23BG	J20BE	N20AK	M15AH	G35AE	N21AC	E22CR	G26CM	N31CK	E12CF	G13CD	B30BL
9	P36BH	U28BF	A28AL	H25AJ	L16AF	A17AD	M29CS	L38CP	A16CL	M37CH	L19CE	N14BM
10	W32BJ	R33BG	Z33AM	S35AK	D24AG	Z22AE	H34AB	D11CR	Z12CM	H27CJ	D30CF	A23BP
11	Y20BK	X15BH	C15BC	P16AL	J21AH	C29AF	S26AC	J31CS	C37CP	S13CK	J14CH	Z36CD
12	T28BL	B25BJ	K25BD	W24AM	U17AJ	K34CG	P38AD	U18AB	K27CR	P19CL	U23CJ	C32CE
13	G33BM	N35BK	F35BE	Y21BC	R22AK	F26AH	W11AF	R12AC	F13CS	W30CH	R36CK	K20CF
14	L15BP	A16BL	V16BF	T17BD	X29AL	V38AJ	Y31AF	X37AD	V19AB	Y14CL	X32CL	F28CH
15	D25CD	Z24BH	E24BG	G22BE	B34AM	E11AK	T18AG	B27AE	E30AC	T23CR	B20CM	V33CJ
16	J35CE	C21BP	M21BH	L29BF	N26BC	M31AL	G12AH	N13AF	M14AD	G36CS	N28CP	E15CK
17	U16CF	K17CD	H17BJ	D34BG	A38BD	H18AJ	L37AF	A19AG	H23AE	L32AB	A33CR	M25CL
18	R24CH	F22CE	S22BM	J26BH	Z11BE	S12BC	D27AK	Z30AH	S36AF	D20AC	Z15CS	H35CM
19	X21CJ	V29CF	P29BL	U38BJ	C31BF	P37BD	J13AL	C14AJ	P32AG	J28AD	C25AB	S16CP
20	B17CK	E34CH	W34BM	R11BK	K18BG	W27BE	U19AM	K23AK	W20AH	U33AE	K35AC	P24CR
21	N22CL	M26CJ	Y26BP	X31BL	F12BH	Y13BF	R30BC	F36AL	Y28AJ	R15AF	F16AD	W21CS
22	A29CM	H38CK	T38CD	B18BM	V37BJ	T19BG	X14BD	V32AM	T33AK	X25AG	V24AE	Y17AB
23	Z34CP	S11CL	G11CE	N12BP	E27BK	G30BH	B23BE	E20BC	G15AL	B35AH	E21AF	T22AC
24	C26CR	P31CM	L31CF	A37CD	M13BL	L14BJ	N36BF	M28BD	L25AM	N16AJ	M17AG	G29AD
25	K38CS	W18CP	D18CH	Z27CE	H19BM	D23BK	A32BG	H33BE	D35BC	A24AK	H22AH	U11AF
26	F11AB	Y12CR	J12CJ	C13CF	S30BP	J36BL	Z20BH	S15BF	J16BD	Z21AL	S29AJ	D26AF
27	V31AC	T37CS	U37CK	K19CH	P14CD	U32BM	C28BJ	P25BG	U24BE	C17AM	P34AK	J38AG
28	E18AD	G27AB	R27CL	F30CJ	W23CE	R20BP	K33BK	W35BH	R21BF	K22BC	W26AL	U11AH
29	M12AE		X13CH	V14CK	Y36CF	X28CD	F15BL	Y16BJ	X17BG	F29BD	Y38AM	R31AJ
30	H37AF		B19CP	E23CL	T32CH	B33CE	V25BM	T24BK	B22BH	V34BE	T11BC	X18AK
31	S27AG		N30CR		G20CJ		E35BP	G21BL		E26BF		B12AL

FIRST LETTER = YOUR PHYSICAL CODE, MIDDLE NUMBER = SENSITIVITY CODE, LAST TWO LETTERS = INTELLECTUAL CODE

1872	JAN	FEB	MAR	APR	MAY	JUN	JUL	AUG	SEP	OCT	NOV	DEC
1	N37AM	M19AK	T30AF	B36AD	V20CS	T15CP	X35CK	V21CH	T29CE	X26BM	V31BK	Y12BG
2	A27BC	H30AL	G14AG	N32AE	E28AB	G25CR	B16CL	E17CJ	G34CF	B38BP	E18BL	T37BH
3	Z13BD	S14AM	L23AH	A20AF	M33AC	L35CS	N24CH	M22CK	L26CH	N11CD	M12BM	G27BJ
4	C19BE	P23BC	D36AJ	Z28AG	H15AD	D16AB	A21CP	H29CL	D38CJ	A31CE	H37BP	L13BK
5	K30BF	W36BD	J32AK	C33AM	S25AE	J24AC	Z17CR	S34CM	J11CK	Z18CF	S27CD	D19BL
6	F14BG	Y32BE	U20AL	K15AJ	P35AF	U21AD	C22CS	P26CP	U31CL	C12CH	P13CE	J30BM
7	V23BH	T20BF	R28AM	F25AK	H16AG	R17AE	K29AB	M38CR	R18CM	K37CJ	W19CF	U14BP
8	E36BJ	G28BG	X33BP	V35AL	Y24AH	X22AF	F34AC	Y11CS	X12CP	F27CK	Y30CH	R23CD
9	M32BK	L33BH	B15BD	E16AM	T21AJ	B29AG	V26AD	T31AB	B37CR	V13CL	T14CJ	X36CE
10	H20BL	D15BJ	N25BE	M24BC	G17AK	N34AH	E38AC	G18AC	N27CS	E19CM	G23CK	B32CF
11	S28BM	J25BK	A35BF	H21BD	L22AL	A26AJ	M11AF	L12AD	A13AB	M30CP	L36CL	N20CH
12	P33BP	U35BL	Z16BG	S17BE	D29AM	Z38AK	H31AG	D37AE	Z19AC	H14CR	D32CM	A28CJ
13	W15CD	R16BM	C24BH	P22BF	J34BC	C11AL	S18AH	J27AF	C30AD	S23CS	J20CP	Z33CK
14	Y25CE	X24BP	K21BJ	W29BG	U26BD	K31AM	P12AJ	U13AG	K14AE	P36AB	U28CR	C15CL
15	T35CF	B21CD	F17BK	Y34BH	R38BE	F18BC	W37AK	R19AH	F23AF	W32AC	R33CS	K25CM
16	G16CH	N17CE	V22BL	T26BJ	X11BF	V12BD	Y27AL	X30AJ	V36AG	Y20AD	X15AB	F35CP
17	L24CJ	A22CF	E29BM	G38BK	B31BG	E37BE	T13AH	B14AK	E32AH	T28AE	B25AC	V16CR
18	D21CK	Z29CH	M34BP	L11BL	N18BH	M27BF	G19BC	N23AL	M20AJ	G33AF	N35AD	E24CS
19	J17CL	C34CJ	H26CD	D31BM	A12BJ	H13BG	L30BD	A36AM	H28AK	L15AG	A16AE	M21AB
20	U22CM	K26CK	S38CE	J18BP	Z37BK	S19BH	D14BE	Z32BC	S33AL	D25AH	Z24AF	H17AC
21	R29CP	F38CL	P11CF	U12CD	C27BL	P30BJ	J23BF	C20BD	P15AM	J35AJ	C21AG	S22AD
22	X34CR	V11CM	W31CH	R37CE	K13BM	W14BK	U36BG	K28BE	W25BC	U16AK	K17AH	P29AE
23	B26CS	E31CP	Y18CJ	X27CF	F19BP	Y23BL	R32BH	F33BF	Y35BD	R24AL	F22AJ	W34AF
24	N38AB	M18CR	T12CK	B13CH	V30CD	T36BM	X20BJ	V15BG	T16BE	X21AM	V29AK	Y26AG
25	A11AC	H12CS	G37CL	N19CJ	E14CE	G32BP	B28BK	E25BH	G24BF	B17BC	E34AL	T38AH
26	Z31AD	S37AB	L27CM	A30CK	M23CF	L20CD	N33BL	M35BJ	L21BG	N22BD	M26AM	G11AJ
27	C18AE	P27AC	D13CP	Z14CL	H36CH	D28CE	A15BM	H16BK	D17BH	A29BE	H38BC	L31AK
28	K12AF	W13AD	J19CR	C23CM	S32CJ	J33CF	Z25BP	S24BL	J22BJ	Z34BF	S11BD	D18AL
29	F37AG	Y19AE	U30CS	K36CP	P20CK	U15CH	C35CD	P21BM	U29BK	C26BG	P31BE	J12AM
30	V27AH		R14AB	F32CR	W28CL	R25CJ	K16CE	W17BP	R34BL	K38BH	W18BF	U37BC
31	E13AJ		X23AC		Y33CM		F24CF	Y22CD		F11BJ		R27BD

1873	JAN	FEB	MAR	APR	MAY	JUN	JUL	AUG	SEP	OCT	NOV	DEC
1	X13BE	V14BC	P14AH	U32AF	C28AC	P25CS	J16CH	C17CK	P34CH	J36CD	C18BM	S37BJ
2	B19BF	E23BD	W23AJ	R20AG	K33AD	W35AB	U24CP	K22CL	W26CJ	U11CE	K12BP	P27BK
3	N30BG	M36BE	Y36AK	X28AH	F15AE	Y16AC	R21CR	F29CH	Y38CK	R31CF	F37CD	W13BL
4	A14BH	H32BF	T32AL	B33AJ	V25AF	T24AD	X17CS	V34CP	T11CL	X18CH	V27CE	Y19BM
5	Z23BJ	S20BG	G20AM	N15AK	E35AG	G21AE	B22AB	E26CR	G31CH	B12CJ	E13CF	T30BP
6	C36BK	P28BH	L28BC	A25AL	M16AH	L17AF	N29AC	M38CS	L18CP	N37CK	M19CH	G14CD
7	K32BL	W33BJ	D33BD	Z35AM	H24AJ	D22AG	A34AD	H11AB	D12CR	A27CL	H30CJ	L23CE
8	F20BM	Y15BK	J15BE	C16BC	S21AK	J29AH	Z26AE	S31AC	J37CS	Z13CH	S14CK	D36CF
9	V28BP	T25BL	U25BF	K24BD	P17AL	U34AJ	C38AF	P18AD	U27AB	C19CP	P23CL	J32CH
10	E33CD	G35BM	R35BG	F21BE	W22AM	R26AK	K11AG	W12AE	R13AC	K30CR	W36CH	U20CJ
11	M15CE	L16BP	X16BH	V17BF	Y29BC	X38AL	F31AH	Y37AF	X19AD	F14CS	Y32CP	R28CK
12	H25CF	D24CD	B24BJ	E22BG	T34BD	B11AM	V18AJ	T27AG	B30AE	V23AB	T20CR	X33CL
13	S35CH	J21CE	N21BK	M29BH	G26BE	N31BC	E12AK	G13AH	N14AF	E36AC	G28CS	B15CM
14	P16CJ	U17CF	A17BL	H34BJ	L38BF	A19BD	M37AL	L19AJ	A23AG	M32AD	L33AB	N25CP
15	W24CK	R22CH	Z22BM	S26BK	D11BG	Z12BE	H27AM	D30AK	Z36AH	H20AE	D15AC	A35CR
16	Y21CL	X29CJ	C29BP	P38BL	J31BH	C37BF	S13BC	J14AL	C32AJ	S28AF	J25AD	Z16CS
17	T17CM	B34CK	K34CD	W11BM	U18BJ	K27BG	P19BD	U23AM	K20AK	P33AG	U35AE	C24AB
18	G22CP	N26CL	F26CE	Y31BP	R12BK	F13BH	W30BE	R36BC	F28AL	W15AH	R16AF	K21AC
19	L29CR	A38CM	V38CF	T18CD	X37BL	V19BJ	Y14BF	X32BD	V33AM	Y25AJ	X24AG	F17AD
20	D34CS	Z11CP	E11CH	G12CE	B27BM	E30BK	T23BG	B20BE	E15BC	T35AK	B21AH	V22AE
21	J26AB	C31CR	M31CJ	L37CF	N13BP	M14BL	G36BH	N28BF	M25BD	G16AL	N17AJ	E29AF
22	U38AC	K18CS	H18CK	D27CH	A19CD	H23BM	L32BJ	A33BG	H35BE	L24AM	A22AK	M34AG
23	R11AD	F12AB	S12CL	J13CJ	Z30CE	S36BP	D20BK	Z15BH	S16BF	D21BC	Z29AL	H26AH
24	X31AE	V37AC	P37CM	U19CK	C14CF	P32CD	J28BL	C25BJ	P24BG	J17BD	C34AM	S38AJ
25	B18AF	E27AD	W27CP	R30CL	K23CH	W20CE	U33BM	K35BK	W21BH	U22BE	K26BC	P11AK
26	N12AG	M13AE	Y13CR	X14CH	F36CJ	Y28CF	R15BP	F16BL	Y17BJ	R29BF	F38BD	W31AL
27	A37AH	H19AF	T19CS	B23CP	V32CK	T33CH	X25CD	V24BM	T22BK	X34BG	V11BE	Y18AM
28	Z27AJ	S30AG	G30AB	N36CR	E20CL	G15CJ	B35CE	E21BP	G29BL	B26BH	E31BF	T12BC
29	C13AK		L14AC	A32CS	M28CH	L25CK	N16CF	M17CD	L34BM	N38BJ	M18BG	G37BD
30	K19AL		D23AD	Z20AB	H33CP	D35CL	A24CH	H22CE	D26BP	A11BK	H12BH	L27BE
31	F30AM		J36AE		S15CR		Z21CJ	S29CF		Z31BL		D13BF

1874	JAN	FEB	MAR	APR	MAY	JUN	JUL	AUG	SEP	OCT	NOV	DEC
1	J19BG	C23BE	M23AK	L20AH	N33AE	M35AC	G24CR	N22CM	M26CK	G11CF	N12CD	E27BL
2	U30BH	K36BF	F36AL	D28AJ	A15AF	H16AD	L21CS	A29CP	H38CL	L31CH	A37CE	M13BM
3	R14BJ	F32BG	S32AM	J33AK	Z25AG	S24AC	D17AB	Z34CR	S11CH	D18CJ	Z27CF	H19BP
4	X23BK	V20BH	P20BC	U15AL	C35AH	P21AF	J22AC	C26CS	P31CP	J12CK	C13CH	S30CD
5	B36BL	E28BJ	W28BD	R25AM	K16AJ	W17AG	U29AD	K38AB	W18CR	U37CL	K19CJ	P14CE
6	N32BM	M33BK	Y33BE	X35BC	F24AK	Y22AH	R34AE	F11AC	Y12CS	R27CM	F30CH	W23CF
7	A20BP	H15BL	T15BF	B16BD	V21AL	T29AJ	X26AF	V31AD	T37AB	X13CP	V14CL	Y36CH
8	Z28CD	S25BM	G25BG	N24BE	E17AM	G34AK	B38AG	E18AE	G27AC	B19CR	E23CH	T32CJ
9	C33CE	P35BP	L35BH	A21BF	M22BC	L26AL	N11AH	M12AF	L13AD	N30CS	M36CP	G28CK
10	K15CF	W16CD	D16BJ	Z17BG	H29BD	D38AM	A31AJ	H37AG	D19AE	A14AB	H32CR	L28CL
11	F25CH	Y24CE	J24BK	C22BH	S34BE	J11BC	Z18AK	S27AH	J30AF	Z23AC	S20CS	D33CM
12	V35CJ	T21CF	U21BL	K29BJ	P26BF	U31BD	C12AL	P13AJ	U14AG	C36AD	P28AB	J15CP
13	E16CK	G17CH	R17BM	F34BK	W38BG	R18BE	K37AM	W19AK	R23AH	K32AE	W33AC	U25CR
14	M24CL	L22CJ	X22BP	W26BL	Y11BH	X12BF	P27BD	Y30AL	X36AJ	P20AF	Y15AD	R35CS
15	H21CM	D29CK	B29CD	E38BM	T31BJ	B37BG	V13BD	T14AM	B32AK	V28AG	T25AE	X16AB
16	S17CP	J34CL	N34CE	M11BP	G18BK	N27BH	E19BC	G23BF	N20AL	E33AH	G35AF	B24AC
17	P22CR	U26CM	A26CF	H31CD	L12BL	A13BJ	M30BF	L36BD	A28AM	M15AJ	L16AG	N21AD
18	W29CS	R38CP	Z38CH	S18CE	D37BM	Z19BK	H14BG	D32BE	Z33BC	H25AK	D24AH	A17AE
19	Y34AB	X11CR	C11CJ	P12CF	J27BP	C30BL	S23BH	J20BF	C15BD	S35AL	J21AJ	Z22AF
20	T26AC	B31CS	K31CK	W37CH	U13CD	K14BM	P36BJ	U28BG	K25BE	P16AM	U17AK	C29AG
21	G38AD	N18AB	F18CL	Y27CJ	R19CE	F23BP	W32BH	R33BF	F35BD	W22AL	R22AJ	K34AH
22	L11AE	A12AC	V12CM	T13CK	X30CF	V36CD	Y20BL	X15BJ	V16BG	Y21BD	X29AM	F26AJ
23	D31AF	Z37AD	E37CP	G19CL	B14CH	E32CE	T28BM	B25BK	E24BH	T17BE	B34BC	V38AK
24	J18AG	C27AE	M27CR	L30CM	N23CJ	M20CF	G33BP	N35BL	M21BJ	G22BF	N26BD	E11AL
25	U12AH	K13AF	H13CS	D14CP	A36CK	H28CH	L15CD	A17AM	H17BK	L29BG	A38BE	M31AM
26	R37AJ	F19AG	S19AB	J23CR	Z32CL	S33CJ	D25CE	Z24BP	S22BL	D34BH	Z11BF	H18BC
27	X27AK	V30AH	P30AC	U36CS	C20CM	P15CK	J35CD	C21BM	P29BM	J26BJ	C31BG	S12BD
28	B13AL	E14AJ	W14AD	R32AB	K28CP	W25CL	U16CH	K17CE	W34BP	U38BK	K18BH	P37BE
29	N19AM		Y23AE	X20AC	F33CR	Y35CM	R24CJ	F22CF	Y26CD	R11BL	F12BJ	W27BF
30	A30BC		T36AF	B28AD	V15CS	T16CP	X21CK	V29CH	T38CE	X31BM	V37BK	Y13BG
31	Z14BD		G32AG		E25AB		B17CL		E34CJ		B18BP	T19BH

FIRST LETTER = YOUR PHYSICAL CODE, MIDDLE NUMBER = SENSITIVITY CODE, LAST TWO LETTERS = INTELLECTUAL CODE

1875	JAN	FEB	MAR	APR	MAY	JUN	JUL	AUG	SEP	OCT	NOV	DEC
1	G30BJ	N36BG	F36AM	Y28AK	R15AG	F16AE	W21AB	R29CR	F38CM	H31CJ	R37CF	K13BP
2	L14BK	A32BH	V32BC	T33AL	X25AH	V24AF	Y17AC	X34CS	V11CP	Y18CK	X27CH	F19CD
3	D23BL	Z20BJ	E20BD	G15AM	B35AJ	E21AG	T22AD	B26AB	E31CR	T12CL	B13CJ	V30CE
4	J36BM	C28BK	M28BE	L25BC	N16AK	M17AH	G29AE	N38AC	M18CS	G37CM	N19CK	E14CF
5	U32BP	K33BL	H33BF	D35BD	A24AL	H22AJ	L34AF	A11AD	H12AB	L27CP	A30CL	M23CH
6	R20CD	F15BM	S15BG	J16BE	Z21AM	S29AK	D26AG	Z31AE	S37AC	D13CR	Z14CM	H36CJ
7	X28CE	V25BP	P25BH	U24BF	C17BC	P34AL	J38AH	C18AF	P27AD	J19CS	C23CP	S32CK
8	B33CF	E35CD	W35BJ	R21BG	K22BD	W26AM	U11AJ	K12AG	W13AE	U30AB	K36CR	P20CL
9	N15CH	M16CE	Y16BK	X17BH	F29BE	Y38BC	R31AK	F37AH	Y19AF	R14AC	F32CS	W28CM
10	A25CJ	H24CF	T24BL	B22BJ	V34BF	T11BD	X18AL	V27AJ	T30AG	X23AD	V20AB	Y33CP
11	Z35CK	S21CH	G21BM	N29BK	E26BG	G31BE	B12AM	E13AK	G14AH	B36AE	E28AC	T15CR
12	C16CL	P17CJ	L17BP	A34BL	M38BH	L18BF	N37BC	M19AL	L23AJ	N32AF	M33AD	G25CS
13	K24CM	W22CK	D22CD	Z26BM	H11BJ	D12BG	A27BD	H30AM	D36AK	A20AG	H15AE	L35AB
14	F21CP	Y29CL	J29CE	C38BP	S31BH	J37BH	Z13BE	S14BC	J32AL	Z28AH	S25AF	D16AC
15	V17CR	T34CM	U34CF	K11CD	P18BL	U27BJ	C19BF	P23BD	U20AM	C33AJ	P35AG	J24AD
16	E22CS	G26CP	R26CH	F31CE	W12BM	R13BK	K30BG	W36BE	R28BC	K15AK	W16AH	U21AE
17	M29AB	L38CR	X38CJ	V18CF	Y37BP	X19BL	F14BH	Y32BF	X33BD	F25AL	Y24AJ	R17AF
18	H34AC	D11CS	B11CH	E12CH	T27CD	B30BM	V23BJ	T20BG	B15BE	V35AM	T21AK	X22AG
19	S26AD	J31AB	N31CL	M37CJ	G13CE	N14BP	E36BK	G28BH	N25BF	E16BC	G17AL	B29AH
20	P38AE	U18AC	A18CM	H27CK	L19CF	A23CD	M32BL	L33BJ	A35BG	M24BD	L22AM	N34AJ
21	W11AF	R12AD	Z12CP	S13CL	D30CH	Z36CE	H20BM	D15BK	Z16BH	H21BE	D29BC	A26AK
22	Y31AG	X37AE	C37CR	P19CM	J14CJ	C32CF	S28BP	J25BL	C24BJ	S17BF	J34BD	Z38AL
23	T18AH	B27AF	K27CS	W30CP	U23CK	K20CH	P33CD	U35BM	K21BK	P22BG	U26BE	C11AM
24	G12AJ	N13AG	F13AB	Y14CR	R36CL	F28CJ	W15CE	R16BP	F17BL	W29BH	R38BF	K31BC
25	L37AK	A19AM	V19AC	T23CS	X32CH	V33CK	Y25CF	X24CD	V22BM	Y34BJ	X11BG	F18BD
26	D27AL	Z30AJ	E30AD	G36AB	B20CP	E15CL	T35CH	B21CE	E29BP	T26BK	B31BH	V12BE
27	J13AM	C14AK	M14AE	L32AC	N28CR	M25CM	G16CJ	N17CE	M34CD	G38BL	N18BJ	E37BG
28	U19BC	K23AL	H23AF	D20AD	A33CS	H35CP	L24CK	A22CH	H26CE	L11BM	A12BK	M27BG
29	R30BD		S36AG	J28AE	Z15AB	S16CR	D21CL	Z29CJ	S38CF	D31BP	Z37BL	H13BH
30	X14BE		P32AH	U33AF	C25AC	P24CS	J17CM	C34CK	P11CH	J18CD	C27BM	S19BJ
31	B23BF		W20AJ		K35AD		U22CP	K26CL		U12CE		P30BK

1876	JAN	FEB	MAR	APR	MAY	JUN	JUL	AUG	SEP	OCT	NOV	DEC
1	W14BL	R32BJ	C20BE	P15BC	J35AK	C21AH	S22AE	J26AC	C31CS	S12CM	J13CK	Z30CF
2	Y23BM	X20BK	K28BF	W25BD	U16AL	K17AJ	P29AF	U38AD	K18AB	P37CP	U19CL	C14CH
3	T36BP	B28BL	F33BG	Y35BE	R24AH	F22AK	W34AG	R11AE	F12AC	W27CR	R30CH	K23CJ
4	G32CD	N33BM	V15BH	T16BF	X21BC	V29AL	Y26AH	X31AF	V37AD	Y13CS	X14CP	F36CK
5	L20CE	A15BP	E25BJ	G24BG	B17BD	E34AM	T38AJ	B18AG	E27AE	T19AB	B23CR	V32CL
6	D28CF	Z25CD	M35BK	L21BH	M26BC	G11AK	N12AH	M13AF	G30AC	N36CS	E20CM
7	J33CH	C35CE	H16BL	D17BJ	A29BF	H38BD	L31AL	A37AJ	H19AG	L14AD	A32AB	M28CP
8	U15CJ	K16CF	S24BM	J22BK	Z34BG	S11BE	D18AM	Z27AK	S30AH	D23AE	Z20AC	H33CR
9	R25CK	F24CH	P21BP	U29BL	C26BH	P31BF	J12BC	C13AL	P14AJ	J36AF	C28AD	S15CS
10	X35CL	V21CJ	M17CD	R34BM	K38BJ	M18BG	U37BD	K19AM	M23AK	U32AG	K33AE	P25AB
11	B16CM	E17CK	Y22CE	X26BP	F11BK	Y12BH	R27BE	F30BC	Y36AL	R20AH	F15AF	W35AC
12	N24CP	M22CL	T29CF	B33CD	V31BL	T37BJ	X13BF	V14BD	T32AM	X28AJ	V25AG	Y16AD
13	A21CR	H29CM	G34CH	N11CE	E18BM	G27BK	B19BG	E23BE	G20BC	B33AK	E35AH	T24AE
14	Z17CS	S34CP	L26CJ	A31CF	M12BP	L13BL	N30BH	M36BF	L28BD	N15AL	M16AJ	G21AF
15	C22AB	P26CR	D38CK	Z18CH	H37CD	D19BM	A14BJ	H32BG	D33BE	A25AM	H24AK	L17AG
16	K29AC	W38CS	J11CL	C12CJ	S27CE	J30BP	Z23BK	S20BH	J15BF	Z35BC	S21AL	D22AH
17	F34AD	Y11AB	U31CM	K37CK	P13CF	U14CD	C36BL	P28BJ	U25BG	C16BD	P17AM	J29AJ
18	V26AE	T31AC	R18CP	F27CL	W19CH	R23CE	K32BH	W33BK	R35BH	K24BE	W22BC	U34AK
19	E38AF	G18AD	X12CR	V13CM	Y30CJ	X36CF	F20BP	Y15BL	X16BJ	F21BF	Y29BD	R26AL
20	M11AG	L12AE	B37CS	E19CP	T14CK	B32CH	V28CD	T25BM	B24BK	V17BG	T34BE	X38AM
21	H31AH	D37AF	N27AB	M30CR	G23CL	N20CJ	E33CE	G35BP	N21BL	E22BH	G26BF	B11BC
22	S18AJ	J27AG	A13AC	H14CS	L36CM	A28CK	M15CF	L16CD	A17BM	M29BJ	L38BG	N31BD
23	P12AK	U13AH	Z19AD	S23AB	D32CP	Z33CL	H25CH	D24CE	Z22BP	H34BK	D11BH	A18BE
24	W37AL	R19AJ	C30AE	P36AC	J20CR	C15CM	S35CJ	J21CF	C29CD	S26BL	J31BJ	Z12BF
25	Y27AM	X30AK	K14AF	W32AD	U28CS	K25CP	P16CK	U17CH	K34CE	P38BH	U18BK	C37BG
26	T13BC	B14AL	F23AG	Y20AE	R33AB	F35CR	W24CL	R22CJ	F26CF	W11BP	R12BL	K27BH
27	G19BD	N23AM	V36AH	T28AF	X15AC	V16CS	Y21CM	X29CK	V38CH	Y31CD	X37BH	F13BJ
28	L30BE	A36BC	E32AJ	G33AG	B25AD	E24AB	T17CP	B34CL	E11CJ	T18CE	B27BP	V19BK
29	D14BF	Z32BD	M20AK	L15AH	N35AE	M21AC	G22CR	N26CM	M31CK	G12CF	N13CD	E30BL
30	J23BG		H28AL	D25AJ	A16AF	H17AD	L29CS	A38CP	H18CL	L37CH	A19CE	M14BM
31	U36BH		S33AM		Z24AG		D34AB	Z11CR		D27CJ		H23BP

1877	JAN	FEB	MAR	APR	MAY	JUN	JUL	AUG	SEP	OCT	NOV	DEC
1	S36CD	J28BM	N28BG	M25BE	G16AM	N17AK	E29AG	G38AE	N18AC	E37CR	G19CM	B14CJ
2	P32CE	U33BP	A33BH	H35BF	L24BC	A22AL	M34AH	L11AF	A12AD	M27CS	L30CP	N23CK
3	W20CF	R15CD	Z15BJ	S16BG	D21BD	Z29AM	H26AJ	D31AG	Z37AE	H13AB	D14CR	A36CL
4	Y28CH	X25CE	C25BK	P24BH	J17BC	C34BC	S38AK	J18AH	C27AF	S19AC	J23CS	Z32CM
5	T33CJ	B35CF	K35BL	W21BJ	U22BF	K26BD	P11AL	U12AJ	K13AG	P30AD	U36AB	C20CP
6	G15CM	N16CH	F16BM	Y17BK	R29BG	F38BE	W31AM	R37AK	F19AH	W14AE	R32AC	K28CR
7	L25CL	A24CJ	V24BP	T22BL	X34BH	V11BF	Y18BC	X27AL	V30AJ	Y23AF	X20AD	F33CS
8	D35CH	Z21CK	E21CD	G29BM	B26BJ	E31BG	T12BD	B13AM	E14AK	T36AG	B28AE	V15AB
9	J16CP	C17CL	M17CE	L34BP	N38BK	M18BH	G37BE	N19BC	M23AL	G32AH	N33AF	E25AC
10	U24CR	K22CH	H22CF	D26CD	A11BL	H28BJ	L27BF	A30BD	H36AM	L20AJ	A15AG	M35AD
11	R21CS	F29CP	S29CH	J38CE	Z31BM	S37BK	D13BG	Z14BE	S32BC	D28AK	Z25AH	H16AE
12	X17AB	V34CR	P34CJ	U11CF	C18BP	P27BL	J19BH	C23BF	P20BD	J33AL	C35AJ	S24AF
13	B22AC	E26CS	W26CK	R31CH	K12CD	W13BM	U30BJ	K36BG	W28BE	U15AM	K16AK	P21AG
14	N29AD	M38AB	Y38CL	X16CJ	F37CE	Y19BP	R14BK	F32BH	Y33BF	R25BC	F24AL	W17AH
15	A34AE	H11AC	T11CM	B12CK	V27CF	T30CD	X23BL	V20BJ	T15BG	X35BD	V21AM	Y22AJ
16	Z26AF	S31AD	G31CP	N37CL	E13CH	G14CE	B36BM	E28BK	G25BH	B16BE	E17BC	T29AK
17	C38AG	P18AE	L18CR	A27CM	M19CJ	L23CF	N32BP	M33BF	L35BJ	N24BF	M22BD	G34AL
18	K11AH	W12AF	D12CS	Z13CP	H30CK	D36CH	A20CD	H15BM	D16BK	A21BG	H29BE	L26AM
19	F31AJ	Y37AG	J37AB	C19CR	S14CL	J32CJ	Z28CE	S12BP	J24BL	Z17BH	S34BF	D38BG
20	V18AK	T27AH	U27AC	K30CS	P23CM	U20CK	C33CF	P35CD	U21BM	C22BJ	P26BG	J11BD
21	E12AL	G13AJ	R13AD	F14AB	W36CP	R28CL	K15CH	W16CF	R17BP	K29BK	W38BH	U31BE
22	M37AH	L19AK	X19AE	V23AC	Y32CH	X33CM	F25CJ	Y24CF	X22CD	F34BL	Y11BJ	R18BF
23	H27BC	D30AL	B30AF	E36AD	T20CS	B15CP	V35CK	T21CH	B29CE	V26BM	T31BK	X12BG
24	S13BD	J14AM	N14AG	M32AE	G28AB	N25CR	E16CL	G17CJ	N34CF	E38BP	G18BL	B37BH
25	P19BE	U23BC	A23AH	H20AF	L33AC	A35CS	M24CH	L22CC	A26CM	M11CD	L12BM	N27BJ
26	W30BF	R36BD	Z36AJ	S28AG	D15AD	Z16AB	H21CP	D29CL	Z38CK	H31CE	D37BP	A13BK
27	Y14BG	X32BE	C32AK	P33AH	J25AE	C24AC	S17CR	J34CM	C11CK	S18CF	J27CD	Z19BL
28	T23BH	B20BF	K20AL	W15AJ	U35AF	K21AD	P22CS	U26CP	K31CL	P12CH	U13CE	C30BM
29	G36BJ		F28AM	Y25AK	R16AG	F17AE	W29AB	R38CH	F18CM	W37CJ	R19CF	K14BP
30	L32BK		V33BC	T35AL	X24AH	V22AF	Y34AC	X11CS	V12CP	Y27CK	X30CH	F23CD
31	D20BL		E15BD		B21AJ		T26AD	B31AB		T13CL		V36CE

FIRST LETTER = YOUR PHYSICAL CODE, MIDDLE NUMBER = SENSITIVITY CODE, LAST TWO LETTERS = INTELLECTUAL CODE

1878

	JAN	FEB	MAR	APR	MAY	JUN	JUL	AUG	SEP	OCT	NOV	DEC
1	E32CF	G33CD	R33BJ	F35BG	W24BD	R22AM	K34AJ	N11AG	R12AE	K27AB	W30CR	U23CL
2	M20CH	L15CE	X15BK	V16BH	Y21BE	X29BC	F26AK	Y31AH	X37AF	F13AC	Y14CS	R36CM
3	H28CJ	D25CF	B25BL	E24BJ	T17BF	B34BD	V38AL	T18AJ	B27AG	V19AD	T23AB	X32CP
4	S33CK	J35CH	N35BM	M21BK	G22BG	N26BE	E11AM	G12AK	N13AH	E30AE	G36AC	B20CR
5	P15CL	U16CJ	A16BP	H17BL	L29BH	A38BF	M31BC	L37AL	A19AJ	M14AF	L32AD	N28CS
6	W25CM	R24CK	Z24CD	S22BM	D34BJ	Z11BG	H18BD	D27AM	Z30AK	H23AG	D20AE	A33AB
7	Y35CP	X21CL	C21CE	P29BP	J26BK	C31BH	S12BE	J13BC	C14AL	S36AH	J28AF	Z15AC
8	T16CR	B17CM	K17CF	W34CD	U38BL	K18BJ	P37BF	U19BD	K23AH	P32AJ	U33AG	C25AD
9	G24CS	N22CP	F22CH	Y26CE	R11BM	F12BK	W27BG	R30BE	F36BC	W20AK	R15AH	K35AE
10	L21AB	A29CR	V29CJ	T38CF	X31BP	V37BL	Y13BH	X14BF	V32BD	Y28AL	X25AJ	F16AF
11	D17AC	Z34CS	E34CR	G11CH	B18CD	E27BM	T19BJ	B23BG	E20BE	T33AM	B35AK	V24AG
12	J22AD	C26AB	M26CL	L31CJ	N12CE	M13BP	G30BK	N36BH	M28BF	G15BC	N16AL	E21AH
13	U29AE	K38AC	H38CM	D18CK	A37CF	H19CD	L14BL	A32BJ	H33BG	L25BD	A24AM	M17AJ
14	R34AF	F11AD	S11CP	J12CL	Z27CH	S30CE	D23BM	Z20BK	S15BH	D35BE	Z21BG	H22AK
15	X26AG	V31AE	P31CR	U37CM	C13CJ	P14CF	J36BP	C28BL	P25BJ	J16BF	C17BD	S29AL
16	B38AH	E18AF	W18CS	R27CP	K19CK	W23CH	U32CD	K33BM	W35BK	U24BG	K22BE	P34AM
17	N11AJ	M12AG	Y12AB	X13CR	F30CL	Y36CJ	R20CE	F15BP	Y16BL	R21BH	F29BF	W26BC
18	A31AK	H37AH	T37AC	B19CS	V14CM	T32CK	X28CF	V25CD	T24BM	X17BJ	V34BG	Y38BD
19	Z18AL	S27AJ	G27AD	N30AB	E23CP	G20CL	B33CH	E35CE	G21BP	B22BK	E26BH	T11BE
20	C12AM	P13AK	L13AE	A14AC	M36CR	L28CH	N15CJ	M16CF	L17CD	N29BL	M38BJ	G31BF
21	K37BC	W19AL	D19AF	Z23AD	H32CS	D33CP	A25CK	H24CH	D22CE	A34BM	H11BK	L18BG
22	F27BD	Y30AM	J30AG	C36AE	S20AB	J15CR	Z35CL	S21CJ	J29CF	Z26BP	S31BL	D12BH
23	V13BE	T14BC	U14AH	K32AF	P28AC	U25CS	C16CM	P17CK	U34CH	C38CD	P18BM	J37BJ
24	E19BF	G23BD	R23AJ	F20AG	W33AD	R35AB	K24CP	W22CL	R26CJ	K11CE	W28BP	U27BK
25	M30BG	L36BE	X36AK	V28AH	Y15AE	X16AC	F21CR	Y29CH	X38CK	F31CF	Y37CD	R13BL
26	H14BH	D32BF	B32AL	E33AJ	T25AF	B24AD	V17CS	T34CP	B11CL	V18CH	T27CE	X19BM
27	S23BJ	J20BG	N20AM	M15AK	G35AG	N21AE	E22AB	G26CR	N31CM	E12CJ	G13CF	B30BP
28	P36BK	U28BH	A28BC	H25AL	L16AH	A17AF	M29AC	L38CS	A18CP	M37CK	L19CH	N14CD
29	W32BL		Z33BD	S35AM	D24AJ	Z22AG	H34AD	D11AB	Z12CR	H27CL	D30CJ	A23CE
30	Y20BM		C15BE	P16BC	J21AK	C29AH	S26AE	J31AC	C37CS	S13CM	J14CK	Z36CF
31	T28BP		K25BF		U17AL		P38AF	U18AD		P19CP		C32CH

1879

	JAN	FEB	MAR	APR	MAY	JUN	JUL	AUG	SEP	OCT	NOV	DEC
1	K20CJ	W15CF	D15BL	Z16BJ	H21BF	D29BD	A26AL	H31AJ	D37AG	A13AD	H14AB	L36CP
2	F28CK	Y25CH	J25BK	C24BK	S17BG	J34BE	Z38AM	S18AK	J27AH	Z19AE	S23AC	D32CR
3	V33CL	T35CJ	U35BP	K21BL	P22BH	U26BF	C11BC	P12AL	U13AJ	C30AF	P36AD	J20CS
4	E15CM	G16CK	R16CD	F17BM	W29BJ	R38BG	K31BD	W37AM	R19AK	K14AG	W32AF	U28AB
5	M25CP	L24CL	X24CE	V22BP	Y34BK	X11BH	F18BE	Y27BC	X30AL	F23AH	Y20AF	R33AC
6	H35CR	D21CM	B21CF	E29CD	T26BL	B31BJ	V12BF	T13BD	B14AM	V36AJ	T28AG	X15AD
7	S16CS	J17CP	N17CH	M34CE	G38BM	N18BK	E37BG	G19BE	N23BC	E32AK	G33AH	B25AE
8	P24BH	U22CR	A22CJ	H26CF	L11BP	A12BL	M27BH	L30BF	A36BD	M20AL	L15AJ	N35AF
9	W21AC	R29CS	Z29CK	D31CD	Z37BM	H13BJ	D14BG	Z28BE	H28AM	D25AK	A16AG	
10	Y17AD	X34AB	C34CL	P11CJ	J18CE	C27BP	S19BK	J23BH	G20BF	S33BC	J35AL	Z24AH
11	T22AE	B26AC	K26CH	W31CK	U12CF	K13CD	P30BL	U36BJ	K28BG	P15BD	U16AM	C21AJ
12	G29AF	N38AD	F38CP	Y18CL	R37CH	F19CE	W14BM	R32BK	F33BH	W25BE	R24BC	K17AK
13	L34AG	A11AF	V11CR	T12CM	X27CJ	V30CF	Y23BP	X20BL	V15BJ	Y35BF	X21BD	F22AL
14	D26AH	Z31AF	E31CS	G37CP	B13CK	E14CH	T36CD	B28BM	E25BK	T16BG	B17BE	V29AM
15	J38AJ	C18AG	M18AB	L27CR	N19CL	M23CJ	G32CE	N33BP	M35BL	G24BH	N22BF	E34BC
16	U11AK	K12AH	H12AC	D13CS	A38CM	H36CK	L20CF	A15CD	H16BM	L21BJ	A29BG	M26BD
17	R31AL	F37AJ	S37AD	J19AB	Z14CP	S32CL	D28CH	Z25CE	S24BP	D17BK	Z34BH	H38BE
18	X18AM	V27AK	P27AE	U30AC	C23CR	P20CH	J33CJ	C35CF	P21CD	J22BL	C26BJ	S11BF
19	B12BC	E13AL	W13AF	R14AD	K36CS	W28CP	U15CK	K16CH	W17CE	U29BM	K38BK	P31BG
20	N37BD	M19AM	Y19AG	X23AE	F32AB	Y33CR	R25CL	F24CJ	Y22CF	R34BP	F11BL	W18BH
21	A27BE	H30BC	T30AH	B36AF	V20AC	T15CS	X35CM	V21CK	T29CH	X26CD	V31BM	Y12BJ
22	Z13BF	S14BD	G14AJ	N32AG	E28AD	G25AB	B16CP	E17CL	G34CJ	B38CE	E18BP	T37BK
23	C19BG	P23BE	L23AK	A20AH	M33AE	L35AC	N24CR	M22CH	L26CK	N11CF	M12CD	G27BL
24	K30BH	W36BF	D36AL	Z28AJ	H15AF	D16AD	A21CS	H29CP	D38CL	A31CH	H37CE	L13BM
25	F14BJ	Y32BG	J32AM	C33AK	S25AG	J24AE	Z17AB	S34CR	J11CH	Z16CJ	S27CF	D19BP
26	V23BK	T28BH	U20BC	K15AL	P35AH	U21AF	C22AC	P26CS	U31CP	C12CK	P13CH	J30CD
27	E36BL	G28BJ	R28BD	F25AM	W16AJ	R17AG	K29AD	W38AB	R18CR	K37CL	W19CJ	U14CE
28	M32BM	L33BK	X33BE	V35BC	Y24AK	X22AH	F34AE	Y11AC	X12CS	F27CM	Y30CK	R23CF
29	H20BP		B15BF	E16BD	T21AL	B29AJ	V26AF	T31AD	B37AB	V13CP	T14CL	X36CH
30	S28CD		N25BG	M24BE	G17AM	N34AK	E38AG	G18AE	N27AC	E19CR	G23CM	B32CJ
31	P33CE		A35BH		L22BC		M11AH	L12AF		M30CS		N20CK

1880

	JAN	FEB	MAR	APR	MAY	JUN	JUL	AUG	SEP	OCT	NOV	DEC
1	A28CL	H25CJ	G35CD	N21BM	E22BJ	G26BG	B11BD	E12AM	G13AK	B30AG	E36AE	T20AB
2	Z33CM	S35CK	L16CD	A17BP	M29BK	L38BH	N31BE	M37BC	L19AL	N14AH	M32AF	G28AC
3	C15CP	P16CL	D24CF	Z22CD	H34BL	D11BJ	A18BF	H27BD	D30AM	A23AJ	H20AG	L33AD
4	K25CR	W24CK	J21CH	O29CE	S26BM	J31BK	Z12BG	S13BE	J14BC	Z36AK	S28AH	D15AE
5	F35CS	Y21CP	U17CJ	K34CF	P38BP	U18BL	C37BH	P19BF	U23BD	C32AL	P33AJ	J25AF
6	V16AB	T17BG	R22CK	F26CH	W11CD	R28BM	K27BJ	W30BG	R36BE	K20AM	W15AK	U35AG
7	E24AC	G22CS	X29CL	V38CJ	Y31CE	X37BP	F13BM	Y14BH	X32BF	F28BC	Y25AL	R16AH
8	M21AD	L29AB	B34CM	E11CK	T18CF	B27CD	V19BL	T23BJ	B20BG	V33BD	T35AM	X24AJ
9	H17AE	D34AC	N26CP	M31CL	G12CE	N13CE	E30BM	G36BK	N28BH	E15BE	G16BC	B21AK
10	S22AF	J26AD	A38CR	H18CH	L37CJ	A19CF	M14BP	L32BM	A33BK	M25BF	L24BD	N17AL
11	P29AG	U38AE	Z11CS	S12CP	D27CK	Z30CH	H23CD	D20BM	Z15BK	H35BG	D21BE	A22AM
12	W34AH	R11AF	C31AB	P37CR	J13CL	C14CJ	S36CE	J28BP	C25BL	S16BH	J17BF	Z29BC
13	Y26AJ	X31AG	K18AC	W27CS	U19CM	K23CK	P32CF	U33BD	K35BM	P24BJ	U22BG	C34BD
14	T38AK	B18AH	F12AD	Y13AB	R30CP	F36CL	W20CH	R15CE	F16BP	W21BK	R29BH	K26BE
15	G11AL	N12AJ	V37AE	T19AC	X14CR	V32CH	Y28CP	X25CF	V24CD	Y17BL	X34BJ	F38BF
16	L31AM	A37AK	E27AF	G30AD	B23CS	E20CP	T33CK	B35CH	E21CE	T22BM	B26BK	V11BG
17	D18BC	Z27AL	M13AG	L14AE	N36AB	M28CR	G15CL	N16CJ	M17CF	G29BP	N38BL	E31BH
18	J12BD	C13AM	H19AH	D23AF	A32AC	H33CS	L25CM	A24CK	H22CH	L34CD	A11BM	M18BJ
19	U37BE	K19BC	S30AJ	J36AG	Z20AD	S15AB	D35CP	Z21CL	S29CJ	D26CE	Z31BP	H12BH
20	R27BF	F30BD	P14AK	U32AH	C28AE	P25AC	J16CR	C17CH	P34CK	J38CF	C18CD	S37BL
21	X13BG	V14BE	W23AL	R20AJ	K33AF	W35AD	U24CS	K22CP	W26CL	U11CH	K12CE	P27BM
22	B19BH	E23BF	Y36AM	X28AK	F15AG	Y16AE	T24AF	F29CR	Y38CH	T11CP	X18CK	W13BD
23	N30BJ	M36BG	T32BC	B33AL	V25AH	T24AF	X17AC	V34CS	T11CP	X18CK	V27CH	Y19CD
24	A14BK	H32BH	G20BD	N15AM	E35AG	G21AE	B22AD	E26AB	G31CR	B12CL	E13CJ	T30CE
25	Z23BL	S20BJ	L28BE	A25BC	M16AK	L17AH	N29AE	M38AC	L18CS	N37CH	M19CK	G14CF
26	C36BM	P28BK	D33BF	Z35BD	H24AL	D22AJ	A34AF	H11AD	D12AB	A27CP	H30CL	L23CH
27	K32BP	W33BL	J15BG	C16BE	S21AM	J29AG	Z26AG	S31AE	J37AC	Z13CR	S14CM	D36CJ
28	F20CD	Y15BM	U25BH	K24BF	P17BC	U34AL	C38AH	P18AF	U27AD	C19CS	P23CP	J32CK
29	V28CE	T25BP	R35BJ	F21BG	W22BD	R26AK	K11AJ	M12AG	R13AE	K30AB	W36CR	U20CL
30	E33CF		X16BK	V17BH	Y29BE	X38BC	F31AK	Y37AH	X19AF	F14AC	Y32CS	R28CM
31	M15CH		B24BL		T34BF		V18AL	T27AJ		V23AD		X33CP

FIRST LETTER = YOUR PHYSICAL CODE, MIDDLE NUMBER = SENSITIVITY CODE, LAST TWO LETTERS = INTELLECTUAL CODE

1881	JAN	FEB	MAR	APR	MAY	JUN	JUL	AUG	SEP	OCT	NOV	DEC
1	B15CR	E16CM	W16CF	R17CD	K29BL	M38BJ	U31BF	K37BD	W19AM	U14AJ	K32AG	P28AD
2	N25CS	M24CP	Y24CH	X22CE	F34BM	Y11BK	R18BG	F27BE	Y30BC	R23AK	F20AH	W33AE
3	A35AB	H21CR	T21CJ	B29CF	V26BP	T31BL	X12BH	V13BF	T14BD	X36AL	V28AJ	Y15AF
4	Z16AC	S17CS	G17CK	N34CH	E38CD	G18BM	B37BJ	E19BG	G23BE	B32AM	E33AK	T25AG
5	C24AD	P22AB	L22CL	A26CJ	M11CE	L12BP	N27BK	M30BH	L36BF	N20BC	M15AL	G35AH
6	K21AE	W29AC	D29CM	Z38CK	H31CF	D37CD	A13BL	H14BJ	D32BG	A28BD	H25AM	L16AJ
7	F17AF	Y34AD	J34CP	C11CL	S18CH	J27CE	Z19BM	S23BK	J20BH	Z33BE	S35BC	D24AK
8	V22AG	T26AE	U26CR	K31CH	P12CJ	U13CF	C30BP	P36BL	U28BJ	C15BF	P16BD	J21AL
9	E29AH	G38AF	R38CS	F18CP	W37CK	R19CH	K14CD	W32BM	R33BK	K25BG	W24BE	U17AM
10	M34AJ	L11AG	X11AB	V12CR	Y27CL	X30CJ	F23CE	Y20BP	X15BL	F35BH	Y21BF	R22BC
11	H26AK	D31AH	B31AC	E37CS	T13CM	B14CK	V36CF	T28CD	B25BM	V16BJ	T17BG	X29BD
12	S38AL	J18AJ	N18AD	M27AB	G19CP	N23CL	E32CH	G33CE	N35BP	E24BK	G22BH	B34BE
13	P11AM	U12AK	A12AE	H13AC	L30CR	A36CM	M20CJ	L15CF	A16CD	M21BL	L29BJ	N26BF
14	W31BC	R37AL	Z37AF	S19AD	D14CS	Z32CP	H28CK	D25CH	Z24CE	H17BM	D34BK	A38BG
15	Y18BD	X27AM	C27AG	P30AE	J23AB	C20CR	S33CL	J35CJ	C21CF	S22BP	J26BL	Z11BH
16	T12BE	B13BC	K13AH	W14AF	U36CK	K28CS	P15CM	U16CK	K17CH	P29CD	U38BM	C31BJ
17	G37BF	N19BD	F19AJ	Y23AG	R32AD	F33AB	W25CP	R24CL	F22CJ	W34CE	R11BP	K18BK
18	L27BG	A30BE	V30AK	T36AH	X20AE	V15AC	Y35CR	X21CM	V29CK	Y26CF	X31CD	F12BL
19	D13BH	Z14BF	E14AL	G32AJ	B28AF	E25AD	T16CS	B17CP	E34CL	T38CH	B18CE	V37BM
20	J19BJ	C23BG	M23AH	L20AK	N33AG	M35AE	G24AB	N22CR	M26CM	G11CJ	N12CF	E27BP
21	U30BK	K36BH	H36BG	D28AL	A15AH	H16AF	L21AC	A29CS	H38CP	L31CK	A37CH	M13CD
22	R14BL	F32BJ	S32BD	J33AM	Z25AJ	S24AG	D17AD	Z34AB	S11CR	D18CL	Z27CJ	H19CE
23	X23BM	V20BK	P20BE	U15BC	C35AK	P21AH	J22AE	C26AC	P31CS	J12CM	C13CK	S30CF
24	B36BP	E28BL	W28BF	R25BD	K16AL	W17AJ	U29AF	K38AD	W18AB	U37CP	K19CL	P14CH
25	N32CD	M33BM	Y33BG	X35BE	F24AM	Y22AK	R34AG	F11AF	Y12AC	R27CR	F30CM	W23CJ
26	A20CE	H15BP	T15BH	B16BF	V21BC	T29AL	X26AH	V31AF	T37AD	X13CS	V14CP	Y36CK
27	Z28CF	S25CD	G25BJ	N24BG	E17BD	G34AM	B38AJ	E18AG	G27AE	B19AB	E23CR	T32CL
28	C33CH	P35CE	L35BK	A21BH	M22BE	L26BC	N11AK	M12AH	L13AF	N30AC	M36CS	G20CM
29	K15CJ		D16BL	Z17BJ	H29BF	D38BD	A31AL	H37AJ	D19AG	A14AD	H32AB	L28CP
30	F25CK		J24BM	C22BK	S34BG	J11BE	Z18AM	S27AK	J30AH	Z23AE	S20AC	D33CR
31	V35CL		U21BP		P26BH		C12BC	P13AL		C36AF		J15CS

1882	JAN	FEB	MAR	APR	MAY	JUN	JUL	AUG	SEP	OCT	NOV	DEC
1	U25AB	K24CR	H24CJ	D22CF	A34BP	H11BL	L18BH	A27BF	H30BD	L23AL	A20AJ	M33AF
2	R35AC	F21CS	S21CK	J29CH	Z26CD	S31BM	D12BJ	Z13BG	S14BE	D36AM	Z28AK	H15AG
3	X16AD	V17AB	P17CL	U34CJ	C38CE	P18BP	J37BK	C19BH	P23BF	J32BC	C33AL	S25AH
4	B24AE	E22AC	W22CM	R26CK	K11CF	W12CD	U27BL	K30BJ	W36BG	U20BD	K15AM	P35AJ
5	N21AF	M29AD	Y29CP	X38CL	F31CH	Y37CE	R13BM	F14BK	Y32BH	R28BE	F25BC	W16AK
6	A17AG	H34AE	T34CR	B11CH	V18CJ	T27CF	X19BP	V23BL	T20BJ	X33BF	V35BD	Y24AL
7	Z22AH	S28AF	G26CS	N31CP	E12CK	G13CH	B30CD	E36BM	G28BK	B15BG	E16BE	T21AM
8	C29AJ	P38AG	L38AB	A18CR	M37CL	L19CJ	N14CE	M32BP	L33BL	N25BH	M24BF	G17BC
9	K34AK	W11AH	D11AC	Z12CS	H27CM	D30CK	A23CF	H20CD	D15BM	A35BJ	H21BG	L22BD
10	F26AL	Y31AJ	J31AD	C37AB	S13CP	J14CL	Z36CH	S28CE	J25BP	Z16BK	S17BH	D29BE
11	V38AM	T18AK	U18AE	K27AC	P19CR	U23CM	C32CJ	P33CF	U35CD	C24BL	P22BJ	J34BF
12	E11BC	G12AL	R12AF	F13AD	W30CS	R36CP	K20CK	W15CH	R16CE	K21BM	W29BK	U26BG
13	M31BD	L37AM	X37AG	V19AE	Y14AB	X32CR	F28CL	Y25CJ	X24CF	F17BP	Y34BL	R38BH
14	H18BE	D27BC	B27AH	E30AF	T23AC	B29CS	V33CM	T35CK	B21CH	V22CD	T26BM	X11BJ
15	S12BF	J13BD	N13AJ	M14AG	G36AD	N28AB	E15CP	G16CL	N17CJ	E29CE	G38BP	B31BK
16	P37BG	U19BE	A19AK	H23AH	L32AE	A33AC	M25CR	L24CM	A22CK	M34CF	L11CD	N18BL
17	W27BH	R30BF	Z30AL	S36AJ	D20AF	Z15AD	H35CS	D21CP	Z29CL	H26CH	D31CE	A12BM
18	Y13BJ	X14BG	C14AM	P32AK	J28AG	C25AE	S16AB	J17CR	C34CM	S38CJ	J18CF	Z37BP
19	T19BK	B23BH	K23BG	W20AL	U33AH	K35AF	P24AC	U22CS	K26CP	P11CK	U12CH	C27CD
20	G30BL	N36BJ	F36BD	Y28AH	R15AJ	F16AG	W21AD	R29AB	F38CR	W31CL	R37CJ	K13CE
21	L14BM	A32BH	V32BE	T33BC	X25AK	V24AH	Y17AE	X34AC	V11CS	Y18CM	X27CK	F19CF
22	D23BP	Z20BL	E20BF	G15BD	B35AL	E21AJ	T22AF	B26AD	E31AB	T12CP	B13CL	V30CH
23	J36CD	C28BM	K28BK	L25BE	N16AM	M17AK	G29AG	N38AE	M18AC	G37CR	N19CM	E14CJ
24	U32CE	K33BP	H33BH	D35BF	A24BC	H22AL	L34AH	A11AF	H12AD	L27CS	A30CP	M23CK
25	R20CF	F15CD	S18BJ	J16BG	Z21BD	S29AM	D26AJ	Z31AG	S37AE	D13AB	Z14CR	H36CL
26	X28CH	V25CE	P25BK	U24BH	C17BE	P34BC	J38AK	C18AH	P27AF	J19AC	C23CS	S32CM
27	B33CJ	E35CF	W35BL	R21BJ	W26BD	W26BD	U11AL	K12AJ	W13AG	U30AD	K36AB	P20CP
28	N15CK	M16CH	Y16BM	X17BK	F29BG	Y38BE	R31AM	F37AK	Y19AH	R14AE	F32AC	W28CR
29	A25CL		T24BP	B22BL	V34BH	T11BF	X18BC	V27AL	T30AJ	X23AF	V20AD	Y33CS
30	Z35CM		G21CD	N29BH	E26BJ	G31BG	B12BD	E13AM	G14AK	B36AG	E28AE	T15AB
31	C16CP		L17CE		M38BK		N37BE	M19BC		N32AH		G25AC

1883	JAN	FEB	MAR	APR	MAY	JUN	JUL	AUG	SEP	OCT	NOV	DEC
1	L35AD	A21AB	V21CL	T29CJ	X26CE	V31BP	Y12BK	X13BH	V14BF	Y36BC	X28AL	F15AH
2	D16AE	Z17AC	E17CM	G34CK	B38CF	E16CD	T37BL	B19BJ	E23BG	T32BD	B33AM	V25AJ
3	J24AF	C22AD	M22CP	L26CL	N11CH	M12CE	G27BM	N30BK	M36BH	G20BE	N15BC	E35AK
4	U21AG	K29AE	H29CR	D38CH	A31CJ	H37CF	L13BP	A14BL	H32BJ	L28BF	A25BD	M16AL
5	R17AH	F34AF	S34CS	J11CP	Z18CK	S27CH	D19CD	Z23BM	S20BK	D33BG	Z35BE	H24AM
6	X22AJ	V26AG	P26AB	U31CR	C12CL	P13CJ	J30CE	C36BP	P28BL	J15BH	C16BF	S21BC
7	B29AK	E38AH	W38CS	R18CS	K37CH	W19CK	U14CF	K32CD	W33BM	U25BJ	K24BG	P17BD
8	N34AL	M11AJ	Y11AD	X12AB	F27CP	Y30CL	R23CH	F20CF	Y15BP	R35BK	F21BH	W28BE
9	A26AM	H31AK	T31AE	B37AC	V13CR	T14CM	X36CJ	V28CF	T25CD	X16BL	V17BJ	Y29BF
10	Z38BC	S18AL	G18AF	N27AD	E19CS	G23CP	B32CK	E33CH	G35CE	B24BM	E22BB	T34BG
11	C11BD	P12AM	L12AG	A13AE	M30AB	L36CR	N20CL	M15CJ	L16CF	N21BP	M29BL	G26BH
12	K31BE	W37BC	D37AH	Z19AF	H14AC	D32CS	A28CM	H25CK	D24CH	A17CD	H34BM	L38BJ
13	F18BF	Y27BD	J27AJ	C30AG	S23AD	J20AB	Z33CP	S35CL	J21CJ	Z22CC	S26BP	D11BK
14	V12BG	T13BD	U13AK	K14AH	P36AE	U28AC	C15CR	P16CM	U17CK	C29CF	P38CD	J31BL
15	E37BH	G19BF	R19AL	F23AJ	W32AF	R33AD	K25CS	W24CM	R22CL	K34CH	W11CE	U18BM
16	M27BJ	L30BG	X30AM	W36AK	Y20AG	X15AE	F35AB	Y21CR	X29CM	F26CJ	Y31CF	R12BP
17	H13BK	D14BH	B14BC	E32AL	T28AH	B25AF	V16AC	T17CS	B34CP	V38CL	T18CH	X37CD
18	S19BL	J23BJ	N23BD	M20AM	G33AJ	N35AG	E24AD	G22AB	N26CR	E11CL	G12CJ	B27CE
19	P30BM	U36BK	A36BE	H28BC	L15AK	A16AH	M21AE	L29AC	A38CS	M31CM	L37CK	N13CF
20	W14BP	R32BL	Z32BF	S33BD	D25AL	Z24AJ	H17AF	D34AD	Z11AB	H18CP	D27CL	A19CH
21	Y23CD	X20BM	C20BG	P15BE	J35AM	C21AK	S22AG	J26AE	C31AC	S12CR	J13CM	Z30CJ
22	T36CE	B28BP	K28BH	W25BF	U16BD	K17AL	P29AH	U38AF	K18AD	P37CS	U19CP	C14CK
23	G32CF	N33CD	F33BJ	Y35BG	R24BD	F22BB	W34AJ	R11AG	F12AD	W27AB	R30CR	K23CL
24	L20CH	A15CE	V15BK	T16BH	X28BF	V29BD	Y26AK	X31AH	V37AF	Y13AD	X14CS	F36CH
25	D28CJ	Z25CF	E25BL	G24BJ	B17BF	E34BD	T38AL	B18AJ	E27AG	T19AD	B23AB	V32CP
26	J33CK	C35CH	M35BM	L21BK	N22BH	M26BE	G11AM	N12AK	M13AH	G30AE	N36AC	E20CR
27	U15CL	K16CJ	H16BP	D17BL	A29BH	H38BF	L31BC	A37AL	H19AJ	L14AF	A32AD	M28CS
28	R25CM	F24CK	S24CD	J22BM	Z34BJ	S11BG	D16BD	Z27AM	S14AK	P14AL	J36AH	S15AC
29	X35CP		P21CE	U29BP	C26BK	P31BH	J12BE	C13BC			C28AF	
30	B16CR		W17CF	R34CD	K38BL	W18BJ	U37BF	K19BD	W23AM	U32AJ	K33AG	P25AD
31	N24CS		Y22CH		F11BM		R27BG	F30BE		R20AK		W35AE

FIRST LETTER = YOUR PHYSICAL CODE, MIDDLE NUMBER = SENSITIVITY CODE, LAST TWO LETTERS = INTELLECTUAL CODE

1884	JAN	FEB	MAR	APR	MAY	JUN	JUL	AUG	SEP	OCT	NOV	DEC
1	Y16AF	X17AD	K22CR	W26CM	U11CJ	K12CF	P27BP	U30BL	K36BJ	P20BF	U15BD	C35AL
2	T24AG	B22AE	F29CS	Y38CP	R31CK	F37CH	W13CD	R14BM	F32BK	W28BG	R25BE	K16AM
3	G21AH	N29AF	V34AB	T11CR	X18CL	V27CJ	Y19CE	X23BP	V20BL	Y33BH	X35BF	F24BC
4	L17AJ	A34AG	E26AC	G31CS	B12CM	E13CK	T30CF	B36CD	E28BM	T15BJ	B16BG	V21BD
5	D22AK	Z26AH	M38AD	L18AB	N37CP	M19CL	G14CH	N32CE	M33BP	G25BK	N24BH	E17BE
6	J29AL	C38AJ	H11AE	D12AC	A27CR	H30CM	L23CJ	A20CF	H15CD	L35BL	A21BJ	M22BF
7	U34AM	K11AK	S31AF	J37AD	Z13CS	S14CP	D36CK	Z28CH	S25CE	D16BM	Z17BK	H29BG
8	R26BC	F31AL	P18AG	U27AE	C19AB	P23CR	J32CL	C33CJ	P35CF	J24BP	C22BL	S34BH
9	X38BD	V18AM	W12AH	R13AF	K30AC	W36CS	U20CM	K15CK	W16CH	U21CD	K29BM	P26BJ
10	B11BE	E12BC	Y37AJ	X19AG	F14AD	Y32AB	R28CP	F25CL	Y24CJ	R17CE	F34BP	W38BK
11	N31BF	M37BD	T27AK	B30AH	V23AE	T20AC	X33CR	V35CH	T21CK	X22CF	V26CD	Y11BL
12	A18BG	H27BE	G13AL	N14AJ	E36AF	G28AD	B15CS	E16CP	G17CL	B29CH	E38CE	T31BM
13	Z12BH	S13BF	L19AM	A23AK	M32AG	L33AE	N25AB	M24CR	L22CM	N34CJ	M11CF	G18BP
14	C37BJ	P19BG	D30BC	Z36AL	H20AH	D15AF	A35AC	H21CS	D29CP	A26CK	H31CH	L12CD
15	K27BK	W38BH	J14BD	C32AM	S28AJ	J25AG	Z16AD	S17AB	J34CR	Z38CL	S18CJ	D37CE
16	F13BL	Y14BJ	U23BE	K20BC	P33AK	U35AH	C24AE	P22AC	U26CS	C11CM	P12CK	J27CF
17	V19BM	T23BK	R36BF	F28BD	W15AL	R16AJ	K21AF	W29AD	R38AB	K31CP	W37CL	U13CH
18	E30BP	G36BL	X32BG	V33BE	Y25AM	X24AK	F17AG	Y34AE	X11AC	F18CR	Y27CM	R19CJ
19	M14CD	L32BM	B20BH	E15BF	T35BC	B21AL	V22AH	T26AF	B31AD	V12CS	T13CP	X30CK
20	H23CE	D20BP	N28BJ	M25BG	G16BD	N17AM	E29AJ	G38AG	N18AE	E37AB	G19CR	B14CL
21	S36CF	J28CD	A33BK	H35BH	L24BE	A22BC	M34AK	L11AH	A12AF	M27AC	L30CS	N23CM
22	P32CH	U33CE	Z15BL	S16BJ	D21BF	Z29BD	H26AL	D31AJ	Z37AG	H13AD	D14AB	A36CP
23	W20CJ	R15CF	C25BM	P24BK	J17BG	C34BE	S38AM	J18AK	C27AH	S19AE	J23AC	Z32CR
24	Y28CK	X25CH	K35BP	W21BL	U22BH	K26BF	P11BC	U12AL	K13AJ	P30AF	U36AD	C20CS
25	T33CL	B35CJ	F16CD	Y17BM	R29BJ	F38BG	W31BD	R37AM	F19AK	W14AG	R32AE	K28AB
26	G15CM	N16CK	V24CE	T22BP	X34BK	V11BH	Y18BE	X27BC	V30AL	Y23AH	X20AF	F33AC
27	L25CP	A24CL	E21CF	G29CD	B26BL	E31BJ	T12BF	B13BD	E14AM	T36AJ	B28AG	V15AD
28	D35CR	Z21CH	M17CH	L34CE	N38BM	M18BK	G37BG	N19BE	M23BC	G32AK	N33AH	E25AE
29	J16CS	C17CP	H22CJ	D26CF	A11BP	H12BL	L27BH	A30BF	H36BD	L20AL	A15AJ	M35AF
30	U24AB		S29CK	J38CH	Z31CD	S37BM	D13BJ	Z14BG	S32BE	D28AM	Z25AK	H16AG
31	R21AC		P34CL		C18CE		J19BK	C23BH		J33BC		S24AH

1885	JAN	FEB	MAR	APR	MAY	JUN	JUL	AUG	SEP	OCT	NOV	DEC
1	P21AJ	U29AG	A29AB	H38CR	L31CL	A37CJ	M13CE	L14BP	A32BL	M28BH	L25BF	N16BC
2	W17AK	R34AH	Z34AC	S11CS	D18CM	Z27CK	H19CF	D23CD	Z20BM	H33BJ	D35BG	A24BD
3	Y22AL	X26AJ	C26AD	P31AB	J12CP	C13CL	S30CH	J36CE	C28BP	S15BK	J16BH	Z21BE
4	T29AM	B38AK	K38AE	W18AC	U37CR	K19CM	P14CJ	U32CF	K33CD	P25BL	U24BJ	C17BF
5	G34BC	N11AL	F11AF	Y12AD	R27CS	F30CP	W23CK	R20CH	F15CE	W35BM	R21BK	K22BG
6	L26BD	A31AM	V31AG	T37AE	X13AB	V14CR	Y36CL	X28CJ	V25CF	Y16BP	X17BL	F29BH
7	D38BE	Z18BC	E18AH	G27AF	B19AC	E23CS	T32CM	B33CK	E35CH	T24CD	B22BM	V34BJ
8	J11BF	C12BD	M12AJ	L13AG	N30AD	M36AB	G20CP	N15CL	M16CJ	G21CE	N29BP	E26BK
9	U31BG	K37BE	H37AK	D19AH	A14AE	H32AC	L28CR	A25CM	H24CK	L17CF	A34CD	M38BL
10	R18BH	F27BF	S27AL	J30AJ	Z23AF	S20AD	D33CS	Z35CP	S21CL	D22CH	Z26CE	H11BM
11	X12BJ	V13BG	P13AM	U14AK	C36AG	P28AE	J15AB	C16CR	P17CM	J29CJ	C38CF	S31BP
12	B37BK	E19BH	W19BC	R23AL	K32AH	W33AF	U25AC	K24CS	W22CP	U34CK	K11CH	P18CD
13	N27BL	M30BJ	Y30BD	X36AM	F20AJ	Y15AG	R35AD	F21AB	Y29CR	R26CL	F31CJ	W12CE
14	A13BM	H14BK	T14BE	B32BC	V28AK	T25AH	X16AE	V17AC	T34CS	X38CM	V18CK	Y37CF
15	Z19BP	S23BF	G23BF	N20BD	E33AL	G35AJ	B24AF	E22AD	G26AB	B11CP	E12CL	T27CH
16	C30CD	P36BM	L36BG	A28BE	M15AM	L16AK	N21AG	M29AE	L38AC	N31CR	M37CM	G13CJ
17	K14CE	W32BP	D32BH	Z33BP	H25BC	D24AL	A17AH	H34AF	D11AD	A18CS	H27CP	L19CK
18	F23CF	Y20CD	J20BJ	C15BG	S35BD	J21AM	Z22AJ	S26AG	J31AE	Z12AB	S13CP	D30CL
19	V36CH	T28CE	U28BK	K25BH	P16BE	U17BC	C29AK	P38AH	U18AF	C37AC	P19CS	J14CM
20	E32CJ	G33CF	R33BL	F35BJ	W24BF	R22BD	K34AL	W11AJ	R12AG	K27AD	W30AB	U23CP
21	M20CK	L15CH	X15BM	V16BK	Y21BG	X29BE	F26AM	Y31AK	X37AH	F13AE	Y14AC	R36CR
22	H28CL	D25CJ	B25BP	E24BL	T17BH	B34BF	V38BC	T18AL	B27AJ	V19AF	T23AD	X32CS
23	S33CM	J35CK	N35CD	M21BM	G22BJ	N26BG	E11BD	G12AM	N13AK	E30AG	G36AE	B20AB
24	P15CP	U16CL	A16CE	H17BP	L29BK	A38BD	M31BE	L37BC	A19AL	M14AH	L32AF	N28AC
25	W25CR	R24CM	Z24CF	S22CD	D34BL	Z11BJ	H18BF	D27BD	Z30AM	H23AJ	D20AG	A33AD
26	Y35CS	X21CP	C21CH	P29CE	J26BM	C31BK	S12BG	J13BE	C14BC	S36AK	J28AH	Z15AE
27	T16AB	B17CR	K17CJ	W34CF	U38BP	K18BL	P37BH	U19BF	K23BD	P32AL	U33AJ	C25AF
28	G24AC	N22CS	F22CK	Y26CH	R11CD	F12BM	W27BJ	R30BG	F36BE	W20AM	R15AK	K35AG
29	L21AD		V29CL	T38CJ	X31CE	V37BP	Y13BK	X14BH	V32BF	Y28BC	X25AL	F16AH
30	D17AE		E34CM	G11CK	B18CF	E27CD	T19BL	B23BJ	E20BG	T33BD	B35AM	V24AJ
31	J22AF		M26CP		N12CH		G30BM	N36BK		G15BE		E21AK

1886	JAN	FEB	MAR	APR	MAY	JUN	JUL	AUG	SEP	OCT	NOV	DEC
1	M17AL	L34AJ	X34AD	V11AB	Y16CP	X27CL	F19CH	Y23CE	X20BP	F33BK	Y35BH	R24BE
2	H22AM	D26AK	B26AE	E31AC	T12CR	B13CM	V30CJ	T36CF	B28CD	V15BL	T16BJ	X21BF
3	S29BC	J38AL	N38AF	M18AD	G37CS	N19CP	E14CK	G32CH	N33CE	E25BM	G24BK	B17BG
4	P34BD	U11AM	A11AG	H12AE	L27AB	A30CR	M23CL	L20CJ	A15CF	M35BP	L21BL	N22BH
5	W26BE	R31BC	Z31AH	S37AF	D13AC	Z14CS	H36CM	D28CK	Z25CH	H16CD	D17BH	A29BJ
6	Y38BF	X18BD	C18AJ	P27AG	J19AD	C23AB	S32CP	J33CL	C35CJ	S24CE	J22BP	Z34BK
7	T11BG	B12BE	K12AH	W13AH	U30AE	K36AC	P20CR	U15CM	K16CK	P21CF	U29CD	C26BL
8	G31BH	N37BF	F37AK	Y19AJ	R14AF	F32AD	W28CS	R25CP	F24CL	W17CH	R34CF	K38BM
9	L18BJ	A27BG	V27AM	T30AK	X23AG	V20AE	Y33AB	X35CR	V21CM	Y22CJ	X26CF	F11CD
10	D12BK	Z13BH	E13BC	G14AL	B36AH	E28AD	T15AC	B16CS	E17CP	T29CK	B38CH	V31CD
11	J37BL	C19BJ	M19BD	L23AM	N32AJ	M33AC	S25AD	N24CR	M22CM	S34CJ	N11CF	E18CE
12	U27BM	K30BK	H30BE	D36BC	A20AK	H15AH	L35AE	A21CP	H29CS	L26CM	A31CK	M12CF
13	R13BP	F14BL	S14BF	J32BD	Z28AL	S25AG	D16AF	Z17AD	S34BH	D38CP	Z18CH	H37CJ
14	X19CD	V23BM	P23BG	U20BE	C33AM	P35AH	J24AG	C22AE	P26BC	J11CS	C12CK	S27CL
15	B30CE	E36BP	W36BH	R28BF	K15BC	W16AJ	Y24AH	K29AF	W38AD	U31CS	K37CH	P13CL
16	N14CF	M32CD	Y32BJ	X33BG	F25BD	Y24AM	R17AJ	F34AG	Y11AE	R18AB	F27CR	W19CL
17	A23CH	H20CE	T20BK	B15BH	V35BE	T21BC	X22AK	V26AH	T31AF	X12AC	V13CS	Y30CM
18	Z36CJ	S28CF	G28BL	N25BJ	E16BF	G17BD	B29AL	E38AJ	G18AG	B37AD	E19AB	T14CP
19	C32CK	P33CH	L33BM	A35BK	M24BG	L22BE	N34AM	M11AK	L12AH	N30AE	M30AC	G23CR
20	K20CL	W15CJ	D15BP	Z16BL	H21BH	D29BF	A26AM	H31AL	D37AJ	A13AF	H14AD	L36CS
21	F28CM	Y25CK	J25CD	C24BM	S17BJ	J34BG	Z38BD	S18AB	J27AK	Z19AG	S23AE	D32AB
22	V33CP	T35CL	U35CE	K21BP	P22BK	U26BF	C11BE	P12AD	U13AL	C30AH	P36AF	J20AC
23	E15CR	G16CM	R16CF	F17CD	W29BL	R38BD	K31BF	W37BD	R19AM	K14AJ	W32AG	U28AB
24	M25CS	L24CP	X24CH	V22CE	Y34BP	X11BC	F18BG	Y27BE	X30BC	F23AK	Y20AH	R33AE
25	H35AB	D21CR	B21CJ	E29CF	T26BP	B31BL	V12BH	T13BF	B14BD	V36AL	T28AJ	X15AF
26	S16AC	J17CS	N17CK	M34CH	G38CD	N18BK	S19BG	G19BE	N23BC	E32AM	G33AK	B25AG
27	P24AD	U22AB	A22CL	H26CJ	L11CE	A12BP	M27BK	L30BC	A36BF	M20BD	L15AL	N35AH
28	W21AE	R29AC	Z29CM	S36CK	D31CF	Z37CD	H13BL	D14BJ	Z32BG	H28BD	D25AM	A16AJ
29	Y17AF		C34CP	P11CL	J18CH	C27CE	S19BM	J23BK	C20BF	S33AL	J35BC	Z24AK
30	T22AG		K26CR	W31CH	U12CJ	K13CF	P38BP	U36BL	K28BJ	P15BF	U16BD	C21AL
31	G29AH		F38CS		N31CM		W14CD	R32BM		M25BG		K17AM

FIRST LETTER = YOUR PHYSICAL CODE, MIDDLE NUMBER = SENSITIVITY CODE, LAST TWO LETTERS = INTELLECTUAL CODE

1887	JAN	FEB	MAR	APR	MAY	JUN	JUL	AUG	SEP	OCT	NOV	DEC
1	F22BC	Y26AL	J26AF	C31AD	S12CS	J13CP	Z30CK	S36CH	J28CE	Z15BM	S16BK	D21BG
2	V29BD	T38AM	U38AG	K18AE	P37AB	U19CR	C14CL	P32CJ	U33CF	C25BP	P24BL	J17BH
3	E34BE	G11BC	R11AH	F12AF	H27AC	R30CS	K23CM	W20CK	R15CH	X35CD	H21BM	U22BJ
4	M26BF	L31BD	X31AJ	V37AG	Y13AD	X14AB	F36CP	Y28CL	X25CJ	F16CE	Y17BP	R29BK
5	H38BG	D18BE	B18AK	E27AH	T19AE	B23AC	V32CR	T33CM	B35CK	V24CF	T22CD	X34BL
6	S11BH	J12BF	N12AL	M13AJ	G30AF	N36AD	E20CS	G15CP	N16CL	E21CH	G29CE	B26BM
7	P31BJ	U37BG	A37AM	H19AK	L14AG	A32AE	M28AB	L25CR	A24CM	M17CJ	L34CF	N38BP
8	W18BK	R27BH	Z27BC	S30AL	D23AH	Z20AF	H33AC	D35CS	Z21CP	H22CK	D26CH	A11CD
9	Y12BL	X13BJ	C13BD	P14AM	J36AJ	C28AG	S15AD	J16AB	C17CR	S29CL	J38CJ	Z31CE
10	T37BM	B19BK	K19BE	W23BC	U32AK	K33AH	P25AE	U24AC	K22CS	P34CR	U11CK	C18CF
11	G27BP	N30BL	F30BF	Y36BD	R20AL	F15AJ	W35AF	R21AD	F29AB	W26CP	R31CL	K12CH
12	L13CD	A14BM	V14BG	T32BE	X28AM	V25AK	Y16AG	X17AE	V34AC	Y38CR	X18CM	F37CJ
13	D19CE	Z23BP	E23BH	G20BF	B33BC	E35AL	T24AH	B22AF	E26AD	T11CS	B12CP	V27CK
14	J30CF	C36CD	M36BJ	L28BG	N15BD	M16AM	G21AJ	N29AG	M38AE	G31AB	N37CR	E13CL
15	U14CH	K32CE	H32BK	D33BH	A25BE	H24BC	L17AK	A34AH	H11AF	L18AC	A27CS	M19CM
16	R23CJ	F20CF	S20BL	J15BJ	Z35BF	S21BD	D22AL	Z26AJ	S31AG	D12AD	Z13AB	H30CP
17	X36CK	V28CH	P28BM	U25BK	C16BG	P17BE	J29AM	C38AK	P18AH	J37AE	C19AC	S14CR
18	B32CL	E33CJ	W33BP	R35BL	K24BH	W22BF	U34BC	K11AL	W12AJ	U27AF	K30AD	P23CS
19	N20CM	M15CK	Y15CD	X16BM	F21BJ	Y29BG	R26BD	F31AM	Y37AK	R13AG	F14AE	W36AB
20	A28CP	H25CL	T25CE	B24BP	V17BK	T34BH	X38BE	V18BC	T27AL	X19AH	V23AF	Y32AC
21	Z33CR	S35CM	G35CF	N21CD	E22BL	G26BJ	B11BF	E12BD	G13AM	B30AJ	E36AG	T20AD
22	C15CS	P16CP	L16CH	A17CE	M29BM	L38BK	N31BG	M37BE	L19BC	N14AK	M32AH	G28AE
23	K25AB	W24CR	D24CJ	Z22CF	H34BP	D11BL	A18BH	H27BF	D30BD	A23AL	H20AJ	L33AF
24	F35AC	Y21CS	J21CK	C29CH	S26CS	J31BM	Z12BJ	S13BG	J14BE	Z36AM	S28AK	D15AG
25	V16AD	T17AB	U17CL	K34CJ	P38CE	U18BP	C37BK	P19BH	U23BF	C32BC	P33AL	J25AH
26	E24AE	G22AC	R22CM	F26CK	W11CF	R12CD	K27BL	W30BJ	R36BG	K20BD	W15AM	U35AJ
27	M21AF	L29AD	X29CP	V38CL	Y31CH	X37CE	F13BM	Y14BK	X32BH	F28BE	Y25BC	R16AK
28	H17AG	D34AE	B34CR	E11CM	T18CJ	B27CF	V19BP	T23BL	B20BJ	V33BF	T35BD	X24AL
29	S22AH		N26CS	M31CP	G12CK	N13CH	E30CD	G36BM	N28BK	E15BG	G16BE	B21AM
30	P29AJ		A38AB	H18CR	L37CL	A19CJ	M14CE	L32BP	A33BL	M25BH	L24BF	N17BC
31	W34AK		Z11AC		D27CM		H23CF	D20CD		H35BJ		A22BD

1888	JAN	FEB	MAR	APR	MAY	JUN	JUL	AUG	SEP	OCT	NOV	DEC
1	Z29BE	S38BC	L11AJ	A12AG	M27AD	L30AB	N23CP	M20CL	L15CJ	N35CE	M21BP	G22BK
2	C34BF	P11BD	D31AK	Z37AH	H13AE	D14AC	A36CR	H28CM	D25CK	A16CF	H17CD	L29BL
3	K26BG	W31BE	J18AL	C27AJ	S19AF	J23AD	Z32CS	S33CP	J35CL	Z24CH	S22CE	D34BM
4	F38BH	Y18BF	U12AM	K13AK	P30AG	U36AE	C20AB	P15CR	U16CM	C21CJ	P29CF	J26BP
5	V11BJ	T12BG	R37BC	F19AL	W14AH	R32AF	K28AC	W25CS	R24CP	K17CK	W34CH	U38CD
6	E31BK	G37BH	X27BD	V30AM	Y23AJ	X20AG	F33AD	Y35AB	X21CR	F22CL	Y26CJ	R11CE
7	M18BL	L27BJ	B13BE	E14BC	T16AK	B28AE	V15AE	T16AC	B17CS	V29CH	T38CK	X31CF
8	H12BM	D13BK	N19BF	M23BD	G32AL	N33AJ	E25AF	G24AD	N22AB	E34CP	G11CL	B18CH
9	S37BP	J19BL	A30BG	H36BE	L20AM	A15AK	M35AG	L21AE	A29AC	M26CR	L31CH	N12CJ
10	P27CD	U30BM	Z14BH	S32BF	D28BG	Z25AL	H16AH	D17AF	Z34AD	H38CS	D18CP	A37CK
11	W13CE	R14BP	C23BJ	P20BG	J33BD	C35AM	S24AJ	J22AG	C26AE	S11AB	J12CR	Z27CL
12	Y19CF	X23CD	K36BK	W28BH	U15BE	K16BC	P21AK	U29AH	K38AF	P31AC	U37CS	C13CM
13	T30CH	B36CE	F32BL	Y33BJ	R25BF	F24BD	M17AL	R34AJ	F11AG	M18AD	R27AB	K19CP
14	G14CJ	N32CF	V28BM	T15BK	X35BG	V21BE	Y22AM	X26AK	V31AH	Y12AE	X13AC	F30CR
15	L23CK	A20CH	E28BP	G25BL	B16BH	E17BF	T29BC	B38AL	E18AJ	T37AF	B19AD	V14CS
16	D36CL	Z28CJ	M33CD	L35BM	N24BJ	M22BG	G34BD	N11AM	M12AK	G27AG	N30AE	E23AB
17	J32CM	C33CK	H15CE	D16BP	A21BK	H29BH	L26BE	A31BC	H37AL	L13AH	A14AF	M36AC
18	U20CP	K15CL	S25CF	J24CD	Z17BL	S34BJ	D38BF	Z18BD	S27AM	D19AJ	Z23AG	H32AD
19	R28CR	F25CM	P35CH	U21CE	C22BM	P26BK	J11BG	C12BE	P13BC	J30AK	C36AH	S20AE
20	X33CS	V35CP	W16CJ	R17CF	F34CD	W38BL	U31BH	F37BL	W19BD	U14AL	K32AJ	P28AF
21	B15AB	E16CR	Y24CK	X22CH	F34CD	Y11BM	R18BJ	F27BG	Y30BE	R23AM	F20AK	W33AG
22	N25AC	M24CS	T21CL	B29CJ	V26CE	T31BP	X12BK	V13BH	T14BF	X36BC	V28AL	Y15AH
23	A35AD	H21AB	G17CM	N34CK	E38CF	G18CD	B37BL	E19BJ	G23BG	B32BD	E33AM	T25AJ
24	Z16AE	S17AC	L22CP	A26CL	M11CH	L12CE	N27BM	M30BK	L36BH	N20BE	M15BC	G35AK
25	C24AF	P22AD	D29CR	Z38CM	H31CJ	D37CF	A13BP	H14BL	D32BJ	A28BF	H25BD	L16AL
26	K21AG	W29AE	J34CS	C11CP	S18CK	J27CH	Z19CD	S23BM	J20BK	Z33BG	S35BE	D24AM
27	F17AH	Y34AF	U26BA	K31CR	P12CL	U13CJ	C30CE	P36BP	U28BL	C15BH	P16BF	J21BC
28	V22AJ	T26AG	R38AC	F18CS	W37CM	R19CK	K14CF	W32CD	R33BM	K25BJ	W24BG	U17BD
29	E29AK	G38AH	X11AD	V12AB	Y27CP	X30CL	F23CH	Y20CE	X15BP	F35BK	Y21BH	R22BE
30	M34AL		B31AE	E37AC	T13CR	B14CM	V36CJ	T28CF	B25CD	V16BL	T17BJ	X29BF
31	H26AM		N18AF		G19CS		E32CK	G33CH		E24BM		B34BG

1889	JAN	FEB	MAR	APR	MAY	JUN	JUL	AUG	SEP	OCT	NOV	DEC
1	N26BH	M31BF	Y31AL	X37AJ	F13AF	Y14AD	R36CS	F28CP	Y25CL	R16CH	F17CE	W29BM
2	A38BJ	H18BG	T18AM	B27AK	V19AG	T23AE	X32AB	V33CR	T35CM	X24CJ	V22CF	Y34BP
3	Z11BK	S12BH	G12BC	N13AL	E30AH	G36AF	B20AC	E15CS	G16CP	B21CK	E29CH	T26CD
4	C31BL	P37BJ	L37BD	A19AM	M14AJ	L32AG	N28AD	M25AB	L24CR	N17CL	M34CJ	G38CE
5	K18BM	W27BK	D27BE	Z30BC	H23AK	D20AH	A33AE	H35AC	D21CS	A22CM	H26CK	L11CF
6	F12BP	Y13BL	J13BF	C14BD	S36AL	J28AJ	Z15AF	S16AD	J17AB	Z29CP	S38CL	D31CH
7	V37CD	T19BM	U19BG	K23BE	P32AM	U33AK	C25AG	P24AE	U22AC	C34CR	P11CM	J18CJ
8	E27CE	G30BP	R30BH	F36BF	W20BC	R15AL	K35AH	W21AF	R29AD	K26CS	W31CP	U12CK
9	M13CF	L14CD	X14BJ	V32BG	Y28BD	V24AD	K16AE	T17AG	X34AE	F38AB	Y18CR	R37CL
10	H19CD	D23CE	B23BK	E20BH	T33BE	B35BC	V24AK	T22AH	B26AF	E31AC	G14CS	B13CP
11	S30CJ	J36CF	N36BL	M28BJ	G15BF	N16BD	E21AL	G29AJ	N38AG	M18AD	J31AB	N19CP
12	P14CK	U32CH	A32BM	H33BK	L25BG	A24BE	M17AM	L34AK	A11AH	M18AE	L27AC	N19CR
13	W23CL	R20CJ	Z20BP	S15BL	D35BH	Z21BF	M22AM	R34AK	F11AG	M18AF	D13AD	A30CS
14	Y36CM	X28CK	C28CD	P25BM	J16BJ	C17BG	S29BD	J38AM	C18AK	S37AG	J19AE	Z14AB
15	T32CP	B33CP	K33CE	W35BP	U24BK	K22BG	P34BE	U11BC	K12AL	P27AH	U30AF	C23AC
16	G20CR	N15CM	F15CF	Y16CD	R28BL	F29BJ	W26BF	R31BD	F37AM	W13AJ	R14AG	K36AD
17	L28CS	A25CP	V25CH	T24CE	X17BM	V34BH	Y38BG	X18BC	V27BD	Y19AK	X23AH	F32AE
18	D33AB	Z35CR	E35CJ	G21CF	B22BP	E26BL	T11BH	B12BF	E13BD	T30AL	B36AJ	V20AF
19	J15AC	C16CS	M16CK	L17CH	N29CD	M38BM	G31BJ	N37BG	M19BE	G14AM	N32AK	E28AG
20	U25AD	K24AB	H24CL	D22CJ	A34CE	H11BP	L18BK	A27BE	H30BF	L23BC	A20AL	M33AH
21	R35AE	F21AC	S21CM	J29CK	Z26CF	S31CD	D12BL	Z13BJ	S14BG	D36BD	Z28AM	H15AJ
22	X16AF	V17AD	P17CP	U34CL	H27CH	W12CF	U27BP	K30BL	W36BJ	U20BF	K15BD	P35AL
23	B24AG	E22AE	W22CR	R26CM	K11CJ	W12CF	U27BP	K30BL	W36BJ	U20BF	K15BD	P35AL
24	N21AH	M29AF	T34AB	B11CR	V18CL	T27CJ	X19CE	V23BP	T20BL	X33BH	V35BF	Y24BC
25	A17AJ	H34AG	G26AC	N31CS	E12CM	G13CH	B38CF	E36CD	G28BG	N26BD	B15BJ	T21BD
26	Z22AK	S26AH	L38AD	A18AB	M37CP	L19CJ	N14CH	M32CE	L33BP	N25BK	M24BH	G17BE
27	C29AL	P38AJ	D11AE	Z12AC	H27CR	D30CM	A23CJ	H20CF	D15CD	A35BL	Z16BM	L22BF
28	K34AM	W11AK	J31CF	C37AD	S13CS	J14CP	Z36CK	S28CH	J25CF	Z16BM	S17BK	D29BG
29	F26BC		U18AG	K27AE	P19AB	U23CR	C32CL	P33CJ	U35CF	C24BP	P22BL	J34BH
30	V38BD		R12AH		W30AC		K20CH	W15CK		K21CD		U26BJ
31	E11BE		R12AH		W30AC		K20CH	W15CK		K21CD		U26BJ

FIRST LETTER = YOUR PHYSICAL CODE, MIDDLE NUMBER = SENSITIVITY CODE, LAST TWO LETTERS = INTELLECTUAL CODE

1890

	JAN	FEB	MAR	APR	MAY	JUN	JUL	AUG	SEP	OCT	NOV	DEC
1	R38BK	F18BH	S18BC	J27AL	Z19AH	S23AF	D32AC	Z33CS	S35CP	D24CK	Z22CH	H34CD
2	X11BL	V12BJ	P12BD	U13AM	C30AJ	P36AG	J20AD	C15AB	P16CR	J21CL	C29CJ	S26CE
3	B31BM	E37BK	W37BF	R19BC	K14AK	W32AH	U28AE	K25AC	W24CS	U17CH	K34CK	P38CF
4	N18BP	M27BL	Y27BF	X30BD	F23AL	Y20AJ	R33AF	F35AD	Y21AB	R22CP	F26CL	W11CH
5	A12CD	H13BM	T13BG	B14BE	V36AM	T28AK	X15AG	V16AE	T17AC	X29CR	V38CM	Y31CJ
6	Z37CE	S19BP	G19BH	N23BF	E32BC	G33AL	B25AH	E24AF	G22AD	B34CS	E11CP	T18CK
7	C27CF	P30CD	L30BJ	A36BG	M20BD	L15AM	N35AJ	M21AG	L29AE	N26AB	M31CR	G12CL
8	K13CH	W14CE	D14BK	Z32BH	H28BE	D25BC	A16AK	H17AH	D34AF	A38AC	H16CS	L37CM
9	F19CJ	Y23CF	J23BL	C20BJ	S33BF	J35BD	Z24AL	S22AJ	J26AG	Z11AD	S12AB	D27CP
10	V30CK	T36CH	U36BM	K28BK	P15BG	U16BE	C21AM	P29AK	U38AH	C31AE	P37AC	J13CR
11	E14CL	G32CJ	R32BP	F33BL	W25BH	R24BF	K17BC	W34AL	R11AJ	K18AF	W27AD	U19CS
12	H23CM	L20CK	X20CD	V15BM	Y35BJ	X21BG	F22BD	Y26AM	X31AK	F12AG	Y13AE	R30AB
13	H36CP	D28CL	B28CE	E25BP	T16BK	B17BH	V29BE	T38BC	B18AL	V37AH	T19AF	X14AC
14	S32CR	J33CM	N33CF	M35CD	G24BL	N22BJ	E34BF	G11BD	N12AM	E27AJ	G30AG	B23AD
15	P20CS	U15CP	A15CH	H16CE	L21BM	A29BK	M25BG	L31BE	A37BC	M13AK	L14AH	N36AE
16	W28AB	R25CR	Z25CJ	S24CF	D17BP	Z34BL	H38BH	D18BF	Z27BD	H19AL	D23AJ	A32AF
17	Y33AC	X35CS	C35CK	P21CH	J22CD	C26BM	S11BJ	J12BG	C13BE	S30AM	J36AK	Z20AG
18	T15AD	B16AB	K16CL	W17CJ	U29CE	K38BP	P31BK	U37BH	K19BF	P14BC	U32AL	C28AH
19	G25AE	N24AC	F24CM	Y22CK	R34CF	F11CD	W18BL	R27BJ	F30BG	W23BD	R20AM	K33AJ
20	L35AF	A21AD	V21CP	T29CL	X26CH	V31CE	Y12BM	X13BK	V14BH	Y36BE	X28BC	F15AK
21	D16AG	Z17AE	E17CR	G34CM	B38CJ	E18CF	T37BP	B19BL	E23BJ	T32BF	B33BD	V25AL
22	J24AH	C22AF	M22CS	L26CP	N11CH	M12CH	G27CD	N30BM	M36BK	G20BG	N15BE	E35AM
23	U21AJ	K29AG	H29AB	D38CR	A31CL	H37CJ	L13CE	A14BP	H32BL	L28BH	A25BF	M16BC
24	R17AK	F34AH	S34AC	J11CS	Z18CM	S27CK	D19CF	Z23CD	S20BM	D33BJ	Z35BG	H24BD
25	X22AL	V26AJ	P26AD	U31AB	C12CP	P13CL	J30CH	C36CE	P28BP	J15BK	C16BH	S21BE
26	B29AM	E38AK	W38AE	R18AC	K37CR	W19CM	M19CH	K32CE	W33CD	U25BL	K24BJ	P17BF
27	N34BC	M11AL	Y11AF	X12AD	F27CS	Y30CP	R23CK	F20CH	Y15CE	R35BM	F21BK	H22BG
28	A26BD	H31AM	T31AG	B37AE	V13AB	T14CR	X36CL	V28CJ	T25CF	X16BP	V17BL	Y29BH
29	Z38BE		G18AH	N27AF	E19AC	G23CS	B32CM	E33CK	G35CH	B24CD	E22BM	T34BJ
30	C11BF		L12AJ	A13AG	M30AD	L36AB	N20CP	M15CL	L16CJ	N21CE	M29BP	G26BK
31	K31BG		D37AK		H14AE		A28CR	H25CM		A17CF		L38BL

1891

	JAN	FEB	MAR	APR	MAY	JUN	JUL	AUG	SEP	OCT	NOV	DEC
1	D11BM	Z12BK	E12BE	G13BC	B30AK	E36AH	T20AE	B15AC	E16CS	T21CM	B29CK	V26CF
2	J31BP	C37BL	M37BF	L19BD	N14AL	M32AJ	G28AF	N25AD	M24AB	G17CP	N34CL	E38CH
3	U18CD	K27BM	H27BG	D30BE	A23AM	H20AK	L33AG	A35AE	H21AC	L22CR	A26CM	M11CJ
4	R12CF	F13BP	S13BH	J14BF	Z36BC	S28AL	D15AH	Z16AF	S17AD	D29CS	Z38CP	H31CK
5	X37CF	V19CD	P19BK	U23BG	C32BD	P33AM	J25AJ	C24AG	P22AE	J34AB	C11CR	S18CL
6	B27CH	E38CE	W30BK	R36BH	K20BE	W15BC	U35AK	K21AH	W29AF	U26AC	K31CS	P12CM
7	N13CJ	M14CF	Y14BL	X32BJ	F28BF	Y25BD	R16AL	F17AJ	Y34AG	R38AD	F18AB	W37CP
8	A19CK	H23CH	T23BM	B20BK	V33BG	T35BE	X24AM	V22AK	T26AH	X11AE	V12AC	Y27CR
9	Z30CL	S36CJ	G36BP	N28BL	E15BH	G16BF	B21BC	E29AL	G38AJ	B31AF	E37AD	T13CS
10	C14CM	P32CK	L32CD	A33BM	M25BJ	L24BG	N17BD	M34AM	L11AK	N18AG	M27AE	G19AB
11	K23CP	W20CL	D20CE	Z15BP	H35BK	D21BH	A22BE	H26BC	D31AL	A12AH	H13AF	L30AC
12	F36CR	Y28CM	J28CF	C25CD	S16BL	J17BJ	Z29BF	S38BD	J18AM	Z37AJ	S19AG	D14AD
13	V32CS	T33CP	U33CH	K35CE	P24BH	U22BK	C34BG	P11BE	U12BC	C27AK	P30AH	J23AE
14	E20AB	G15CR	R15CJ	F16CF	W21BP	R29BL	K26BH	W31BF	R37BD	K13AL	W14AJ	U36AF
15	M28AC	L25CS	X25CK	V24CH	Y17CD	X34BM	F38BJ	Y18BG	X27BE	F19AH	Y23AK	R32AG
16	H33AD	D35AB	B35CL	E21CJ	T22CE	B26BP	V11BK	T12BH	B13BF	V30BC	T36AL	X20AH
17	S15AE	J16AC	N16CM	M17CK	G29CF	N38CD	E31BL	G37BJ	N19BG	E14BD	G32AM	B28AJ
18	P25AF	U24AD	A24CP	H22CL	L34CH	A11CE	M18BM	L27BK	A30BH	M23BE	L20BC	N33AK
19	W35AG	R21AE	Z21CR	S29CM	D26CJ	Z31CF	H12BP	D13BL	Z14BJ	H36BF	D28BD	A15AL
20	Y16AH	X17AF	C17CS	P34CP	J38CK	C18CH	S37CD	J19BM	C23BK	S32BG	J33BE	Z25AM
21	T24AJ	B22AG	K22AB	W26CR	U11CL	K12CJ	P27CE	U30BP	K36BL	P20BH	U15BF	C35BC
22	G21AK	N29AH	F29AC	Y38CS	R31CM	F37CK	M13CF	R14CD	F32BM	M28BJ	R25BG	K16BD
23	L17AL	A34AJ	V34AD	T11AB	X18CP	V27CL	Y19CH	X23CE	V20BP	Y33BK	X35BH	F24BE
24	D22AM	Z26AK	E26AE	G31AC	B12CR	E13CM	T30CJ	B36CF	E28CD	T15BL	B16BJ	V21BF
25	J29BC	C38AL	M38AF	L18AD	N37CS	M19CP	G14CK	N32CH	M33CE	G25BM	N24BK	E17BG
26	U34BD	K11AM	H11AG	D12AE	A27AB	H30CR	L23CL	A20CJ	H15CF	L35BP	A21BL	M22BH
27	R26BE	F31BC	S31AH	J37AF	Z13AC	S14CS	D36CM	Z28CK	S25CH	D16CD	Z17BM	H29BJ
28	X38BF	V18BD	P18AJ	U27AG	C19AD	P23AB	J32CP	C33CL	P35CJ	J24CE	C22BP	S34BK
29	B11BG		W12AK	R13AH	K30AE	W36AC	U20CR	K15CH	W16CK	U21CF	K29CD	P26BL
30	N31BH		Y37AL	X19AJ	F14AF	Y32AD	R28CS	F25CP	Y24CL	R17CH	F34CE	W38BM
31	A18BJ		T27AM		V23AG		X33AB	V35CR		X22CJ		Y11BP

1892

	JAN	FEB	MAR	APR	MAY	JUN	JUL	AUG	SEP	OCT	NOV	DEC
1	T31CD	B37BH	F27BF	Y38BF	R23BC	F20AL	W33AH	R35AF	F21AD	W22CS	R26CP	K11CK
2	G18CE	N27BP	V13BJ	T14BG	X36BD	V28AM	Y15AJ	X16AG	V17AE	Y29AB	X38CR	F31CL
3	L12CF	A13CD	E19BK	G23BH	B32BE	E33BC	T25AK	B24AH	E22AF	T34AC	B11CS	V18CM
4	D37CH	Z19CE	M30BL	L36BJ	N20BF	M15BD	G35AL	N21AJ	M29AG	G26AD	N31AB	E12CP
5	J27CJ	C30CF	H14BM	D32BK	A28BG	H25BE	L16AM	A17AK	H34AH	L38AC	A18AC	M37CR
6	U13CK	K14CH	S23BP	J20BL	Z33BH	S35BF	D24BC	Z22AH	S26AJ	D11AF	Z12AD	H27CS
7	R19CL	F23CJ	P36CD	U28BM	C15BJ	P16BG	J21BD	C29AM	P38AK	J31AG	C37AE	S13AB
8	X30CM	V36CK	W32CE	R33BP	K25BE	W24BC	U17BF	K34BH	W11AL	U18AH	K27AF	P19AC
9	B14CP	E32CL	Y20CF	X15CD	F35BL	Y21BJ	R22BF	F26BD	Y31AM	R12AJ	F13AG	W30AD
10	N23CR	M20CM	T28CH	B25CE	V16BM	T17BK	X29BG	V38BE	T18BC	X37AK	V19AH	Y14AE
11	A36CS	H28CP	G33CJ	N35CF	E24BP	G22BH	B34BH	E11BF	G12BD	B27AL	E30AJ	T23AF
12	Z32AB	S33CR	L15CK	A16CH	M21CD	L29BM	N26BJ	M31BG	L37BE	N13AM	M14AK	G36AG
13	C20AC	P15CS	D25CL	Z24CJ	H17CE	D34BP	A38BK	H18BH	D27BF	A19BC	H23AL	L32AH
14	K28AD	W25AB	J35CM	C21CK	S22CF	J26CD	Z11BL	S12BJ	J13BG	Z30BD	S36AM	D20AJ
15	F33AE	Y35AC	U16CP	K17CL	P29CH	U38CE	C31BM	P37BK	U19BH	C14BE	P32BC	J28AK
16	V15AF	T16AD	R24CR	F22CM	W34CJ	R11CF	K18BP	W27BL	R30BJ	K23BF	W20BD	U33AL
17	E25AG	G24AE	X21CS	V29CP	Y26CK	X31CH	F12CD	Y13BM	X14BK	F36BG	Y28BE	R15AM
18	M35AH	L21AF	B17AB	E34CR	T38CL	B18CJ	V37CE	T19BP	B23BL	V32BH	T33BF	X25BC
19	H16AJ	D17AG	N22AC	M26CS	G11CM	N12CK	E27CF	G30CD	N36BM	E20BJ	G15BG	B35BD
20	S24AK	J22AH	A29AD	H38AB	L31CP	A37CL	M13CH	L14CE	A32BP	M28BK	L25BH	N16BF
21	P21AL	U29AJ	Z34AE	S11AC	D16CR	Z27CM	H19CJ	D23CF	Z20CD	H33BL	D35BJ	A24BF
22	W17AM	R34AK	C26AF	P31AD	J12CS	C13CP	S30CK	J36CH	C28CE	S15BM	J16BK	Z21BG
23	Y22BC	X26AL	K36AG	W18AE	U37AB	K19CR	P14CL	U32CJ	K33CF	P25BP	U24BL	C17BH
24	T29BD	B38AM	F11AH	Y12AF	R27AC	F30CS	W23CM	R20CK	F15CH	W35CD	R21BM	K22BJ
25	G34BE	N11BC	V31AJ	T37AG	X13AD	V14AB	Y36CP	X28CJ	V25CJ	Y16CE	X17BP	F29BK
26	L26BF	A31BD	E18AK	G27AH	B19AE	E23AC	T32CR	B33CM	E35CK	T24CF	B22CD	V34BL
27	D38BG	Z18BE	M12AL	L13AJ	N30AF	M36AD	G20CS	N15CP	M16CL	G21CH	N29CE	E26BM
28	J11BH	C12BF	H37AM	D19AK	A14AG	H32AE	L28AB	A25CR	H24CM	L17CJ	A34CF	M38BP
29	U31BJ	K37BG	S27BC	J30AL	Z23AM	S20AF	D33AC	Z35CS	S21CP	D22CK	Z26CH	H11CG
30	R18BK		P13BD	U14AM	C36AJ	P28AG	J15AD	C16AB	P17CR	J29CL	C38CJ	S31CE
31	X12BL		W19BE		K32AK		U25AE	K24AC		U34CM		P18CF

FIRST LETTER = YOUR PHYSICAL CODE, MIDDLE NUMBER = SENSITIVITY CODE, LAST TWO LETTERS = INTELLECTUAL CODE

1893	JAN	FEB	MAR	APR	MAY	JUN	JUL	AUG	SEP	OCT	NOV	DEC
1	M12CH	R13CE	Z13BK	S14BH	D36BE	Z28BC	H15AK	D16AH	Z17AF	H29AC	D38CS	A31CM
2	Y37CJ	X19CF	C19BL	P23BJ	J32BF	C33BD	S25AL	J24AJ	C22AG	S34AD	J11AB	Z18CP
3	T27CK	B30CM	K30BM	W36BK	U20BG	K15BE	P35AM	U21AK	K29AH	P26AE	U31AC	C12CR
4	G13CL	N14CJ	F14BP	Y32BL	R28BH	F25BF	W16BC	R17AL	F34AJ	W38AF	R18AD	K37CS
5	L19CM	A23CK	V23CD	T20BM	X33BJ	V35BG	Y24BD	X22AM	V26AK	Y11AG	X12AE	F27AB
6	D30CP	Z36CL	E36CE	G28BP	B15BK	E16BH	T21BE	B29BC	E38AL	T31AH	B37AF	V13AC
7	J14CR	C32CM	M32CF	L33CD	N25BL	M24BJ	G17BF	N34BD	M11AM	G18AJ	N27AG	E19AD
8	U23CS	K20CP	H20CH	D15CE	A35BM	H21BK	L22BG	A26BE	H31BC	L12AK	A13AH	M30AE
9	R36AB	F28CR	S28CJ	J25CF	Z16BP	S17BL	D29BH	Z38BF	S18BD	D37AL	Z19AJ	H14AF
10	X32AC	V33CS	P33CK	U35CH	C24CD	P22BM	J34BJ	C11BG	P12BE	J27AM	C30AK	S23AG
11	B20AD	E15AB	W15CL	R16CJ	K21CE	W29BP	U26BK	K31BH	W37BF	U13BC	K14AL	P36AH
12	N28AE	M25AC	Y25CM	X24CK	F17CF	Y34CD	R38BL	F18BJ	Y27BG	R19BD	F23AM	W32AJ
13	A33AF	H35AD	T35CP	B21CL	V22CH	T26CE	X11BM	V12BK	T13BH	X30BE	V36BC	Y20AK
14	Z15AG	S16AE	G16CR	N17CM	E29CJ	G38CF	B31BP	E37BL	G19BJ	B14BF	E32BD	T28AL
15	C25AH	P24AF	L24CS	A22CP	M34CK	L11CH	N18CD	M27BM	L30BK	N23BG	M20BE	G33AM
16	K35AJ	W21AG	D21AB	Z29CR	H26CL	D31CJ	A12CE	H13BP	D14BL	A36BH	H28BF	L15BC
17	F16AK	Y17AH	J17AC	C34CS	S38CM	J18CK	Z37CF	S19CD	J23BM	Z32BJ	S33BG	D25BD
18	V24AL	T22AJ	U22AD	K26AB	P11CP	U12CL	C27CH	P30CE	U36BP	C20BK	P15BH	J35BE
19	E21AM	G29AK	R29AC	F38AC	W31CR	R37CM	K13CJ	W14CF	R32CD	K28BL	W25BJ	U16BF
20	M17BC	L34AL	X34AF	V11AD	Y18CS	X27CP	F19CK	Y23CH	X20CE	F33BM	Y35BK	R24BG
21	H22BD	D26AM	B26AG	E31AE	T12AB	B13CR	V30CL	T36CJ	B28CF	V15BP	T16BL	X21BH
22	S29BE	J38BC	N38AH	M18AF	G37AC	N19CS	E14CM	G32CK	N33CH	E25CD	G24BM	B17BJ
23	P34BF	U11BD	A11AJ	H12AG	L27AD	A30AB	M23CP	L20CL	A15CJ	M35CE	L21BP	N22BK
24	W26BG	R31BE	Z31AK	S37AH	D13AE	Z14AC	H36CR	D28CM	Z25CK	H16CF	D17CD	A29BL
25	Y38BH	X18BF	C18AL	P27AJ	J19AF	C23AD	S32CS	J33CP	C35CL	S24CH	J22CE	Z34BM
26	T11BJ	B12BG	K12AM	W13AK	U30AG	K36AE	P20AB	U15CR	K16CM	P21CJ	U29CF	C26BP
27	G31BK	N37BH	F37BC	Y19AL	R14AH	F32AF	W28AC	R25CS	F24CP	W17CK	R34CH	K38CD
28	L18BL	A27BJ	V27BD	T30AM	X23AJ	V20AG	Y33AD	X35AB	V21CR	Y22CL	X26CJ	F11CE
29	D12BM		E13BE	G14BC	B36AK	E28AH	T15AE	B16AC	E17CS	T29CM	B38CK	V31CF
30	J37BP		M19BF	L23BD	N32AL	M33AJ	G25AF	N24AD	M22AB	G34CP	N11CL	E18CH
31	U27CD		H30BG		A20AM		L35AG	A21AE		L26CR		M12CJ

1894	JAN	FEB	MAR	APR	MAY	JUN	JUL	AUG	SEP	OCT	NOV	DEC
1	H37CK	D19CH	B19BM	E23BK	T32BG	B33BE	V25AM	T24AK	B22AH	V34AE	T11AC	X18CR
2	S27CL	J30CJ	N30BP	M36BL	G20BH	N15BF	E35BC	G21AL	N29AJ	E26AF	G31AD	B12CS
3	P13CM	U14CK	A14CD	H32BM	L28BJ	A25BG	M16BD	L17AM	A34AK	M38AG	L18AE	N37AB
4	W19CP	R23CL	Z23CE	S20BP	D33BK	Z35BH	H24BE	D22BC	Z26AL	H11AH	D12AF	A27AC
5	Y30CR	X36CM	C36CF	P28CD	J15BL	C16BJ	S21BF	J29BD	C38AM	S31AJ	J37AG	Z13AD
6	T14CS	B32CP	K32CH	W33CE	U25BM	K24BK	P17BG	U34BE	K11BC	P18AK	U27AH	C19AE
7	G23AB	N20CR	F20CJ	Y15CF	R35BP	F21BL	W22BH	R26BF	F31BD	W12AL	R13AJ	K30AF
8	L36AC	A28CS	V28CK	T25CH	X16CD	V17BM	Y29BJ	X38BF	V18BE	Y37AM	X19AK	F14AG
9	D32AD	Z33AB	E33CL	G35CJ	B24CE	E26BP	T34BK	B11BH	E12BF	T27BC	B30AL	V23AH
10	J20AE	C15AC	M15CM	L16CK	N21CF	M29CD	G26BL	N31BJ	M37BG	G13BD	N14AM	E36AJ
11	U28AF	K25AD	H25CP	D24CL	A17CH	H34CE	L38BM	A18BK	H27BH	L19BE	A23BC	H32AK
12	R33AG	F35AE	S35CR	J21CM	Z22CJ	S26CF	D11BP	Z12BL	S13BJ	D30BF	Z36BD	H20AL
13	X15AH	V16AF	P16CS	U17CP	C29CK	P38CH	J31CD	C37BM	P19BK	J14BG	C32BE	S28AM
14	B25AJ	E24AG	W24AB	R22CR	K34CL	W11CJ	U18CE	K27BP	W38BL	U23BH	K20BF	P33BC
15	N35AK	M21AH	Y21AC	X29CS	F26CM	Y31CK	R12CF	F13CD	Y14BM	R36BJ	F28BG	W15BD
16	A16AL	H17AJ	T17AD	B34AB	V38CP	T18CL	X37CH	V19CE	T23BP	X32BK	V33BH	Y25BE
17	Z24AM	S22AK	G22AE	N26AC	E11CR	G12CM	B27CJ	E30CF	G35CD	B20BL	E15BJ	T35BF
18	C21BC	P29AL	L29AF	A38AD	M31CS	L37CP	N13CK	M14CH	L32CE	N28BM	M25BK	G16BG
19	K17BD	W34AM	D34AB	Z11AE	H18AB	D27CR	A19CL	H23CJ	D20CF	A33BP	H35BL	L24BH
20	F22BE	Y26BC	J26AH	C31AF	S12AC	J13CS	Z30CM	S36CK	J28CH	Z15CD	S16BM	D21BJ
21	V29BF	T38BD	U38AJ	K18AG	P37AD	U19AB	C14CP	P32CL	U33CJ	C25CE	P24BP	J17BK
22	E34BG	G11BE	R11AK	F12AH	W27AE	R30AC	K23CR	W20CM	R15CK	K35CF	W21CD	U22BL
23	M26BH	L31BF	X31AL	V37AJ	Y13AF	X14AD	F36CS	Y28CP	X25CL	F16CH	Y17CE	R29BM
24	H38BJ	D18BG	B18AM	E27AK	T19AG	B23AE	V32AB	T33CR	B35CM	V24CJ	T22CF	X34BP
25	S11BK	J12BH	N12BC	M13AL	G30AH	N36AF	E20AC	G15CS	N16CP	E21CK	G29CH	B26CD
26	P31BL	U37BJ	A37BD	H19AM	L14AJ	A32AG	M28AD	L25AB	A24CR	M17CL	L34CJ	N38CE
27	W18BM	R27BK	Z27BE	S30BC	D23AK	Z20AH	H33AE	D35AC	Z21CS	H22CM	D26CK	A11CF
28	Y12BP	X13BL	C13BF	P14BD	J36AL	C28AJ	S15AF	J16AD	C17AB	S29CP	J38CL	Z31CH
29	T37CD		K19BG	W23BF	U32AM	K33AK	P25AG	U24AE	K22AC	P34CR	U11CM	C18CJ
30	G27CE		F30BH	Y36BF	R20BC	F15AL	W35AH	R21AF	F29AD	W26CS	R31CP	K12CK
31	L13CF		V14BJ		X28BD		Y16AJ	X17AG		Y38AB		F37CL

1895	JAN	FEB	MAR	APR	MAY	JUN	JUL	AUG	SEP	OCT	NOV	DEC
1	V27CH	T30CK	U30CD	K36BM	P20BJ	U15BG	C35BD	P21AM	U29AK	C26AG	P31AE	J12AB
2	E13CP	G14CL	R14CE	F32BP	W28BK	R25BH	K16BE	W17BC	R34AL	K38AH	W18AF	U37AC
3	M19CR	L23CM	X23CF	V20CD	Y33BL	X35BJ	F24BF	Y22BD	X26AM	F11AJ	Y12AG	R27AD
4	H30CS	D36CP	B36CH	E28CE	T15BM	B16BK	V21BG	T29BE	B38BC	V31AK	T37AH	X13AE
5	S14AB	J32CR	N32CJ	M33CF	G25BP	N24BL	E17BH	G34BF	N11BD	E18AL	G27AJ	B19AF
6	P23AC	U20CS	A20CK	H15CH	A21BM	M28BJ	L26BG	A31BE		M12AH	L13AG	N30AG
7	W36AD	R28AB	Z28CL	S25CJ	D16CE	Z17BP	H29BK	D38BH	Z18BF	H37BC	D19AL	A14AH
8	Y32AE	X33AC	C33CM	P35CK	J24CF	C22CD	S34BL	J11BJ	C12BG	S27BD	J30AM	Z23AJ
9	T20AF	B15AD	K15CP	W16CL	U21CH	K29CE	P26BK	U31BJ	K37BH	P13BE	U14BC	C36AK
10	G28AG	N25AE	F25CR	Y24CM	R17CJ	F34CF	W38BP	R18BL	F27BH	W19BF	R23BD	K32AL
11	L33AH	A35AF	V35CS	T21CP	X22CK	V26CH	Y11CD	X12BM	V13BK	Y30BG	X36BE	F20AM
12	D15AJ	Z16AG	E16AB	G17CR	B29CL	E38CJ	T31CE	B37BP	E19BL	T14BH	B32BF	V28BC
13	J25AK	C24AH	M24AC	H24AC	N34CM	M11CK	G18CF	N27CD	M30BM	G23BJ	N20BG	E33BD
14	U35AL	K21AJ	H21AD	D29AB	A26CP	H31CL	L12CH	A13CE	H14BP	L36BK	A28BH	H15BE
15	R16AM	F17AK	S17AE	J34AC	Z38CM	S18CK	D37CJ	Z19CF	S23CD	D32BL	Z33BJ	H25BF
16	X24BC	V22AL	P22AF	U26AD	C11CS	P12CP	J27CK	C30CH	P36CE	J20BH	C15BK	S35BG
17	B21BD	E29AM	W29AG	R38AK	K31AF	W37AH	U13CM	K14CJ	W32CF	U28BP	K25BL	P16BH
18	N17BE	M34BC	Y34AH	X11AF	F18AC	Y27CS	R19CM	F23CK	Y20CH	R33CD	F35BM	W24BJ
19	A22BF	H26BD	T26AJ	B31AG	V12AD	T13AB	X30CP	V36CL	T28CJ	X15CE	V16BP	Y21BK
20	Z29BG	S38BE	G38AK	N18AH	E37AE	G19AC	B14CR	E32CM	G33CK	B25CF	E24CD	T17BL
21	C34BH	P11BF	L11AL	A12AJ	M27AF	L30AD	N23CS	M20CH	L15CL	N35CH	M21CE	G22BM
22	K26BJ	W31BG	D31AM	Z37AK	H13AG	D14AF	A36AB	H28CR	D25CM	A16CJ	H17CF	L29BP
23	F38BK	Y18BH	J18BC	C27AL	S19AH	J23AF	Z32AC	S33CS	J35CP	Z24CK	S22CH	D34CG
24	V11BL	T12AJ	U12BD	K13AM	P30AG	U36AE	C20AD	P15AB	U16CR	C21CL	P29CJ	J26CE
25	E31BM	G37BK	R37BE	F19BC	W14AK	R32AH	K28AE	W25AC	R24CS	K17CM	W34CK	U38CF
26	M18BP	L27BL	X27BF	V30BD	Y23AL	X20AJ	F33AF	Y35AD	X21AB	F22CP	Y26CL	R11CH
27	H12CD	D13BM	B13BG	E14BE	T36AM	B28AK	V15AG	T16AE	B17AC	V29CR	T38CM	X31CJ
28	S37CE	J19BP	N19BH	M23BF	G32BC	N33AL	E25AM	G24AF	N22AD	E34CS	G11CP	B18CK
29	P27CF		A30BJ	H36BG	L20BD	A15AM	M35AJ	L21AG	A29AE	M26AB	L31CR	N12CL
30	W13CH		Z14BK	S32BH	D28BE	Z25BC	H16AK	D17AH	Z34AF	H38AC	D18CS	A37CM
31	Y19CJ		C23BL		J33BF		S24AL	J22AJ		S11AD		Z27CP

FIRST LETTER = YOUR PHYSICAL CODE, MIDDLE NUMBER = SENSITIVITY CODE, LAST TWO LETTERS = INTELLECTUAL CODE

1896

	JAN	FEB	MAR	APR	MAY	JUN	JUL	AUG	SEP	OCT	NOV	DEC
1	C13CR	P14CH	D23CH	Z20CE	H33BM	D35BK	A24BG	H22BE	D26BC	A11AK	H12AH	L27AE
2	K19CS	W23CP	J36CJ	C28CP	S15BP	J16BL	Z21BH	S29BF	J38BD	Z31AL	S37AJ	O13AF
3	F30AB	Y36CR	U32CK	K33CH	P25CO	U24BM	C17BJ	P34BG	U11BE	C18AH	P27AK	J19AG
4	V14AC	T32CL	R20CL	F15CJ	W35CE	R21BP	K22BH	W26BH	R31BF	K12BC	W13AL	U30AH
5	E23AD	G20AB	X28CM	V25CK	Y16CF	X17CD	F29BL	Y38BJ	X18BG	F37BD	Y19AM	R14AJ
6	M36AE	L28AC	B33CP	E35CL	T24CH	B22CE	V34BM	T11BK	B12BH	V27BE	T30BC	X23AK
7	H32AF	D33AD	N15CR	M16CM	G21CJ	N29CF	E26BP	G31BL	N37BJ	E13BF	G14BD	B36AL
8	S20AG	J15AE	A25CS	H24CP	L17CK	A34CH	M38CO	L18BM	A27BK	M19BG	L23BE	N32AM
9	P28AH	U25AF	Z35AB	S21CR	D22CL	Z26CJ	H11CE	D12BP	Z13BL	H30BH	D36BF	A20BC
10	W33AJ	R35AG	C16AC	P17CS	J29CM	C38CK	S31CF	J37CO	C19BM	S14BJ	J32BG	Z28BD
11	Y15AK	X16AH	K24AD	W22AB	U34CP	K11CL	P18CH	U27CE	K30BP	P23BK	U20BH	C33BE
12	T25AL	B24AJ	F21AE	Y29AC	R26CR	F31CM	W12CJ	R13CF	F14CO	W36BL	R28BJ	K15BF
13	G35AM	N21AK	V17AF	T34AD	X38CS	V18CP	Y37CK	X19CH	V23CE	Y32BM	X33BK	F25BG
14	L16BC	A17AL	E22AG	G26AE	B11AB	E12CR	T27CL	B30CJ	E36CF	T20BP	B15BL	V35BH
15	D24BD	Z22AM	M29AH	L38AF	N31AC	M37CS	G13CM	N14CK	M32CH	G28CD	N25BM	E16BJ
16	J21BE	C29BC	H34AJ	D11AG	A18AD	H27AB	L19CP	A23CL	H20CJ	L33CE	A35BP	M24BK
17	U17BF	K34BD	S26AK	J31AH	Z12AE	S13AC	D30CR	Z36CH	S28CK	D15CF	Z16CD	H21BL
18	R22BG	F26BE	P38AL	U18AJ	C37AF	P19AD	J14CS	C32CP	P33CL	J25CH	C24CE	S17BM
19	X29BH	V38BF	W11AM	R12AK	K27AG	M30AE	U23AB	K20CR	M15CM	U35CJ	K21CF	P22BP
20	B34BJ	E11BG	Y31BC	X37AL	F13AH	Y14AF	R36AC	F28CS	Y25CP	R16CK	F17CH	W29CD
21	N26BK	M31BH	T18BD	B27AM	V19AJ	T23AG	X32AD	V33AB	T35CR	X24CL	V22CJ	Y34CE
22	A38BL	H18BJ	G12BE	N13BC	E30AK	G36AH	B20AE	E15AC	G16CS	B21CM	E29CK	T26CF
23	Z11BM	S12BK	L37BF	A19BD	M14AL	L32AJ	N28AF	M25AD	L24AB	N17CP	M34CL	G38CH
24	C31BP	P37BL	D27BG	Z30BE	H23AM	D20AK	A33AG	H35AE	D21AC	A22CR	H26CH	L11CJ
25	K18CD	W27BM	J13BH	C14BF	S36BC	J28AL	Z15AH	S16AF	J17AD	Z29CS	S38CP	D31CK
26	F12CE	Y13BP	U19BJ	K23BG	P32BD	U33AM	C25AJ	P24AG	U22AE	C34AB	P11CR	J18CL
27	V37CF	T19CO	R30BK	F36BH	W20BE	R15BC	K35AK	W21AH	R29AF	K26AC	W31CS	U12CM
28	E27CH	G30CE	X14BL	V32BJ	Y28BF	X25BD	F16AL	Y17AJ	X34AG	F38AD	Y18AB	R37CP
29	M13CJ	L14CF	B23BM	E20BK	T33BG	B35BE	V24AM	T22AK	B26AH	V11AE	T12AC	X27CR
30	H19CK		N36BP	M28BL	G15BH	N16BF	E21BC	G29AL	N38AJ	E31AF	G37AD	B13CS
31	S30CL		A32CD		L25BJ		M17BD	L34AM		M18AG		N19AB

1897

	JAN	FEB	MAR	APR	MAY	JUN	JUL	AUG	SEP	OCT	NOV	DEC
1	A30AC	H36CS	T36CK	B28CH	V15CD	T16BM	X21BJ	V29BG	T38BE	X31AM	V37AK	Y13AG
2	Z14AD	S32AB	G32CL	N33CJ	E25CE	G24BP	B17BK	E34BH	G11BF	B18BC	E27AL	T19AH
3	C23AE	P20AC	L20CM	A15CK	M35CF	L21CD	N22BL	M26BJ	L31BG	N12BD	M13AM	G30AJ
4	K36AF	W28AD	D28CP	Z25CL	H16CH	D17CE	A29BM	H38BK	D18BH	A37BE	H19BC	L14AK
5	F32AG	Y33AE	J33CR	C35CM	S24CJ	J22CF	Z34BP	S11BL	J12BJ	Z27BF	S30BD	D23AL
6	V20AH	T15AF	U15CS	K16CP	P21CK	U29CH	C26CO	P31BM	U37BK	C13BG	P14BE	J36AJ
7	E28AJ	G25AG	R25AB	P24CR	W17CL	R34CJ	K38CE	W18BP	R27BL	K19BH	W23BF	U32BC
8	M33AK	L35AM	X35AC	V21CS	Y22CM	X26CK	F11CF	Y12CO	X13BM	F30BJ	Y36BG	R20BD
9	H15AL	D16AJ	B16AD	E17AB	T29CP	B38CL	V31CH	T37CE	B19BP	V14BK	T32BH	X28BE
10	S25AM	J24AK	N24AE	M22AC	G34CR	N11CM	E18CJ	G27CF	N30CD	E23BL	G20BJ	B33BF
11	P35BC	U21AL	A21AF	H29AD	L26CS	A31CP	M12CK	L13CH	A14CE	M36BH	L28BK	N15BG
12	W16BD	R17AM	Z17AG	S34AE	D38AB	Z18CR	H37CL	D19CJ	Z23CF	H32BP	D33BL	A25BH
13	Y24BE	X22BC	C22AH	P26AF	J11AC	C12CS	S27CM	J30CK	C36CH	S20CD	J15BM	Z35BJ
14	T21BF	B29BD	K29AJ	W38AG	U31AD	K37AB	P13CP	U14CL	K32CJ	P28CE	U25BP	C16BK
15	G17BG	N34BE	F34AK	Y11AH	R18AE	F27AC	W19CR	R23CH	F20CK	W33CF	R35CD	K24BL
16	L22BH	A26BF	V26AL	T31AJ	X12AF	V13AD	Y30CS	X36CP	V28CL	Y15CH	X16CE	F21BM
17	D29BJ	Z38BG	E38AM	G18AK	B37AG	E19AE	T14AB	B32CR	E33CM	T25CJ	B24CF	V17BP
18	J34BK	C11BH	M11BC	L12AL	N27AH	M30AF	G23AC	N20CS	M15CP	G35CK	N21CH	E22CD
19	U26BL	K31BJ	H31BD	D37AM	A13AJ	H14AG	L36AD	A28AB	H25CR	L16CL	A17CJ	M29CE
20	R38BM	F18BK	S18BE	J27BC	Z19AK	S23AH	O32AE	Z33AC	S35CS	D24CM	Z22CK	H34CF
21	X11BP	V12BL	P12BF	U13BD	C30AL	P36AJ	J20AF	C15AD	P16AB	J21CP	C29CL	S26CH
22	B31CD	E37BH	W37BG	R19BE	K14AM	W32AK	U28AG	K25AE	W24AC	U17CR	K34CH	P38CJ
23	N18CE	M27BP	Y27BH	X30BF	F23BC	Y20AL	R33AH	F35AF	Y21AD	R22CS	F26CP	W11CK
24	A12CF	H13CO	T13BJ	B14BG	V36BD	T28AM	X15AJ	V16AG	T17AE	X29AB	V38CR	Y31CL
25	Z37CH	S19CE	G19BK	N23BH	E32BE	G33BC	B25AK	E24AH	G22AF	B34AC	E11CS	T18CM
26	C27CJ	P30CF	L30BL	A36BJ	M20BF	L15BD	N35AL	M21AJ	L29AG	N26AD	M31AB	G12CP
27	K13CK	W14CH	D14BM	Z32BK	H28BH	D25BE	A16AM	H17AK	D34AH	A38AE	H18AC	L37CP
28	F19CL	Y23CJ	J23BP	C20BL	S33BH	J35BF	Z24BC	S22AL	J25AJ	Z11AF	S12AD	D27CS
29	V30CH		U36CO	K28BP	P15BJ	U16BG	C21BD	P29AM	U38AK	C31AG	P37AE	J13AB
30	E14CP		R32CE	F33BP	W25BK	R24BH	K17BE	W34BD	R11AL	K18AH	W27AF	U19AC
31	M23CR		X20CF		Y35BL		F22BF	Y26BD		F12AJ		R30AD

1898

	JAN	FEB	MAR	APR	MAY	JUN	JUL	AUG	SEP	OCT	NOV	DEC
1	X14AE	V32AC	P32CM	U33CK	C25CF	P24CD	J17BL	C34BJ	P11BG	J18BD	C27AM	S19AJ
2	B23AF	E20AD	W20CP	R15CL	K35CH	W21CE	U22BM	K26BK	W31BH	U12BE	K13BC	P30AK
3	N36AG	M28AE	Y28CR	X25CM	F16CJ	Y17CF	R29BP	F38BL	Y18BJ	R37BF	F19BD	W14AL
4	A32AH	H33AF	T33CS	B35CP	V24CK	T22CH	X34CD	V11BM	T12BK	X27BG	V30BE	Y23AM
5	Z20AJ	S15AG	G15AB	N16CR	E21CL	G29CJ	B26CE	E31BP	G37BL	B13BH	E14BF	T36BC
6	C28AK	P25AH	L25AC	A24CS	M17CM	L34CK	N38CO	M18CO	L27BM	N19BJ	M23BG	G32BD
7	K33AL	W35AJ	D35AD	Z21AB	H22CP	D26CL	A11CH	H12CE	D13BP	A30BK	H36BH	L20BE
8	F15AM	Y16AK	J16AE	C17AC	S29CR	J38CH	Z31CJ	S37CF	J19CO	Z14BL	S32BF	D28BP
9	V25BC	T24AL	U24AF	K22AD	P34CS	U11CP	C18CK	P27CH	U30CE	C23BM	P20BK	J33BG
10	E35BD	G21AM	R21AG	F29AE	W26AB	R31CM	K12CJ	W13CJ	R14CF	K36BP	W28BL	U15BH
11	M16BE	L17BC	X17AH	V34AF	Y38AC	X18CS	F37CM	Y19CH	X23CH	F32CD	Y33BM	R25BJ
12	H24BF	D22BD	B22AJ	E26AG	T11AD	B12AB	V27CL	T30CL	B36CJ	V20CE	T15BP	X35BK
13	S21BG	J29BE	N29AK	M38AH	G31AE	N37AC	E13CR	G14CR	N32CH	E28CF	G25CO	B16BL
14	P17BH	U34BF	A34AL	H11AJ	L18AF	A27AD	M19CS	L23CP	A20CL	M33CH	L35CE	N24BM
15	W22BJ	R26BG	Z26AM	S31AK	D12AG	Z13AE	H30AB	D36CO	Z28CM	H15CJ	D16CF	A21BP
16	Y29BK	X38BH	C38BC	P18AL	J37AH	C19AF	S14AC	J32CS	C33CP	S25CK	J24CH	Z17CD
17	T34BL	B11BJ	K11BD	W12AM	U27AJ	K30AG	P23AD	U20AB	K15CR	P35CL	U21CJ	C22CE
18	G26BM	N31BK	F31BE	Y37BC	R13AK	F14AH	W36AE	R28AC	F25CS	W16CM	R17CK	K29CF
19	L38BP	A18BL	V18BF	T27BD	X19AL	V23AH	Y32AF	X33AB	V35AB	Y24CP	X22CL	F34CH
20	D11CD	Z12BM	E12BG	G13BE	B30AM	E36AK	T20AG	B15AE	E16AC	T21CR	B29CM	V26CJ
21	J31CE	C37BP	M37BH	L19BF	N14BD	M32AL	G28AH	N25AF	M24AD	G17CS	N34CP	E38CK
22	U18CF	K27CD	H27BJ	D30BG	A23BD	H20AM	L33AJ	A35AG	H21AE	L22AB	A26CR	M11CL
23	R12CH	F13CE	S13BK	J14BH	Z36BE	S28BC	D15AK	Z16AH	S17AF	D29AC	Z38CS	H31CM
24	X37CJ	V19CF	P19BL	U23BL	C32BF	P33BD	J25AL	C24AJ	P22AG	J34AD	C11AB	S18CP
25	B27CK	E30CH	W30BM	R36BK	K20BG	W15BE	U35AM	K21AH	M29AH	U26AE	K31AC	P12CR
26	N13CL	M14CO	Y14BP	X32BM	F28BH	Y25BF	R16BC	F17AL	Y34AJ	R38AF	F18AD	W37CS
27	A19CM	H23CK	T23CD	B20BM	V33BJ	T35BG	X24BD	V22AM	T26AK	X11AG	V12AE	Y27AB
28	Z30CP	S36CL	G36CE	N28BP	E15BK	G16BH	B21BE	E29BC	G38AL	B31AH	E37AF	T13AC
29	C14CR		L32CF	A33CD	M25BL	L24BJ	N17BF	M34BD	L11AM	N18AJ	M27AG	G19AD
30	K23CS		D28CH	Z15CE	H35BM	D21BK	A22BG	H26BE	D31BC	A12AK	H13AH	L30AE
31	F36AB		J28CJ		S16BP		Z29BH	S38BF		Z37AL		D14AF

FIRST LETTER = YOUR PHYSICAL CODE, MIDDLE NUMBER = SENSITIVITY CODE, LAST TWO LETTERS = INTELLECTUAL CODE

1899	JAN	FEB	MAR	APR	MAY	JUN	JUL	AUG	SEP	OCT	NOV	DEC
1	J23AG	C20AE	M20CR	L15CM	N35CJ	M21CF	G22BP	N26BL	M31BJ	G12BF	N13BD	E30AL
2	U36AH	K28AF	H28CS	D25CP	A16CK	H17CH	L29CD	A38BM	H18BK	L37BG	A19BE	M14AM
3	R32AJ	F33AG	S33AB	J35CR	Z24CL	S22CJ	O34CE	Z11BP	S12BL	D27BH	Z30BF	M23BC
4	X20AK	V15AH	P15AC	U16CS	C21CM	P29CK	J26CF	C31CD	P37BM	J13BJ	C14BG	S36BD
5	B28AL	E25AJ	W25AD	R24AB	K17CP	W34CL	U38CH	K18CE	W27BP	U19BK	K23BH	P32BE
6	N33AM	M35AK	Y35AE	X21AC	F22CR	Y26CM	R11CJ	V37CH	T19CE	X14BM	V32BK	Y28BG
7	A15BC	H16AL	T16AF	B17AD	V29CS	T38CP	X31CK	E27CJ	G30CF	B23BP	E20BL	T33BH
8	Z25BD	S24AM	G24AG	N22AE	E34AB	G11CR	B18CL	E27CJ	G30CF	B23BP	E20BL	T33BH
9	C35BE	P21BC	L21AH	A29AF	M26AC	L31CS	N12CM	M13CK	L14CH	N36CD	M28BM	G15BJ
10	K16BF	W17BD	D17AJ	Z34AG	H38AD	D18AB	A37CP	H19CL	D23CJ	A32CE	H33BP	L25BK
11	F24BG	Y22BE	J22AK	C26AH	S11AE	J12AC	Z27CR	S30CM	J36CK	Z20CF	S15CD	D35BL
12	V21BH	T29BF	U29AL	K38AJ	P31AF	U37AD	C13CS	P14CP	U32CL	C28CH	P25CE	J16BM
13	E17BJ	G34BG	R34AM	F11AK	W18AG	R27AE	K19AB	H23CR	R20CM	K33CJ	W35CF	U24BP
14	M22BK	L26BH	X26BC	V31AL	Y12AH	X13AF	F30AC	Y36CS	X28CP	F15CK	Y16CH	R21CD
15	H29BL	D38BJ	B39BD	E18AM	T37AJ	B19AG	V14AD	T32AB	B33CR	V25CL	T24CJ	X17CE
16	S34BM	J11BK	N11BE	M12BC	G27AK	N30AH	E23AE	G20AC	N15CS	E35CM	G21CK	B22CF
17	P26BP	U31BL	A31BF	H37BD	L13AL	A14AJ	M36AF	L28AD	A25AB	M16CP	L17CL	N29CH
18	W38CD	R18BM	Z18BG	S27BE	D19AM	Z23AK	H32AG	D33AE	Z35AC	H24CR	D22CM	A34CJ
19	Y11CE	X12BP	G12BH	P13BF	J30BC	C36AL	S20AH	J15AF	C16AD	S21CS	J29CP	Z26CK
20	T31CF	B37CD	K37BJ	W19BG	U14BD	K32AM	P28AJ	U25AG	K24AE	P17AB	U34CR	C38CL
21	G18CH	N27CE	F27BK	Y30BH	R23BE	F20BC	W33AK	R35AH	F21AF	W22AC	R26CS	K11CM
22	L12CJ	A13CF	V13BL	T14BJ	X36BF	V28BD	Y15AL	X16AJ	V17AG	Y29AD	X38AB	F31CP
23	D37CK	Z19CH	E19BM	G23BK	B32BG	E33BE	T25AM	B24AK	E22AH	T34AE	B11AC	V18CR
24	J27CL	C30CJ	M30BP	L36BL	N20BH	M15BF	G35BC	N21AL	M29AJ	G26AF	N31AD	E12CS
25	U13CM	K14CK	H14CD	D32BM	A25BG	H25BG	L16BD	A17AM	H34AG	L38AG	A18AE	M37AB
26	R19CP	F23CL	S23CE	J20BP	Z33BK	S35BH	D24BE	Z22BC	S26AL	D11AH	Z12AF	H27AC
27	X30CR	V36CM	P36CF	U28CD	C15BL	P16BJ	J21BF	C29BD	P33AH	J31AJ	C37AG	S13AD
28	B14CS	E32CP	W32CH	R33CE	K25BM	W24BK	U17BG	K34BE	W11BC	U18AK	K27AH	P19AE
29	N23AB		Y20CJ	X15CF	F35BP	Y21BL	R22BH	F26BF	Y31BD	R12AL	F13AJ	W30AF
30	A36AC		T28CK	B25CH	V16CD	T17BM	X29BJ	V38BG	T18BE	X37AM	V19AK	Y14AG
31	Z32AD		G33CL		E24CE		B34BK	E11BH		B27BC		T23AH

1900	JAN	FEB	MAR	APR	MAY	JUN	JUL	AUG	SEP	OCT	NOV	DEC
1	G36AJ	N28AG	F28AB	Y25CR	R16CL	F17CJ	W29CE	R38BP	F18BL	W37BH	R19BF	K14BC
2	L32AK	A33AH	V33AC	T35CS	X24CM	V22CK	Y34CF	X11CD	V12BM	Y27BJ	X30BG	F23BD
3	D20AL	Z15AJ	E15AD	G16AB	B21CP	E29CL	T26CH	B31CE	E37BP	T13BK	B14BH	V36BE
4	J28AM	C25AK	M25AE	L24AC	N17CR	M34CM	G38CJ	N18CF	M27CD	G19BL	N23BJ	E32BF
5	U33BC	K35AL	H35AF	D21AD	A22CS	H26CP	L11CK	A12CH	H13CE	L30BM	A36BK	M20BG
6	R15BD	F16AM	S16AG	J17AE	Z29AB	S38CR	D31CL	Z37CJ	S19CF	D14BP	Z32BL	H28BH
7	X25BE	V24BC	P24AH	U22AF	C34AC	P11CS	J18CM	C27CK	P30CH	J23CD	C20BM	S33BJ
8	B35BF	E21BD	W21AJ	R29AG	K26AD	W31AB	U12CP	K13CL	W14CJ	U36CE	K28BP	P15BK
9	N16BG	M17BE	Y17AK	X34AH	F38AE	Y18AC	R37CR	F19CM	Y23CK	R32CF	F33CD	W25BL
10	A24BH	H22BF	T22AL	B26AJ	V11AF	T12AD	X27CS	V30CP	T36CL	X20CH	V15CE	Y35BM
11	Z21BJ	S29BG	G29AM	N38AK	E31AG	G37AE	B13AB	E14CR	G32CM	B28CJ	E25CF	T16BP
12	C17BK	P34BH	L34BC	A11AL	M18AH	L27AF	N19AC	M23CS	L20CP	N33CK	M35CH	G24CD
13	K22BL	W26BJ	D26BD	Z31AM	H12AJ	D13AG	A30AD	H36AB	D28CR	A15CL	H16CJ	L21CE
14	F29BM	Y38BK	J38BE	C18BC	S37AK	J19AH	Z14AE	S32AC	J33CS	Z25CH	S24CK	D17CF
15	V34BP	T11BL	U11BF	K12BD	P27AL	U30AJ	C23AF	P20AD	U15AB	C35CP	P21CL	J22CH
16	E26CD	G31BM	R31BG	F37BE	W13AM	R14AK	K35AG	W28AE	R25AC	K16CR	W17CM	U29CJ
17	H38CE	L18BP	X18BH	V27BF	Y19BC	X23AL	F32AH	Y33AF	X35AD	F24CS	Y22CP	R34CK
18	H11CF	D12CD	B12BJ	E13BG	T30BD	B36AM	V20AJ	T15AG	B16AE	V21AB	T29CR	X26CL
19	S31CH	J37CE	N37BK	M19BH	G14BE	N32BC	E28AK	G25AH	N24AF	E17AC	G34CS	B38CM
20	P18CJ	U27CF	A27BL	H30BJ	L23BF	A20BD	M33AL	L35AJ	A21AG	M22AD	L26AB	N11CP
21	W12CK	R13CH	Z13BM	S14BK	D36BG	Z28BE	H15AM	D16AK	Z17AH	H29AE	D38AC	A31CR
22	Y37CL	X19CJ	C19BP	P23BL	J32BH	C33BF	S25BC	J24AL	C22AJ	S34AF	J11AD	Z18CS
23	T27CM	B30CK	K30CD	W36BM	U20BJ	K15BG	P35BD	U21AM	K29AK	P26AG	U31AE	C12AB
24	G13CP	N14CL	F14CE	Y32BP	R28BK	F25BH	W16BE	R17BC	F34AL	W38AH	R18AF	K37AC
25	L19CR	A23CH	V23CF	T20CD	X33BL	V35BJ	Y24BF	X22BD	V26AM	Y11AJ	X12AG	F27AD
26	D30CS	Z36CP	E36CH	G28CE	B15BM	E16BK	T21BG	B29BE	E38BC	T31AK	B37AH	V13AE
27	J14AB	C32CM	H32CJ	L33CF	N25BP	H24BJ	G17BH	N34BF	H11BD	G18AL	N27AJ	E19AF
28	U23AC	K20CS	H20CH	D15CH	A35CD	H21BM	L22BJ	A26BG	H31BE	L12AM	A13AK	M30AG
29	R36AD		S28CH	J25CJ	Z16CE	S17BP	D29BK	Z38BH	S18BF	D37BC	Z19AL	H14AH
30	X32AE		P33CM	U35CK	C24CF	P22CD	J34BL	C11BJ	P12BF	J27BD	C30AM	S23AJ
31	B20AF		W15CP		K21CH		U26BM	K31BK		U13BE		P36AK

1901	JAN	FEB	MAR	APR	MAY	JUN	JUL	AUG	SEP	OCT	NOV	DEC
1	W32AL	R33AJ	Z33AD	S35AB	D24CP	Z22CL	H34CH	D11CE	Z12BP	H27BK	D30BH	A23BE
2	Y20AM	X15AK	C15AE	P16AC	J21CR	C29CM	S26CJ	J31CF	C37CD	S13BL	J14BJ	Z36BF
3	T28BC	B25AL	K25AF	W24AD	U17CS	K34CP	P38CK	U16CH	K27CE	P19BM	U23BK	C32BG
4	G33BD	N35AM	F35AG	Y21AE	R22AB	F26CR	W11CL	R12CJ	F13CF	W30BP	R36BL	K20BH
5	L15BE	A16BC	V16AH	T17AF	X29AC	V38CS	Y31CH	X37CK	V19CH	Y14CD	X32BM	F28BJ
6	D25BF	Z24AE	E24AJ	G22AG	B34AD	E11AB	T18CP	B27CL	E30CJ	T23CE	B20BP	V33BK
7	J35BG	C21BE	M21AK	L29AH	N26AE	M31AC	G12CR	N13CM	M14CK	G36CF	N28CD	E15BL
8	U16BH	K17BF	H17AL	D34AJ	A38AF	H18AD	L37CS	A19CP	H23CL	L32CH	A33CE	M25BM
9	R24BJ	F22BG	S22AM	J26AK	Z11AG	S12AE	D27AB	Z30CR	S36CM	D20CJ	Z15CF	H35BP
10	X21BK	V29BH	P29BC	U38AL	C31AH	P37AF	J13AC	C14CS	P32CP	J28CK	C25CH	S16CD
11	B17BL	E34BJ	W34BD	R11AM	K18AJ	W27AD	U19AD	K23AB	W20CR	U33CL	K35CJ	P24CE
12	N22BM	M26BK	Y26BE	X31BC	F12AK	Y13AH	R30AE	F36AC	V32AD	T33AB	X25CP	W21CH
13	A29BP	H38BL	T38BF	B18BD	V37AL	T19AJ	X14AF	V32AD	T33AB	X25CP	V24CL	Y17CH
14	Z34CD	S11BM	G11BG	N12BE	E27AM	G30AK	B23AG	E20AE	G15AC	B35CS	E21CM	T22CJ
15	C26CE	P31BP	L31BH	A37BF	M13BC	L14AL	N36AH	M28AF	L25AD	N16CS	M17CP	G29CK
16	K38CF	W18CD	D18BJ	Z27BG	H19BD	D23AM	A32AJ	H33AG	D35AE	A24AB	H22CR	L34CL
17	F11CH	Y12CE	J12BK	C13BH	S30BE	J36BC	Z20AK	S15AH	J16AF	Z21AC	S29CS	D26CM
18	V31CJ	T37CF	U37BL	K19BJ	P14BF	U32BD	C28AL	P25AJ	U24AG	C17AD	P34AB	J38CP
19	E18CK	G27CH	R27BM	F30BK	W23BG	R20BE	K33AK	W35AH	R21AF	K22AE	W26AC	U11CR
20	M12CL	L13CJ	X13BP	V14BL	Y36BH	X28BF	F15BC	Y16AL	X17AJ	F29AF	Y38AD	R31CS
21	H37CM	D19CK	B19CD	E23BM	T32BJ	B33BF	V25BD	T24AK	B22AH	V34AE	T11AC	X18AB
22	S27CP	J38CL	N38CE	M36BP	G20BK	N15BH	E35BE	G21BC	N29AL	E26AH	G31AF	B12AC
23	P13CR	U16CM	A14CF	H32CD	L28BL	A25BJ	M16BF	L17BD	A34AM	M38AJ	L18AG	N37AD
24	W19CS	R23CP	Z23CH	S28CE	D33BM	Z35BK	H24BG	D22BE	Z26BC	H11AK	D12AH	A27AE
25	Y30AB	X36CR	C36CJ	P28CF	J15BP	C16BL	S21BH	J29BE	C38BD	S31AL	J37AJ	Z13AF
26	T14AC	B32CS	K32CK	W33CH	U25CD	K24BM	P17BJ	U34BG	K11BE	P18AM	U27AK	C19AG
27	G23AD	N20AB	F20CL	Y15CJ	R35CE	F21BH	W22BK	R26BH	F31BF	W12BC	R13AL	K30AH
28	L36AE	A28AC	V28CH	T25CK	X16CF	V17CD	Y29BL	X38BJ	V18BK	Y37BD	X19AM	F14AJ
29	D32AF		E33CP	G35CL	B24CH	E22CE	T34BM	B11BK	E12BH	T27BE	B30BC	V23AK
30	J20AG		M15CR	L16CM	N21CJ	M29CF	G26BP	N31BL	M37BJ	G13BF	N14BD	E36AL
31	U28AH		H25CS		A17CK		L38CD	A18BM		L19BG		M32AM

FIRST LETTER = YOUR PHYSICAL CODE, MIDDLE NUMBER = SENSITIVITY CODE, LAST TWO LETTERS = INTELLECTUAL CODE

1902

	JAN	FEB	MAR	APR	MAY	JUN	JUL	AUG	SEP	OCT	NOV	DEC
1	H20BC	O15AL	B15AF	E16AD	T21CS	B29CP	V26CK	T31CH	B37CE	V13BM	T14BK	X36BG
2	S26BD	J25AM	N25AG	M24AE	G17AB	N34CR	E38CL	G18CJ	N27CF	E19BP	G23BL	B32BH
3	P33BE	U35BC	A35AH	H21AF	L22AC	A26CS	M11CM	L12CK	A13CH	M30CD	L36BM	N20BJ
4	W15BF	R16BD	Z16AJ	S17AG	D29AD	Z38AB	H31CP	D37CL	Z19CJ	H14CE	D32BP	A28BK
5	Y25BG	X24BE	C24AK	P22AH	J34AE	C11AC	S18CR	J27CM	C30CK	S23CF	J20CD	Z33BL
6	G16BJ	B21BF	K21AL	W29AJ	U26AF	K31AD	P12CS	U13CP	K14CL	P36CH	U28CE	C15BM
7	L24BK	N17BG	F17AM	Y34AK	R38AG	F18AE	W37AB	R19CR	F23CM	W32CJ	R33CF	K25BP
8	D21BL	A22BH	V22BC	T26AL	X11AH	V12AF	Y27AC	X30CS	V36CP	Y20CK	X15CH	F35CD
9	J17BM	Z29BD	E29BD	G38AM	B31AJ	E37AG	T13AD	B14AB	E32CR	T28CL	B25CJ	V16CE
10	U22BP	C34BK	M34BE	L11BC	N18AK	M27AH	G19AE	N23AC	M20CS	G33CM	N35CK	E24CF
11	R29CD	K26BL	H26BF	D31BD	A12AL	H13AJ	L30AF	A36AD	H28AB	L15CP	A16CL	M21CH
12	X34CE	F38BM	S38BG	J18BE	Z37AM	S19AK	D14AG	Z32AE	S33AC	D25CR	Z24CM	H17CJ
13	B26CF	V11BP	P11BH	U12BF	C27BC	P30AL	J23AH	C20AF	P15AD	J35CS	C21CP	S22CK
14	N38CH	E31CD	W31BJ	R37BG	K13BD	W14AM	U36AJ	K28AG	W25AE	U16AB	K17CR	P29CL
15	A11CJ	M18CE	Y18BK	X27BH	F19BE	Y23BC	R32AK	F33AH	Y35AF	R24AC	F22CS	W34CM
16	Z31CK	H12CF	T12BL	B13BJ	V30BF	T36BD	X20AL	V15AJ	T16AG	X21AD	V29AB	Y26CP
17	C18CL	S37CH	G37BM	N19BK	E14BG	G32BE	B28AM	E25AK	G24AH	B17AE	E34AC	T38CR
18	K12CM	P27CJ	L27BP	A30BL	M23BH	L20BF	N33BC	M35AL	L21AJ	N22AF	M26AD	G11CS
19	F37CP	W13CK	D13CD	Z14BM	H36BJ	D28BG	A15BD	H16AM	D17AK	A29AG	H38AE	L31AB
20	V27CR	Y19CL	J19CE	C23BP	S32BK	J33BH	Z25BE	S24BC	J22AL	Z34AH	S11AF	D18AC
21	E13CS	T30CM	U30CF	K36CD	P20BL	U15BJ	G35BF	P21BD	U29AM	C26AJ	P31AG	J12AD
22	M19AB	G14CP	R14CH	F32CE	W28BM	R25BK	K16BG	W17BE	R34BC	K38AK	W18AH	U37AE
23	H30AC	L23CR	X23CJ	V20CF	Y33BP	X35BL	F24BH	Y22BF	X26BD	F11AL	Y12AJ	R27AF
24	S14AD	D36CS	B36CK	E28CH	T15CD	B16BM	V21BJ	T29BG	B38BE	V31AM	T37AK	X13AG
25	P23AE	J32AB	N32CL	M33CJ	G25CE	N24BP	E17BK	G34BH	N11BF	E18BC	G27AL	B19AH
26	W36AF	U20AC	A20CM	H15CK	L35CF	A21CD	M22BL	L26BJ	A31BG	M12BD	L13AM	N30AJ
27	Y32AG	R28AD	Z28CP	S25CL	D16CH	Z17CE	H29BM	D38BK	Z18BH	H37BE	D19BC	A14AK
28	T20AH	X33AE	C33CR	P35CM	J24CJ	C22CF	S34BP	J11BL	C12BJ	S27BF	J30BD	Z23AL
29	G28AJ		K15CS	W16CP	U21CK	K29CH	P26CD	U31BM	K37BK	P13BG	U14BE	C36AM
30	L33AK		F25BH	Y24CR	R17CL	F34CJ	W38CE	R18BP	F27BL	W19BH	R23BF	K32BC
31			V35AC		X22CM		Y11CF	X12CD		Y30BJ		F20BD

1903

	JAN	FEB	MAR	APR	MAY	JUN	JUL	AUG	SEP	OCT	NOV	DEC
1	V28BE	T25BC	U25AH	K24AF	P17AC	U34CS	C38CH	P18CK	U27CH	C19CD	P23BM	J32BJ
2	E33BF	G35BD	R35AJ	F21AG	W22AD	R26AB	K11CP	M12CL	R13CJ	K30CE	W36BP	U20BK
3	M15BG	L16BE	X16AK	V17AH	Y29AE	X38AC	F31CR	Y37CH	X19CK	F14CF	Y32CD	R28BL
4	H25BH	D24BF	B24AL	E22AJ	T34AF	B11AD	V18CS	T27CP	B30CL	V23CH	T20CE	X33BM
5	S35BJ	J21BG	N21AM	M29AK	G26AG	N31AE	E12AB	G13CR	N14CM	E36CJ	G28CF	B15BP
6	P16BK	U17BH	A17BC	H34AL	L38AH	A18AF	M37AC	L19CS	A23CP	M32CK	L33CH	N25CD
7	W24BL	R22BJ	Z22BD	S26AM	D11AJ	Z12AG	H27AD	D30AB	Z36AF	H20CL	D15CJ	A35CE
8	Y21BM	X29BK	C29BE	P38BC	J31AK	C37AH	S13AE	J14AC	C32CS	S28CH	J25CK	Z16CF
9	T17BP	B34BL	K34BF	W11BD	U18AL	K27AJ	P19AF	U23AD	K20AB	P33CP	U35CL	C24CH
10	G22CD	N26BM	F26BG	Y31BE	R12AM	F13AK	W30AG	R36AE	F28AC	W15CR	R16CM	K21CJ
11	L29CE	A38BP	V38BH	T16BF	X37BC	V19AL	Y14AH	X32AF	V33AD	Y25CS	X24CP	F17CK
12	D34CF	Z11CD	E11BJ	G12BG	B27BD	E30AM	T23AJ	B20AG	E15AE	T35AB	B21CR	V22CL
13	J26CH	C31CE	M31BK	L37BH	N13BE	M16AK	G36AK	N28AH	M25AF	G16AC	N17CS	E29CM
14	U38CJ	K18CF	H18BL	D27BJ	A19BF	H23BD	L32AL	A33AJ	H35AG	L24AD	A22AB	M34CP
15	R11CK	F12CH	S12BM	J13BK	Z30BG	S36BE	D20AM	Z15AK	S16AH	D21AE	Z29AC	H26CR
16	X31CL	V37CJ	P37BP	U19BL	C14BH	P32BF	J28BC	C25AL	P24AJ	J17AF	C34AD	S38CS
17	B18CM	E27CK	W27CD	R30BM	K23BJ	W20BG	U33BD	K35AM	W21AK	U22AG	K26AE	P11AB
18	N12CP	M13CL	Y13CE	X14BP	F36BK	Y28BH	R15BE	F16BD	Y17AL	R29AH	F38AF	W31AC
19	A37CR	H19CM	T19CF	B23CD	V32BL	T33BJ	X25BF	V24BD	T22AM	X34AJ	V11AG	Y18AD
20	Z27CS	S30CP	G30CH	N36CE	E20BM	G15BK	B35BG	E21BE	G29AP	B26AK	E31AH	T12AE
21	C13AB	P14CR	L14CJ	A32CF	M28BP	L25BL	N16BH	M17BF	L34BD	N38AL	M18AJ	G37AF
22	K19AC	W23CS	D23CK	Z20CH	H33CD	D35BM	A24BJ	H22BG	D26BE	A11AM	H12AK	L27AG
23	F30AD	Y36AB	J36CL	C28CJ	S15CE	J16BP	Z21BK	S29BH	J38BF	Z31BL	S37AL	D13AH
24	V14AE	T32AC	U32CM	K33CK	P25CF	U24CD	C17BL	P34BJ	U11BG	C18BD	P27AM	J19AJ
25	E23AF	G20AD	R20CP	V25CH	W35CH	R21CE	K22BK	W26BK	R31BH	K12BD	W13BC	N11BJ
26	M36AG	L28AE	X28CR	Y16CJ	X17CF	F29BP	Y38BL	X18BJ	F37BF	Y19BD	R14AG	U30AK
27	H32AH	D33AF	B33CS	E35CP	T24CK	B22CH	V34CD	T11BM	B12BC	V27BG	T30BE	X23AM
28	S20AJ	J15AG	N15AB	M16CR	G21CL	N29CJ	E26CE	G31BP	N37BL	E13BH	G14BF	B36BC
29	P28AK		A25AE	H24CS	L17CH	A34CK	M38CF	L18CD	A27BM	M19BJ	L23BG	N32BD
30	W33AL		Z35AD	S21AB	D22CP	Z26CL	H11CH	D12CD	Z13BP	H30BH	D36BH	A20BE
31	Y15AM		C16AE		J29CR		S31CJ	J37CF		S14BL		Z28BF

1904

	JAN	FEB	MAR	APR	MAY	JUN	JUL	AUG	SEP	OCT	NOV	DEC
1	C33BG	P35BE	D16AL	Z17AJ	H29AF	D38AD	A31CS	H37CP	D19CL	A14CH	H32CE	L28BM
2	K15BH	W16BF	J24AM	C22AK	S34AG	J11AE	Z18AB	S27CR	J30CM	Z23CJ	P28CH	D33BP
3	F25BJ	Y24BG	U21BC	K29AL	P26AH	U31AF	C12AC	P13CS	U14CP	C36CK	U24CL	J15CD
4	W35BK	T21BH	R17BD	F34AM	W38AJ	R18AG	K37AD	W19AB	R23CH	K32CL	W33CJ	U25CE
5	E16BL	G17BJ	X22BE	V26BC	Y11AK	X12AH	F27AE	Y30AC	X35CS	F20CH	Y15CK	R35CD
6	M24BM	L22BK	B29BF	E38BD	T31AL	B37AJ	V13AF	T14AD	B32AB	V28CR	T25CD	X16CH
7	H21BP	D29BL	N34BG	M11BE	G18AM	N27AH	E19AG	G23AE	N20AC	E33CP	G35CM	B24CJ
8	S17CD	J34BM	A26BH	H31BF	L12BC	A13AL	M30AG	L36AF	A28AD	M15CS	L16CP	N21CK
9	P22CE	U26BP	Z38BJ	S18BG	D37BD	Z19AM	H14AJ	D32AG	Z33AE	H25AB	D24CR	A17CL
10	W29CF	R38CD	C11BK	P12BH	J27BD	C30BC	S23AK	J20AH	C15AF	S35AC	J21CS	Z22CM
11	Y34CH	X11CE	K31BL	W37BJ	U13BF	K14BD	P36AL	U24AJ	K25AG	P16AD	U17AB	H18AE
12	T26CJ	B31CF	F18BM	Y27BK	R19BG	F23BE	W32AH	R33AK	F35AH	W24AE	R22AC	C29CP
13	G38CK	N18CH	V12BP	T13BL	X30BH	V36BF	Y20BC	X15AL	V16AJ	Y21AF	X29AD	K34CR
14	L11CL	A12CJ	E37CD	G19BM	B14BJ	E32BG	T28BD	B25AM	E24AK	T17AG	B34AE	V38AB
15	D31CM	Z37CK	M27CE	L30BP	N23BK	M20BH	G33BE	N35AP	M21AH	G22AD	N26AF	E11AC
16	J18CP	C27CL	H13CF	D14CD	A36BL	H28BJ	L15BF	A16BD	H17AM	L29AJ	A38AG	M31AD
17	U12CR	K13CM	S19CH	J23CE	Z32BM	S33BH	D25BG	Z24BE	S22BL	D34AM	Z11AH	H18AE
18	R37CS	F19CP	P30CJ	U36CF	C20BP	P15BL	J35BH	C21BF	P29BD	J26AL	C31AJ	S12AF
19	X27AB	V30CR	W14CK	R32CH	K28CD	W25BM	U16BJ	K17BG	W34BE	U38AM	K18AK	P37AG
20	B13AC	E14CS	Y23CL	X20CJ	F33CE	Y35BP	R24BK	F22BH	Y36BG	R11BC	F12AL	W27AH
21	N19AD	M24CH	T36CM	B28CK	V15CF	T16CD	X21BL	V29BJ	T38BG	X31BD	V37AM	Y13AJ
22	A30AE	H36AC	G32CP	N33CL	E25CH	G24CF	B17BM	E34BH	G11BF	B18BD	E27BC	T19AK
23	Z14AF	S32AD	L20CR	A15CM	M35CJ	L21CF	N22BP	M26BL	L31BJ	N12BF	M13BD	G30AL
24	C23AG	P20AE	D28CS	Z25CP	H16CK	D17CH	A29CD	H38BM	D18BK	A37BG	H19BE	L14AM
25	K36AH	W28AF	J33CM	C35CR	S24CL	J22CJ	Z34CE	S11BP	J12BL	Z27BH	S30BF	D23BC
26	F32AJ	Y33AG	U15AC	K16CS	P21CM	U29CH	C26CF	P31CG	U37BH	C13BJ	P14BG	J36BD
27	V20AK	T15AH	R25AD	F24AB	W17CP	R34CL	K38CH	W18CE	R27BP	K19BK	W23BH	U32BE
28	E28AL	G25AJ	X35AE	V21AC	Y22CR	X26CH	F11CJ	Y12CE	X13CD	F30BL	Y36BJ	R20BP
29	M33AM	L35AK	B16AF	E17AD	T29CS	B38CP	V31CK	T37CH				
30	H15BC		N24AG	M22AE	G34AB	N11CR	E18CL	G27CJ	N30CF	E23BP	G20BL	B33BH
31	S25BD		A21AH		L26AC		M12CH	L13CK		M36CD		N15BJ

FIRST LETTER = YOUR PHYSICAL CODE, MIDDLE NUMBER = SENSITIVITY CODE, LAST TWO LETTERS = INTELLECTUAL CODE

1905

Day	JAN	FEB	MAR	APR	MAY	JUN	JUL	AUG	SEP	OCT	NOV	DEC
1	A25BK	H24BH	T24BC	B22AL	V34AH	T11AF	X18AC	V27CS	T30CP	X23CK	V20CH	Y33CD
2	Z35BL	S21BJ	G21BD	N29AM	E26AJ	G31AG	B12AD	E13AB	G14CR	B36CL	E28CJ	T15CE
3	C16BM	P17BK	L17BE	A34BC	M38AK	L18AH	N37AE	M19AC	L23CS	N32CM	M33CK	G25CF
4	K24BP	W22BL	D22BF	Z26BD	H11AL	D12AJ	A27AF	H30AD	D36AB	A20CP	H15CL	L35CM
5	F21CD	Y29BM	J29BG	C38BE	S31AM	J37AK	Z13AG	S14AE	J32AC	Z28CR	S25CM	D16CJ
6	V17CE	T34BP	U34BH	K11BF	P18BC	U27AL	C19AH	P23AF	J20AD	C33CS	P35CP	J24CK
7	E22CF	G26CD	R26BJ	F31BG	W12BD	R13AM	K30AJ	W36AG	R28AE	K15AB	W16CR	U21CL
8	M29CH	L38CE	X38BK	V18BH	Y37BK	X19BC	F14AK	Y32AH	X33AF	F25AC	Y24CS	R17CM
9	H34CJ	D11CF	B11BL	E12BJ	T27BF	B30BD	V23AL	T20AJ	B15AG	V35AD	T21AB	X22CP
10	S26CK	J31CH	N31BM	M37BK	G13BG	N14BE	E36AM	G28AK	N25AH	E16AE	G17AC	B29CR
11	P38CL	U18CJ	A18BP	H27BL	L19BH	A23BF	M32BC	L33AL	A35AJ	M24AF	L22AD	N34CS
12	W11CM	R12CK	Z12CD	S13BM	D30BJ	Z36BG	H20BD	D15AM	Z16AK	H21AG	D29AE	A26AB
13	Y31CP	X37CL	C37CE	P19BP	J14BK	C32BH	S28BE	J25BC	C24AL	S17AH	J34AF	Z38AC
14	T18CR	B27CM	K27CF	W30CD	K20BJ	P33BF	U35BD	K21AM	P22AJ	U26AG		C11AD
15	G12CS	N13CP	F13CH	Y14CE	R36BM	F28BK	W15BG	R16BE	F17BC	W29AK	R38AH	K31AE
16	L37AB	A19CR	V19CJ	T23CF	X32BP	V33BL	Y25BH	X24BF	V22BD	Y34AL	X11AJ	F18AF
17	D27AC	Z30CS	E30CK	G36CH	B20CD	E15BM	T35BJ	B21BG	E29BE	T26AM	B31AK	V12AG
18	J13AD	C14AB	M14CL	L32CJ	N28CE	M25BP	G16BK	N17BH	M34BF	G38BC	N18AL	E37AH
19	U19AE	K23AC	H23CM	D20CK	A33CF	K35CD	L24BL	A22BJ	H26BE	L11BD	A12AM	M27AJ
20	R30AF	F36AD	S36CP	J28CL	Z15CH	S16CE	D21BM	Z29BK	S38BH	D31BE	Z37BC	H13AK
21	X14AG	V32AE	P32CR	U33CM	K25CP	P24CE	J17BP	C34BL	P11BJ	J18BF	C27BD	S19AL
22	B23AH	E20AF	H20CS	R15CP	K35CK	W21CH	U22CD	K26BM	W31BK	U12BG	K13BE	P30AM
23	N36AJ	M28AG	Y28AB	X25CR	F16CL	Y17CJ	R29CE	F38BP	Y18BL	R37BH	F19BF	W14BC
24	A32AK	H33AH	T33AC	B35CS	V24CM	T22CK	X34CF	V11CD	T12BM	X27BJ	V30BG	Y23BD
25	Z20AL	S15AJ	G15AD	N16AB	E21CP	G29CL	B26CH	E31CE	G37BP	B13BK	E14BH	T36BE
26	C28AM	P25AK	L25AE	A24AC	M17CR	L34CM	N38CJ	M18CF	L27CD	N19BL	M23BJ	G32BF
27	K33BC	W35AL	D35AF	Z21AD	H22CS	D26CP	A11CK	H12CH	D13CE	A30BM	H36BK	L20BG
28	F15BD	Y16AM	J16AG	C17AE	S28CF	J38CR	Z31CL	S37CJ	J19CF	Z14BP	S32BL	D28BH
29	V25BE		U24AH	K22AF	P34AC	U11CS	C18CM	P27CK	U30CH	C23CD	P20BM	J33BJ
30	E35BF		R21AJ	F29AG	W26AD	R31AB	K12CP	W13CL	R14CJ	K36CE	W28BP	U15BK
31	M16BG		X17AK		Y38AE		F37CR	Y19CM		F32CF		R25BL

1906

Day	JAN	FEB	MAR	APR	MAY	JUN	JUL	AUG	SEP	OCT	NOV	DEC
1	X35BM	V21BK	P21BE	U29BC	C26AK	P31AH	J12AE	C13AC	P14CS	J36CH	C28CK	S15CF
2	B16BP	E17BL	H17BF	R34BD	K38AL	H18AJ	U37AF	K19AD	N23AB	U32CP	K33CL	P25CH
3	N24CD	H22BH	Y22BG	X26BE	F11AM	Y12AK	R27AG	F30AE	Y36AC	R20CR	F15CM	W35CJ
4	A21CE	H29BP	I29BH	B38BF	V31BC	I37AL	X13AH	V14AF	I32AD	X28CS	V25CP	Y16CK
5	Z17CF	S34CD	G34BJ	N11BG	E18BD	G27AM	B19AJ	E23AG	G20AE	B33AB	E35CR	T24CL
6	C22CH	P26CE	L26BK	A31BH	M12BE	L13BC	N30AK	M36AH	L28AF	N15AC	M16CS	G21CM
7	K29CJ	Y11CH	D38BL	Z18BJ	H37BF	D19BD	A14AL	H32AJ	D33AG	A25AD	H24AB	L17CP
8	F34CK	Y11CH	J11BM	C12BK	S27BG	J30BE	Z23AM	S20AK	J15AH	Z35AE	S21AC	D22CR
9	V26CL	T31CJ	U31BP	K37BL	P13BH	U14BF	C36BC	P28AL	U25AJ	C16AF	P17AD	J29CS
10	E38CM	G18CK	R18CD	F27BM	W19BJ	R23BG	K32BD	W33AM	R15AK	K24AG	W22AE	U34AB
11	M11CR	L12CL	X12CE	V13BP	Y30BK	X36BH	F20BE	Y15BC	X16AL	F21AH	Y29AF	R26AC
12	H31CR	D37CM	B37CF	E19CD	T14BL	B32BJ	V28BF	T25BD	B24AM	V17AJ	T34AG	X38AD
13	S18CS	J27CP	N27CH	M30CE	G23BM	N20BK	E33BG	G59BE	N21BC	E22AK	G26AH	B11AE
14	P12AB	U13CR	A13CJ	H14CF	L36BP	A28BL	M15BH	L16BF	A17BD	M29AL	L38AJ	N31AF
15	W37AC	R19CS	Z19CK	S23CH	D32CD	Z33BM	H25BJ	D24BG	Z22BE	H34AM	D11AK	A18AG
16	Y27AD	X30AB	C30CL	P36CJ	J20CE	C15BP	S35BK	J21BH	C29BF	S26BC	J31AL	Z12AH
17	T13AE	B14AC	K14CM	W32CK	K25CD	P16BL	U17BJ	K34BG	P38BD	U18AM		C37AJ
18	G19AF	N23AD	F23CP	Y20CL	R33CH	F35CE	W24BM	R22BK	F26BH	W11BE	R12BC	K27AK
19	L30AG	A36AE	V36CR	T28CM	X15CJ	V16CF	Y21BP	X29BL	V38BJ	Y31BF	X37BD	F13AL
20	D14AH	Z32AF	E32CS	G33CP	B25CK	E24CH	T17CD	B34BM	E11BK	T18BG	B27BE	V19AM
21	J23AJ	C20AG	M20AB	L15CR	N35CL	M21CJ	G22CE	N26BP	M31BL	G12BH	N13BF	E30BC
22	U36AK	K28AM	H28AC	D25CS	A16CM	H17CK	L29CF	A38CD	H18BM	L37BJ	A19BG	M14BD
23	R32AL	F33AJ	S33AD	J35AB	Z24CP	S22CL	D34CH	Z11CE	S12BP	D27BK	Z30BH	H23BE
24	X20AM	V15AK	P15AE	U16AC	C21CR	P29CM	J26CJ	C31CF	P37CD	J13BL	C14BJ	S36BF
25	B28BC	E25AL	W25AF	R24AD	K17CS	W34CP	U38CK	K18CH	W27CE	U19BM	K23BK	P32BG
26	N33BD	M35AM	Y35AG	X21AE	F22AB	Y26CR	R11CL	F12CJ	Y13CF	R30BP	F36BL	W20BH
27	A15BE	H16BC	T16AH	B17AF	V29AC	T38CS	X31CH	V37CK	T19CH	X14CD	V32BH	Y28BJ
28	Z25BF	S24BD	G24AJ	N22AG	E34AD	G11AB	B18CP	E27CL	G30CJ	B23CE	E20BP	T33BK
29	C35BG		L21AK	A29AH	M26AE	L31AC	N12CR	M13CH	L14CK	N36CF	M28CD	G15BL
30	K16BH		D17AL	Z34AJ	H38AF	D18AD	A37CS	H19CP	D23CL	A32CH	H33CE	L25BH
31	F24BJ		J22AM		S11AG			S30CR		Z20CJ		D35BP

1907

Day	JAN	FEB	MAR	APR	MAY	JUN	JUL	AUG	SEP	OCT	NOV	DEC
1	J16CD	C17BM	M17BG	L34BE	N38AM	M18AK	G37AG	N19AE	M23AC	G32CH	N33CM	E25CJ
2	U24CE	K22BP	S29BJ	D26BF	A11BC	H12AL	L27AH	A30AF	H36AD	L20CS	A15CP	M35CK
3	R21CF	F29CD	S29BJ	J38BG	Z31BD	S37AM	D13AJ	Z14AG	S32AE	D28AB	Z25CR	H16CL
4	X17CH	V34CE	P34BK	U11BH	C18BE	P27BC	J19AK	C23AH	P20AF	J33AC	C35CS	S24CH
5	B22CJ	E26CF	W26BL	R31BJ	K12BF	W13BD	U30AL	K36AJ	W28AG	U15AD	K16AB	P21CP
6	N29CK	M38CH	Y38BM	X18BK	F37BG	Y19BE	R14AM	F32AK	Y33AH	R25AE	F24AC	W17CR
7	A34CL	H11CJ	T11BP	B12BL	V27BH	T30BF	X23BC	V20AL	T15AJ	X35AF	V21AD	Y22CS
8	Z26CM	S31CK	G31CD	N37BM	E13BJ	G14BG	B36BD	E28AM	G25AK	B16AG	E17AE	T29AB
9	C38CP	P18CL	L18CE	M27BP	M19BK	L23BH	N32BE	M33BC	L35AL	N24AH	M22AF	G34AC
10	K11CR	W12CM	D12CF	Z13CD	H30BL	D36BJ	A20BF	H15BD	D15AM	A21AJ	H29AG	L26AD
11	F31CS	Y37CP	J37CH	C19CE	S14BM	J32BK	Z28BG	S25BE	J24BC	Z17AK	S34AH	D38AE
12	V18AB	T27CR	U27CJ	K30CF	P23BP	U20BL	C33BH	P35BF	U21BD	C22AL	P26AJ	J11AF
13	E12AC	G13CS	R13CK	F14CH	W36CD	R28BM	K15BJ	W16BG	R17BE	K29AM	W38AK	X12AJ
14	M37AD	L19AB	X19CL	V23CJ	Y32CE	X33BP	F25BK	Y24BH	X22BF	F34BC	Y11AL	R18AH
15	H27AE	D30AC	B30CM	E36CK	T20CF	B15CD	V35BL	T21BJ	B29BG	V26BD	T31AM	X12AJ
16	S13AF	J14AD	N14CP	M32CL	G28CH	N25CE	E16BM	G17BK	N34BH	E38BE	G18BC	B37AK
17	P19AG	U23AE	A23CR	H20CM	L33CJ	A35CF	M24BP	L22BJ	A26BH	M11BF	L12BD	N27AL
18	W30AH	R36AF	Z36CS	S28CP	D15CK	Z16CH	H21CD	D29BM	Z38BK	H31BG	D37BE	A13AM
19	Y14AJ	X32AG	C32AB	P33CR	J25CL	C24CJ	S17CE	J34BF	C11BL	S18BH	J27BF	Z19BC
20	T23AK	B20AM	K20AH	W15CS	U35CM	K21CK	P22CF	U26CD	K31BM	P12BJ	U13BG	C30BD
21	G36AL	N28AJ	F28AD	Y25AB	R16CP	F17CL	W29CH	R38CE	F18BP	W37BK	R19BH	K14BC
22	L32AM	A33AK	V33AE	T35AC	X24CR	V22CP	Y34CJ	X11CF	V12CD	Y27BL	X30BJ	F23BF
23	D20BC	Z15AL	E15AF	G16AD	B21CS	E29CP	T26CK	B31CH	E37CE	T13BM	B14BK	V36BG
24	J28BD	C25AM	M25AG	L24AE	N17AB	M34CR	G38CL	N16CJ	C31CF	M27CF	G19BP	N23BL
25	U33BE	K35BC	H35AH	D21AF	A22AC	H26CS	L11CH	A12CK	H13CH	L30CD	A36BM	M20BJ
26	R15BF	F15BD	S16AJ	J17AG	Z29AD	S38AB	D31CP	Z37CL	S19CJ	D14CE	Z32BP	H28BK
27	X25BG	V24BE	P16AK	W22AH	C34AE	P11AC	J18CR	C27CH	P30CK	J23CF	C20CD	S33BL
28	B35BH	E21BF	W21AL	R29AJ	K26AF	W31AD	U12CS	K13CP	W14CL	U36CH	K28CE	P15BM
29	N16BJ		Y17AM	P29AJ	F38AG	Y18AE	R37AB	F19CR	Y23CM	R32CJ	F33CF	N25BP
30	A24BK		T22BC	B26AL	V11AH	T12AF	X27AC	V30CS	T36CP	X20CK	V15CH	Y35CD
31	Z21BL		G29BD		E31AJ		B13AD	E14AB		B28CL		T16CE

FIRST LETTER = YOUR PHYSICAL CODE, MIDDLE NUMBER = SENSITIVITY CODE, LAST TWO LETTERS = INTELLECTUAL CODE

1908	JAN	FEB	MAR	APR	MAY	JUN	JUL	AUG	SEP	OCT	NOV	DEC
1	G24CF	N22CD	V29BK	T38BH	X31BE	V37BC	Y13AK	X14AH	V32AF	Y28AC	X25CS	F16CM
2	L21CH	A29CE	E34BL	G11BJ	B18BF	E27BD	T19AL	B23AJ	E20AG	T33AD	B35AB	V24CP
3	D17CJ	Z34CF	M26BM	L31BK	N12BG	M13BE	G30AM	N36AK	M28AH	G15AE	N16AC	E21CR
4	J22CK	C26CH	H38BP	D18BL	A37BH	H19BF	L14BC	A32AL	H33AJ	L25AF	A24AD	M17CS
5	U29CL	K38CJ	S11CD	J12BM	Z27BJ	S30BG	D23BD	Z20AM	S15AK	D35AG	Z21AE	H22AB
6	R34CM	F11CK	P31CE	U37BP	C13BK	P14BH	J36BE	C28BC	P25AL	J16AH	C17AF	S29AC
7	X26CP	V31CL	W18CF	R27CD	K19BL	W23BJ	U32BF	K33BD	W35AM	U24AJ	K22AG	P34AD
8	B38CR	E18CM	Y12CH	X13CE	F30BM	Y36BK	R20BG	F15BE	Y16BC	R21AK	F29AH	W26AE
9	N11CS	M12CP	T37CJ	B19CF	V14BP	T32BL	X28BH	V25BF	T24BD	X17AL	V34AJ	Y38AF
10	A31AB	H37CR	G27CK	N30CH	E23CD	G20BM	B33BJ	E35BG	G21BE	B22AM	E26AK	T11AG
11	Z18AC	S27CS	L13CL	A14CJ	M36CE	L28BP	N15BK	M16BH	_17BF	N29BC	M38AL	G31AH
12	C12AD	P13AB	D19CM	Z23CK	H32CF	D33CD	A25BL	H24BJ	D22BG	A34BD	H11AM	L18AJ
13	K37AE	W19AC	J30CP	C36CL	S20CH	J15CE	Z35BM	S21BK	J29BH	Z26BE	S31BC	D12AK
14	F27AF	Y30AD	U14CR	K32CM	P28CJ	U25CF	C16BP	P17BL	U34BJ	C38BF	P18BD	J37AL
15	V13AG	T14AE	R23CS	F20CP	W33CK	R35CH	K24CD	W22BM	R26BK	K11BG	W12BE	U27AM
16	E19AH	G23AF	X36AB	V28CR	Y15CL	X16CJ	F21CE	Y29BP	X38BL	F31BH	Y37BF	R13BC
17	M30AJ	L36AG	B32AC	E33CS	T25CM	B24CK	V17CF	T34CD	B11BM	V18BJ	T27BG	X19BD
18	H14AK	D32AH	N20AD	M15AB	G35CP	N21CL	E22CH	G26CE	N31BP	E12BK	G13BH	B30BE
19	S23AL	J20AJ	A28AE	H25AC	L16CR	A17CM	M29CJ	L38CF	A18CD	M37BL	L19BJ	N14BF
20	P36AM	U28AK	Z33AF	S35AD	D24CS	Z22CP	H34CK	D11CH	Z12CE	H27BM	D30BK	A23BG
21	W32BC	R33AL	C15AG	P16AE	J21AB	C29CR	S26CL	J31CJ	C37CF	S13BP	J14BL	Z36BH
22	Y20BD	X15AM	K25AH	W24AF	U17AC	K34CS	P38CM	U18CK	K27CH	P19CD	U23BM	C32BJ
23	T28BE	B25BC	F35AJ	Y21AG	R22AD	F26AH	H11CP	R12CL	F13CJ	W30CE	R36BP	K20BK
24	G33BF	N35BD	V16AK	T17AH	X29AE	V38AC	Y31CR	X37CM	V19CK	Y14CF	X32CD	F28BL
25	L15BG	A16BE	E24AL	G22AJ	B34AF	E11AD	T18CS	B27CP	E30CL	T23CH	B20CE	V33BM
26	D25BH	Z24BF	M21AM	L29AK	N26AG	M31AE	G12AB	N13CR	M14CM	G36CJ	N28CF	E15BP
27	J35BJ	C21BG	H17BC	D34AL	A38AH	H18AF	L37AC	A19CS	H23CP	L32CK	A33CH	M25CD
28	U16BK	K17BH	S22BD	J26AM	Z11AJ	S12AG	D27AD	Z30AB	S36CR	D20CL	Z15CJ	H35CE
29	R24BL	F22BJ	P29BE	U36BC	C31AK	P37AH	J13AE	C14AC	P32CS	J28CM	C25CK	S16CF
30	X21BM		W34BF	R11BD	K18AL	W27AJ	U19AF	K23AD	W20AB	U33CP	K35CL	P24CH
31	B17BP		Y26BG		F12AM		R30AG	F36AE		R15CR		W21CJ

1909	JAN	FEB	MAR	APR	MAY	JUN	JUL	AUG	SEP	OCT	NOV	DEC
1	Y17CK	X34CH	C34BM	P11BK	J18BG	C27BE	S19AM	J23AK	C20AH	S33AE	J35AC	Z24CR
2	T22CL	B26CJ	K26BP	W31BL	U12BH	K13BF	P30BC	U36AL	K28AJ	P15AF	U16AD	C21CS
3	G29CM	N38CK	F38CD	Y18BM	R37BJ	F19BG	W14BD	R32AM	F33AK	W25AG	R24AE	K17AB
4	L34CP	A11CL	V11CE	T12BP	X27BK	V30BH	Y23BE	X20BC	V15AL	Y35AH	X21AF	F22AC
5	D26CR	Z31CM	E31CF	G37CD	B13BL	E14BJ	T36BF	B28BD	E25AM	T16AJ	B17AG	V29AD
6	J38CS	C18CP	M18CH	L27CE	N19BM	M23BK	G32BG	N33BE	M35BC	G24AK	N22AH	E34AE
7	U11AB	K12CR	H12CJ	D13CF	A30BP	H36BL	L20BH	A15BF	H16BD	L21AL	A29AJ	M26AF
8	R31AC	F37CS	S37CK	J19CH	Z14CD	S32BM	D28BJ	Z25BG	S24BE	D17AM	Z34AK	H38AG
9	X18AD	V27AB	P27CL	U30CJ	C23CE	P20BP	J33BK	C35BH	P21BF	J22BC	C26AL	S11AH
10	B12AE	E13AC	W13CM	R14CK	K36CF	W28CD	U15BL	K16BJ	W17BG	U29BD	K38AM	P31AJ
11	N37AF	M19AD	Y19CP	X23CL	F32CH	Y33CE	R25BM	F24BK	Y22BH	R34BE	F11BC	W18AK
12	A27AG	H38AE	T30CR	B36CM	V20CJ	T15CF	X35BP	V21BL	T29BJ	X26BF	V31BD	Y12AL
13	Z13AH	S14AF	G14CS	N32CP	E28CK	G25CH	B16CD	E17BM	G34BK	B38BG	E18BE	T37AM
14	C19AJ	P23AG	L23AB	A20CR	M33CL	L35CJ	N24CE	M22BP	L26BL	N11BH	M12BF	G27BC
15	K30AK	W36AH	D36AC	Z28CS	H15CM	D16CK	A21CF	H29CG	D38BM	A31BJ	H37BG	L13BD
16	F14AL	Y32AJ	J32AD	C33AB	S25CP	J24CL	Z17CH	S34CE	J11BP	Z18BK	S27BH	D19BE
17	V23AM	T20AK	U20AE	K15AC	P35CR	U21CM	C22CJ	P26CF	U31CD	C12BL	P13BJ	J30BF
18	E36BC	G28AL	R28AF	F25AD	W16CS	R17CP	K29CK	W38CH	R18CE	K37BM	W19BK	U14BG
19	M32BD	L33AM	X33AG	W35AE	Y24AB	X22CR	F34CL	Y11CJ	X12CF	F27BP	Y30BL	R23BH
20	H20BE	D15BC	B15AH	E16AF	T21AC	B29CS	V26CM	T31CK	B37CH	V13CD	T14BM	X36BJ
21	S28BF	J25BD	N25AJ	M24AG	G17AD	N34AB	E38CP	G16CL	N27CJ	E19CE	G23BP	B32BK
22	P33BG	U35BE	A35AK	H21AH	L22AE	A26AC	M11CR	L12CM	A13CK	M30CF	L36CD	N20BL
23	W15BH	R16BF	Z16AL	S17AJ	D29AF	Z38AD	H31CS	D37CP	Z19CL	H14CH	D32CE	A28BM
24	Y25BJ	X24BG	C24AM	P22AK	J34AG	C11AE	S18AB	J27CR	C30CM	S23CJ	J20CF	Z33BP
25	T35BK	B21BH	K21BC	W29AL	U26AH	K31AF	P12AC	U13CS	K14CP	P36CK	U28CH	C15CD
26	G16BL	N17BJ	F17BD	Y34AM	R38AJ	F18AG	W37AD	R19AB	F23CR	W32CL	R33CJ	K25CE
27	L24BM	A22BK	V22BE	T26BC	X11AK	V12AH	Y27AE	X30AC	V36CS	Y20CM	X15CK	F35CF
28	D21BP	Z29BL	E29BF	G38BD	B31AL	E37AJ	T13AF	B14AD	E32AB	T28CP	B25CL	V16CH
29	J17CD		M34BG	L11BE	N18AM	M27AK	G19AG	N23AE	M20AC	G33CR	N35CM	E24CJ
30	U22CE		H26BH	D31BF	A12BC	H13AL	L30AH	A36AF	H28AD	L15CS	A16CP	M21CL
31	R29CF		S38BJ		Z37BD		D14AJ	Z32AG		D25AB		H17CL

1910	JAN	FEB	MAR	APR	MAY	JUN	JUL	AUG	SEP	OCT	NOV	DEC
1	S22CM	J26CK	N26CD	M31BH	G12BJ	N13BG	E30BD	G36AM	N28AK	E15AG	G16AE	B21AB
2	P29CP	U38CL	A38CE	H18BP	L37BK	A19BH	M14BE	L32BC	A33AL	M25AH	L24AF	N17AC
3	W34CR	R11CM	Z11CF	S12CD	D27BL	Z30BJ	H23BF	D20BD	Z15AM	H35AJ	D21AG	A22AD
4	Y26CS	X31CP	C31CH	P37CE	J13BM	C14BL	S36BG	J28BE	C25BC	S16AK	J17AH	Z29AE
5	T38AB	B18CR	K18CJ	W27CF	U19BP	K23BL	P32BH	U33BF	K35BD	P24AL	U22AJ	C34AF
6	G11AC	N12CS	F12CK	Y13CL	R30CD	F36BM	W20BJ	R15BG	F16BE	W21AK	R29AK	K26AG
7	L31AD	A37AB	V37CL	T19CJ	X14CE	V32BP	Y28BK	X25BH	V24BF	Y17BC	X34AL	F38AM
8	D18AE	Z27AC	E27CM	G30CH	B23CF	E20CD	T33BL	B35BJ	E21BG	T22BD	B26AM	V11AJ
9	J12AF	C13AD	M13CP	L14CL	N36CH	M28CE	G15BM	N16BK	M17BH	G29BE	N38BC	E31AK
10	U37AG	K19AE	H19CR	D23CM	A32CJ	H33CF	L25BP	A24BL	H22BJ	L34BF	A11BD	M18AL
11	R27AH	F30AF	S30CS	J36CP	Z20CK	S15CH	D35CD	Z21BM	S29BK	D26BG	Z31BE	H12AM
12	X13AJ	V14AG	P14AB	U32CR	C28CL	P25CJ	J16CE	C17BP	P34BL	J38BH	C18BF	S37BC
13	B19AK	E23AH	W23AC	R20CS	K33CM	W35CK	U24CF	K22CD	W26BM	U11BG	K12BE	P27BD
14	N30AL	M36AJ	Y36AD	X28AB	F15CP	Y16CL	R21CH	F29CE	Y38BP	R31BK	F37BH	W13BE
15	A14AM	H32AK	T32AE	B33AC	V25CR	T24CM	X17CJ	V34CF	T11CD	X18BL	V27BJ	Y19BF
16	Z23BC	S20AL	G20AF	N15AD	E35CS	G21CP	B22CK	E26CH	G31CE	B12BH	E13BG	G14BD
17	C36BD	P28AM	L28AG	A25AE	M16AB	L17CR	N29CL	M38CJ	L18CF	N37BP	M19BL	T30BG
18	K32BE	W33BC	D33AH	Z35AF	H24AC	D22CS	A34CM	H11CK	D12CH	A27CD	H30BM	L23BJ
19	F20BF	Y15BD	J15AJ	C16AG	S21AD	J29AB	Z26CP	S31CL	J37CJ	Z13CE	S14BP	D36BK
20	V28BG	T25BE	U25AK	K24AH	P17AE	U34AC	C38CR	P18CM	U27CK	C19CF	P23CD	J32BL
21	E33BH	G35BF	R35AL	F21AJ	W22AF	R26AD	K11CS	W12CP	R13CL	K30CH	W36CE	U20BM
22	M15BJ	L16BG	X16AM	V17AK	Y29AG	X38AE	F31AB	Y37CR	X19CM	F14CJ	Y32CF	R28BP
23	H25BK	D24BM	B24BC	E22AL	T34AH	B11AF	V18AC	T27CS	B30CP	V23CK	T20CH	X33CD
24	S35BL	J21BJ	N21BD	M29AM	G26AJ	N31AG	E12AD	G13AB	N14CR	E36CL	G28CJ	B15CE
25	P16BM	U17BK	A17BE	H34AC	L38AK	A18AH	M37AE	L19AC	A23CS	M32CH	L33CK	N25CF
26	W24BP	R22BL	Z22BF	M26AB	D11AG	Z12AE	H27AF	D30AD	Z36AB	H20CP	D15CL	A35CM
27	Y21CD	X29BM	C29BG	P38BE	J31AM	C37AK	S13AG	J14AE	C32AC	S28CR	J25CM	M24CD
28	T17CE	B34BP	K34BH	W11BF	U18BC	K27AL	P19AH	U23AF	K20AD	P33CS	U35CP	C24CL
29	G22CF		F26BJ	Y31BG	R12BD	F13AL	W30AH	R36AF	F28AE	W15AB	R16CR	K21CL
30	L29CH		V38BK	T18BH	X37BE	V19BC	Y14AK	X32AH	V33AF	Y25AC	X24CS	F17CM
31	D34CJ		E11BL		B27BF		T23AL	B20AJ		T35AD		V22CP

FIRST LETTER = YOUR PHYSICAL CODE, MIDDLE NUMBER = SENSITIVITY CODE, LAST TWO LETTERS = INTELLECTUAL CODE

1911	JAN	FEB	MAR	APR	MAY	JUN	JUL	AUG	SEP	OCT	NOV	DEC
1	E29CR	G38CM	R38CF	F18CD	W37BL	R19BJ	K14BF	W32BD	R33AM	K25AJ	W24AG	U17AD
2	M34CS	L11CP	X11CH	V12CE	Y27BM	X30BK	F23BG	Y20BE	X15BC	F35AK	Y21AH	R22AE
3	H26AB	D31CR	B31CJ	E37CF	T13BP	B14BL	V36BH	T28BF	B25BD	V16AL	T17AJ	X29AF
4	S38AC	J18CS	N18CK	M27CH	G19CD	N23BM	E32BJ	G33BG	N35BE	E24AM	G22AK	B34AG
5	P11AD	U12AB	A12CL	H13CJ	L30CE	A36BP	M20BK	L15BH	A16BF	M21BC	L29AL	N26AH
6	W31AE	R37AC	Z37CM	S19CK	D14CF	Z32CD	H28BL	D25BJ	Z24BG	H17BD	D34AM	A38AJ
7	Y18AF	X27AD	C27CP	P30CL	J23CH	C20CE	S33BM	J35BK	C21BH	S22BE	J26BC	Z11AK
8	T12AG	B13AE	K13CR	W14CM	U36CJ	K28CF	P15BP	U16BL	K17BJ	P29BF	U38BD	C31AL
9	G37AH	N19AF	F19CS	Y23CP	R32CK	F33CH	W25CD	R24BM	F22BK	W34BG	R11BE	K18AM
10	L27AJ	A30AG	V30AB	T36CR	X20CL	V15CJ	Y35CE	X21BP	V29BL	Y26BH	X31BF	F12BC
11	D13AK	Z14AH	E14AC	G32CS	B28CM	E25CK	T16CF	B17CD	E34BM	T38BJ	B18BG	V37BD
12	J19AL	C23AJ	M23AD	L20AB	N33CP	M35CL	G24CH	N22CE	M26BP	G11BK	N12BH	E27BE
13	U30AM	K36AK	H36AE	D28AC	A15CR	H16CM	L21CJ	A29CF	H38CD	L31BL	A37BJ	M13BF
14	R14BC	F32AL	S32AF	J33AD	Z25CS	S24CP	D17CK	Z34CH	S11CE	D18BM	Z27BK	H19BG
15	X23BD	V20AM	P20AG	U15AE	C35AB	P21CR	J22CL	C26CJ	P31CF	J12BP	C13BL	S30BH
16	B36BE	E28BC	W28AH	R25AF	K16AC	W17CS	U29CH	K38CK	W18CH	U37CD	K19BM	P14BJ
17	N32BF	M33BD	Y33AJ	X35AG	F24AD	Y22AB	R34CP	F11CL	Y12CJ	R27CE	F30BP	W23BK
18	A20BG	H15BE	T15AK	B16AH	V21AE	T29AC	X26CR	V31CM	T37CK	X13CF	V14CD	Y36BL
19	Z28BH	S25BF	G25AL	N24AJ	E17AF	G34AD	B38CS	E18CP	G27CL	B19CH	E23CE	T32BM
20	C33BJ	P35BG	L35AM	A21AK	M22AG	L26AE	N11AB	M12CR	L13CM	N30CJ	M36CF	G20BP
21	K15BK	W16BH	D16BC	Z17AL	H29AH	D38AF	A31AC	H37CS	D19CP	A14CK	H32CH	L28CD
22	F25BL	Y24BJ	J24BD	C22AM	S34AJ	J11AG	Z18AD	S27AB	J30CR	Z23CL	S20CJ	D33CE
23	V35BM	T21BK	U21BE	K29BC	P26AK	U31AH	C12AE	P13AC	U14CS	C36CM	P28CK	J15CF
24	E16BP	G17BL	R17BF	F34BD	W38AL	R18AJ	K37AF	W19AD	R23AB	K32CP	W33CL	U25CH
25	M24CD	L22BM	X22BG	V26BE	Y11AM	X12AK	F27AG	Y30AE	X36AC	F20CR	Y15CM	R35CJ
26	H21CE	D29BP	B29BH	E38BF	T31BC	B37AL	V13AH	T14AF	B32AD	V28CS	T25CP	X16CK
27	S17CF	J34CD	N34BJ	M11BG	G18BD	N27AM	E19AJ	G23AG	N20AE	E33AB	G35CR	B24CL
28	P22CH	U26CE	A26BK	H31BH	L12BE	A13BC	M30AK	L36AH	A28AF	M15AC	L16CS	N21CM
29	W29CJ		Z38BL	S18BJ	D37BF	Z19BD	H14AL	D32AJ	Z33AG	H25AD	D24AB	A17CP
30	Y34CK		C11BM	P12BK	J27BG	C30BE	S23AM	J20AK	C15AH	S35AE	J21AC	Z22CR
31	T26CL		K31BP		U13BH		P36BC	U28AL		P16AF		C29CS

1912	JAN	FEB	MAR	APR	MAY	JUN	JUL	AUG	SEP	OCT	NOV	DEC
1	K34AB	W11CR	J31CK	C37CH	S13CD	J14BM	Z36BJ	S28BG	J25BE	Z16AM	S17AK	D29AG
2	F26AC	Y31CS	U18CL	K27CJ	P19CE	U23BP	C32BK	P33BH	U35BF	C24BC	P22AL	J34AH
3	V38AD	T18AB	R12CM	F13CK	W30CF	R36CD	K20BL	W15BJ	R16BG	K21BD	W29AM	U26AJ
4	E11AE	G12AC	X37CP	V19CL	Y14CH	X32CE	F28BM	Y25BK	X24BH	F17BE	Y34BC	R38AK
5	M31AF	L37AD	B27CR	E30CM	T23CJ	B20CF	V33BP	T35BL	B21BJ	V22BF	T26BD	X11AL
6	H18AG	D27AE	N13CS	M14CP	G36CK	N28CH	E15CD	G16BM	N17BK	E29BG	G38BE	B31AM
7	S12AH	J13AF	A19AB	H23CR	L32CL	A33CJ	M25CE	L24BP	A22BL	M34BH	L11BF	N18BC
8	P37AJ	U19AG	Z30AC	S36CS	D20CM	Z15CK	H35CF	D21CD	Z29BM	H26BJ	D31BG	A12BD
9	W27AK	R30AH	C14AD	P32AB	J28CP	C25CE	S16CH	J17CE	C34BP	S38BK	J18BH	Z37BE
10	Y13AL	X14AJ	K23AE	W20AG	U33CR	K35CM	P24CJ	U22CF	K26CD	P11BL	U12BJ	C27BF
11	T19AM	B23AK	F36AF	Y28AD	R15CS	F16CP	W21CK	R29CH	F38CD	W31BM	R37BK	K13BG
12	G30BC	N36AL	V32AG	T33AE	X25AB	V24CR	Y17CL	X34CJ	V11CF	Y18BP	X27BL	F19BH
13	L14BD	A32AM	E20AH	G15AF	B35AC	E21CS	T22CM	B26CK	E31CH	T12CD	B13BM	V30BJ
14	D23BE	Z20BC	M28AJ	L25AG	N16AD	M17AB	G29CP	N38CL	M18CJ	G37CE	N19BP	E14BK
15	J36BF	C28BD	H33AK	D35AH	A24AE	H22AC	L34CR	A11CM	H12CK	L27CF	A30CD	M23BL
16	U32BG	K33BE	S15AL	J16AJ	Z21AF	S29AD	D26CS	Z31CP	S37CL	D13CH	Z14CE	H36BM
17	R20BH	F15BF	P25AM	U24AK	C17AG	P34AE	J38AB	C18CR	P27CM	J19CJ	C23CF	S32BP
18	X28BJ	V25BG	W35BC	R21AL	K22AH	W24AF	U11AC	K12CS	W13CP	U30CK	K36CH	P20CD
19	B33BK	E35BH	Y16BD	X17AM	F29AJ	Y38AG	R31AD	F37AB	Y19CR	R14CL	F32CJ	W28CE
20	N15BL	M16BJ	T24BE	B22BC	V34AK	T11AH	X18AE	V27AC	T30CS	X23CH	V20CK	Y33CF
21	A25BM	H24BK	G21BF	N29BD	E26AL	G31AJ	B12AF	E13AD	G14AB	B36CP	E28CL	T15CH
22	Z35BP	S21BL	L17BG	A34BE	M38AM	L18AK	N37AG	M19AE	L23AC	N32CR	M33CM	G25CJ
23	C16CD	P17BM	D22BH	Z26BF	H11BC	D12AL	A27AH	H30AF	D36AD	A20CS	H15CP	L35CK
24	K24CE	W22BP	J29BJ	C38BG	S31BD	J37AM	Z13AJ	S14AG	J32AE	Z28AB	S25CR	D16CL
25	F21CF	Y29CD	U34BK	K11BH	P18BE	U27BC	C19AK	P23AH	U20AF	C33AC	P35CS	J24CM
26	V17CH	T34CE	R26BL	F31BJ	W12BF	R13BD	K30AL	W36AJ	R28AG	K15AD	W16AB	U21CP
27	E22CJ	G26CF	X38BM	V18BK		X19BE	F14AM	Y32AK	X33AH	F25AE	Y24AC	R17CR
28	M29CK	L38CM	B11BP	E12BL	T27BH	B30BF	V23BC	T20AL	B15AJ	V35AF	T21AD	X22CS
29	H34CL	D11CJ	N31CD	M37BM	N14BG	E36BD	G28AM	N25AK				
30	S26CM		A18CE	H27BP	L19BK	A23BH	M32BE	L33BC	A35AL	M24AH	L22AF	N34AC
31	P38CP		Z12CF		D30BL		H20BF	D15BD		H21AJ		A26AD

1913	JAN	FEB	MAR	APR	MAY	JUN	JUL	AUG	SEP	OCT	NOV	DEC
1	Z38AE	S18AC	G18CM	N27CK	E19CF	G23CD	B32BL	E33BJ	G35BG	B24BD	E22AM	T34AJ
2	C11AF	P12AD	L12CP	A13CL	M30CH	L36CE	N20BM	M15BK	L16BH	N21BE	M29BC	G26AK
3	K31AG	W37AE	D37CP	Z19CM	H14CJ	D32CF	A28BP	H25BL	D24BJ	A17BF	H34BD	L38AL
4	F18AH	Y27AF	J27CS	C30CP	S23CK	J20CH	Z33CD	S35BM	J21BK	Z22BG	S26BE	D11AM
5	V12AJ	T13AG	U13AB	K14CR	P36CL	U28CJ	C15CE	P16BP	U17BL	C29BH	P38BF	J31BL
6	E37AK	G19AH	R19AC	F23CS	W32CH	R25CF	W24CD	R22BM	K34BJ	W11BG	U18BD	
7	M27AL	L30AJ	X30AD	V36AB	Y20CP	X15CL	F35CH	Y21CE	X29BP	F26BK	Y31BH	R12BE
8	H13AM	D14AK	B14AE	E32AC	T25CH	B25CH	V16CJ	T17CF	B34CD	V38BL	T18BJ	X37BF
9	S19BC	J23AL	N23AF	M20AD	G33CS	N35CP	E24CK	G22CH	N26CE	E11BM	G12BK	B27BG
10	P30BD	U36AM	A36AG	H28AE	L15AB	A16CR	M21CL	L29CJ	D34CK	M31BP	L37BL	N13BH
11	W14BE	R32BC	Z32AH	S33AF	D25AC	Z24CS	H17CM	D34CJ	Z11CH	H18CD	D27BM	A19BJ
12	Y23BF	X20BD	C20AJ	P15AG	J35AD	C21AB	S22CP	J26CL	C31CJ	S12CE	J13BP	Z30BK
13	T36BG	B28BE	K28AK	W25AH	U16AE	K17AC	P29CR	U38CM	K18CK	P37CF	U19CD	C14BL
14	G32BH	N33BF	F33AL	Y35AJ	R24AF	F22AD	W34CS	R11CP	F12CL	W27CH	R30CE	K23BM
15	L20BJ	A15BG	V15AM	T16AK	X21AG	V29AE	Y26AB	X31CH	V37CM	Y13CJ	X14CF	F36BP
16	D28BK	Z25BH	E25BC	G24AL	B17AH	E34AF	T38AC	B18CS	E27CP	T19CK	B23CH	V32CD
17	J33BL	C35BJ	M35BD	L21AM	N12AD	M26AB	G11AD	N12AD	M13CR	G30CL	N36CJ	E20CE
18	U15BM	K16BK	H16BE	D17BC	A29AK	H38AH	L31AE	A37AC	H19CS	L14CM	A32CK	M28CF
19	R25BP	F24BL	S24BF	J22BD	Z34AL	S11AJ	O18AF	Z27AD	S30AB	O23CP	Z20CL	H33CH
20	X35CD	B16CE	P21BG	U29BE	C26AM	P31AK	J12AG	C13AE	P14AC	J36CR	C28CM	S15CJ
21	B16CE	E17BP	W17BH	R34BF	K38AG	W18AL	U37AH	K19AF	W23AD	U32CS	K33CP	P25CK
22	N24CF	M22CD	Y22BJ	X26BG	F11BD	Y12AM	R27AJ	F30AG	Y36AE	R20AB	F15CR	W35CL
23	A21CH	H29CE	T29BC	B38BH	V31BE	T37BC	X13AK	V14AH	T32AF	X28AC	V25CS	Y16CM
24	Z17CJ	S34CF	G34BL	N11BJ	A31BK	G27BD	B19AL	E23AJ	G20AG	B33AD	E35AB	T24CP
25	C22CK	P26CH	L26BM	A31BK	M12BG	L13BE	N30AM	M36AK	L28AH	N15AE	M16AC	G21CR
26	K29CL	W38CJ	D38BP	H37BF		D19BF	A14BC	H32AL	D33AJ	A25AF	H24AD	L17CS
27	F34CM	Y11CK	J11CD	C12BM	S27BJ	J30BG	Z23BD	S20AM	J15AK	Z35AG	S21AE	D22AB
28	V26CP	T31CL	U31CE	K37BP	P13BK	U14BH	C36BE	P28BC	U25AL	C16AH	P17AF	J29AC
29	E38CR		R18CF	F27CD	W19BL	R23BJ	K32BF	W33BD	R35AM	K24AJ	W22AG	U34AD
30	M11CS		X12CH	V13CE	Y30BM	X36BK	F20BG	Y15BE	X16BC	F21AK	Y29AH	R26AE
31	H31AB		B37CJ		T14BP		V28BH	T25BF		V17AL		X38AF

FIRST LETTER = YOUR PHYSICAL CODE, MIDDLE NUMBER = SENSITIVITY CODE, LAST TWO LETTERS = INTELLECTUAL CODE

1914	JAN	FEB	MAR	APR	MAY	JUN	JUL	AUG	SEP	OCT	NOV	DEC
1	B11AG	E12AE	H12CR	R13CM	K30CJ	W36CF	U20BP	K15BL	W16BJ	U21BF	K29BD	P26AL
2	N31AH	M37AF	Y37CS	X19CP	F14CK	Y32CH	R28CD	F25BM	Y24BK	R17BG	F34BE	W38AM
3	A18AJ	H27AG	T27AB	B30CR	V23CL	T20CJ	X33CE	V35BP	T21BL	X22BH	V26BF	Y11BG
4	Z12AK	S13AH	G13AC	N14CP	E36CM	G28CK	B15CF	E16CD	G17BM	B29BJ	E38BG	T31BD
5	C37AL	P19AJ	L19AD	A23AB	M32CP	L33CL	N25CH	M24CE	L22BP	N34BK	M11BH	G18BE
6	K27AM	W30AK	D30AE	Z36AC	H20CR	D15CM	A35CJ	H21CF	D29CD	A26BL	H31BJ	L12BF
7	F13BC	Y14AL	J14AF	C32AD	S28CS	J25CP	Z16CK	S17CH	J34CE	Z38BM	S18BK	D37BG
8	W19BD	T23AM	U23AG	K20AE	P33AB	U35CR	G24CL	P22CJ	U25CF	C11BP	P12BL	J27BH
9	E30BE	G36BC	R36AH	F28AF	W15AC	R16CS	K21CH	W29CK	R38CH	K31CD	W37BM	U13BJ
10	M14BF	L32BD	X32AJ	V33AG	Y25AD	X24AB	F17CP	Y34CL	X11CJ	F18CE	Y27BP	R19BK
11	H23BG	D20BE	B20AK	E15AH	T35AE	B21AC	V22CR	T26CM	B31CK	V12CF	T13CD	X30BL
12	S36BH	J28BF	N28AL	M25AJ	G29AF	N17AD	E29CS	G38CP	N18CL	E37CH	G19CE	B14BH
13	P32BJ	U33BG	A33AM	H35AK	L24AG	A22AE	M34AB	L11CR	A12CM	M27CJ	L30CF	N23BP
14	W20BK	R15BH	Z15BC	S16AL	D21AH	Z29AF	H26AC	D31CS	Z37CP	H13CK	D14CH	A36CD
15	Y28BL	X25BJ	C25BD	P24AM	J17AJ	C34AG	S36AD	J18AB	C27CR	S19CL	J23CJ	Z32CE
16	T33BM	B35BK	K35BE	W21BC	U22AK	K26AH	P11AE	U12AC	K13CS	P30CH	U36CK	C20CF
17	G15BP	N16BL	F16BF	Y17BD	R29AL	F38AJ	W31AF	R37AD	F19AB	W14CP	R32CL	K28CH
18	L25CD	A24BM	V24BG	T22BE	X34AM	V11AK	Y18AG	X27AE	V30AC	Y23CR	X20CM	F33CJ
19	D35CE	Z21BP	E21BH	G29BF	B26BC	E31AL	T12AH	B13AF	E14AD	T36CS	B28CP	V15CK
20	J16CF	C17CD	M17BJ	L34BG	N38BD	M18AM	G37AJ	N19AG	M23AE	G32AB	N33CR	E25CL
21	U24CH	K22CE	H22BK	D26BH	A11BE	H12BC	L27AK	A30AH	H36AF	L20AC	A15CS	M35CM
22	R21CJ	F29CF	S29BL	J38BJ	Z31BF	S37BD	D13AL	Z14AJ	S32AG	D28AD	Z25AB	H16CP
23	X17CK	V34CH	P34BM	U11BK	C18BG	P27BE	J19AM	C23AK	P20AH	J33AE	C35AC	S24CR
24	B22CL	E26CJ	W26BP	R31BL	K12BH	W13BF	U30BC	K36AL	W28AJ	U15AF	K16AD	P21CS
25	N29CM	M38CK	Y38CD	X18BM	F37BJ	Y19BG	R14BD	F32AM	Y33AK	R25AG	F24AE	W17AB
26	A34CP	H11CE	T11CE	B12BP	V27BK	T30BH	X23BE	V20BC	T15AL	X35AH	V21AF	Y22AC
27	Z26CR	S31CM	G31CF	N37CD	E13BL	G14BJ	B36BF	E28BD	G25AM	B16AJ	E17AG	T29AD
28	C30CS	P18CP	L18CH	A27CE	M19BM	L23BL	N32BG	M33BE	L35BC	N24AK	M22AH	G34AE
29	K11AB		D12CJ	Z13CF	H30BP	D36BL	A20BH	H15BF	D16BD	A21AL	H29AJ	L26AF
30	F31AC		J37CK	C19CH	S14CD	J32BM	Z28BJ	S25BG	J24BE	Z17AM	S34AK	D38AG
31	V18AD		U27CL		P23CE		C33BK	P35BH		C22BC		J11AH

1915	JAN	FEB	MAR	APR	MAY	JUN	JUL	AUG	SEP	OCT	NOV	DEC
1	U31AJ	K37AG	H37AB	D19CR	A14CL	H32CJ	L28CE	A25BP	H24BL	L17BH	A34BF	M38BC
2	R18AK	F27AH	S27AC	J30CS	Z23CM	S20CK	D33CF	Z35CD	S21BM	D22BJ	Z26BG	H11BD
3	X12AL	V13AJ	P13AD	U14AB	C36CP	P28CL	J15CH	C16CE	P17BP	J29BK	C38BH	S31BE
4	B37AM	E19AK	W19AE	R23AC	K32CR	W33CM	U25CJ	K24CF	W22CD	U34BL	K11BJ	P18BF
5	N27BC	M30AL	Y30AF	X36AD	F20CS	Y15CP	R35CK	F21CH	Y29CE	R26BM	F31BK	W12BG
6	A13BD	H14AM	T14AG	B32AE	V28AB	T25CR	X16CL	V17CJ	T34CF	X38BP	V18BL	Y37BH
7	Z19BE	S23BC	G23AH	W20AF	E33AC	G35CS	B24CH	E22CK	G26CH	B11CD	E12BM	T27BJ
8	C30BF	P36BD	L36AJ	A28AG	M15AD	L16AB	N21CP	M29CL	L38CJ	N31CE	M37BP	G13BK
9	K14BG	W32BE	D32AK	Z33AH	H25AE	D24AC	A17CR	H34CM	D11CK	A18CF	H27CD	L19BL
10	F23BH	Y20BF	J20AL	C15AJ	S35AF	J21AD	Z22CS	S26CP	J31CL	Z12CH	S13CE	D30BM
11	V36BJ	T28BG	U28AM	K25AK	P16AG	U17AE	C29AB	P38CR	U18CM	C37CJ	P19CF	J14BP
12	E32BK	G33BH	R33BC	F35AL	W24AH	R22AF	K34AC	W11CS	R12CP	K27CK	W30CH	U23CD
13	M20BL	L15BJ	X15BD	V16AM	Y21AJ	X29AG	F26AD	Y31AB	X37CR	F13CL	Y14CJ	R36CE
14	H28BM	D25BK	B25BE	E24BC	T17AK	B34AH	W38AE	T18AC	B27CS	W19CM	T23CK	X32CF
15	S33BP	J35BL	N35BF	M21BD	G22AL	N26AJ	E11AF	G12AD	N13AB	E30CP	G36CL	B20CH
16	P15CD	U16BM	A16BG	H17BE	L29AM	A38AK	M31AG	L37AE	A19AC	M14CR	L32CH	N28CJ
17	W25CE	R24BP	Z24BH	S22BF	D34BC	Z11AL	H18AH	D27AF	Z30AD	H23CS	D20CP	A33CK
18	Y35CF	X21CD	C21BJ	P29BG	J26BD	C31AM	S12AJ	J13AG	C14AE	S36AB	J28CR	Z15CL
19	T16CH	B17CE	K17BK	W34BH	U38BE	K18BC	P37AK	U19AH	K23AF	P32AC	U33CS	C25CM
20	G24CJ	N22CF	F22BL	Y26BJ	R11BF	F12BD	W27AM	R30AJ	F36AG	W20AD	R15AB	K35CP
21	L21CK	A29CH	V29BM	T38BK	X31BG	V37BE	Y13AM	X14AK	V32AH	Y28AE	X25AC	F16CR
22	D17CL	Z34CJ	E34BP	G11BL	B18BH	E27BC	T19BC	B23AL	E20AJ	T33AF	B35AD	V24CS
23	J22CM	C26CK	M26CD	L31BM	N28BJ	M13BG	G30BD	N36AM	M28AK	G15AG	N16AE	E21AB
24	U29CP	K38CL	H38CE	D18BP	A37BK	H19BH	L14BC	A32BC	H33AL	L25AH	A24AF	M17AC
25	R34CR	F11CM	S11CF	J12CD	Z27BL	S30BJ	D23BF	Z20BD	S15AM	D35AJ	Z21AG	H22AD
26	X26CS	V31CP	P31CH	U37CE	C13BM	P14BK	J36BG	C28BE	P25BC	J16AK	C17AH	S29AE
27	B38AB	E18CR	W18CJ	R27CF	K19BP	W23BL	U32BH	K33BF	W35BD	U24AL	K22AJ	P34AF
28	N11AC	M12CS	Y12CK	X13CH	F30CD	Y36BM	R20BJ	F15BG	Y16BE	R21AM	F29AK	W26AG
29	A31AD		T37CL	B19CJ	V14CE	T32BP	X28BK	V25BH	T24BF	X17BC	V34AL	Y38AH
30	Z18AE		G27CM	N30CK	E23CF	G20CD	B33BL	E35BJ	G21BG	B22BD	E26AM	T11AJ
31	C12AF		L13CP		M36CH		N15BH	M16BK		N29BE		G31AK

1916	JAN	FEB	MAR	APR	MAY	JUN	JUL	AUG	SEP	OCT	NOV	DEC
1	L18AL	A27AJ	E13AE	G14AC	B36CR	E28CM	T15CJ	B16CF	E17CD	T29BL	B38BJ	V31BF
2	D12AM	Z13AK	M19AF	L23AD	N32CS	M33CP	G25CK	N24CH	M22CE	G34BM	N11BK	E18BG
3	J37BC	C19AL	H30AG	D36AE	A20AB	H15CR	L35CL	A21CJ	H29CF	L26BP	A31BL	M12BH
4	U27BD	K30AM	S14AH	J32AF	Z28AC	S25CS	D16CM	Z17CK	S34CH	D38CD	Z18BM	H37BJ
5	R13BE	F14BC	P23AJ	U20AG	C33AD	P35AB	J24CP	C22CL	P26CJ	J11CE	C12BP	S27BK
6	X19BF	V23BD	W36AK	R28AH	K15AF	W16CL	U21CH	K29CK	W38CK	U31CF	K37BD	P13BL
7	B30BG	E36BE	Y32AL	X33AJ	F25AE	Y24AD	R17CS	F34CE	Y11CL	R18CH	F27CE	W19BM
8	N14BH	M32BF	T20AM	B15AK	V35AG	T21AE	X22AB	V26CR	T31CM	X12CJ	V13CF	Y30BP
9	A23BJ	H20BG	G28BC	N25AL	E16AH	G17AF	B29AC	E38CS	G18CP	B37CH	E19CH	T14CD
10	Z36BK	S28BH	L33BD	A35AM	H24AJ	L22AG	N34AD	M11AB	L12CR	N27CL	M30CJ	G23CE
11	C32BL	P33BJ	D15BE	Z16BG	H21AK	D29AH	A26AE	H31AL	D37CS	A13CM	H14CK	L36CF
12	K20BM	W15BK	J25BF	O24BD	S17AL	J34AJ	Z38AF	S18AD	J27BE	Z19CP	S23CL	D32CH
13	F28BP	Y25BL	U35BG	K21BE	P22AM	U26AK	C11AG	P12AE	U13AC	C30CR	P36CH	J20CJ
14	V33CD	T35BM	R16BH	F17BF	W29BC	R38AL	K31AH	W37AF	R19AD	K14CS	W32CP	U28CK
15	E15CE	G16BP	X24BJ	V22BG	Y34BD	X11AM	F18AG	Y27AE	X30AB	F23AB	Y20CR	R33CL
16	M25CF	L24CD	B21BK	E29BH	T26BE	B31BC	V12AK	T13AK	B14AF	V36AC	T28CS	X15CM
17	H35CH	D21CE	N17BL	M34BJ	G38BF	N18BD	E37AL	G19AJ	N23AG	E32AD	G33AB	B25CP
18	S16CJ	J17CF	A22BM	H26BK	L11BG	A12BE	M27AH	L30AK	A35AF	M20AD	L15AC	N35CR
19	P24CK	U22CH	Z29BP	S38BL	D31BH	Z37BF	H13BC	D14AL	Z32AJ	H28AF	D25AD	A16CS
20	W21CL	R29CJ	C34CD	P11BM	J18BJ	C27BE	S19BD	J23AM	C20AK	S33AG	J35AE	
21	Y17CM	X34CK	K26CE	W31BP	U12BK	K13BH	P30BE	U36BC	K28AL	P15AH	U16AF	C21AC
22	T22CP	B26CL	F38CF	Y18CD	R37BL	F19BJ	W14BF	R32BD	F33AM	W25AJ	R24AG	K17AD
23	G29CR	N38CM	V11CH	T12CE	X27BM	V30BK	Y23BG	X20BE	V15BC	Y35AK	X21AH	F22AE
24	L34CS	A11CP	E31CJ	G37CF	B13BP	E14BL	T36BH	B28BF	E25BD	T16AL	B17AJ	V29AF
25	D26AB	Z31CR	M18CK	L27CH	N19CD	M23BM	G32BJ	N33BG	M35BE	G24AM	N22AK	E34AG
26	J38AC	C18CS	H12CL	D13CJ	A30CE	H36BP	L20BK	A15BH	H16BF	L21BC	A29AL	M26AH
27	U11AD	K12AB	S37CM	J19CK	Z14CF	S32CD	D28BL	Z25BJ	S24BG	D17BD	Z34AM	H38AJ
28	R31AE	F3FAC	P27CP	U30CL	C23CH	P20CE	J33BM	C35BK	P21BH	J22BE	C26BC	S11AK
29	X18AF	V27AD	W13CR	R14CM	K36CJ	W28CF	U15BP	K16BL	W17BJ	U29BF	K38BD	P31AL
30	B12AG		Y19CS	X23CP	F32CK	Y33CK	R25CD	F24BM	Y22BK	R34BG	F11BE	W18AM
31	N37AH		T30AB		V20CL		X35CE	V21BP		X26BH		Y12BC

FIRST LETTER = YOUR PHYSICAL CODE, MIDDLE NUMBER = SENSITIVITY CODE, LAST TWO LETTERS = INTELLECTUAL CODE

1917	JAN	FEB	MAR	APR	MAY	JUN	JUL	AUG	SEP	OCT	NOV	DEC
1	T37BD	B19AM	K19AG	W23AE	U32AB	K33CR	P25CL	U24CJ	K22CF	P34BP	U11BL	C18BH
2	G27BE	N30BC	F30AH	Y36AF	R20AC	F15CS	W35CH	R21CK	F29CH	W26CD	R31BM	K12BJ
3	L13BF	A14BD	V14AJ	T32AG	X28AD	V25AB	Y16CP	X17CL	V34CJ	Y38CE	X18BP	F37BK
4	D19BG	Z23BE	E23AK	G20AH	B33AE	E35AC	T24CR	B22CM	E26CK	T11CF	B12CD	V27BL
5	J30BH	C36BF	M36AL	L28AJ	N15AF	M16AD	G21CS	N29CP	M38CL	G31CH	Z13CH	M13BM
6	U14BJ	K32BG	H32AM	D33AK	A25AG	H24AE	L17AB	A34CR	H11CM	L18CJ	A27CF	M19BP
7	R23BK	F20BH	S29BC	J15AL	Z35AH	S21AF	D22AC	Z26CS	S31CP	D12CK	Z13CM	H30CD
8	X36BL	V28BJ	P28BD	U25AM	C16AJ	P17AG	J29AD	C38AB	P18CR	J37CL	C19CJ	S14CE
9	B32BM	E33BK	W33BE	R35BC	K24AK	W22AH	U34AE	K11AC	W12CS	U27CM	K30CK	P23CF
10	N20BP	M15BL	Y15BF	X16BD	F21AL	Y29AJ	R26AF	F31AD	Y37AB	R13CP	F14CL	W36CH
11	A28CD	H25BM	T25BG	B24BE	V17AM	T34AK	X38AG	V18AE	T27AC	X19CR	V23CM	Y32CJ
12	Z33CE	S35BP	G35BH	N21BF	E22BC	G26AL	B11AH	E12AF	G13AD	B30CS	E36CP	T20CK
13	C15CF	P16CD	L16BJ	A17BG	M29BD	L38AM	N31AJ	M37AG	L19AE	N14AB	M32CR	G28CL
14	K25CH	W24CE	D24BK	Z22BH	H34BE	D11BC	A18AK	H27AH	D30AF	A23AC	H20CS	L33CM
15	F35CJ	Y21CF	J21BL	C29BJ	S26BF	J31BD	Z12AL	S13AJ	J14AG	Z36AD	S28AB	D15CP
16	V16CK	T17CH	U17BH	K34BK	P38BG	U18BE	C37AM	P19AK	U23AH	C32AE	P33AC	J25CR
17	E24CL	G22CJ	R22BP	F26BL	W11BH	R12BF	K27BC	W30AL	R36AJ	K20AF	W15AD	U35CS
18	M21CM	L29CK	X29CD	V38BM	T18BK	X37BG	F13BD	Y14AM	X32AK	F28AG	Y25AE	R16AB
19	H17CP	D34CL	B34CE	E11BP	G12BL	B27BH	V19BE	T23BC	B20AL	V33AH	T35AF	X24AC
20	S22CR	J26CM	N26CF	M31CD	G12BL	N13BJ	E30BF	G36BD	N28AM	E15AJ	G16AG	B21AD
21	P29CS	U36CP	A38CH	H18CE	L37BM	A19BK	M14BG	L32BE	A33BC	M25AK	L24AH	N17AE
22	W34AB	R11CR	Z11CJ	S12CF	D27BP	Z30BL	H23BH	D20BF	Z15BD	H35AL	D21AJ	A22AF
23	Y26AC	X31CS	C31CK	P37CH	J13CD	C14BM	S36BJ	J28BG	C25BE	S16AM	J17AK	Z29AG
24	T38AD	B18AB	K18CL	W27CJ	U19CE	K23BP	P32BK	U33BH	K35BF	P24BC	U22AL	C34AH
25	G11AE	N12AC	F12CM	Y13CK	R30CF	F36CD	M20BL	R15BJ	F16BG	W21BD	R29AM	K26AJ
26	L31AF	A37AD	V37CP	T19CL	X14CH	V32CE	Y28BM	X25BK	V24BH	Y17BE	X34BC	F38AK
27	D18AG	Z27AE	E27CR	G30CM	B23CJ	E20CF	T33BP	B35BL	E21BJ	T22BF	B26BD	V11AL
28	J12AH	C13AF	M13CS	L14CP	N36CK	N28CH	G15CD	N16BM	M17BK	L34BH	A11BF	E31AM
29	U37AJ		H19AB	D23CR	A32CL	H33CJ	L25CE	A24BP	H22BL	L34BH	A11BF	M18BC
30	R27AK		S30AC	J36CS	Z20CM	S15CK	D35CF	Z21CD	S29BM	D26BJ	Z31BG	H12BD
31	X13AL		P14AD		C28CP		J16CH	C17CE		J38BK		S37BE

1918	JAN	FEB	MAR	APR	MAY	JUN	JUL	AUG	SEP	OCT	NOV	DEC
1	P27BF	U30BD	A30AJ	H36AG	L20AD	A15AB	M35CP	L21CL	A29CJ	M26CE	L31BP	N12BK
2	W13BG	R14BE	Z14AK	S32AH	D28AE	Z25AC	H16CR	D17CM	Z34CK	H38CF	D18CD	A37BL
3	Y19BH	X23BF	C23AL	P20AJ	J33AF	C35AD	S24CS	J22CP	C26CL	S11CH	J12CE	Z27BM
4	T30BJ	B36BG	K36AM	W28AK	U15AG	K16AE	P21AB	U29CR	K38CM	P31CJ	U37CF	C13BP
5	G14BK	N32BH	F32BC	Y33AL	R25AH	F24AF	W17AC	R34CS	F11CP	W18CK	R27CH	K19CD
6	L23BL	A20BJ	V20BD	T15AM	X35AJ	V21AG	Y22AD	X26AB	V31CR	Y12CL	X13CJ	F30CE
7	D36BM	Z28BK	E28BE	G25BC	B16AK	E17AH	T29AE	B38AC	E18CS	T37CM	B19CK	V14CF
8	J32BP	C33BL	M33BF	L35BD	N24AL	M22AJ	G34AF	N11AD	M12AB	G27CP	N30CL	E23CH
9	U20CD	K15BM	H15BG	D16BE	A21AM	H29AK	L26AG	A31AB	H37AC	L13CR	A14CM	M36CJ
10	R28CE	F25BP	S25BH	J24BF	Z17BG	S34AL	D38AH	Z18AF	S27AD	D19CS	Z23CP	H32CK
11	X33CF	V35CD	P35BJ	U21BG	C22BD	P26AM	J11AJ	C12AG	P13AE	J30AB	C36CR	S20CL
12	B15CH	E16CE	H16BK	R17BH	K29BE	W38BC	U31AK	K37AH	W19AF	U14AC	K32CS	P28CM
13	N25CJ	M24CF	Y24BL	X22BJ	F34BF	Y11BD	R18AL	F27AJ	Y30AG	R23AD	F20AB	W33CP
14	A35CK	H21CH	T21BM	B29BK	V26BG	T31BE	X12AM	V13AK	T14AH	X36AF	V28AC	Y15CR
15	Z16BL	S17CJ	G17BP	N34BL	E38BH	G18BF	B37BC	E19AL	G23AJ	B32AF	E33AD	T25CS
16	C24CM	P22CK	L22CD	A26BM	M11BJ	L12BG	N27BD	M30AM	L36AK	N20AG	M15AE	G35AB
17	K21CP	W29CL	D29CE	Z38BP	H31BK	D37BH	A13BE	H14BC	D32AL	A28AH	H25AF	L16AC
18	F17CR	Y34CM	J34CF	G11CD	S18BL	J27BJ	V19BF	H14BC	J33BC	U28BC	Z19BD	O24AD
19	V22CS	T26CP	U26CH	K31CE	P12BH	U13BK	C30BG	P36BE	U28BC	C15AK	P16AH	J21AE
20	E29AB	G38CR	R38CJ	F16CF	R19BL	K14BH	W32BF	R24CD	R33BD	K25AL	W24AJ	U17AF
21	M34CG	L11CS	X11CK	V12CH	Y27CD	X30BM	F23BJ	Y20BG	X15BE	F35AM	Y21AK	R22AG
22	H26AD	D31AB	B31CL	E37CJ	T13CE	B14BP	V36BK	T28BH	B25BF	V16BC	T17AL	X29AH
23	S38AE	J18AG	N18CH	M27CK	G19CF	N23CD	E32BL	G33BJ	N35BG	E24BD	G22AM	B34AJ
24	P11AF	U12AD	A12CP	H13CL	L30CH	A36CE	M20BM	L15BK	A16BH	M21BE	L29BC	N26AK
25	W31AG	R37AE	Z37CR	S19CH	D14CJ	Z32CF	H28BP	D25BL	Z24BJ	H17BF	D34BD	A38AL
26	Y18AH	X27AF	C27CS	P30CP	J23CK	C20CH	S33CD	J35BM	C21BK	S22BG	J26BE	Z11AM
27	T12AJ	B13AG	K13AB	W14CH	U36CL	K28CJ	P15CE	U16BP	K17BL	P29BH	U38BF	C31BC
28	G37AK	N19AH	F19AC	Y23CS	R32CM	F33CK	W25CF	R24CD	W34BJ	R11BG	K18BD	
29	L27AL		V30AD	T36AB	X20CP	V15CL	Y35CH	X21CE	V29BP	Y26BK	X31BH	F12BE
30	D13AM		E14CE	G32AC	E25CM	T16CJ	B17CF	E34CD	G11BM	B18BJ	V37BF	
31	J19BC		H23AF		N33CS		G24CK	N22CH				E27BG

1919	JAN	FEB	MAR	APR	MAY	JUN	JUL	AUG	SEP	OCT	NOV	DEC
1	M13BH	L14BF	X14AL	V32AJ	Y28AF	X25AD	F16CS	Y17CP	X34CL	F38CH	Y16CE	R37BM
2	H19BJ	D23BG	B23AM	E20AK	T33AG	B35AE	V24AB	T22CR	B26CM	V11CJ	T12CF	X27BP
3	S30BK	J36BH	N36BC	M28AL	G15AH	N16AF	E21AC	G29CS	N38CP	E31CK	G37CH	B13CD
4	P14BL	U32BJ	A33AM	H33AM	L25AJ	A24AG	M17AD	L34AB	A11CR	M18CL	L27CJ	N19CE
5	W23BM	R20BK	Z20BE	S15BC	D35AK	Z21AH	H22AE	D26AC	Z31CS	H12CM	D13CK	A30CF
6	Y36BP	X28BL	C28BF	P25BD	J16AL	C17AJ	S29AF	J38AD	C18AB	S37CP	J19CL	Z14CH
7	T32CD	B33BM	K33BG	W35BE	U24AM	K22AK	P34AG	U11AE	K12AC	P27CR	U30CM	C23CJ
8	G20CE	N15BP	F15BH	Y16BF	R21BC	F29AL	W26AH	R31AF	F37AD	W13CS	R14CP	K36CK
9	L28CF	A25CD	V25BJ	T24BG	X17BD	V34AM	Y38AJ	X18AG	V27AE	Y19AB	X23CR	F32CL
10	D33CH	Z35CE	E35BK	G21BH	B22BE	E26BC	T11AK	B12AH	E13AF	T30AC	B36CS	V20CM
11	J15CJ	C16CF	M16BL	L17BJ	N29BF	M38BD	G31AL	N37AJ	M19AG	G14AD	N32AB	E28CP
12	U25CK	K24CH	H24BM	D22BK	A34BG	H11BE	L18AM	A27AK	H30AH	L23AE	A20AC	M33CR
13	R35CL	F21CJ	S21BP	Z26BL	M38BH	S31BF	D12BC	Z13AL	S14AJ	D36AF	Z28AD	H15CS
14	X16CM	V17CK	P17CD	U34BM	C38BJ	P18BG	J37BD	C19AM	P23AK	J32AG	C33AE	S25AB
15	B24CP	E22CL	W22CE	K11BP	K11BK	W12BH	H27BE	K38BC	U20AL	K15AH	P35AF	P35AC
16	N21CR	M29CN	Y29CF	X38CD	F31BL	Y37BJ	R13BF	F14BD	Y32AM	R28AJ	F25AG	W16AD
17	A17CS	H34CP	T34CH	B11CE	V18BP	T27BK	X19BG	V23BE	T28AK	X33AH	V35AF	U24AE
18	Z22AB	S26CR	G26CJ	N31CF	E12BP	G13BL	B30BH	E36BF	G28BD	B15AL	E16AJ	T21AF
19	C29AC	P38CS	L36CK	A18CH	M37CF	L19BM	N14BJ	M32BG	L33BE	N25AM	M24AH	G17AG
20	K34AD	W11AB	D11CL	Z12CJ	H27CE	D38BP	A23BK	H20BH	D15BF	A35BC	H21AL	L22AH
21	F26AE	Y31AC	J31CM	C37CK	S13CF	J14CD	Z36BL	S28BJ	P33BK	U35BH	P22BC	J34AK
22	V38AF	T18AD	U18CR	F13CH?	W30CJ	R36CF	K28BP	W15BL	R16BJ	K21BF	W29BD	U26AJ
23	E11AG	G12AE	R12CS	F13CM	W30CJ	R36CF	K28BP	W15BL	R16BJ	K21BF	W29BD	R38AM
24	M31AH	L37AF	X37CS	V19CP	Y14CK	X32CH	F28CD	Y25BM	X24BK	F17BG	Y34BE	X11BC
25	H18AJ	D27AG	B27AB	E38CH	T23CL	B20CJ	V33CE	T35BP	B21BL	V22BH	T26BF	X11BC
26	S12AK	J13AH	N13AC	M14CS	L32CP	N28CK	E15CF	G16CD	N17BM	E29BJ	G38BG	B31BD
27	P37AL	U19AJ	A19AD	H23AB	L32CP	M25CH	L24CE	A22BP	M34BL	A22BP	L11BH	N18BE
28	W27AM	R30AK	Z30AE	S36AC	D20CR	Z15CH	H35CJ	D21CF	Z29CD	H26BL	D31BJ	A12BF
29	Y13BC		C14AF	P32AD	J28CS	C25CP	S16CK	J17CH	C34CE	Z29CD	J18BK	Z37BG
30	T19BD		K23AG	W20AE	U33AB	K35CR	P24CL	U22CJ	K26CF	P11BP	U12BL	C27BH
31	G30BE		F36AH		R15AC		N21CH	R29CK				K13BJ

(141)

FIRST LETTER = YOUR PHYSICAL CODE, MIDDLE NUMBER = SENSITIVITY CODE, LAST TWO LETTERS = INTELLECTUAL CODE

1920	JAN	FEB	MAR	APR	MAY	JUN	JUL	AUG	SEP	OCT	NOV	DEC
1	F19BK	Y23BH	U36BD	K28AM	P15AJ	U16AG	C21AD	P29AB	J38CR	C31CL	P37CJ	J13CE
2	V30BL	T36BJ	R32BE	F33BC	W25AK	R24AH	K17AE	W34AC	R11CS	K18CM	W27CK	U19CF
3	E14BM	G32BK	X20BF	V15BD	Y35AL	X21AJ	F22AF	Y26AD	X31AB	F12CP	Y13CL	R30CH
4	M23BP	L20BL	B28BG	E25BE	T16AM	B17AK	V29AG	T38AE	B18AC	V37CR	T19CM	X14CJ
5	H36CD	D28BM	N33BH	M35BF	G24BC	N22AL	E34AH	G11AF	N12AD	E27CS	G30CP	B23CK
6	S32CE	J33BP	A15BJ	H16BG	L21BD	A29AM	M26AJ	L31AG	A37AE	M13AB	L14CR	N36CL
7	P20CF	U15CD	Z25BK	S24BH	D17BE	Z34BC	H38AK	D18AH	Z27AF	H19AC	D23CS	A32CM
8	W28CH	R25CE	C35BL	P21BJ	J22BF	C26BD	S11AL	J12AJ	C13AG	S30AD	J36AB	Z20CP
9	Y33CJ	X35CF	K16BM	W17BK	U29BG	K38BE	P31AM	U37AK	K19AH	P14AE	U32AC	C28CR
10	T15CK	B16CH	F24BP	Y22BL	R34BH	F11BF	W18BC	R27AL	F30AJ	W23AF	R20AD	K33CS
11	G25CL	N24CJ	V21CD	T29BM	X26BJ	V31BG	Y12BD	X13AM	V14AK	Y36AG	X28AE	F15AB
12	L35CM	A21CK	E17CE	G34BP	B38BK	E18BH	T37BE	B19BC	E23AL	T32AH	B33AF	V25AC
13	D16CP	Z17CL	M22CF	L26CD	N11BL	M12BJ	G27BF	N30BD	M36AM	G20AJ	N15AG	E35AD
14	J24CR	C22CM	H29CH	D38CE	A31BM	H37BK	L13BG	A14BE	H32BC	L28AK	A25AH	M16AE
15	U21CS	K29CP	S34CJ	J11CF	Z18BP	S27BL	D19BH	Z23BF	S20BD	D33AL	Z35AJ	H24AF
16	R17AB	F34CR	P26CK	U31CH	C12CD	P13BM	J30BJ	C36BG	P28BE	J15AM	C16AK	S21AG
17	X22AC	V26CS	W38CL	R18CJ	K37CE	W19BP	U14BK	K32BH	W33BF	U25BC	K24AL	P17AH
18	B29AD	E38AB	Y11CM	X12CK	F27CF	Y30CD	R23BL	F20BJ	Y15BG	R35BD	F21AM	W22AJ
19	N34AE	M11AC	T31CP	B37CL	V13CH	T14CE	X36BM	V28BK	T25BH	X16BE	V17BC	Y29AK
20	A26AF	H31AD	G18CR	N27CM	E19CJ	G23CF	B32BP	E33BL	G35BJ	B24BF	E22BD	T34AL
21	Z38AG	S18AE	L12CS	A13CP	M30CK	L36CH	N20CD	M15BM	L16BK	N21BG	M29BE	G26AM
22	C11AH	P12AF	D37AB	Z19CR	H14CL	D32CJ	A28CE	H25BP	D24BL	A17BH	H34BF	L38BC
23	K31AJ	W37AG	J27AC	C30CS	S23CM	J20CK	Z33CF	S35CD	J21BM	Z22BJ	S26BG	D11BD
24	F18AK	Y27AH	U13AD	K14AB	P36CP	U28CL	C15CH	P16CE	J17BP	C29BK	P38BH	J31BE
25	V12AL	T13AJ	R19AE	F23AC	W32CR	R33CM	K25CJ	W24CF	Z22BE	K34BL	W11BJ	U18BF
26	E37AM	G19AK	X30AF	V36AD	Y20CS	X15CP	F35CK	Y21CH	X29CE	F26BM	Y31BK	R12BG
27	M27BC	L30AL	B14AG	E32AE	T28AB	B25CR	V16CL	T17CJ	B34CF	V38BP	T18BL	X37BH
28	H13BD	D14AM	N23AH	M20AF	G33AC	N35CS	E24CH	G22CK	N25CH	E11CD	G12BM	B27BJ
29	S19BE	J23BC	A36AJ	H28AG	L15AD	A16AB	M21CP	L29CL	A38CJ	M31CE	L37BP	N13BK
30	P30BF		Z32AK	S33AH	D25AE	Z24AC	H17CR	D34CM	Z11CK	H18CF	D27CD	A19BL
31	W14BG		C20AL		J35AF		S22CS	J26CP		S12CH		Z30BM

1921	JAN	FEB	MAR	APR	MAY	JUN	JUL	AUG	SEP	OCT	NOV	DEC
1	C14BP	P32BL	L32BF	A33BD	M25AL	L24AJ	N17AF	M34AD	L11AB	N18CP	M27CL	G19CH
2	K23CD	W20BM	D29BG	Z15BE	H35AM	D21AK	A22AG	H26AE	D31AC	A12CR	H13CM	L30CJ
3	F36CE	Y28BP	J28BH	C25BF	S16BC	J17AL	Z29AH	S38AF	J18AD	Z37CS	S19CP	D14CK
4	V32CF	T33CD	U33BJ	K35BG	P24BD	U22AM	C34AJ	P11AG	U12AE	C27AB	P30CR	J23CL
5	E20CH	G15CE	R15BK	F16BH	W21BE	R29BC	K26AK	W31AH	R37AF	K13AC	W14CS	U36CM
6	M28CJ	L25CF	X25BL	V24BJ	Y17BF	X34BD	F38AL	Y18AJ	X27AG	F19AD	Y23AB	R32CP
7	H33CK	D35CH	B35BM	E21BK	T22BG	B26BE	V11AM	T12AK	B13AH	V30AE	T36AC	X20CR
8	S15CL	J16CJ	N16BP	M17BL	G29BH	N38BF	E31BC	G37AL	N19AJ	E14AF	G32AD	B28CS
9	P25CM	U24CK	A24CD	H22BM	L34BJ	A11BG	M18BD	L27AM	A30AK	M23AG	L20AE	N33AB
10	W35CP	R21CL	Z21CE	S29BP	D26BK	Z31BH	H12BE	D13BC	Z14AL	H36AH	D28AF	A15AC
11	Y16CR	X17CM	C17CF	P34CD	J38BL	C18BJ	S37BF	J19BD	C23AM	S32AJ	J33AG	Z25AD
12	T24CS	B22CP	K22CH	W26CE	U11BM	K12BK	P27BG	U30BE	K36BC	P20AK	U15AH	C35AE
13	G21AB	N29CR	F29CJ	Y38CF	R31BH	F37BD	W13BH	R14BF	F32BD	W28AL	R25AJ	K16AF
14	L17AC	A34CS	V34CK	T11CH	X18CD	V27BM	Y19BJ	X23BG	V20BE	Y33AM	X35AK	F24AG
15	D22AD	Z26AB	E26CL	G31CJ	B12CE	E13BP	T30BK	B36BH	E28BF	T15BC	B16AL	V21AH
16	J29AE	C38AC	M38CM	L18CK	N37CF	M19CD	G14BL	N32BJ	M33BG	G25BD	N24AM	E17AJ
17	U34AF	K11AD	H11CP	D12CL	A27CH	H30CE	L23BM	A20BK	H15BH	L35BE	A21BC	M22AK
18	R26AG	F31AE	S31CR	J37CM	Z13CJ	S14CF	D36BP	Z28BL	S25BJ	D16BF	Z17BD	H29AL
19	X38AH	V18AF	P18CS	U27CP	C19CK	P23CH	J32CD	C33BM	P35BK	J24BG	C22BE	S34AM
20	B11AJ	E12AG	W12AB	R13CR	K30CL	W36CJ	U20CE	K15BP	W16BL	U21BH	K29BF	P26BC
21	N31AK	M37AH	Y37AC	X19CS	F14CM	Y32CK	R28CF	F25CD	Y24BM	R17BJ	F34BG	W38BD
22	A18AL	H27AJ	T27AD	B30AB	V23CP	T20CL	X33CH	V35CE	T21BP	X22BK	V26BH	Y11BE
23	Z12AM	S13AK	G13AE	N14AC	E36CR	G28CM	B15CJ	E16CF	G17CD	B29BL	E38BJ	T31BF
24	C37BC	P19AL	L19AF	A23AD	M32CS	L33CP	N25CK	M24CH	L22CE	N34BM	M11BK	G18BG
25	K27BD	W30AM	D30AG	Z36AE	H20AB	D15CR	A35CL	H21CJ	D29CF	A26BP	H31BL	L12BH
26	F13BE	Y14BC	J14AH	C32AF	S28AC	J25CS	Z16CM	S17CK	J34CH	Z38CD	S18BM	D37BJ
27	V19BF	T23BD	U23AJ	K20AG	P33AD	U35AB	C24CP	P22CL	U26CJ	C11CE	P12BP	J27BK
28	E30BG	G36BE	R36AK	F28AH	W15AE	R16AC	K21CR	W29CM	R38CK	K31CF	W37CD	U13BL
29	M14BH		X32AL	V33AJ	Y25AF	X24AD	F17CS	Y34CF	X11CL	F18CH	Y27CE	R19BM
30	H23BJ		B20AM	E15AK	T35AG	B21AE	V22AB	T26CR	B31CM	V12CJ	T13CF	X30BP
31	S36BK		N28BC		G16AH		E29AC	G38CS		E37CK		B14CD

1922	JAN	FEB	MAR	APR	MAY	JUN	JUL	AUG	SEP	OCT	NOV	DEC
1	N23CE	M20BP	Y20BH	X15BF	F35BC	Y21AL	R22AH	F26AF	Y31AD	R12CS	F13CP	W30CH
2	A36CF	H28CD	T28BJ	B25BG	V16BD	T17AM	X29AJ	V38AG	T18AE	X37AB	V19CR	Y14CL
3	Z32CH	S33CE	G33BK	N35BH	E24BE	G22BC	B34AK	E11AH	G12AF	B27AC	E30CS	T23CM
4	C20CJ	P15CF	L15BL	A16BJ	M21BF	L29BD	N26AL	M31AJ	L37AG	N13AD	M14AB	G36CP
5	K28CK	W25CH	D25BM	Z24BK	H17BG	D34BE	A38AM	H18AK	D27AH	A19AE	H23AC	L32CR
6	F33CL	Y35CJ	J35BP	C21BL	S22BH	J26BF	Z11BC	S12AL	J13AJ	Z30AF	S36AD	D20CS
7	V15CM	T16CK	U16CD	K17BM	P29BJ	U38BG	C31BD	P37AM	U19AK	C14AG	P32AE	J28AB
8	E25CP	G24CL	R24CE	F22BP	W34BK	R11BH	K18BE	W27BC	R30AL	K23AH	W20AF	U33AC
9	M35CR	L21CM	X21CF	V29CD	Y26BL	X31BJ	F12BF	Y13BD	X14AM	F36AJ	Y28AG	R15AD
10	H16CS	D17CD	B17CH	E34CE	T38BM	B18BK	V37BG	T19BE	B23BC	V32AK	T33AH	X25AE
11	S24AB	J22CR	N22CJ	M26CF	G11BP	N12BL	E27BH	G30BF	N36BD	E20AL	G15AG	B35AF
12	P21AC	U29CS	A29CK	H38CH	L31CD	A37BM	M13BJ	L14BG	A32BE	M28AH	L25AK	N16AG
13	W17AD	R34CM	Z34CL	S11CJ	D18CE	Z27BP	H19BK	D23BH	Z20BF	H33BC	D35AL	A24AH
14	Y22AE	X26AC	C26CM	P31CK	J12CF	C13BD	S30BL	J36BJ	C28BG	S15BD	J16AM	Z21AJ
15	T29AF	B38AD	K38CP	W18CL	U37CH	K19CE	P14BM	U32BK	K33BH	P25BE	U24BC	C17AK
16	G34AG	N11AE	F11CR	Y12CM	R27CJ	F30CF	W23BP	R20BL	F15BG	W35BF	R21BD	K22AL
17	L26AH	A31AF	V31CS	T37CP	X13CK	V14CH	Y36CD	X28BM	V25BK	Y16BG	X17BE	F29AM
18	D38AJ	Z18AG	E18AB	G27CR	B19CL	E23CJ	T32CE	B33BP	E35BL	T24BH	B22BF	V34BC
19	J11AK	C12AH	M12AC	L13CS	N38CK	M36CK	G28CF	N15CD	M16BH	G21BJ	N29BG	E26BD
20	U31AL	K37AJ	H37AD	D19AB	A14CP	H32CL	L28CH	A25CE	H24BP	L17BK	A34BH	M38BE
21	R18AM	F27AK	S27AE	J30AC	Z23CR	S20CM	D33CJ	Z35CF	S21CD	D22BL	Z26BJ	H11BF
22	X12BC	V13AL	P13AF	U14AD	C36CS	P28CP	J15CK	C16CH	P17CE	J29BH	C38BK	S31BG
23	B37BD	E19AM	W19AG	R23AE	K32AB	W33CR	U25CL	K24CJ	W22CF	U34BP	K11BL	P18BH
24	N27BE	M30BC	Y30AH	X36AF	F20AC	Y15CS	R35CH	F21CK	Y29CH	R26CD	F31BH	W12BJ
25	A13BF	H14BD	T14AJ	B32AG	V28AD	T25AB	X16CP	V17CL	T34CJ	X38CE	V18BP	Y37BK
26	Z19BG	S23BE	G23AK	N20AH	E33AE	G35AC	B24CR	E22CH	G26CK	B11CF	E12CD	T27BL
27	C30BH	P36BF	L36AL	A28AJ	M15AF	L16AD	N21CS	M29CP	L38CL	N31CH	M37CE	G13BM
28	K14BJ	W32BG	D32AM	Z33AK	H25AG	D24AE	A17AB	H34CR	D11CM	A18CJ	H27CF	L19BM
29	F23BK		J20BC	C15AL	S35AH	J21AF	Z22AC	S26CS	J31CE	Z12CK	S13CH	O30CD
30	V36BL		U28BD	K25AM	P16AJ	U17AG	C29AD	P38AB	U18CR	C37CL	P19CJ	J14CE
31	E32BM		R33BE		W24AK		K34AE	W11AC		K27CH		U23CF

FIRST LETTER = YOUR PHYSICAL CODE, MIDDLE NUMBER = SENSITIVITY CODE, LAST TWO LETTERS = INTELLECTUAL CODE

1923	JAN	FEB	MAR	APR	MAY	JUN	JUL	AUG	SEP	OCT	NOV	DEC
1	R36CH	F28CE	S28BK	J25BH	Z16BE	S17BC	D29AK	Z38AH	S18AF	D37AC	Z19CS	H14CM
2	X32CJ	V33CF	P33BL	U35BJ	C24BF	P22BD	J34AL	C11AJ	P12AG	J27AD	C30AB	S23CP
3	B2OCK	E15CH	H15BM	R16BK	K21BG	W29BE	U26AM	K31AK	N37AH	U13AE	K14AC	P36CR
4	N28CL	M25CJ	Y25BP	X24BL	F17BH	Y34BF	R38BC	F18AL	Y27AJ	R19AF	F23AD	W32CS
5	A33CM	H35CK	T35CD	B21BM	V22BJ	T26BG	X11BD	V12AM	T13AK	X30AG	V36AE	Y20AB
6	Z15CP	S16CL	G16CE	N17BP	E29BK	G38BH	B31BE	E37BC	S19AL	B14AH	E32AF	T28AC
7	C25CR	P24CM	L24CF	A22CD	M34BL	L11BJ	N18BF	M27BD	L30AM	N23AJ	M2OAG	G33AD
8	K35CS	W21CP	D21CH	Z29CE	H26BM	D31BK	A12BG	H13BE	D14BC	A36AK	H28AH	L15AE
9	F16AB	Y17CR	J17CJ	C34CF	S38BP	J18BL	Z37BH	S19BF	J23BD	Z32AL	S33AJ	D25AF
10	V24AC	T22CS	U22CK	K26CH	P11CD	U12BM	C27BJ	P3OBG	U36BE	C20AM	P15AK	J35AG
11	E21AD	G29AB	R29CL	F38CJ	W31CE	R37BP	K13BK	W14BH	R32BF	K28BC	W25AL	U16AH
12	M17AE	L34AC	X34CM	V11CK	Y18CF	X27CD	F19BL	Y23BJ	X2OBG	F33BD	Y35AM	R24AJ
13	H22AF	D26AD	B26CP	E31CL	T12CH	B13CE	V30BM	T36BK	B28BH	V15BE	T16BC	X21AK
14	S29AG	J38AE	N38CR	M18CM	G37CJ	N19CF	E14BP	G32BL	N33BJ	E25BF	G24BD	B17AL
15	P34AH	U11AF	A11CS	H12CP	L27CK	A30CH	M23CD	L20BM	A15BK	M35BG	L21BE	N22AM
16	W26AJ	R31AG	Z31AB	S37CR	D13CL	Z14CJ	H36CE	D28BP	Z25BL	H16BH	D17BF	A29BC
17	Y38AK	X18AH	C18AC	P27CS	J19CM	C23CK	S32CF	J33CD	C35BM	S24BJ	J22BG	Z34BD
18	T11AL	B12AJ	K12AD	W13AB	U3OCP	K36CL	P2OCH	U15CE	K16BP	P21BK	U29BH	C26BE
19	G31AM	N37AK	F37AE	Y19AC	R14CR	F32CM	W28CJ	R25CF	F24CD	W17BL	R34BJ	K38BF
20	L18BC	A27AL	V27AF	T3OAD	X23CS	V2OCP	Y33CK	X35CH	V21CE	Y22BM	X26BK	F11BG
21	D12BD	Z13AM	E13AG	G14AE	B36AB	E28CR	T15CL	B16CJ	E17CF	T29BP	B38BL	V31BH
22	J37BE	C19BC	M19AH	L23AF	N32AC	M33CS	G25CM	N24CK	M22CH	G34CD	N11BH	E18BJ
23	U27BF	K30BD	H30AJ	D36AG	A20AD	H15AB	L35CP	A21CL	H29CJ	L26CE	A31BP	M12BK
24	R13BG	F14BE	S14AK	J32AH	Z28AE	S25AC	D16CR	Z17CM	S34CK	D38CF	Z18CD	H37BL
25	X19BH	V23BF	P23AL	U2OAJ	C33AF	P35AD	J24CS	C22CP	P26CL	J11CH	C12CE	S27BM
26	B3OBJ	E36BG	W36AM	R28AK	K15AG	W16AE	U21AB	K29CR	W38CM	U31CJ	K37CF	P13BP
27	N14BK	M32BH	Y32BG	X33AL	F25AH	Y24AF	R17AC	F34CS	Y11CP	R18CK	F27CH	W19CD
28	A23BL	H2OBJ	T2OBD	B15AM	V35AJ	T21AG	X22AD	V26AB	T31CR	X12CL	V13CJ	Y3OCE
29	Z36BM		G28BE	N25BC	E16AK	G17AH	B29AE	E38AC	G18CS	B37CM	E19CK	T14CF
30	C32BP		L33BF	A35BD	M24AL	L22AJ	N34AF	M11AD	L12AB	N27CP	M3OCL	G23CH
31	K20CD		D15BG		H21AM		A26AG	H31AE		A13CR		L36CJ

1924	JAN	FEB	MAR	APR	MAY	JUN	JUL	AUG	SEP	OCT	NOV	DEC
1	D32CK	Z33CH	M15BP	L16BL	N21BH	M29BF	G26BC	N31AL	M37AJ	G13AF	N14AD	E36CS
2	J2OCL	C15CJ	H25CD	D24BM	A17BJ	H34BG	L38BD	A18AM	H27AK	L19AG	A23AE	M32AB
3	U28CM	K25CK	S35CE	J21BP	Z22BK	S26BH	D11BE	Z12BC	S13AL	D3OAH	Z36AF	H2OAC
4	R33CP	F35CL	P16CF	U17CD	C29BL	P38BJ	J31BF	C37BD	P19AM	J14AJ	C32AG	S28AD
5	X15CR	V16CM	W24CH	R22CE	K34BM	W11BK	U18BG	K27BE	W3OBC	U23AK	K2OAH	P33AE
6	B25CS	E24CP	Y21CJ	X29CF	F26BP	Y31BL	R12BH	F13BF	Y14BD	R36AL	F28AJ	W15AF
7	N35AB	M21CR	T17CK	B34CH	V38CD	T18BM	X37BJ	V19BG	T23BE	X32AM	V33AK	Y25AG
8	A16AC	H17CS	G22CL	N26CJ	E11CE	G12BP	B27BK	E3OBH	G36BF	B2OBC	E15AL	T35AH
9	Z24AD	S22AB	L29CM	A38CK	M31CF	L37CD	N13BL	M14BJ	L32BG	N28BD	M25AM	G16AJ
10	C21AE	P29AC	D34CP	Z11CL	H18CH	D27CE	A19BM	H23BK	D2OBH	A33BE	H35BC	L24AK
11	K17AF	W34AD	J26CR	C31CM	S12CJ	J13CF	Z3OBP	S36BL	J28BJ	Z15BF	S16BD	D21AL
12	F22AG	Y26AE	U38CS	K18CP	P37CK	U19CH	C14CD	P32BM	U33BK	C25BG	P24BE	J17AM
13	V29AH	T38AF	R11AB	F12CR	W27CL	R3OCJ	K23CE	W2OBP	R15BL	K35BH	W21BF	U22BC
14	E34AJ	G11AG	X31AC	V37CS	Y13CM	X14CK	F36CF	Y28BD	X25BM	F16BJ	Y17BG	R29BD
15	M26AK	L31AH	B18AD	E27AB	T19CP	B23CL	V32CH	T33CE	B35BP	V24BK	T22BH	X34BE
16	H38AL	D18AJ	N12AE	M13AC	G3OCR	N36CM	E2OCJ	G15CF	N16CD	E21BL	G29BJ	B26BF
17	S11AM	J12AK	A37AF	H19AD	L14CS	A32CP	M28CK	L25CH	A24CE	M17BM	L34BK	N38BG
18	P31BC	U37AL	Z27AG	S3OAE	D23AB	Z2OCR	H33CL	D35CJ	Z21CF	H22BP	D26BL	A11BH
19	W18BD	R27AM	C13AH	P14AF	J36AC	C28CS	S15CM	J16CK	C17CH	S29CD	J38BH	Z31BJ
20	Y12BE	X13BC	K19AJ	W23AG	U32AD	K33AB	P25CP	U24CL	K22CJ	P34CE	U11BP	C18BK
21	T37BF	B19BD	F3OAK	X36AH	R2OAE	F15AC	W35CR	R21CM	F29CK	W26CF	R31CD	K12BL
22	G27BG	N3OBE	V14AL	T32AJ	X28AF	V25AD	Y16CS	X17CP	V34CL	Y38CH	X18CE	F37BM
23	L13BH	A14BF	E23AM	G2OAK	B33AG	E35AE	T24AB	B22CR	E26CM	T11CJ	B12CF	V27BP
24	D19BJ	Z23BG	M36BC	L28AL	N15AH	M16AF	G21AC	N29CS	M38CP	G31CK	N37CH	E13CD
25	J30BK	C36BM	H32BD	D33AM	A25AJ	H24AG	L17AD	A34AB	H11CR	L18CL	A27CJ	M19CE
26	U14BL	K32BJ	S2OBE	J15BC	Z35AK	S21AH	D22AE	Z26AC	S31CS	D12CM	Z13CK	H3OCF
27	R23BM	F2OBK	P28BF	U25BD	C16AL	P17AJ	J29AF	C38AD	P18AB	J37CP	C19CL	S14CH
28	X36BP	V28BL	W33BG	R35BE	K24AM	W22AK	U34AG	K11AE	W12AC	U27CR	K3OCM	P23CJ
29	B32CD	E33BM	Y15BH	X16BF	F21BC	Y29AL	R26AH	F31AF	Y37AD	R13CS	F14CP	W36CK
30	N2OCE		T25BJ	B24BG	V17BD	T34AM	X38AJ	V18AG	T27AE	X19AB	V23CR	Y32CL
31	A28CF		G35BK		E22BE		B11AK	E12AH		B3OAC		T2OCM

1925	JAN	FEB	MAR	APR	MAY	JUN	JUL	AUG	SEP	OCT	NOV	DEC
1	G28CP	N25CL	F25CE	Y24BP	R17BK	F34BH	W38BE	R18BC	F27AL	W19AH	R23AF	K32AC
2	L33CR	A35CM	W35CF	T21CD	X22BL	V26BJ	Y11BF	X12BD	V13AM	Y3OAJ	X36AG	F2OAD
3	D15CS	Z16CP	E16CH	G17CE	B29BM	E38BK	T31BG	B37BE	E19BC	T14AK	B32AH	V28AE
4	J25AB	C24CR	M24CJ	L22CF	N34BP	M11BL	G18BH	N27BF	M3OBD	G23AL	N2OAJ	E33AF
5	U35AC	K21CS	H21CK	D29CH	A26CD	H31BM	L12BJ	A13BG	H14BE	L36AM	A28AK	M15AG
6	R16AD	F17AB	S17CL	J34CJ	Z38CE	S18BP	D37BK	Z19BH	S23BF	D32BC	Z33AL	H25AH
7	X24AE	V22AC	P22CM	U26CK	C11CF	P12BD	J27BJ	C3OBJ	P36BG	J2OBD	C15AM	S35AJ
8	B21AF	E29AD	W29CP	R38CL	K31CH	W37CE	U13BM	K14BK	W32BH	U28BE	K25BC	P16AK
9	N17AG	M34AE	Y34CR	X11CM	F18CJ	Y27CF	R19BP	F23BL	Y2OBJ	R33BF	F35BD	M24AL
10	A22AH	H26AF	T26CS	B31CP	V12CK	T13CH	X3OCD	V36BM	T28BK	X15BG	V16BE	Y21AM
11	Z29AJ	S38AG	G38AB	N18CR	E37CL	G19CJ	B14CE	E32BP	G33BL	B25BH	E24BF	T17BC
12	C34AK	P11AH	L11AC	A12CS	M27CM	L30CK	N23CD	M2OCO	L15BM	N35BJ	M21BG	G22BD
13	K26AL	W31AJ	D31AD	Z37AB	H13CP	D14CL	A36CH	H28CE	D25BP	A16BK	H17BH	L29BE
14	F38AM	Y18AK	J18AE	C27AD	S19CR	J23CF	Z32CJ	S33CF	J35CD	Z24BL	S22BJ	D34BF
15	V11BC	T12AL	U12AF	K13AD	P3OCS	U36CP	C20CK	P15CH	U16CE	C21BM	P29BK	J26BG
16	E31BD	G37AM	R37AG	F19AE	W14AB	R32CR	K26CL	W25CJ	R24CF	K17BP	W34BL	U38BH
17	M18BE	L27BC	X27AH	V30AF	Y23AC	X20CS	F33CM	Y35CK	X21CH	F22CD	Y26BM	R11BJ
18	H12BF	D13BD	B13AJ	E14AG	T36AD	B28AB	V15CP	T16CL	B17CJ	V29CE	T38BP	X31BK
19	S37BG	J19BE	N19AK	M23AH	G32AE	N33AC	E25CR	G24CM	N22CH	E34CF	G11CD	B18BL
20	P27BH	U3OBF	A3OAL	H36AJ	L2OAF	A15AD	M35CS	L21CP	A29CL	M26CH	L31CE	N12BM
21	W13BJ	R14BG	Z14AM	S32AK	D28AG	Z25AE	H16AB	D17CR	Z34CK	H38CJ	D18CF	A37BP
22	Y19BK	X23BH	C23BC	P2OAL	J33AH	C35AF	S24AC	J22CS	C26CM	S11CK	J12CH	Z27CD
23	T3OBL	B36BJ	K36BD	W28AM	U15AJ	K16AG	P21AD	U29CS	K38CP	P31CL	U37CJ	C13CE
24	G14BM	N32BK	F32BE	Y33BC	R25AK	F24AH	W17AE	R34AC	F11CS	W18CM	R27CK	K19CF
25	L23BP	A2OBL	V2OBF	T15BD	X35AL	V21AJ	Y22AF	X26AD	V31AB	Y12CP	X13CL	F3OCH
26	D36CD	Z28BM	E28BG	G25BE	B16AM	E17AK	T29AG	B38AE	E18AC	T37CR	B19CH	V14CJ
27	J32CE	C33BP	M33BH	L35BF	N24BC	M22AL	G34AH	N11AF	M12AD	G27CS	N3OCP	E23CK
28	U2OCF	K15CD	H15BJ	D16BG	A21BD	H29AM	L26AJ	A31AG	H37AE	L13AB	A14CR	M36CL
29	R28CH		S25BK	J24BH	Z17BE	S34BC	D38AK	Z18AH	S27AF	D19AG	Z23CS	H32CM
30	X33CJ		P35BL	U21BJ	C22BF	P26BD	J11AL	C12AJ	P13AG	J3OAD	C36AB	S2OCP
31	B15CK		W16BH		K29BG		U31AM	K37AK		U14AE		P28CR

FIRST LETTER = YOUR PHYSICAL CODE, MIDDLE NUMBER = SENSITIVITY CODE, LAST TWO LETTERS = INTELLECTUAL CODE

1926	JAN	FEB	MAR	APR	MAY	JUN	JUL	AUG	SEP	OCT	NOV	DEC
1	H33CS	R35CP	Z35CH	S21CE	D22BM	Z26BK	H11BG	D12BE	Z13BC	H30AK	D36AH	A20AE
2	Y15AB	X16CR	C16CJ	P17CF	J29BP	C38BL	S31BH	J37BF	C19BD	S14AL	J32AJ	Z28AF
3	T25AC	B24CK	K24CM	W22CH	U34CD	K11BM	P18BJ	U27BG	K30BE	P23AM	U20AK	C33AG
4	G35AD	N21AB	F21CL	Y29CJ	R26CE	F31BP	W12BK	R13BH	F14BF	W36BC	R28AL	K15AH
5	L16AE	A17AC	V17CM	T34CK	X38CF	V18CD	Y37BL	X19BJ	V23BG	Y32BD	X33AM	F25AJ
6	D24AF	Z22AD	E22CP	G26CL	B11CH	E12CE	T27BM	B30BK	E36BH	T20BE	B15BC	V35AK
7	J21AG	C29AE	M29CR	L36CM	N31CJ	M37CF	G13BP	N14BL	M32BJ	G28BF	N25BD	E16AL
8	U17AH	K34AF	H34CS	D11CP	A18CK	H27CH	L19CD	A23BM	H20BK	L33BG	A35BE	M24AM
9	R22AJ	F26AG	S26AB	J31CR	Z12CL	S13CJ	D30CE	Z36BP	S28BL	D15BH	Z16BF	H21BC
10	X29AK	V38AH	P38AC	U18CS	C37CM	P19CK	J14CF	C32CD	P33BM	J25BJ	C24BG	S17BD
11	B34AL	E11AJ	H11AD	R12AB	K27CP	W30CL	U23CH	K20CE	W15BP	U35BK	K21BH	P22BE
12	N26AM	M31AK	Y31AE	X37AC	F13CR	Y14CM	R36CJ	F28CF	Y25CD	R16BL	F17BJ	W29BF
13	A38BC	H18AL	T18AF	B27AD	V19CS	T23CP	X32CK	V33CH	T35CE	X24BM	V22BK	Y34BG
14	Z11BD	S12AM	G12AG	N13AE	E30AB	G36CR	B20CL	E15CJ	G16CF	B21BP	E29BL	T26BH
15	C31BE	P37BC	L37AH	A19AF	M14AC	L32CS	N28CM	M25CK	L24CH	N17CD	M34BM	G38BJ
16	K18BF	W27BD	D27AJ	Z30AG	H23AD	D20AB	A33CP	H35CL	D21CJ	A22CE	H26BP	L11BK
17	F12BG	Y13BE	J13AK	C14AH	S36AE	J28AC	Z15CR	S16CM	J17CK	Z29CF	S38CD	D31BL
18	V37BH	T19BF	U19AL	K23AJ	P32AF	U33AD	C25CS	P24CP	U22CL	C34CH	P11CE	J18BM
19	E27BJ	G30BG	R30AM	F36AK	W20AG	R15AE	K35AB	W21CR	R29CM	K26CJ	W31CF	U12BP
20	M13BK	L14BH	X14BC	V32AL	Y28AH	X25AF	F16AC	Y17CS	X34CP	F38CK	Y18CH	R37CD
21	H19BL	D23BJ	B23BD	E20AM	T33AJ	B35AG	V24AD	T22AB	B26CR	V11CL	T12CJ	X27CE
22	S30BH	J36BK	N36BE	M28BC	J25AK	N16AH	E21AE	G29AC	N38CS	E31CH	G37CK	B13CF
23	P14BP	U32BL	A32BF	H33BD	L25AL	A24AJ	M17AF	L34AD	A11AB	M18CP	L27CL	N19CH
24	W23CD	R20BM	Z20BG	S15BE	D35AM	Z21AK	H22AG	D26AE	Z31AC	H12CR	D13CM	A30CJ
25	Y36CE	X28BP	C28BH	P25BF	J16BC	C17AL	S29AH	J38AF	C18AD	S37CS	J19CP	Z14CK
26	T32CF	B33CD	K33BJ	W35BG	U24BD	K22AM	P34AJ	U11AG	K12AE	P27AB	U30CR	C23CL
27	G20CH	N15CE	F15BK	Y16BH	R21BE	F29BC	M26AK	R31AH	E37AF	W13AC	R14CS	K36CM
28	L28CJ	A25CF	V25BL	T24BJ	X17BF	V34BD	Y38AL	X18AJ	V27AG	Y19AD	X23AB	F32CP
29	D33CK		E35BM	G21BK	B22BG	E26BE	T11AM	B12AK	E13AH	T30AE	B36AC	V20CR
30	J15CL		M16BP	L17BL	N29BH	M38BF	G31BC	N37AL	M19AJ	G14AF	N32AD	E28CS
31	U25CM		H24CD		A34BJ		L18BD	A27AM		L23AG		M33AB

1927	JAN	FEB	MAR	APR	MAY	JUN	JUL	AUG	SEP	OCT	NOV	DEC
1	H15AC	D16CS	B16CK	E17CH	T29CD	B38BM	V31BJ	T37BG	B19BE	V14AM	T32AK	X28AG
2	S25AD	J24AB	N24CL	M22CJ	G34CE	N11BP	E18BK	G27BH	N30BF	E23BC	G20AL	B33AH
3	P35AE	U21AC	A21CM	H29CK	L26CF	A31CD	M12BL	L13BJ	A14BG	M36BD	L28AM	N15AJ
4	W16AF	R17AD	Z17CP	S34CL	D38CH	Z18CE	H37BM	D19BK	Z23BH	H32BE	D33BC	A25AK
5	Y24AG	X22AE	C22CR	P26CH	J11CJ	C12CF	S27BP	J30BL	C36BJ	S20BF	J15BD	Z35AL
6	T21AH	B29AF	K29CS	W38CP	U31CK	K37CH	P13CD	U14BM	K32BK	P28BG	U25BE	C16AM
7	G17AJ	N34AG	F34AB	Y11CR	R18CL	F27CJ	W19CE	R23BP	F20BL	W33BH	R35BF	K24BC
8	L22AK	A26AH	V26AC	T31CS	X12CM	V13CK	Y30CF	X36CD	V28BM	Y15BJ	X16BG	F21BD
9	D29AL	Z38AJ	E38AD	G18AB	B37CP	E19CL	T14CH	B32CE	E33BP	T25BK	B24BH	V17BE
10	J34AM	C11AK	M11AE	L12AC	N27CR	M30CM	G23CJ	N20CE	M15CD	G35BL	N21BJ	E22BF
11	U26BC	K31AL	H31AF	D37AD	A13CS	H14CP	L36CK	A28CH	H25CE	L16BM	A17BK	M29BG
12	R38BD	F18AM	S18AG	J27AE	Z19AB	S23CR	D32CL	Z33CJ	S35CF	D24BP	Z22BL	H34BH
13	X11BE	V12BC	P12AH	U13AF	C30AC	P36CS	J20CM	C15CK	P16CH	J21CD	C29BM	S26BJ
14	B31BF	E37BD	M37AJ	R19AG	K14AD	M32AB	U28CP	K25CL	M24CJ	U17CE	K34BP	P38BK
15	N18BG	M27BE	Y27AK	X30AH	F23AE	Y20AC	R33CR	F35CM	Y21CK	R22CF	F26CD	W11BL
16	A12BH	H13BF	T13AL	B14AJ	V36AF	T28AD	X15CS	V16CP	T17CL	X29CH	V38CE	Y31BM
17	Z37BJ	S19BG	G19AM	N23AK	E32AG	G33AE	B25AB	E24CR	G22CM	B34CJ	E11CF	T18BP
18	C27BK	P30BH	L30BC	A36AL	M20AH	L15AF	N35AC	M21CS	L29CP	N26CK	M31CH	G12CD
19	K13BL	W14BJ	D14BD	Z32AM	H28AJ	D25AG	A16AD	H17AB	D34CR	A38CL	H18CJ	L37CE
20	F19BM	Y23BK	J23BE	C20BC	S33AK	J35AH	Z24AE	S22AC	J26CS	Z11CH	S12CK	D27CF
21	V30BP	T36BL	U36BF	K28BD	P15AL	U16AJ	C21AF	P29AD	U38AB	C31CP	P37CL	J13CH
22	E14CD	G32BM	R32BG	F33BE	W25AM	R24AK	K17AG	W34AE	R11AC	K18CR	W27CM	U19CJ
23	M23CE	L20BP	X20BH	V15BF	Y35BC	X21AL	F22AH	Y26AF	X31AD	F12CS	Y13CP	R30CK
24	H36CF	D28CD	B28BJ	E25BG	T16BD	B17AM	V29AJ	T38AG	B18AE	V37AB	T19CR	X14CL
25	S32CH	J33CE	N33BK	M35BH	G24BE	N22BC	E34AK	G11AH	N12AF	E27AC	G30CS	B23CM
26	P20CJ	U15CF	A15BL	H16BJ	Z21AD	A29BP	M26AL	Z31AJ	A37AG	M13AD	L14AB	N36CP
27	W28CK	R25CH	Z25BM	S24BK	D17BG	Z34BE	H38AM	D18AK	Z27AH	H19AE	D23AC	A32CR
28	Y33CL	X35CJ	C35BP	P21BL	J22BH	C26BF	S11BC	J12AL	C13AJ	S30AF	J36AD	Z20CS
29	T15CM		K16CD	M17BM	U29BJ	K38BG	P31BD	U37AM	K19AK	P14AG	U32AE	C28AB
30	G25CP		F24CE	Y22BP	R34BK	F11BH	W18BE	R27BC	F30AL	W23AH	R20AF	K33AC
31	L35CR		V21CF		X26BL		Y12BF	X13BD		Y36AJ		F15AD

1928	JAN	FEB	MAR	APR	MAY	JUN	JUL	AUG	SEP	OCT	NOV	DEC
1	V25AE	T24AC	R21CP	F29CL	W26CH	R31CE	K12BM	W13BK	R14BH	K36BE	W28BC	U15AK
2	E35AF	G21AD	X17CR	V34CM	Y38CJ	X18CF	F37BP	Y19BL	X23BJ	F32BF	Y33BD	R25AL
3	M16AG	L17AE	B22CS	E26CP	T11CK	B12CH	V27CD	T30BM	B36BK	V20BG	T15BE	X35AM
4	H24AH	D22AF	N29AB	M38CR	G31CL	N37CJ	E13CE	G14BP	N32BL	E28BH	G25BF	B16BC
5	S21AJ	J29AG	A34AC	H11CS	L18CM	A27CK	M19CF	L23CD	A20BM	M33BJ	L35BG	N24BD
6	P17AK	U34AH	Z26AD	S31AB	D12CP	Z13CL	H30CH	D36CE	Z28BP	H15BK	D16BH	A21BE
7	W22AL	R26AJ	C38AE	P18AC	J37CJ	C19CM	S14CJ	J32CF	C33CD	S25BL	J24BJ	Z17BF
8	Y29AM	X38AK	K11AF	W12AD	U27CS	K30CP	P23CH	U20CH	K15CE	P35BM	U21BK	C22BG
9	T34BC	B11AL	F31AG	Y37AE	R13AH	F14AC	W36CL	R28CJ	F25CF	W16BP	R17BL	K29BH
10	G26BD	N31AM	V18AH	T27AF	X19AC	V23CS	Y32CM	X33CK	V35CH	Y24CD	X22BM	F34BJ
11	L38BE	A18BG	E12AJ	G13AG	B30AD	E36AB	T20CP	B15CL	E16CJ	T21CH	B29BP	V26BK
12	D11BF	Z12BD	M37AL	L19AH	N14AE	M32AC	G28CR	N25CM	M24CK	G17CF	N34CD	E38BL
13	J31BG	C37BE	H27AL	D30AJ	A23AF	H20AD	L33CS	A35CP	H21CL	L22CH	A26CE	M11BM
14	U18BH	K27BF	S13AM	J14AK	Z36AG	S28AE	D15AB	Z16CR	S17CM	D29CJ	Z38CF	H31BP
15	R12BJ	F13BG	P19BC	U23AL	C32AH	P33AF	J25AC	C24CS	P22CP	J34CK	C11CH	S18CD
16	X37BK	V19BH	W30BD	R36AM	K20AJ	W15AG	U35AD	K21AB	W29CR	U26CL	K31CJ	P12CE
17	B27BL	E38BJ	Y14BE	X32BC	F28AK	Y25AH	R16AC	F17AC	Y34CS	R38CM	F16CK	M37CF
18	N13BM	M14BK	T23BF	B20BD	V33AL	T35AJ	X24AF	V22AD	T26AB	X11CP	V12CL	Y27CH
19	A19BP	H23BL	G36BG	N28BE	E15AM	G16AH	B21AE	E29AE	G38AC	B31CR	E37CM	T13CJ
20	Z30CD	S36BM	L32BH	A33BF	M25BC	L24AL	N17AH	M34AF	L11AD	N18CS	M27CP	G19CK
21	C14CE	P32BP	D20BJ	Z15BG	H35BD	D21AM	A22AJ	H26AG	D31AE	A12AB	H13CR	L30CL
22	K23CF	W20CD	J28BK	C25BH	S16BE	J17BC	Z29AK	S38AH	J18AF	Z37AC	S19CS	D14CM
23	F36CH	Y28CE	U33BL	K35BJ	P24BF	U22BD	C34AL	P11AJ	U12AG	C27AD	P30AB	J23CR
24	V32CJ	T33CF	R15BM	F16BP	W21BG	R29BE	K26AM	W31AK	R37AH	K13AE	W14AC	U36CR
25	E20CK	G15CH	X25BP	V24BL	Y17BH	X34BF	F38BC	Y18AL	X27AJ	F19AF	Y23AD	R32CS
26	M28CL	L25CJ	B35CD	E21BM	T22BJ	B26BG	V11BD	T12AM	B13AK	V30AG	T36AE	X20AB
27	H33CM	D35CK	N16CE	M17BP	W29BK	N38BH	E31BE	G37BC	N19AL	E14AH	G32AF	B28AC
28	S15CP	J16CL	A24CF	H22CD	L34BL	A11BJ	M18BF	L27BD	A30AM	M23AJ	L20AG	N33AD
29	P25CR	U24CM	Z21CH	S29CE	D26BM	Z31BK	H12BG	D13BE	Z14BC	H36AK	D28AH	A15AM
30	W35CS		C17CJ	P34CF	J38BP	C18BL	S37BH	J19BF	C23BD	S32AL	J33AJ	Z25AF
31	Y16AB		K22CK		U11CD		P27BJ	U30BG		P20AH		C35AG

FIRST LETTER = YOUR PHYSICAL CODE, MIDDLE NUMBER = SENSITIVITY CODE, LAST TWO LETTERS = INTELLECTUAL CODE

1929	JAN	FEB	MAR	APR	MAY	JUN	JUL	AUG	SEP	OCT	NOV	DEC
1	K16AH	H17AF	D17CS	Z34CP	H38CK	D18CH	A37CD	H19BM	D23BK	A32BG	H33BE	L25AM
2	F24AJ	Y22AG	J22AB	C26CR	S11CL	J12CJ	Z27CE	S30BP	J35BL	Z20BH	S15BF	D35BC
3	V21AK	T29AH	U29AC	K38CS	P31CM	U37CK	C13CF	P14CD	U32BM	C28BJ	P25BG	J16BD
4	E17AL	G34AJ	R34AD	F11AB	W18CP	R27CL	K19CH	W23CE	R20BP	K33BK	W35BH	U24BE
5	M22AM	L26AK	X26AE	V31AC	Y12CR	X13CH	F30CJ	Y36CF	K28CD	F15BL	Y16BJ	R21BF
6	H29BC	D38AL	B38AF	E18AD	T37CS	B19CP	V14CK	T32CH	B33CE	V25BM	T24BK	X17BG
7	S34BD	J11AM	N11AG	M12AE	G27AB	N30CR	E23CL	G20CJ	N15CF	E35BP	G21BL	B22BH
8	P26BE	U31BC	A31AH	H37AF	L13AC	A14CS	M36CH	L28CK	A25CH	M16CD	L17BM	N29BJ
9	W38BF	R18BD	Z18AJ	S27AG	D19AD	Z23AB	H32CP	D33CL	Z35CJ	H24CE	D22BP	A34BK
10	Y11BG	X12BE	C12AK	P13AH	J30AE	C36AC	S20CR	J15CM	C16CK	S21CF	J29CD	Z26BL
11	T31BH	B37BF	K37AL	W19AJ	U14AF	K32AD	P28CS	U25CP	K24CL	P17CH	U34CE	C38BM
12	G18BJ	N27BG	F27AM	Y30AK	R23AG	F20AE	W33AB	R35CR	F21CM	W22CJ	R26CF	K11BP
13	L12BK	A13BH	V13BC	T14AL	X36AH	V28AF	Y15AC	X16CS	V17CP	Y29CK	X38CH	F31CD
14	D37BL	Z19BJ	E19BD	G23AM	B32AJ	E33AG	T25AD	B24AB	E22CR	T34CL	B11CJ	V18CE
15	J27BM	C30BK	M30BE	L36BC	N20AK	M15AH	G35AE	N21AC	M29CS	G26CM	N31CK	E12CF
16	U13BP	K14BL	H14BF	D32BD	A28AL	H25AJ	L16AF	A17AD	H34AB	L38CP	A18CL	M37CH
17	R19CD	F23BM	S23BG	J20BE	Z33AM	S35AK	D24AG	Z22AE	S26AC	D11CR	Z12CM	H27CJ
18	X30CE	V36BP	P36BH	U28BF	C15BC	P16AL	J21AH	C29AF	P38AD	J31CS	C37CP	S13CK
19	B14CF	E32CD	W32BJ	R33BG	K25BD	W24AM	U17AJ	K34AG	W11AE	U18AB	K27CR	P19CL
20	N23CH	M20CE	Y20BK	X15BH	F35BE	Y21BC	R22AK	F26AH	Y31AF	R12AC	F13CS	W30CH
21	A36CJ	H28CF	T28BL	B25BJ	V16BF	T17BD	X29AL	V38AJ	T18AG	X37AD	V19AB	Y14CP
22	Z32CK	S33CH	G33BM	N35BK	E24BG	G22BE	B34AM	E11AK	G12AH	B27AE	E30AC	T23CR
23	C20CL	P15CJ	L15BP	A16BL	M21BH	L29BF	N26BC	M31AL	L37AJ	N13AF	M14AD	G36CS
24	K28CM	W25CK	D25CD	Z24BM	H17BJ	D34BG	A38BD	H18AM	D27AK	A19AG	H23AE	L32AB
25	F33CP	Y35CL	J35CE	C21BP	S22BK	J26BH	Z11BE	S12BC	J13AL	Z30AH	S36AF	D20AC
26	V15CR	T16CM	U16CF	K17CD	P29BL	U38BJ	C31BF	P37BD	U19AM	C14AJ	P32AG	J28AD
27	E25CS	G24CP	R24CH	F22CE	W34BM	R11BK	K18BG	W27BE	R30BC	K23AK	W20AH	U33AE
28	M35AB	L21CR	X21CJ	V29CF	Y26BP	X31BL	F12BH	Y13BF	X14BD	F36AL	Y28AJ	R15AF
29	H16AC		B17CK	E34CH	T38CD	B18BM	V37BJ	T19BG	B23BE	V32AM	T33AK	X25AG
30	S24AD		N22CL	M26CJ	G11CE	N12BP	E27BK	G30BH	N36BF	E20BC	G15AL	B35AH
31	P21AE		A29CM		L31CF		M13BL	L14BJ		M28BD		N16AJ

1930	JAN	FEB	MAR	APR	MAY	JUN	JUL	AUG	SEP	OCT	NOV	DEC
1	A24AK	H22AM	T22AC	B26CS	V11CM	T12CK	X27CF	V30CD	T36BM	X20BJ	V15BG	Y35BD
2	Z21AL	S29AJ	G29AD	N38AB	E31CP	G37CL	B13CH	E14CE	G32BP	B28BK	E25BH	T16BE
3	C17AM	P34AK	L34AE	A11AC	M18CR	L27CM	N19CJ	M23CF	L20CD	N33BL	M35BJ	G24BF
4	K22BC	W26AL	D26AF	Z31AD	H12CS	D13CP	A30CK	H36CH	D28CE	A15BM	H16BK	L21BG
5	F29BD	Y38AM	J38AG	C18AE	S37AB	J19CR	Z14CL	S32CJ	J33CF	Z25BP	S24BL	D17BH
6	V34BE	T11BC	U11AH	K12AF	P27AC	U30CS	C23CM	P20CK	U15CH	C35CD	P21BM	J22BJ
7	E26BF	G31BD	R31AJ	F37AG	W13AD	R14AB	K36CP	W28CL	R25CJ	K16CE	W17BP	U29BK
8	M38BG	L18BE	X18AK	V27AH	Y19AE	X23AC	F32CR	Y33CH	X35CK	F24CF	Y22CD	R34BL
9	H11BH	D12BF	B12AL	E13AJ	T30AF	B36AD	V20CS	T15CP	B16CL	V21CH	T29CE	X26BM
10	S31BJ	J37BG	N37AM	M19AK	G14AG	N32AE	E28AB	G25CR	N24CM	E17CJ	G34CF	B38BP
11	P18BK	U27BH	A27BC	H30AL	L23AH	A20AF	M33AC	L35CS	A21CP	M22CK	L26CH	N11CD
12	W12BL	R13BJ	Z13BD	S14AM	D36AJ	Z28AG	H15AD	D16AB	Z17CR	H29CL	D38CJ	A31CE
13	Y37BM	X19BK	C19BE	P23BC	J32AK	C33AH	S25AE	J24AC	C22CS	S34CM	J11CK	Z18CF
14	I27BP	B30BL	K30BF	W36BD	U20AL	K15AJ	P35AF	U21AD	K29AB	P26CP	U31CL	C12CH
15	G13CD	N14BM	F14BG	Y32BE	R28AM	F25AK	W16AG	R17AE	F34AC	W38CR	R18CM	K37CJ
16	L19CE	A23BP	V23BH	T20BF	X33BC	V35AL	Y24AH	X22AF	V26AD	Y11CS	X12CP	F27CK
17	D30CF	Z36CD	E36BJ	G28BG	B15BD	E16AM	T21AJ	B29AG	E38AE	T31AB	B37CR	V13CL
18	J14CH	C32CE	M32BK	L33BH	N25BE	M24BC	G17AK	N34AH	M11AF	G18AC	N27CS	E19CM
19	U23CJ	K20CF	H20BL	D15BJ	A35BF	H21BD	L22AL	A26AJ	H31AG	L12AD	A13AB	H30CP
20	R36CK	F28CH	S28BM	J25BK	Z16BG	S17BE	D29AM	Z38AK	S18AH	D37AE	Z19AC	H14CR
21	X32CL	V33CJ	P33BP	U35BL	C24BH	P22BF	J34BC	C11AL	P12AJ	J27AF	C30AD	S23CS
22	B20CM	E15CK	W15CD	R16BM	K21BJ	W29BG	U26BD	K31AM	W37AK	U13AG	K14AE	P36AB
23	N28CP	M25CL	Y25CE	X24BP	F17BK	Y34BH	R38BE	F18BC	Y27AL	R19AH	F23AF	W32AC
24	A33CR	H35CM	T35CF	B21CD	V22BL	T26BJ	X11BF	V12BD	T13AM	X30AJ	V36AG	Y20AD
25	Z15CS	S16CM	G16CH	N17CE	E29BM	G38BK	M34BP	L11BL	G199C	B14AK	E32AH	T28AE
26	C25AB	P24CR	L24CJ	A22CF	M34BP	L11BL	N18BH	M27BF	L30BD	N23AL	M20AJ	G33AF
27	K35AC	W21CS	D21CK	Z29CH	H26CD	D31BM	A12BJ	H13BG	D14BE	A36AM	H28AK	L15AG
28	F16AD	Y17AB	J17CL	C34CJ	S38CE	J18BP	Z37BK	S19BH	J23BF	Z32BC	S33AL	D25AH
29	V24AE		U22CM	K26CK	P11CF	U12CD	C27BL	P30BJ	U35BG	C20BD	P15AM	J35AJ
30	E21AF		R29CP	F38CL	W31CH	R37CE	K13BH	W14BK	R32BH	K28BE	W25BC	U16AK
31	M17AG		X34CR		Y18CJ		F19BP	Y23BL		F33BF		R24AL

1931	JAN	FEB	MAR	APR	MAY	JUN	JUL	AUG	SEP	OCT	NOV	DEC
1	X21AM	V29AK	P29AE	U38AC	C31CR	P37CM	J13CJ	C14CF	P32CD	J28BL	C25BJ	S16BF
2	B17BC	E34AL	W34AF	R11AD	K18CS	W27CP	U19CK	K23CH	W20CE	U33BM	K35BK	P24BG
3	N22BD	M26AM	Y26AG	X31AE	F12AB	Y13CR	R30CL	F36CJ	Y28CF	R15BP	F16BL	W21BH
4	A29BE	H38BC	T38AH	B18AF	V37AC	T19CS	X14CH	V32CK	T33CH	X25CD	V24BM	Y17BJ
5	Z34BF	S11BD	G11AJ	N12AG	E27AD	G30AB	B23CP	E20CL	G15CJ	B35CE	E21BP	T22BK
6	C26BG	P31BE	L31AK	A37AH	M13AE	L14AC	N36CR	M28CM	L25CK	N16CF	M17CD	G29BL
7	K38BH	W18BF	D18AL	Z27AJ	H19AF	D23AD	A32CS	H33CH	D35CL	A24CH	H22CC	L34BM
8	F11BJ	Y12BG	J12AM	C13AK	S30AG	J36AE	Z20AB	S15CR	J16CM	Z21CJ	S29CF	D26BP
9	V31BK	T37BH	U37BC	K19AL	P14AH	U32AF	C28AC	P25CS	U24CP	C17CK	P34CH	J38CD
10	E18BL	G27BJ	R27BD	F30AM	W23AJ	R20AG	K33AD	W35AB	R21CR	K22CL	W26CJ	U11CE
11	M12BM	L13BK	X13BE	V14BC	Y36AK	X28AH	P15AF	Y16AC	X17CS	P18AD	U27AB	R31CL
12	H37BP	D19BL	B19BF	E23BD	T32AL	B33AJ	V25AF	T24AD	B22AB	V34CP	T11CL	X18CH
13	S27CD	J30BM	N30BG	M36BE	G20AM	N15AK	E35AG	G21AE	N29AC	E26CR	G31CM	B12CJ
14	P13CE	U14BP	A14BH	H32BF	L28BC	A25AL	M16AH	L17AF	A34AD	M38CS	L18CP	N37CK
15	W19CF	R23CD	Z23BJ	S20BG	D33BD	Z35AM	H24AJ	D22AG	Z26AE	H11AB	D12CR	A27CL
16	Y30CH	X36CF	C36BK	P28BH	J15BC	C16BC	S21AK	J29AH	C38AF	S31AC	J37CS	Z13CH
17	T14CJ	B32CF	K32BL	W33BJ	U25BF	K24BD	P17AL	U34AJ	K11AG	P18AD	U27AB	C19CP
18	G23CK	N20CH	F20BM	Y15BK	R35BG	F21BE	W22AM	R26AK	F31AH	W14AE	R13AC	K30CR
19	L36CL	A28CJ	V28BP	T25BL	X16BH	V17BF	Y29BC	X38AL	V18AJ	Y37AF	X19AD	F14CS
20	D32CH	Z33CH	E33CD	G35BH	B24BJ	E22BG	T34BD	B11AM	E12AK	T27AG	B30AE	V23AB
21	J20CP	C15CL	M15CE	L16BP	N21BK	M29BH	G26BE	N31BG	M37AL	G13AH	N14AF	E36AC
22	U28CR	K25CM	H25CF	D24CD	A17BL	H34BJ	L38BF	A18BD	H27AM	L19AJ	A23AG	M32AD
23	R33CS	F35CP	S35CH	J21CE	Z22BM	S26BK	D11BG	Z12BE	S13BC	D30AK	Z36AH	H20AE
24	X15AB	V16CR	P16CJ	U17CF	C29BP	P38BL	J31BH	C37BF	P19BD	J14AL	C32AJ	S28AF
25	B25AC	E24CS	W24CK	R22CH	K34CD	W11BM	U18BJ	K27BG	W30BE	U23AH	K20AK	P33AG
26	N35AD	M21AB	Y21CL	X29CJ	F26CE	Y31BD	R12BK	F13BH	Y14BF	R36BC	F28AL	W15AH
27	A16AE	H17AC	T17CM	B34CK	V38CF	T18BD	X37BL	V19BJ	T23BG	X32BD	V33AM	Y25AJ
28	Z24AF	S22AD	G22CP	N26CL	E11CH	G12CE	B27BM	E38BK	G36BH	B20BE	E15BC	T35AK
29	C21AG		L29CR	A38CH	M31CJ	L37CF	N13BP	M14BL	L32BJ	N28BF	M25BD	G16AL
30	K17AH		D34CS	Z11CP	H18CK	D27CH	A19CD	H23BM	D20BK	A33BG	H35BE	L24AM
31	F22AJ		J26AB		S12CL		Z30CE	S36BP		Z15BH		D21BC

FIRST LETTER = YOUR PHYSICAL CODE, MIDDLE NUMBER = SENSITIVITY CODE, LAST TWO LETTERS = INTELLECTUAL CODE

1932	JAN	FEB	MAR	APR	MAY	JUN	JUL	AUG	SEP	OCT	NOV	DEC
1	J17BD	C34AM	H26AH	D31AF	A12AC	H13CS	L30CH	A36CK	H28CH	L15CD	A16BM	M21BJ
2	U22BE	K26BC	S38AJ	J18AG	Z37AD	S19AB	D14CP	Z32CL	S33CJ	D25CE	Z24BP	H17BK
3	R29BF	F38BD	P11AK	U12AH	C27AE	P30AC	J23CR	C20CM	P15CK	J35CF	C21CD	S22BL
4	X34BG	V11BE	W31AL	R37AJ	K13AF	W14AD	U36CS	K28CP	W25CL	U16CH	K17CE	P29BM
5	B26BH	E31BF	Y18AM	X27AK	F19AG	Y23AE	R32AB	F33CR	Y35CM	R24CJ	F22CF	W34BP
6	N38BG	M18BG	T12BC	B13AL	V30AH	T36AF	X20AC	V15CS	T16CP	X21CK	V29CH	Y26CD
7	A11BK	H12BH	G37BD	N19AM	E14AJ	G32AG	B28AD	E25AB	G24CR	B17CL	E34CJ	T38CE
8	Z31BL	S37BJ	L27BE	A30BC	M23AK	L20AH	N33AE	M35AC	L21CS	N22CM	M26CK	G11CF
9	C18BM	P27BK	D13BF	Z14BD	H36AL	D28AJ	A15AF	H16AD	D17AB	A29CP	H38CL	L31CH
10	K12BP	W13BL	J19BG	C23BE	S32AM	J33AK	Z25AG	S24AE	J22AC	Z34CR	S11CH	D18CJ
11	F37CD	Y19BM	U30BH	K36BF	P20BC	U15AL	C35AH	P21AF	U29AD	C26CS	P31CP	J12CK
12	V27CE	T30BP	R14BJ	F32BG	W28BD	R25AM	K16AJ	W17AG	R34AE	K38AB	W18CR	U37CL
13	E13CF	G14CD	X23BK	V20BH	Y33BE	X35BC	F24AK	Y22AH	X26AF	F11AC	Y12CS	R27CM
14	M19CH	L23CE	B36BL	E28BJ	T15BF	B16BD	V21AL	T29AJ	B38AG	V31AD	T37AB	X13CP
15	H30CJ	D36CF	N32BM	M33BK	G25BG	N24BE	E17AM	G34AK	N11AH	E18AE	G27AC	B19CR
16	S14CK	J32CH	A20BP	H15BL	L35BH	A21BF	M22BC	L26AL	A31AJ	M12AF	L13AD	N30CS
17	P23CL	U20CJ	Z28CD	S25BM	D16BJ	Z17BG	H29BD	D38AM	Z18AK	H37AG	D19AE	A14AB
18	W36CM	R28CK	C33CE	P35BP	J24BK	C22BH	S34BE	J11BC	C12AL	S27AH	J30AF	Z23AC
19	Y32CP	X33CL	K15CF	W16CD	U21BL	K29BJ	P26BF	U31BD	K37AM	P13AJ	U14AG	C36AD
20	T20CR	B15CM	F25CH	Y24CE	R17BM	F34BK	W38BG	R18BE	F27BC	W19AK	R23AH	K32AE
21	G28CS	N25CP	V35CJ	T21CF	X22BP	V26BL	Y11BH	X12BF	V13BD	Y30AL	X36AJ	F20AF
22	L33AB	A35CR	E16CK	G17CH	B29CD	E38BM	T31BJ	B37BG	E19BE	T14AM	B32AK	V28AG
23	D15AC	Z16CS	M24CL	L22CJ	N34CE	M11BP	G18BK	N27BH	M30BF	G23BC	N20AL	E33AH
24	J25AD	C24AB	H21CM	D29CK	A26CF	H31CD	L12BL	A13BJ	H14BG	L36BD	A28AM	M15AJ
25	U35AE	K21AC	S17CP	J34CL	Z38CH	S18CE	D37BM	Z19BK	S23BH	D32BE	Z33BC	H25AK
26	R16AF	F17AD	P22CR	U26CM	C11CJ	P12CF	J27BP	C30BL	P36BJ	J20BF	C15BD	S35AL
27	X24AG	V22AE	N29CS	R38CP	K31CK	N37CH	U13CD	K14BM	N32BK	U28BG	K25BE	P16AM
28	B21AH	E29AF	Y34AB	X11CH	F18CL	Y27CJ	R19CE	F23BP	Y20BL	R33BH	F35BF	W24BC
29	N17AJ	M34AG	T26AC	B31CS	V12CM	T13CK	X30CF	V36CD	T28BM	X15BJ	V16BG	Y21BD
30	A22AK		G38AD	N18AB	E37CP	G19CL	B14CH	E32CE	G33BP	B25BK	E24BH	T17BE
31	Z29AL		L11AE		M27CR		N23CJ	M20CF		N35BL		G22BF

1933	JAN	FEB	MAR	APR	MAY	JUN	JUL	AUG	SEP	OCT	NOV	DEC
1	L29BG	A38BE	V38AK	T18AH	X37AE	V19AC	Y14CR	X32CM	V33CK	Y25CF	X24CD	F17BL
2	D34BH	Z11BF	E11AL	G12AJ	B27AF	E30AD	T23CS	B20CP	E15CL	T35CH	B21CE	V22BM
3	J26BJ	C31BG	H31AM	L37AK	N13AG	M14AE	G36AB	N28CR	M25CH	G16CJ	N17CF	E29BP
4	U38BK	K18BM	H18BC	D27AL	A19AH	H23AF	L32AC	A33CS	H35CP	L24CK	A22CH	M34CD
5	R11BL	F12BJ	S12BD	J13AM	Z30AJ	S36AG	D20AD	Z15AB	S16CR	D21CL	Z29CJ	H26CE
6	X31BM	V37BK	P37BE	U19BC	C14AK	P32AH	J28AE	C25AC	P24CS	J17CH	C34CK	S38CF
7	B18BP	E27BL	W27BF	R30BD	K23AL	W20AJ	U33AF	K35AD	W21AB	U22CP	K26CL	P11CH
8	N12CD	M13BM	Y13BG	X14BE	F36AM	Y28AK	R15AG	F16AE	Y17AC	R29CR	F38CM	M31CJ
9	A37CF	H19BP	T19BH	B23BF	V32BC	T33AL	X25AH	V24AF	T22AD	X34CS	V11CP	Y18CK
10	Z27CF	S30CD	G30BJ	N36BG	E20BD	G15AM	B35AJ	E21AG	G29AE	B26AB	E31CR	T12CL
11	C13CH	P14CE	L14BK	A32BH	M28BE	L25BG	N16AK	M17AH	L34AF	N38AC	M18CS	G37CM
12	K19CJ	W23CF	D23BL	Z20BJ	H33BF	D35BD	A24AL	H22AJ	D26AG	A11AD	H12AB	L27CP
13	F30CK	Y36CM	J36BM	C28BK	S15BG	J16BE	Z21AM	S29AK	J38AH	Z31AE	S37AC	D13CR
14	V14CL	T32CJ	U32BP	K33BL	P25BH	U24BF	C17BC	P34AL	U11AJ	C18AF	P27AD	J19CS
15	E23CM	G20CR	R20CD	F15BM	W35BJ	R21BG	K22BD	W26AM	R31AK	K12AG	W13AE	U30AB
16	M36CP	L28CL	X28CE	V25BP	Y16BK	X17BH	F29BE	Y38BC	X18AL	F37AH	X19AF	R14AC
17	H32CR	O33CM	B33CF	E35CD	T24BL	B22BJ	V34BF	T11BD	B12AM	V27AJ	T30AG	X23AD
18	S20CS	J15CP	N15CH	M16CE	G21BM	N29BK	E26BG	G31BE	N37BC	E13AK	G14AH	B36AE
19	P28AB	U25CR	A25CJ	H24CF	L17BP	A34BL	M38BH	L18BF	A27BD	M19AL	L23AJ	N32AF
20	W33AC	R35CS	Z35CK	S21CH	D22CD	Z26BM	H11BJ	D12BG	Z13BE	H30AM	D36AK	A20AG
21	Y15AD	X16AB	C16CL	P17CJ	J29CE	C38BP	S31BK	J37BH	C19BF	S14BC	J32AL	Z28AH
22	T25AE	B24CU	K24CM	W22CK	U34CF	K11CD	P18BL	U27BJ	K30BG	P23BD	U20AM	C33AJ
23	G35AF	N21AD	F21CP	Y29CL	R26CH	F31CE	W12BM	R13BL	F14BJ	W36BE	R28BC	K15AK
24	L16AG	A17AE	V17CR	T34CM	X38CJ	V18CF	Y37BP	X19BL	V23BJ	Y32BF	X33BD	F25AL
25	D24AH	Z22AF	E22CS	G26CP	B11CK	E12CH	T27CD	B30BM	E36BK	T20BG	B15BE	V35AM
26	J21AJ	C29AG	M29AB	L38CM	N31CL	M37CJ	G13CE	N14BP	M32BL	G28BH	N25BF	E16BC
27	U17AK	K34AM	H34AC	D11CS	A18CM	H27CK	L19CF	A23CD	H20BM	L33BJ	A35BG	M24BD
28	R22AL	F26AJ	S26AD	J31AB	Z12CP	S13CL	D30CH	Z36CE	S28BP	D15BK	Z16BH	H21BE
29	X29AM		P38AE	U18AC	C37CR	P19CM	J14CJ	C32CF	P33CD	J25BL	C24BJ	S17BF
30	B34BC		W11AF	R12AD	K27CS	W30CP	U23CK	K20CH	W15CE	U35BM	K21BK	P22BG
31	N26BD		Y31AG		F13AB		R36CL	F28CJ		R16BP		W29BH

1934	JAN	FEB	MAR	APR	MAY	JUN	JUL	AUG	SEP	OCT	NOV	DEC
1	Y34BJ	X11BG	C11AM	P12AK	J27AG	C30AE	S23AB	J20CR	C15CM	S35CJ	J21CF	Z22BP
2	T26BK	B31BH	K31BC	W37AL	U13AH	K14AF	P36AC	U28CS	K25CP	P16CK	U17CH	C29CD
3	G38BL	N18BJ	F18BD	Y27AM	R19AJ	F23AG	W32AD	R33AB	F35CR	W24CL	R22CJ	K34CE
4	L11BM	A12BK	V12BE	T13BC	X30AK	V36AH	Y20AE	X15AC	V16CS	Y21CH	X29CK	F26CF
5	D31BP	Z37BL	E37BF	G19BD	B14AL	E32AJ	T28AF	B25AD	E24AB	T17CP	B34CL	V38CH
6	J18CD	C27BM	W27BG	L30BE	N23AM	W20AK	G33AG	N35AE	W21AC	G22CR	N26CM	E11CJ
7	U12CE	K13BP	H38BH	D14BF	A36BC	H28AL	L15AH	A16AF	H17AD	L29CS	A38CP	M31CH
8	R37CH	F19CD	S19BJ	J23BG	Z32BD	S33AM	D25AJ	Z24AG	S22AE	D34AB	Z11CR	H18CL
9	X27CH	V30CE	P30BK	U36BH	C20BE	P15BC	J35AK	C21AH	P29AF	J26AC	C31CS	S12CM
10	B13CJ	E14CF	W14BL	R32BJ	K28BF	W25BD	U16AL	K17AJ	W34AG	U38AD	K18AB	P37CP
11	N19CK	M23CH	Y23BM	X20BK	F33BG	Y35BE	R24AM	F22AK	Y26AH	R11AE	F12AC	W27CR
12	A30CL	H36CJ	T36BP	B28BL	V15BH	T16BF	X21BC	V29AL	T38AJ	X31AF	V37AD	Y13CS
13	Z14CM	S32CK	G32CD	N33BM	E25BJ	G24BG	B17BD	E34AM	G11AK	B18AG	E27AE	T19AB
14	C23CP	P20CL	L20CE	Z25CD	M35BK	L21BH	N22BE	M26BC	L31AL	N12AH	M13AF	G30AC
15	K36CR	W28CM	D20CF	C35BF	H16BL	D17BJ	A29BF	H38BD	D18AM	A37AJ	H19AG	L14AD
16	F32CS	Y33CP	J33CH	C35CE	S24BM	J22BK	Z34BG	S11BE	J12BC	Z27AK	S30AH	D23AE
17	V20AB	T15CR	U15CJ	K16CF	P21BP	U29BL	C26BH	P31BF	U37BD	C13AL	P14AJ	J36AF
18	E28AC	G25CS	R25CK	F24CH	W17CD	R34BM	K38BK	W18BG	R27BE	K19AM	W23AK	U32AG
19	M33AD	L35AB	X35CL	V21CJ	Y22CE	X26BP	F11BK	Y12BH	X13BF	F30BC	Y36AL	R20AH
20	H15AE	D15AC	B16CM	E17CK	T29CF	B38BK	V31BH	T37BG	B19BF	V14BD	T32AM	X28AJ
21	S25AF	J24AD	N24CP	M22CL	G34CH	N11BE	E18BM	G27BK	N30BH	E23BE	G20BC	B33AK
22	P35AG	U21AE	A21CR	H29CM	L26CJ	A31BF	M12BP	L13BL	A14BJ	M36BF	L28BD	N15AL
23	W16AH	R17AF	Z17CS	S34CP	D38CK	Z18BD	H37BN	D19BM	Z23BK	H32BE	D33BE	A25AM
24	Y24AJ	X22AG	C22AB	P26CR	J11CL	C12BJ	S27CE	J30BP	C36BL	S20BH	J15BF	Z35BC
25	T21AK	B29AH	K29AC	W38CS	U31CM	K37BK	P13CF	U14BD	K32BM	P28BJ	U25BG	C16BD
26	G17AL	N34AJ	F34AD	Y11AB	R18CP	F27CL	M19CH	R23CE	F28BP	W33BK	R35BH	K24BE
27	L22AM	A29AK	E38AF	T31AC	X12CR	V13CM	Y30CJ	X36CO	V26CD	Y15BL	X16BJ	F21BF
28	D29BC	Z38AL	E38AF	G18AD	B37CS	E19CP	T14CK	B32CH	E33CE	T25BM	B24BK	V17BG
29	J34BD		M11AG	L12AE	N27AB	M30CR	G23CL	N20CJ	M15CF	G35BP	N21BL	E22BH
30	U26BE		H31AH	D37AF	A13AC	H14CS	L36CM	A28CK	H25CH	L16CD	A17BM	M29BJ
31	R38BF		S18AJ		Z19AD		D32CP	Z33CL		D24CE		H34BK

FIRST LETTER = YOUR PHYSICAL CODE, MIDDLE NUMBER = SENSITIVITY CODE, LAST TWO LETTERS = INTELLECTUAL CODE

1935	JAN	FEB	MAR	APR	MAY	JUN	JUL	AUG	SEP	OCT	NOV	DEC
1	S26BL	J31BJ	N31BD	M37AM	G13AJ	N14AG	E36AD	G28AB	N25CR	E16CL	G17CJ	B29CE
2	P38BM	U18BK	A18BE	H27BC	L19AK	A23AH	M32AE	L33AC	A35CS	M24CM	L22CK	N34CF
3	W11BP	R12BL	Z12BF	S13BD	D30AL	Z36AJ	H20AF	D15AD	Z16AB	H21CP	D29CL	A26CH
4	Y31CD	X37BM	C37BG	P19BE	J14AM	C32AK	S28AG	J25AE	C24AC	S17CR	J34CM	Z38CJ
5	T18CE	B27BP	K27BH	W30BF	U23BC	K20AL	P33AH	U35AF	K21AD	P22CS	U26CP	C11CK
6	G12CF	N13CD	F13BJ	Y14BG	R36BD	F28AM	W15AJ	R16AG	F17AE	W29AB	R38CR	K31CL
7	L37CH	A19CE	V19BK	T23BH	X32BE	V33BC	Y25AK	X24AH	V22AF	Y34AC	X11CS	F18CM
8	D27CJ	Z30CF	E30BL	G36BJ	B20BF	E15BD	T35AL	B21AJ	E29AG	T26AD	B31AB	V12CP
9	J13CK	C14CH	M14BM	L32BK	N28BG	M25BE	G16AM	N17AK	M34AH	G38AE	N18AC	E37CR
10	U19CL	K23CJ	H23BP	D20BL	A33BH	H35BF	L24BC	A22AL	H26AJ	L11AF	A12AD	M27CS
11	R30CM	F36CK	S36CD	J28BM	Z15BJ	S16BG	D21BD	Z29AM	S38AK	D31AG	Z37AE	H13AB
12	X14CP	V32CL	P32CE	U33BP	C25BK	P24BH	J17BE	C34BC	P11AL	J18AH	C27AF	S19AC
13	B23CR	E20CM	W20CF	R15CD	K35BL	W21BJ	U22BF	K26BD	W31AM	U12AJ	K13AG	P30AD
14	N36CS	M28CP	Y28CH	X25CE	F16BM	Y17BK	R29BG	F38BE	Y18BC	R37AK	F19AH	W14AE
15	A32AB	H33CR	T33CJ	B35CF	V24BP	T22BL	X34BH	V11BF	T12BD	X27AL	V30AJ	Y23AF
16	Z20AC	S15CS	G15CM	N16CH	E21CD	G29BM	B26BJ	E31BG	G37BE	B13AM	E14AK	T36AG
17	C28AD	P25AB	L25CL	A24CJ	M17CE	L34BP	N38BK	M18BH	L27BF	N19BC	M23AL	G32AH
18	K33AE	W35AC	D35CM	Z21CK	H22CF	D26CD	A11BL	H12BJ	D13BG	A30BD	H36AM	L20AJ
19	F15AF	Y16AD	J16CP	C17CL	S29CH	J38CE	Z31BM	S37BK	J19BH	Z14BE	S32BC	D28AK
20	V25AG	T24AE	U24CR	K22CM	P34CJ	U11CF	C18BP	P27BL	U30BJ	C23BF	P20BD	J33AL
21	E35AH	G21AF	R21CS	F29CP	W26CK	R31CH	K12CD	W13BM	R14BK	K36BG	W28BE	U15AM
22	M16AJ	L17AG	X17AB	V34CR	Y38CL	X18CJ	F37CE	Y19BP	X23BL	F32BH	Y33BF	R25BC
23	H24AK	D22AH	B22AC	E26CS	T11CM	B12CK	V27CF	T30CD	B36BM	V20BJ	T15BG	X35BD
24	S21AL	J29AJ	N29AD	M38AB	G31CP	N37CL	E13CH	G14CE	N32BP	E28BK	G25BH	B16BE
25	P17AM	U34AK	A34AE	H11AC	L18CR	A27CM	M19CJ	L23CF	A20CD	M33BL	L35BJ	N24BF
26	W22BC	R26AL	Z26AF	S31AD	D12CS	Z13CP	H30CK	D36CH	Z28CE	H15BM	D16BK	A21BG
27	Y29BD	X38AM	C38AG	P18AE	J37AB	C19CR	S14CL	J32CJ	C33CF	S25BP	J24BL	Z17BH
28	T34BE	B11BC	K11AH	W12AF	U27AC	K30CS	P23CM	U20CK	K15CH	P35CD	U21BM	C22BJ
29	G26BF		F31AJ	Y37AG	R13AD	F14AB	W36CP	R28CL	F25CJ	W16CE	R17BP	K29BK
30	L38BG		V18AK	T27AH	X19AE	V23AC	Y32CR	X33CM	V35CK	Y24CF	X22CD	F34BL
31	D11BH		E12AL		B30AF		T20CS	B15CP		T21CH		V26BM

1936	JAN	FEB	MAR	APR	MAY	JUN	JUL	AUG	SEP	OCT	NOV	DEC
1	E38BP	G18BL	X12BG	V13BE	Y30AM	X36AK	F20AG	Y15AE	X16AC	F21CR	Y29CM	R26CJ
2	M11CD	L12BM	B37BH	E19BF	T14BC	B32AL	V28AH	T25AF	B24AD	V17CS	T34CP	X38CK
3	H31CE	D37BP	N27BJ	M30BG	G23BD	N20AM	E33AJ	G35AG	N21AE	E22AB	G26CR	B11CL
4	S18CF	J27CD	A13BK	H14BH	L36BE	A28BC	M15AK	L16AH	A17AF	M29AC	L38CS	N31CM
5	P12CH	U13CE	Z19BL	S23BJ	D32BF	Z33BD	H25AL	D24AJ	Z22AG	H34AD	D11AB	A18CP
6	W37CJ	R19CF	C30BM	P36BK	J20BG	C15BE	S35AM	J21AK	C29AH	S26AE	J31AC	Z12CR
7	Y27CK	X30CH	K14BP	W32BL	U28BH	K25BF	P16BC	U17AL	K34AJ	P38AF	U18AD	C37CS
8	T13CL	B14CJ	F23CD	Y20BM	R33BJ	F35BG	W24BD	R22AM	F26AK	W11AG	R12AE	K27AB
9	G19CM	N23CK	V36CE	T28BP	X15BK	V16BH	Y21BE	X29BC	V38AL	Y31AH	X37AF	F13AC
10	L30CP	A36CL	E32CF	G33CD	B25BL	E24BJ	T17BF	B34BD	E11AM	T18AJ	B27AG	V19AD
11	D14CR	Z32CM	M20CH	L15CE	N35BM	M21BK	G22BG	N26BE	M31BD	G12AK	N13AH	E30AE
12	J23CS	C20CP	H28CJ	D25CF	A16BP	H17BL	L29BH	A38BF	H18BD	L37AL	A19AJ	M14AF
13	U36AB	K28CR	S33CK	J35CH	Z24CD	S22BM	D34BJ	Z11BG	S12BE	D27AM	Z30AK	H23AG
14	R32AC	F33CS	P15CL	U16CJ	C21CE	P29BP	J26BK	C31BH	P37BF	J13BC	C14AL	S36AH
15	X20AD	V15AB	W25CM	R24CK	K17CF	W34CD	U38BL	K18BJ	W27BG	U19BD	K23AM	P32AJ
16	B28AE	E25AC	Y35CP	X21CL	F22CH	Y26CE	R11BM	F12BK	Y13BH	R39BE	F36BC	W20AK
17	N33AF	M35AD	T16CR	B17CM	V29CJ	T38CF	X31BP	V37BL	T19BJ	X14BF	V32BD	Y28AL
18	A15AG	H16AE	G24CS	N22CP	E34CK	G11CH	B18CD	E27BM	G30BK	B23BG	E20BE	T33AM
19	Z25AH	S24AF	L21AB	A29CR	M26CL	L31CJ	N12CE	M38BP	L14BL	N36BH	M28BF	G15BC
20	C35AJ	P21AG	D17AC	Z34CS	H38CM	D18CK	A37CF	H19CD	D23BM	A32BJ	H33BG	L25BD
21	K16AK	W17AH	J22AD	C26AB	S11CP	J12CL	Z27CH	S30CE	J36BP	Z20BK	S15BH	D35BE
22	F24AL	Y22AJ	U29AE	K38AC	P31CR	U37CM	C13CJ	P14CF	U32CD	C28BL	P25BJ	J16BF
23	V21AM	T29AK	R34AF	F11AD	W18CS	R27CP	K19CK	W23CH	R20CE	K33BM	W35BK	U24BG
24	E17BC	G34AL	X26AG	V31AE	Y12AB	X13CR	F30CL	Y36CJ	X28CF	F15BP	Y16BL	R21BH
25	M22BD	L26AM	B38AH	E18AF	T37AC	B19CS	V14CM	T32CK	B33CH	V25CD	T24BM	X17BJ
26	H29BE	D38BC	N11AJ	M12AG	G27AD	N30AB	E23CP	G20CL	N15CJ	E35CE	G21BP	B22BK
27	S34BF	J11BD	A31AK	H37AH	L13AE	A14AC	M36CR	L28CM	A25CK	M16CF	L17CD	N29BL
28	P26BG	U31BE	Z18AL	S27AJ	D19AF	Z23AD	H32CS	D33CP	Z35CL	H24CH	D22CE	A34BM
29	W38BH	R18BF	C12AM	P13AK	J30AG	C36AE	S20AB	J15CR	C16CM	S21CJ	J29CF	Z26BP
30	Y11BJ		K37BC	W19AL	U14AH	K32AF	P28AC	U25CS	K24CP	P17CK	U34CH	C38CD
31	T31BK		F27BD		R23AJ		W33AD	R35AB		W22CL		K11CE

1937	JAN	FEB	MAR	APR	MAY	JUN	JUL	AUG	SEP	OCT	NOV	DEC
1	F31CF	Y37CD	J37BJ	C19BG	S14BD	J32AM	Z28AJ	S25AG	J24AE	Z17AB	S34CR	D38CL
2	V18CH	T27CE	U27BK	K30BH	P23BE	U20BC	C33AK	P35AH	U21AF	C22AC	P26CS	J11CM
3	E12CJ	G13CF	R13BL	F14BJ	W36BF	R28BD	K15AL	W16AJ	R17AG	K29AD	W38AB	U31CP
4	M37CK	L19CH	X19BM	V23BK	Y32BG	X33BE	F25AM	Y24AK	X22AH	F34AE	Y11AC	R18CR
5	H27CL	D30CJ	B30BP	E36BL	T20BH	B15BF	V35BC	T21AL	B29AJ	V26AF	T31AD	X12CS
6	S13CM	J14CK	N14CD	M32BM	G28BJ	N25BG	E16BD	G17AM	N34AK	E38AG	G18AE	B37AB
7	P19CP	U23CL	A23CE	H20BP	L33BK	A35BH	M24BE	L22BC	A26AL	M11AH	L12AF	N27AC
8	W30CR	R36CM	Z36CF	S28BD	D15BL	Z16BJ	H21BF	D29BD	Z38AM	H31AJ	D37AG	A13AD
9	Y14CS	X32CP	C32CH	P33CE	J25BM	C24BK	S17BG	J34BE	C11BC	S18AK	J27AH	Z19AE
10	T23AB	B20CR	K20CJ	W15CF	U35BP	K21BL	P22BH	U26BF	K31BD	P12AL	U13AJ	C30AF
11	G36AC	N28CS	F28CK	Y25CH	R16CD	F17BM	W29BJ	R38BG	F18BE	W37AK	R19AH	K14AG
12	L32AD	A33AB	V33CL	T35CJ	X24CE	V22BP	Y34BK	X11BH	V12BF	Y27BC	X30AL	F23AH
13	D20AE	Z15CAC	E15CM	G16CK	B21CF	E29CD	T26BL	B31BD	E37BG	T13BD	B14AM	V36AJ
14	J28AF	C25AD	M25CP	L24CL	N17CH	M34CE	G38BM	N18BK	M27BH	G19BE	N23BC	E32AK
15	U33AG	K35AE	H35CR	D21CH	A22CJ	H26CF	L11BP	A12BL	H13BJ	L30BF	A36BD	M20AL
16	R15AH	F16AF	S16CS	J17CP	Z29CK	S38CH	D31CD	Z37BM	S19BK	D14BG	Z32BE	H28AM
17	X25AJ	V24AG	P24AB	U22CR	C34CL	P11CJ	J18CE	C27BP	P30BL	J23BH	C20BF	S33BC
18	B35AK	E21AH	W21AC	R29CS	K26CM	W31CK	U12CF	K13CD	W14BM	U36BJ	K28BG	P15BD
19	N16AL	M17AJ	Y17AD	X34AB	F38CP	Y18CL	R37CH	F19CE	Y23BP	R32BK	F33BH	W25BE
20	A24AM	H22AK	T22AE	B26AC	V11CR	T12CM	X27CJ	V30CF	T36CD	X20BL	V15BJ	Y35BF
21	Z21BC	S29AL	G29AF	N38AD	E31CS	G37CP	B13CK	E14CD	G32CE	B28BM	E25BK	T16BG
22	C17BD	P34AM	L34AG	A11AE	M18AB	L27CR	N19CL	M23CM	L20CF	N33BP	M35BL	G24BH
23	K22BE	W26BC	D26AH	Z31AF	H12AC	D13CS	A30CM	H36CH	D28CD	A15CP	H16BM	L21BJ
24	F29BF	Y38BD	J38AJ	C18AG	S37AD	J19AB	Z14CP	S32CL	J33CJ	Z25CE	S24BP	D17BK
25	V34BG	T11BE	U11AK	K12AH	P27AE	U30AC	C23CR	P20CM	U15CK	C35CF	P21CD	J22BL
26	E26BH	G31BF	R31AL	F37AJ	W13AF	R14AD	K36CS	W28CP	R25CL	K16CH	W17CE	U29BM
27	M38BJ	L18BG	X18AM	W27AK	Y19AG	X23AE	F32AB	Y33CR	X35CM	F24CJ	Y22CF	R34BP
28	H11BK	D12BH	B12BC	E13AL	T30AH	B36AF	V20AC	T15CS	B16CP	V21CK	T29CH	X26CD
29	S31BL		N37BD	M19AM	G14AJ	N32AG	E28AD	G25AB	N24CR	E17CL	G34CJ	B38CE
30	P18BM		A27BE	H30BC	L23AK	A20AH	M33AE	L35AC	A21CS	M22CM	L26CK	N11CF
31	W12BP		Z13BF		D36AL		H15AF	D16AD		H29CP		A31CH

(147)

FIRST LETTER = YOUR PHYSICAL CODE, MIDDLE NUMBER = SENSITIVITY CODE, LAST TWO LETTERS = INTELLECTUAL CODE

1938

	JAN	FEB	MAR	APR	MAY	JUN	JUL	AUG	SEP	OCT	NOV	DEC
1	Z18CJ	S27CF	G27BL	N30BJ	E23BF	G20BD	B33AL	E35AJ	G21AG	B22AD	E26AB	T11CP
2	C12CK	P13CM	L13BM	A14BK	M36BG	L28BE	N15AM	M16AK	L17AH	N29AE	M38AC	G31CR
3	K37CL	W19CJ	D19BP	Z23BL	H32BH	D33BF	A25BC	H24AL	D22AJ	A34AF	H11AD	L18CS
4	F27CM	Y30CK	J30CD	C36BM	S20BJ	J15BG	Z35BD	S21AM	J29AK	Z26AG	S31AE	D12AB
5	V13CP	T14CL	U14CE	K32BP	P28BK	U25BH	C16BE	P17BC	J34AL	C38AH	P18AF	J37AC
6	E19CR	G23CM	R23CF	F20CO	W33BL	R35BJ	K24BF	W22BD	R26AM	K11AJ	W12AG	U27AD
7	M30CS	L36CP	X36CH	V28CE	Y15BM	X16BK	F21BG	Y29BE	X38BC	F31AK	Y37AH	R13AE
8	H14AB	D32CR	B32CJ	E33CF	T25BP	B24BL	V17BH	T34BF	B11BD	V18AL	T27AJ	X19AF
9	S23AC	J20CS	N20CK	M15CH	G35CD	N21BM	E22BJ	G26BG	N31BE	E12AM	G13AK	B30AG
10	P36AD	U28AB	A28CL	H25CJ	L16CE	A17BP	M29BK	L38BH	A18BF	M37BC	L19AL	N14AH
11	W32AE	R33AC	Z33CM	S35CK	D24CF	Z22CD	H34BL	D11BJ	Z12BG	H27BD	D30AM	A23AJ
12	Y20AF	X15AD	C15CP	P16CL	J21CH	C29CE	S26BM	J31BK	C37BH	S13BE	J14BC	Z36AK
13	T28AG	B25AE	K25CR	W24CM	U17CJ	K34CF	P38BP	U18BL	K27BJ	P19BF	U23BD	C32AL
14	G33AH	N35AF	F35CS	Y21CP	R22CK	F26CH	W11CD	R12BM	F13BK	W30BG	R36BE	K20AM
15	L15AJ	A16AG	V16AB	T17CR	X29CL	V38CJ	Y31CE	X37BP	V19BL	Y14BH	X32BF	F28BC
16	D25AK	Z24AH	E24AC	G22CS	B34CM	E11CK	T18CF	B27CD	E30BM	T23BJ	B20BG	V33BD
17	J35AL	C21AJ	M21AD	L29AB	N26CP	M31CL	G12CH	N13CE	M14BP	G36BK	N28BH	E15BE
18	U16AM	K17AK	H17AE	D34AC	A38CR	H18CM	L37CJ	A19CF	H23CD	L32BL	A33BJ	M25BF
19	R24BC	F22AL	S22AF	J26AD	Z11CS	S12CP	D27CK	Z30CH	S36CE	D20BM	Z15BK	H35BG
20	X21BD	V29AM	P29AG	U38AF	C31AB	P37CR	J13CL	C14CJ	P32CF	J28BP	C25BL	S16BH
21	B17BE	E34BC	W34AH	R11AF	K18AC	W27CS	U19CM	K23CK	W20CH	U33CD	K35BM	P24BJ
22	N22BF	M26BD	Y26AJ	X31AG	F12AD	Y13AB	R30CP	F36CL	Y28CJ	R15CE	F16BP	W21BK
23	A29BG	H38BE	T38AK	B18AH	V37AE	T19AC	X14CH	V32CM	T33CK	X25CF	V24CD	Y17BL
24	Z34BH	S11BF	G11AL	N12AJ	E27AF	G30AD	B23CS	E20CP	G15CL	B35CH	E21CE	T22BM
25	C26BJ	P31BG	L31AM	A37AK	M13AG	L14AE	N36AB	M28CR	L25CM	N16CJ	M17CF	G29BP
26	K38BK	W18BH	D18BC	Z27AL	H19AH	D23AF	A32AC	H33CS	D35CP	A24CK	H22CH	L34CD
27	F11BL	Y12BJ	J12BD	C13AM	S30AJ	J36AG	Z20AD	S15AB	J16CR	Z21CL	S29CJ	D26CE
28	V31BM	T37BK	U37BE	K19BC	P14AK	U32AH	C28AE	P25AC	U24CS	C17CM	P34CK	J38CF
29	E18BP		R27BF	F30BD	W23AL	R20AJ	K33AF	W35AD	R21AB	K22CP	W26CL	U11CH
30	M12CO		X13BG	V14BE	Y36AM	X28AK	F15AG	Y16AE	X17AC	F29CR	Y38CM	R31CJ
31	H37CE		B19BH		T32BC		V25AH	T24AF		V34CS		X18CK

1939

	JAN	FEB	MAR	APR	MAY	JUN	JUL	AUG	SEP	OCT	NOV	DEC
1	B12CL	E13CJ	W13BP	R14BL	K36BH	W28BF	U15BC	K16AL	W17AJ	U29AF	K38AD	P31CS
2	N37CM	M19CR	Y19CD	X23BM	Z33BG	Y33BG	R25BD	F24AM	Y22AK	R34AG	F11AE	H18AB
3	A27CP	H30CL	T30CE	B36BP	V20BK	T15BH	X35BE	V21BC	T29AL	X26AH	V31AF	Y12AC
4	Z13CR	S14CM	G14CF	N32CD	E28BL	G25BJ	B16BF	E17BD	G34AM	B38AJ	E18AG	T37AD
5	C19CS	P23CP	L23CH	A20CE	M33BM	L35BK	N24BG	M22BE	L26BC	N11AK	M12AH	G27AE
6	K30AB	M36CR	D36CJ	Z28CF	H15BP	D16BL	A21BH	H29BF	D38BD	A31AL	H37AJ	L13AF
7	F14AC	Y32CS	J32CK	C33CH	S25CD	J24BM	Z17BJ	S34BG	J11BE	Z18AM	S27AK	D19AG
8	V23AD	T20AB	U20CL	K15CJ	P35CE	U21BP	C22BK	P26BH	U31BF	C12BC	P13AL	J30AH
9	E36AE	G28AC	R28CM	V28CE	W16CF	R17CD	K29BL	W38BJ	R18BG	K37BD	W19AM	U14AJ
10	M32AF	L33AD	X33CP	V35CL	Y24CH	X22CE	F34BM	Y11BK	X12BH	F27BE	Y30BC	R23AK
11	H20AG	D15AE	B15CR	E16CM	T21CJ	B29CF	V26BP	T31BL	B37BJ	V13BF	T14BD	X36AL
12	S28AH	J25AF	N25CS	M24CP	G17CK	N34CH	E38CD	G18BM	N27BK	E19BG	G23BE	B32AM
13	P33AJ	U35AG	A35AB	H21CR	L22CL	A26CJ	M11CE	L28BP	A13BL	M30BH	L36BF	N20BC
14	W15AK	R16AH	Z16AC	S17CS	D29CM	Z38CK	N31CF	D37CD	Z19BM	H14BJ	D32BG	A28BD
15	Y25AL	X24AJ	C24AD	P22AB	J34CP	C11CL	S18CH	J27CE	C30BP	S23BK	J20BH	Z33BE
16	T35AM	B21AK	K21AE	W29AC	U26CR	K31CM	P12CJ	U13CF	K14CD	P36BL	U28BJ	C15BF
17	G16BC	N17AL	F17AF	Y34AD	R38CS	F18CP	W37CK	R19CH	F23CE	W32BM	R33BK	K25BG
18	L24BD	A22AM	V22AG	T26AE	X11AB	V12CR	Y27CL	X30CJ	V36CF	Y20BP	X15BL	F35BH
19	D21BE	Z29BC	E29AH	G38AF	B31AC	E37CS	T13CM	B14CK	E32CH	T28CD	B25BM	V16BJ
20	J17BF	C34BD	M34AJ	L11AG	N18AD	M27AB	G19CP	N23CL	M20CJ	G33CE	N35BP	E24BK
21	U22BG	K26BE	H26AK	D31AH	A12AE	H13AC	L30CN	A36CM	H28CK	L15CF	A16CD	M21BL
22	R29BH	F38BF	S38AL	J18AJ	Z37AF	S19AD	D14CS	Z32CP	S33CL	D25CH	Z24CE	H17BM
23	X34BJ	V11BG	P11AM	U12AK	C27AG	P30AE	J23AB	C20CR	P15CM	J35CJ	C21CF	S22BP
24	B26BK	E31BH	W31BC	R37AL	K13AH	W14AF	U36AC	K28CS	W25CP	U16CK	K17CH	P29CD
25	N38BL	M18BJ	Y18BD	X27AM	F19AJ	Y23AG	R32AD	F33AB	Y35CR	R24CL	F22CJ	W34CE
26	A11BM	H12BK	T12BE	B13BC	V30AK	T36AH	X20AE	V15AC	T16CS	X21CH	V29CK	Y26CF
27	Z31BP	S37BL	G37BF	N19BD	E14AL	G32AJ	B28AF	E25AD	G24AB	B17CP	E34CL	T38CH
28	C18CD	P27BM	L27BG	A30BE	M23AM	L20AK	N33AG	M35AE	L21AC	N22CR	M26CL	G11CJ
29	K12CE		D13BH	Z14BF	H36BC	D28AL	A15AH	H16AF	D17AD	A29CS	H38CP	L31CK
30	F37CF		J19BJ	C23BG	S32BD	J33AM	Z25AJ	S24AG	J22AE	Z34AB	S11CR	D18CL
31	V27CH		U30BK		P20BE		C35AK	P21AH		C26AC		J12CM

1940

	JAN	FEB	MAR	APR	MAY	JUN	JUL	AUG	SEP	OCT	NOV	DEC
1	U37CP	K19CL	S30CF	J36CD	Z20BL	S15BJ	D35BF	Z21BD	S29AM	D26AJ	Z31AG	H12AD
2	R27CR	F30CM	P14CH	U32CE	C28BM	P25BK	J16BG	C17BE	P34BC	J38AK	C18AH	S37AE
3	X13CS	V14CP	N23CJ	R20CF	K33BP	W35BL	U24BH	K22BF	W26BD	U11AL	K12AJ	P27AF
4	B19AB	E23CR	Y36CK	X28CH	F15CO	Y16BM	R21BJ	F29BG	Y38BE	R31AM	F37AK	W13AG
5	N30AC	M36CS	T32CL	B33CJ	V25CE	T24BP	X17BK	V34BH	T11BF	X18BC	V27AL	Y19AH
6	A14AD	H32AB	G20CM	N15CK	E35CF	G21CD	B22BL	E26BJ	G31BG	B12BD	E13AM	T30AJ
7	Z23AE	S20AC	L28CP	A25CL	M16CH	L17CD	N29BM	M38BK	L18BG	N37BE	M19BC	G14AK
8	C36AF	P28AD	D33CR	Z35CM	H24CJ	D22CF	A34BH	H11BL	D12BJ	A27BF	H30BD	L23AL
9	K32AG	W33AE	J15CS	C16CP	S21CK	J29CH	Z26CD	S31BH	J37BK	Z13BG	S14BE	D36AM
10	F20AH	Y15AF	U25AB	K24CR	P17CL	U34CJ	C38CE	P18BP	U27BL	C19BH	P23BF	J32BD
11	V28AJ	T25AG	R35AC	F21CS	W22CM	R26CK	K11CF	W12CD	R13BM	K30BJ	W36BG	U20BD
12	E33AK	G35AH	X16AD	V17AB	Y29CP	X38CL	F31CH	Y37CE	X19BP	F14BK	Y32BH	R28BE
13	M15AL	L16AJ	B24AE	E22AC	T34CR	B11CH	V18CJ	T27CF	B30CD	V23BL	T20BJ	X33BF
14	H25AM	D24AK	N21AF	M29AD	G26CS	N31CP	E12CK	G13CH	N14CE	E36BM	G28BH	B15BG
15	S35BC	J21AL	A17AG	H34AE	L38AB	A18CR	M37CL	L19CJ	A23CF	M32BP	L33BL	N25BH
16	P16BD	U17AM	Z22AH	S26AF	D11AC	Z12CS	H27CM	D30CL	Z36CJ	H20CD	D15BM	A35BJ
17	W24BE	R22BC	C29AJ	P38AG	J31AD	C37AB	S13CP	J14CL	C32CJ	S28CE	J25BP	Z16BK
18	Y21BF	X29BD	K34AK	W11AH	U18AE	K27AC	P19CR	U23CM	K20CK	P33CF	U35CD	C24BL
19	T17BG	B34AE	F26AL	Y31AJ	R12AF	F13AD	W30CS	R36CP	F28CL	W15CH	R16CE	K21BM
20	G22BH	N26BF	V38AM	T18AK	X37AG	V19AE	Y14AB	X32CR	V33CN	Y25CJ	X24CF	F17BP
21	L29BJ	A38BG	E11BC	G12AL	B27AH	E30AF	T23AC	B20CS	E15CP	T35CK	B21CH	V22CD
22	D34BK	Z11BH	M31BD	L37AM	N13AJ	M14AG	G26AD	N28AB	M25CS	G16CL	N17CJ	E29CE
23	J26BL	C31BJ	H18BE	D27BC	A19AK	H23AH	L32AC	A33AB	H35CS	L24CH	A22CK	M34CF
24	U38BM	K18BK	S12BF	J13BD	Z30AL	S36AJ	D20AF	Z15AD	S16AB	D21CP	Z29CL	H26CH
25	R11BP	F12BL	P37BG	U19BE	C14AM	P32AK	J28AG	C25AE	P24AC	J17CR	C34CM	S38CJ
26	X31CO	V37BM	W27BH	R30BF	K23BC	W20AL	U33AH	K35AF	W21AD	U22CS	K26CP	P11CK
27	B18CE	E27BP	Y13BJ	X14BG	F36BD	Y28AM	R15AJ	F16AG	Y17AE	R29AB	F38CR	W31CL
28	N12CF	M13CO	T19BK	B23BH	V32BE	T33BC	X25AK	V24AH	T22AF	X34AC	V11CS	Y18CM
29	A37CH	H19CE	G30BL	N36BJ	E20BF	G15BD	B35AL	E21AJ	G29AG	B26AD	E31AB	T12CP
30	Z27CJ		L14BM	A32BK	M28BG	L25BE	N16AM	M17AK	L34AH	N38AE	M18AC	G37CR
31	C13CK		D23BP		H33BH		A24BC	H22AL		A11AF		L27CS

FIRST LETTER = YOUR PHYSICAL CODE, MIDDLE NUMBER = SENSITIVITY CODE, LAST TWO LETTERS = INTELLECTUAL CODE

1941	JAN	FEB	MAR	APR	MAY	JUN	JUL	AUG	SEP	OCT	NOV	DEC
1	D13AB	Z14CR	E14CJ	G32CF	B28BP	E25BL	T16BH	B17BF	E34BD	T38AL	B18AJ	V37AF
2	J19AC	C23CS	M23CK	L20CH	N33CD	M35BM	G24BJ	N22BG	M26BE	G11AM	N12AK	E27AG
3	U30AD	K36AB	H36CL	D28CJ	A15CE	H16BP	L21BK	A29BH	H38BF	L31BC	A37AL	M13AH
4	R14AE	F32AC	S32CM	J33CK	Z25CF	S24CD	D17BL	Z34BJ	S11BG	D18BD	Z27AM	H19AJ
5	X23AF	V20AD	P20CP	U15CL	C35CH	P21CE	J22BM	C26BK	P31BH	J12BE	C13BC	S30AK
6	B36AG	E28AE	W28CR	R25CM	K16CJ	W17CF	U29BP	K38BL	W18BJ	U37BF	K19BD	P14AL
7	N32AH	M33AF	Y33CS	X35CP	F24CK	Y22CH	R34CD	F11BM	Y12BK	R27BG	F30BE	W23AM
8	A20AJ	H15AG	T15AB	B16CR	V21CL	T29CJ	X26CE	V31BP	T37BL	X13BH	V14BF	Y36BC
9	Z28AK	S25AM	G25AC	N24CS	E17CM	G34CK	B38CF	E18CD	G27BM	B19BJ	E23BG	T32BD
10	C33AL	P35AJ	L35AD	A21AB	M22CP	L26CL	N11CH	M12CE	L13BP	N30BK	M36BH	G20BE
11	K15AM	H16AK	D16AE	Z17AC	H29CR	D38CM	A31CJ	H37CF	D19CD	A14BL	H32BJ	L28BF
12	F25BC	Y24AL	J24AF	C22AD	S34CS	J11CP	Z18CK	S27CH	J30CE	Z23BM	S20BK	D33BG
13	V35BD	T21AM	U21AG	K29AE	P26AB	U31CR	C12CL	P13CJ	U14CF	C36BP	P28BL	J15BH
14	E16BE	G17BC	R17AH	F34AF	W38AC	R18CS	K37CM	W19CK	R23CH	K32CD	W33BM	U25BJ
15	M24BF	L22BD	X22AJ	V26AG	Y11AD	X12AB	F27CP	Y30CL	X36CJ	F20CE	Y15BP	R35BK
16	H21BG	D29BE	B29AK	E38AH	T31AE	B37AC	V13CR	T14CM	B32CK	V28CF	T25CD	X16BL
17	S17BH	J34BF	N34AL	M11AJ	G18AF	N27AD	E19CS	G23CP	N20CL	E33CH	G35CE	B24BM
18	P22BJ	U26BG	A26AM	H31AK	L12AG	A13AE	M30AB	L36CR	A28CM	M15CJ	L16CF	N21BP
19	W29BK	R38BH	Z38BC	S18AL	D37AH	Z19AF	H14AC	D32CS	Z33CP	H25CK	D24CH	A17CD
20	Y34BL	X11BJ	C11BD	P12AM	J27AJ	C30AG	S23AD	J20AB	C15CR	S35CL	J21CJ	Z22CE
21	T26BM	B31BK	K31BE	W37BC	U13AK	K14AH	P36AE	U28AC	K25CS	P16CM	U17CK	C29CF
22	G38BP	N16BL	F18BF	Y27BD	R19AL	F23AJ	W32AF	R33AD	F35AB	W24CP	R22CL	K34CH
23	L11CD	A12BM	V12BG	T13BE	X30AM	V36AK	Y20AG	X15AE	V16AC	Y21CR	X29CM	F26CJ
24	D31CE	Z37BP	E37BH	G19BF	B14BC	E32AL	T26AH	B25AF	E24AD	T17CS	B34CP	V38CK
25	J18CF	C27CD	M27BJ	L30BG	N23BD	M20AM	G33AJ	N35AG	M21AE	G22AB	N26CR	E11CL
26	U12CH	K13CE	H13BK	D14BH	A36BE	H28BC	L15AK	A16AH	H17AF	L29AC	A38CS	M31CM
27	R37CJ	F19CF	S19BL	J23BJ	Z32BF	S33BD	D25AL	Z24AJ	S22AG	D34AD	Z11AB	H18CP
28	X27CK	V30CH	P30BM	U36BK	C20BG	P15BE	J35AM	C21AK	P29AH	J26AE	C31AC	S12CR
29	B13CL		W14BP	H34BP	R32BL	K28BH	W25BF	U16BC	K17AL	W34AJ	U38AF	P37CS
30	N19CM		Y23CD	X20BM	F33BJ	Y35BG	R24BD	F22AM	Y26AK	R11AG	F12AE	W27AB
31	A30CP		T36CE		V15BK		X21BE	V29BC		X31AH		Y13AC

1942	JAN	FEB	MAR	APR	MAY	JUN	JUL	AUG	SEP	OCT	NOV	DEC
1	T19AD	B23AB	K23CL	W20CJ	U33CE	K35BP	P24BK	U22BH	K26BF	P11BC	U12AL	C27AH
2	G30AE	N36AC	F36CM	Y28CK	R15CF	F16CD	W21BL	R29BJ	F38BG	W31BD	R37AM	K13AJ
3	L14AF	A32AD	V32CP	T33CL	X25CH	V24CE	Y17BM	X34BK	V11BH	Y18BE	X27BC	F19AK
4	D23AG	Z20AE	E20CR	G15CM	B35CJ	E21CF	T22BP	B26BL	E31BJ	T12BF	B13BD	V30AL
5	J36AH	C28AF	M28CS	L25CP	N16CK	M17CH	G29CD	N38BM	M18BK	G37BG	N19BE	E14AM
6	U32AJ	K33AG	H33AB	D35CR	A24CL	H22CJ	L34CE	A11BP	H12BL	L27BH	A30BF	M23BG
7	R20AK	F15AM	S15AC	J16CS	Z21CM	S29CK	D26CF	Z31CD	S37BM	D13BJ	Z14BG	H36BD
8	X28AL	V25AJ	P25AD	U24AB	C17CP	P34CL	J38CH	C18CE	P27BP	J19BK	C23BH	S32BE
9	B33AM	E35AK	W35AE	R21AC	K22CR	W26CM	U11CJ	K12CF	W13CD	U30BL	K36BJ	P20BF
10	N15BC	M16AL	Y16AF	X17AD	F29CS	Y38CP	R31CK	F37CH	Y19CE	R14BM	F32BK	H28BG
11	A25BD	H24AM	T24AG	B22AE	V34AB	T11CR	X18CL	V27CJ	T30CF	X23BP	V20BL	Y33BH
12	Z35BE	S21BC	G21AH	N29AF	E26AC	G31CS	B12CM	E13CK	G14CH	B36CD	E28BM	T15BJ
13	C16BF	P17BD	L17AJ	A34AG	M38AD	L18AB	N37CP	M19CL	L23CJ	N32CE	M33BP	G25BK
14	K24BG	W22BF	D22AK	Z26AH	H11AE	D12AC	A27CR	H30CM	D36CK	A20CF	H15CD	L35BL
15	F21BH	Y29BF	J29AL	C38AJ	S31AF	J37AD	Z13CS	S14CP	J32CL	Z28CH	S25CE	D16BM
16	V17BJ	T34BG	U34AM	K11AK	P18AG	U27AE	C19AB	P23CR	U20CM	C33CJ	P35CF	J24BP
17	E22BK	G26BM	R26BC	F31AL	W12AH	R13AF	K30AC	W36CS	R28CP	K15CK	W16CH	U21CD
18	M29BL	L38BJ	X36BD	V18AM	Y37AJ	X19AG	F14AD	Y32AB	X33CR	F25CL	Y24CJ	R17CE
19	H34BM	D11BK	B11BE	E12BC	T27AK	B30AH	V23AE	T20AC	B15CS	V35CM	T21CK	X22CF
20	S26BP	J31BL	N31BF	M37BD	G13AL	N14AJ	E36AF	G28AD	N25AB	E16CP	G17CL	B29CH
21	P38CD	U18BM	A18BG	H27BE	L19AM	A23AK	M32AG	L33AE	A35AC	M24CR	L22CM	N34CJ
22	W11CE	R12BP	Z12BH	S13BF	D30BC	Z36AL	H20AH	D15AF	Z16AD	H21CS	D29CP	A26CK
23	Y31CF	X37CD	C37BJ	P19BG	J14BD	C32AM	S28AJ	J25AG	C24AE	S17AB	J34CR	Z38CL
24	T18CH	B27CE	K27BK	W30BH	U23BE	K20BC	P33AK	U35AH	K21AF	P22AC	U26CS	C11CM
25	G12CJ	N13CF	F13BL	Y14BJ	R36BF	F28BD	W15AL	R16AJ	F17AG	W29AD	R38AB	K31CP
26	L37CK	A19CH	V19BM	T23BK	X32BG	V33BE	Y25AM	X24AK	V22AH	Y34AE	X11AC	F18CR
27	D27CL	Z30CJ	E30BP	G36BL	B20BH	E15BF	T35BC	B21AL	E29AJ	T26AF	B31AD	V12CS
28	J13CM	C14CK	M14CD	L32BM	N28BJ	M25BG	G16BD	N17AM	M34AK	G38AG	N18AE	E37AB
29	U19CP		H23CE	D20BP	A33BH	H35BH	L24BE	A22BC	H26AL	L11AH	A12AF	M27AC
30	R30CR		S36CF	J28CD	Z15BL	S16BJ	D21BF	Z29BD	S38AM	D31AJ	Z37AG	H13AD
31	X14CS		P32CH		C25BM		J17BG	C34BE		J18AK		S19AE

1943	JAN	FEB	MAR	APR	MAY	JUN	JUL	AUG	SEP	OCT	NOV	DEC
1	P30AF	U36AD	A36CP	H28CL	L15CH	A16CE	M21BH	L29BK	A38BH	M31BE	L37BC	N13AK
2	W14AG	R32AE	Z32CR	S33CM	D25CJ	Z24CF	H17BP	D34BL	Z11BJ	H18BF	D27BD	A19AL
3	Y23AH	X20AF	C20CS	P15CP	J35CK	C21CH	S22CD	J26BM	C31BK	S12BG	J13BE	Z30AM
4	T36AJ	B28AG	K28AB	W25CR	U16CL	K17CJ	P29CE	U38BP	K18BL	P37BH	U19BF	C14BC
5	G32AK	N33AH	F33AC	Y35CS	R24CM	F22CK	W34CF	R11CD	F12BM	W27BJ	R30BG	K23BD
6	L20AL	A15AJ	V15AD	T16AB	X21CP	V29CL	Y26CH	X31CE	V37BP	Y13BK	X14BH	F36BE
7	D28AM	Z25AK	E25AE	G24AC	B17CR	E34CM	T38CJ	B16CF	E27CD	T19BL	B23BJ	V32BF
8	J33BC	C35AL	M35AF	L21AD	N22CS	M26CP	G11CK	N12CH	M13CE	G30BM	N36BK	E20BG
9	U15BD	K16AM	H16AG	D17AE	A29AB	H38CR	L31CL	A37CJ	H19CF	L14BP	A32BL	H28BH
10	R25BE	F24BC	S24AH	J22AF	Z34AC	S11CS	D18CM	Z27CK	S30CH	D23CD	Z20BM	H33BJ
11	X35BF	V21BD	P21AJ	U29AG	C26AD	P31AB	J12CP	C13CL	P14CJ	J36CE	C28BP	S15BK
12	B16BG	E17BE	W17AK	R34AH	K38AE	W18AC	U37CR	K19CM	W23CK	U32CF	K33CD	P25BL
13	N24BH	M22BF	Y22AL	F11AF	V31AG	Y12AD	R27CS	F30CP	Y36CL	R20CH	F15CE	W35BM
14	A21BJ	H29BC	T29AM	B38AK	V31AG	T37AE	X13AB	V14CR	T32CM	X28CJ	V25CF	Y16BP
15	Z17BK	S34BH	G34BC	N11AL	E18AH	G27AD	B19AF	E23CS	G20CP	B33CK	E35CH	T24CD
16	C22BL	P26BJ	L26BD	A31AM	M12AJ	L13AG	N30AD	M36AB	L28CR	N15CL	M16CJ	G21CE
17	K29BM	W38BK	D38BE	Z18BC	H37AK	D19AH	A14AE	H32AC	D33CS	A25CM	H24CH	L17CF
18	F34BP	Y11BL	J11BF	C12BD	S27AL	J30AJ	Z23AF	S20AD	J15AB	Z35CP	S21CL	D22CH
19	V26CD	T31BM	U31BG	K37BE	P13AM	U14AK	C36AG	P28AE	U25AC	C16CR	P17CM	J29CJ
20	E38CE	G18BP	R18BH	F27BF	W19BG	R23AL	K32AH	W33AF	R35AD	K24CS	W22CP	U34CK
21	M11CF	L12CD	X12BJ	V13BG	Y30BD	X36AM	F20AJ	Y15AG	X16AE	F21AB	Y29CR	R26CL
22	H31CH	D37CE	B37BK	E19BH	T14BE	B32AG	V28AK	T25AH	B24AF	V17AC	T34CS	X38CM
23	S18CJ	J27CF	N27BL	M30BJ	G23BF	N20BD	E33AL	G35AJ	N21AG	E22AD	G26AB	B11CP
24	P12CK	U13CH	A13BM	H14BK	L36BG	A28BE	M15AM	L16AK	A17AH	M29AE	L38AC	N31CR
25	W37CL	R19CJ	Z19BP	S23BL	D32BH	Z33BF	H25AM	D24AL	Z22AJ	H34AF	D11AD	A18CS
26	Y27CM	X30CK	C30CD	P36BM	J20BJ	C15BG	S35BD	J21AM	C29AK	S26AG	J31AE	Z12AB
27	T13CP	B14CL	K14CE	W32BP	U28BK	K25BH	P16BE	U17BC	K34AL	P38AH	U18AF	C37AC
28	G19CR	N23CM	F23CF	Y20CD	R33BL	F35BJ	W24BF	R22BD	F26AM	W11AJ	R12AG	K27AD
29	L30CS		V36CH	T28CE	X15BM	V16BK	Y21BG	X29BD	V38BC	Y31AK	X37AH	F13AE
30	D14AB		E32CJ	G33CF	B25BP	E24BL	T17BH	B34BF	E11BD	T18AL	B27AJ	V19AF
31	J23AC		M28CK		N35CD		G22BJ	N26BG		G12AM		E30AG

FIRST LETTER = YOUR PHYSICAL CODE, MIDDLE NUMBER = SENSITIVITY CODE, LAST TWO LETTERS = INTELLECTUAL CODE

1944	JAN	FEB	MAR	APR	MAY	JUN	JUL	AUG	SEP	OCT	NOV	DEC
1	M14AH	L32AF	B20AB	E15CR	T35CL	B21CJ	V22CE	T26BP	B31BL	V12BH	T13BF	X30BC
2	H23AJ	D20AG	N28AC	M25CS	G16CM	N17CK	E29CF	G38CD	N18BM	E37BJ	G19BG	B14BD
3	S36AK	J28AH	A33AD	H35AB	L24CP	A22CL	M34CH	L11CE	A12BP	M27BK	L30BH	N23BE
4	P32AL	U33AJ	Z15AE	S16AC	D21CR	Z29CM	H26CJ	D31CF	Z37CD	H13BL	D14BJ	A36BF
5	W20AM	R15AK	C25AF	P24AD	J17CS	C34CP	S36CK	J16CH	C27CE	S19BM	J23BK	Z32BG
6	Y28BC	X25AL	K35AG	W21AE	U22AB	K26CR	P11CL	U12CJ	K13CF	P30BP	U36BL	C20BH
7	T33BD	B35AM	F16AH	Y17AF	R29AC	F38CS	W31CM	R37CK	F19CH	W14CD	R32BM	K28BJ
8	G15BE	N16BC	V24AJ	T22AG	X34AD	V11AB	Y18CP	X27CL	V30CJ	Y23CE	X20BP	F33BK
9	L25BF	A24BD	E21AK	G29AH	B26AE	E31AC	T12CS	B13CM	E14CK	T36CF	B28CO	V15BL
10	D35BG	Z21BE	M17AL	L34AJ	N38AF	M18AD	G37CS	N19CP	M23CL	G32CH	N33CE	E25BM
11	J16BH	C17BF	H22AM	D26AK	A11AG	H12AE	L27AB	A30CR	H35CM	L20CJ	A15CF	M35BP
12	U24BJ	K22BG	S29BC	J38AL	Z31AH	S37AF	D13AC	Z14CS	S32CP	D28CK	Z25CH	H16CD
13	R21BK	F29BH	P34BD	U11AM	C18AJ	P27AG	J19AD	C23AB	P20CR	J33CL	C35CJ	S24CE
14	X17BL	V34BJ	W26BE	R31BC	K12AK	W13AH	U30AE	K36AC	W28CS	U15CM	K16CK	P21CF
15	B22BM	E26BK	Y38BF	X18BD	F37AL	Y19AJ	R14AF	F32AD	Y33AB	R25CP	F24CL	W17CH
16	N29BP	M38BL	T11BG	B12BE	V27AM	T30AK	X23AG	V20AE	T15AC	X35CR	V21CH	Y22CJ
17	A34CD	H11BM	G31BH	N37BF	E13BC	G14AL	B36AH	E28AF	G25AD	B16CS	E17CP	T29CK
18	Z26CE	S31BP	L18BJ	A27BG	M19BD	L23AM	N32AJ	M33AG	L35AE	N24AB	M22CR	G34CL
19	C38CF	P18CD	D12BK	Z13BH	H30BE	D36BC	A20AK	H15AH	D16AF	A21AC	H29CS	L26CM
20	K11CH	W12CE	J37BL	C19BJ	S14BF	J32BD	Z28AL	S25AJ	J24AG	Z17AD	S34AB	D38CP
21	F31CJ	Y37CF	U27BM	K30BK	P23BG	U20BE	C33AM	P35AK	U21AH	C22AE	P26AC	J11CR
22	V18CK	T27CH	R13BP	F14BL	W36BH	R28BF	K15BC	W16AL	R17AJ	K29AF	W38AD	U31CS
23	E12CL	G13CJ	X19CD	V23BM	Y32BJ	X33BG	F25BD	Y24AM	X22AK	F34AG	Y11AE	R18AB
24	M37CM	L19CK	B30CE	E36BP	T20BK	B15BH	V35BE	T21BC	B29AL	V26AH	T31AF	X12AC
25	H27CP	D30CL	N14CF	M32CD	G28BL	N25BJ	E16BF	G17BD	N34AM	E38AJ	G18AG	B37AD
26	S13CR	J14CM	A23CH	H20CE	L33BM	A35BK	M24BG	L22BE	A26BC	M11AK	L12AH	N27AE
27	P19CS	U23CP	Z36CJ	S28CF	D15BP	Z16BL	H21BH	D29BF	Z38BD	H31AL	D37AJ	A13AF
28	W30AB	R36CH	C32CK	P33CH	J25CD	C24BM	S17BJ	J34BG	C11BE	S18AM	J27AK	Z19AG
29	Y14AC	X32CS	K20CL	W15CJ	U35CE	K21BP	P22BK	U26BH	K31BF	P12BC	U13AL	C30AH
30	T23AD		F28CM	Y25CK	R16CF	F17CD	W29BL	R38BJ	F18BG	W37BD	R19AM	K14AJ
31	G36AE		V33CP		X24CH		Y34BM	X11BK		Y27BE		F23AK

1945	JAN	FEB	MAR	APR	MAY	JUN	JUL	AUG	SEP	OCT	NOV	DEC
1	V36AL	T28AJ	U28AD	K25AB	P16CP	U17CL	C29CH	P38CE	U18BP	C37BK	P19BH	J14BE
2	E32AM	G33AK	R33AE	F35AC	W24CR	R22CM	K34CJ	W11CF	R12CD	K27BL	W30BJ	U23BF
3	M20BC	L15AL	X15AF	V16AD	Y21CS	X29CP	F26CK	Y31CH	X37CE	F13BM	Y14BK	R36BG
4	H28BD	D25AM	B25AG	E24AE	T17AB	B34CR	V38CL	T18CJ	B27CF	V19BP	T23BL	X32BH
5	S33BE	J35BC	N35AH	M21AF	G22AC	N26CS	E11CM	G12CK	N13CH	E30CD	G36BM	B20BJ
6	P15BF	U16BD	A16AJ	H17AG	L29AD	A38AB	M31CP	L37CL	A19CJ	M14CE	L32BP	N28BK
7	W25BG	R24BE	Z24AK	S22AH	D34AE	Z11AC	H18CR	D27CM	Z30CK	H23CF	D20CD	A33BL
8	Y35BH	X21BF	C21AL	P29AJ	J26AF	C31AD	S12CS	J13CP	C14CL	S36CH	J28CE	Z15BM
9	T16BJ	B17BG	K17AM	W34AK	U38AG	K18AE	P37AB	U19CR	K23CH	P32CJ	U33CF	C25BP
10	G24BK	N22BH	F22BC	Y26AL	R11AH	F12AF	W27AC	R30CS	F36CP	W20CK	R15CH	K35CD
11	L21BL	A29BJ	V29BD	T38AM	X31AJ	V37AG	Y13AD	X14AB	V32CR	Y28CL	X25CJ	F16CE
12	D17BM	Z34BK	E34BE	G11BC	B18AK	E27AH	T19AE	B23AC	E20CS	T33CM	B35CK	V24CF
13	J22BP	C26BL	M26BF	L31BD	N12AL	M13AJ	G30AF	N36AD	M28AB	G15CP	N16CL	E21CH
14	U29CD	K38BM	H38BG	D18BE	A37AM	H19AH	L14AG	A32AE	H33AC	L25CR	A24CM	M17CJ
15	R34CE	F11BP	S11BH	J12BF	Z27BC	S30AL	D23AH	Z20AF	S15AD	D35CS	Z21CP	H22CK
16	X26CF	V31CD	P18BJ	U37BG	K19BE	P14AM	J36AJ	C28AG	P25AE	J16AB	C17CR	S29CL
17	B38CH	E18CE	W18BK	R27BH	K19BE	W23BC	U32AK	K33AH	W35AF	U24AC	K22CS	P34CM
18	N11CJ	M12CF	Y12BL	X13BJ	F30BF	Y36BD	R20AL	F15AJ	Y16AG	R21AD	F29AB	W26CP
19	A31CK	H37CH	T37BM	B19BK	V14BG	T32BE	X28AM	V25AK	T24AH	X17AE	V34AC	Y38CR
20	Z18CL	S27CJ	G27BP	N30BL	E23BH	G20BF	B33BC	E35AL	G21AJ	B22AF	E26AD	T11CS
21	C12CM	P13CK	L13CD	A14BM	M36BJ	L28BG	N15BD	M16AM	L17AK	N29AG	M38AE	G31AB
22	K37CP	W19CL	D19CE	Z23BP	H32BK	D33BH	A25BE	H24BC	D22AL	A34AH	H11AF	L18AC
23	F27CR	Y30CM	J30CF	C36CD	S20BL	J15BJ	Z35BF	S21BD	J29AM	Z26AJ	S31AG	D12AD
24	V13CS	T14CP	U14CH	K32CE	P28BM	U25BK	C16BG	P17BE	U34BC	C38AK	P18AH	J37AE
25	E19AB	G23CR	R23CJ	F20CF	W33BP	R35BL	K24BH	W22BE	R26BD	K11AL	W12AJ	U27AF
26	M30AC	L36CS	X36CK	V28CH	Y15CD	X16BM	F21BJ	Y29BG	X38BE	F31AM	Y37AK	R13AG
27	H14AD	D32AB	B32CL	N25CJ	T25CE	B24BP	V17BK	T34BH	B11BF	V18BD	T27AL	X19AH
28	S23AE	J20AC	N20CM	M15CK	G35CF	N21CD	E22BL	G26BJ	N31BG	E12BD	G13AM	B30AJ
29	P36AF		A28CP	H25CL	L16CH	A17CE	M29BM	L38BK	A18BH	M37BE	L19BC	N14AK
30	W32AG		Z33CM	S35CM	D24CJ	Z22CF	H34BP	D11BL	Z12BJ	H27BF	D30BD	A23AL
31	Y20AH		C15CS		J21CK		S26CD	J31BH		S13BG		Z36AM

1946	JAN	FEB	MAR	APR	MAY	JUN	JUL	AUG	SEP	OCT	NOV	DEC
1	C32BC	P33AL	L33AF	A35AD	M24CS	L22CP	N34CK	M11CH	L12CE	N27BM	M30BK	G23BG
2	K20BD	W15AM	D15AG	Z16AE	H21AB	D29CR	A26CL	H31CJ	D37CF	A13BP	H14BL	L36BH
3	F28BF	Y25BC	J25AH	C24AF	S17AC	J34CS	Z38CM	S18CK	J27CH	Z19CD	S23BM	D32BJ
4	V33BF	T35BD	U35AJ	K21AG	P22AD	U26AB	C11CP	P12CL	U13CJ	C30CE	P36BP	J20BK
5	E15BG	G16BE	R16AK	F17AH	W29AE	R38AC	K31CR	W37CM	R19CK	K14CF	W32CD	U28BL
6	M25BH	L24BF	X24AL	V22AJ	Y34AF	X11AD	F18CS	Y27CP	X30CL	F23CH	Y20CE	R33BH
7	H35BJ	D21BG	B21AM	E29AK	T26AG	B31AE	V12AB	T13CR	B14CM	V36CJ	T28CF	X15BP
8	S16BK	J17BH	N17BC	M34AL	N18AF	E37CH	G19CS	N23CP	E32CK	G33CH	N35CD	B25CD
9	P24BL	U22BG	A22BD	H26AM	L11AJ	A12AG	M27AD	L30AB	A35CR	M20CL	L15CJ	N35CD
10	W21BM	R29BK	Z29BE	S38BC	D31AK	Z37AH	H13AE	D14AC	Z32CS	H28CM	D25CK	A16CF
11	Y17BP	X34BL	C34BF	P11BD	J18AL	C27AJ	S19AF	J23AD	C20AB	S33CP	J35CL	Z24CH
12	T22CD	B26BM	K26BG	W31BE	U12AM	K13AK	P30AG	U36AE	K28AC	P15CR	U16CM	C21CJ
13	G29CE	N38BP	F38BH	Y18BF	R37BC	F19AL	W14AH	R32AF	F33AD	W25CS	R24CP	K17CK
14	L34CF	A11CD	V11BJ	T12BG	X27BD	V30AM	Y23AJ	X20AG	V15AE	Y35AB	X21CR	F22CL
15	D26CH	Z31CE	E31BK	G37BH	B13BE	E14BC	T36AK	B28AF	E25AD	T16AC	B17CS	V29CM
16	J38CJ	C18CF	M18BL	L27BJ	N19BF	M23BD	G32AL	N33AJ	M35AG	G24AD	N22AB	E34CP
17	U11CK	K12CH	H12BM	D13BK	A30BC	H36BE	L20AM	A15AK	H16AH	L21AE	A29AC	M26CR
18	R31CL	F37CJ	S37BP	J19BL	Z14BH	S32BF	D28BC	Z25AL	S24AJ	D17AF	Z34AD	H38CS
19	X18CM	V27CK	P27CD	U30BM	C23BJ	P20BG	J33BD	C35AM	P21AK	J22AG	C26AE	S11AB
20	B12CP	E13CL	W13CE	R14BP	K36BK	W28BH	U15BE	K16BC	W17AL	U29AH	K38AF	P31AC
21	N37CR	M19CM	Y19CF	X23BD	F32BL	Y33BJ	R25BF	F24BD	Y22AM	R34AJ	F11AG	W18AD
22	A27CS	H30CP	T30CH	B36BC	V20BH	T15BF	X35BG	V21BE	T29BC	X26AK	V31AH	Y12AE
23	Z13AB	S14CR	G14CJ	N32BF	E28BP	G25BL	B16BH	E17BF	G34BD	B38AL	E18AJ	T37AF
24	C19AC	P23CS	L23CK	A20CH	M33CD	L35BH	N24BJ	M22BG	L26BE	N11AH	M12AK	G27AG
25	K30AD	W36BM	D36CL	Z28CJ	H15CE	D16BP	A21BK	H29BH	D38BF	A31BC	H37AL	L13AH
26	F14AE	Y32AD	J32CM	O33CK	S25CF	J24CD	Z17BL	S34BJ	J11BG	Z18BD	S27AM	D19AJ
27	V23AF	T20AB	U20CP	K15CL	P35CH	U21CE	C22BM	P26BK	U31BH	C12BE	P13BC	J30AK
28	E36AG	G28AE	R28CM	F25CM	W16CJ	R17CF	K29BP	W38BL	R18BJ	K37BF	W19BD	U14AL
29	M32AH		X33CS	V35CP	Y24CK	X27CD	F34BL	Y11BH	X12BK	F27BG	Y30BE	R23AM
30	H20AJ		B15AB	E16CR	T21CL	B29CJ	V26CE	T31BP	B37BL	V13BH	T14BF	X36BC
31	S28AK		N25AC		G17CM		E38CF	G18CD		E19BJ		B32BD

FIRST LETTER = YOUR PHYSICAL CODE, MIDDLE NUMBER = SENSITIVITY CODE, LAST TWO LETTERS = INTELLECTUAL CODE

1947	JAN	FEB	MAR	APR	MAY	JUN	JUL	AUG	SEP	OCT	NOV	DEC
1	N20BE	M15BC	Y15AH	X16AF	F21AC	Y29CS	R26CH	F31CK	Y37CH	R13CD	F14BM	W36BJ
2	A28BF	H25BD	T25AJ	B24AG	V17AD	T34AB	X38CP	V18CL	T27CJ	X19CE	V23BP	Y32BK
3	Z33BG	S35BE	G35AK	N21AH	E22AE	G26AC	B11CR	E12CM	G13CK	B30CF	E36CD	T20BL
4	C15BH	P16BF	L16AL	A17AJ	M29AF	L38AD	N31CS	M37CP	L19CL	N14CH	M32CE	G28BM
5	K25BJ	W24BG	D24AM	Z22AK	H34AG	D11AE	A18AB	H27CR	D30CM	A23CJ	H20CF	L33BP
6	F35BK	Y21BH	J21BC	C29AL	S26AH	J31AF	Z12AC	S13CS	J14CP	Z36CK	S28CH	D15CD
7	V16BL	T17BJ	U17BD	K34AM	P38AJ	U18AG	C37AD	P19AB	U23CR	C32CL	P33CJ	J25CE
8	E24BM	G22BK	R22BE	F26BC	W11AK	R12AH	K27AE	W30AC	R36CS	K20CM	W15CK	U35CF
9	M21BP	L29BL	X29BF	V38BD	Y31AL	X37AJ	F13AF	Y14AD	X32AB	F28CP	Y25CL	R16CH
10	H17CD	D34BM	B34BG	E11BE	T18AM	B27AK	V19AG	T23AE	B20AC	V33CR	T35CM	X24CJ
11	S22CE	J26BP	N26BH	M31BF	G12BC	N13AL	E30AH	G36AF	N28AD	E15CS	G16CP	B21CK
12	P29CF	U38CD	A38BJ	H18BG	L37BD	A19AM	M14AJ	L32AG	A33AE	M25AB	L24CR	N17CL
13	W34CH	R11CE	Z11BK	S12BH	D27BE	Z30BG	H23AK	D20AH	Z15AF	H35AC	D21CS	A22CM
14	Y26CJ	X31CF	C31BL	P37BJ	J13BF	C14BD	S36AL	J28AJ	C25AG	S16AD	J17AB	Z29CP
15	T38CK	B18CH	K18BM	W27BK	U19BG	K23BE	P32AM	U33AK	K35AH	P24AE	U22AC	C34CR
16	G11CL	N12CJ	F12BP	Y13BL	R30BH	F36BF	W20BC	R15AL	F16AJ	W21AF	R29AD	K26CS
17	L31CM	A37CK	V37CD	T19BM	X14BJ	V32BG	Y28BD	X25AM	V24AK	Y17AG	X34AE	F38AB
18	D18CP	Z27CL	E27CE	G30BP	B23BK	E20BH	T33BE	B35BC	E21AL	T22AH	B26AF	V11AC
19	J12CR	C13CM	M13CF	L14CD	N36BL	M28BJ	G15BF	N16BD	M17AM	G29AJ	N38AG	E31AC
20	U37CS	K19CP	H19CH	D23CE	A32BM	H33BK	L25BG	A24BE	H22BC	L34AK	A11AH	M18AE
21	R27AB	F30CR	S30CJ	J36CF	Z20BP	S15BL	D35BH	Z21BF	S29BD	D26AL	Z31AJ	H12AF
22	X13AC	V14CS	P14CK	U32CH	C28CP	P25BM	J16BJ	C17BG	P34BE	J38AM	C18AK	S37AG
23	B19AD	E23AB	W23CL	R20CJ	K33CE	W35BP	U24BK	K22BH	W26BF	U11BC	K12AL	P27AH
24	N30AE	M36AC	Y36CM	X28CK	F15CF	Y16CD	R21BL	F29BJ	Y38BG	R31BD	F37AM	W13AJ
25	A14AF	H32AD	T32CP	B33CL	V25CH	T24CE	X17BM	V34BK	T11BH	X18BE	V27BC	Y19AK
26	Z23AG	S20AE	G20CR	N15CM	E35CJ	G21CF	B22BP	E26BL	G31BJ	B12BF	E13BD	T30AL
27	C36AH	P28AF	L28CS	A25CP	M16CK	L17CH	N29CD	M38BM	L18BK	N37BG	M19BE	G14AM
28	K32AJ	W33AG	D33AB	Z35CR	H24CL	D22CJ	A34CE	H11BP	D12BL	A27BH	H30BF	L23BC
29	F20AK		J15AC	C16CS	S21CM	J29CK	Z26CF	S31CD	J37BM	Z13BJ	S14BG	D36BD
30	V28AL		U25AD	K24AB	P17CP	U34CL	C38CH	P18CE	U27BP	C19BK	P23BH	J32BE
31	E33AM		R35AE		W22CR		K11CJ	W12CF		K30BL		U20BF

1948	JAN	FEB	MAR	APR	MAY	JUN	JUL	AUG	SEP	OCT	NOV	DEC
1	R28BG	F25BE	P35AL	U21AJ	C22AF	P26AD	J11CS	C12CP	P13CL	J30CH	C36CE	S20BM
2	X33BH	V35BF	M16AM	R17AK	K29AG	W38AE	U31AB	K37CR	M19CM	U14CJ	K32CF	P28BP
3	B15BJ	E16BG	Y24BC	X22AL	F34AH	Y11AF	R18AC	F27CS	Y30CP	R23CK	F20CH	W33CD
4	N25BK	M24BH	T21BD	B29AM	V26AJ	T31AG	X12AD	V13AB	T14CR	X36CL	V28CJ	Y15CE
5	A35BL	H21BJ	G17BE	N34BC	E38AK	G18AH	B37AE	E19AC	G23CS	B32CM	E33CK	T25CF
6	Z16BM	S17BK	L22BF	A26BD	M11AL	L12AJ	N27AF	M30AD	L36AB	N20CP	M15CL	G35CH
7	C24BP	P22BL	D29BG	Z38BE	H31AM	D37AK	A13AG	H14AE	D32AC	A28CR	H25CM	L16CJ
8	K21CD	W29BM	J34BH	C11BF	S18BC	J27AL	Z19AH	S23AF	J20AD	Z33CS	S35CP	D24CK
9	F17CE	Y34BP	U26BJ	K31BG	P12BD	U13AM	C30AJ	P36AG	U28AE	C15AB	P16CR	J21CL
10	V22CF	T26CD	R38BK	F18BH	W37BE	R19BG	K14AK	W32AH	R33AF	K25AC	W24CS	U17CM
11	E29CH	G38CE	X11BL	V12BJ	Y27BF	X30BD	F23AL	Y20AJ	X15AG	F35AD	Y21AB	R22CP
12	M34CJ	L11CF	B31BM	E37BK	T13BG	B14BE	V36AM	T28AK	B25AH	V16AE	T17AC	X29CR
13	H26CK	D31CM	N18BP	M27BL	G19BH	N23BF	E32BC	G33AL	N35AJ	E24AF	G22AD	B34CS
14	S38CL	J18CJ	A12CD	H13BM	L30BJ	A36BG	M28BD	L15AM	A16AK	M21AG	L29AE	N26AB
15	P11CM	U12CK	Z37CE	S19BP	D14BK	Z32BH	H28BE	D25BC	Z24AL	H17AH	D34AF	A38AC
16	W31CP	R37CL	C27CF	P30CD	J23BL	C20BJ	S33BF	J35BD	C21AM	S22AJ	J26AG	Z11AD
17	Y18CR	X27CM	K13CH	W14CE	U36BM	K28BK	P15BG	U16BE	K17BC	P29AK	U38AH	C31AE
18	T12CS	B13CP	F19CJ	Y23CF	R32BP	F33BL	W25BH	R24BF	F22BD	W34AL	R11AJ	K18AF
19	G37AB	N19CR	V30CK	T36CH	X20CD	V15BM	Y35BJ	X21BG	V29BE	Y26AM	X31AK	F12AG
20	L27AC	A30CS	E14CL	G32CJ	B28CE	E25BP	T16BK	B17BH	E34BF	T38BC	B18AL	V37AH
21	D13AD	Z14AB	M23CM	L28CK	N33CD	M35CD	G24BL	N22BJ	M26BG	G11BD	N12AM	E27AJ
22	J19AE	C23AC	H36CP	D28CL	A15CH	H16CE	L21BM	A29BK	H38BH	L31BE	A37BC	M13AK
23	U30AF	K36AD	S32CR	J33CM	Z25CJ	S24CF	D17BP	Z34BL	S11BJ	D18BF	Z27BD	H19AL
24	R14AG	F32AE	P20CS	U15CP	C35CK	P21CH	J22CD	C26BM	P31BK	J12BG	C13BE	S30AM
25	X23AH	V20AF	W28AB	R25CR	K16CL	W17CJ	U29CE	K38BP	W18BL	U37BH	K19BF	P14BC
26	B36AJ	E28AG	Y33AC	X35CS	F24CM	Y22CK	R34CF	F11CD	Y28BM	R27BJ	F30BG	W23BD
27	N32AK	M33AH	T15AD	B16AB	V21CP	T29CL	X26CH	V31CE	T37BP	X13BK	V14BH	Y36BE
28	A20AL	H15AJ	G25AE	N24AC	E17CR	G34CM	B38CJ	E16CF	G27CD	B19BL	E23BJ	T32BF
29	Z28AM	S25AK	L35AF	A21AD	M22CS	L26CP	N11CK	M12CH	L13CE	N30BM	M36BK	G20BG
30	C33BC		D16AG	Z17AE	H29AB	D38CR	A31CL	H37CJ	D19CF	A14BP	H32BL	L28BH
31	K15BD		J24AH		S34AC		Z18CH	S27CK		Z23CD		D33BJ

1949	JAN	FEB	MAR	APR	MAY	JUN	JUL	AUG	SEP	OCT	NOV	DEC
1	J15BK	C16BH	M16BC	L17AL	N29AH	M38AF	G31AC	N37CS	M19CP	G14CK	N32CH	E28CD
2	U25BL	K24BJ	H24BD	D22AM	A34AJ	H11AG	L18AD	A27AB	H30CR	L23CL	A20CJ	M33CE
3	R35BM	F21BK	S21BE	J29BC	Z26AK	S31AH	D12AE	Z13AC	S14CS	D36CM	Z28CK	H15CF
4	X16BP	V17BL	P17BF	U34BD	C38AL	P18AJ	J37AF	C19AD	P23AB	J32CP	C33CL	S25CH
5	B24CD	E22BM	W22BG	R26BE	K11AM	W12AK	U27AG	K30AE	W36AC	U20CR	K15CM	P35CJ
6	N21CE	M29BP	Y29BH	X38BF	F31BC	Y37AL	R13AH	F14AF	Y32AD	R28CS	F25CP	W16CK
7	A17CF	H34CD	T34BJ	B11BG	V18BD	T27AM	X19AJ	V23AG	T20AE	X33AB	V35CR	Y24CL
8	Z22CH	S26CE	G26BK	N31BH	E12BE	G13BC	B30AK	E36AH	G28AF	B15AC	E16CS	T21CM
9	C29CJ	P36CF	L38BL	A18BJ	M37BF	L19BD	N14AL	M32AJ	L33AG	N25AD	M24AB	G17CP
10	K34CK	W17CH	D11BM	Z12BK	H27BG	D30BE	A23AM	H20AK	D15AH	A35AE	H21AC	L22CR
11	F26CL	Y31CJ	J31BP	G37BL	S13BH	J14BF	Z36BC	S28AL	J25AJ	Z16AF	S17AD	D29CS
12	V38CM	T18CL	U18BJ	K27BM	P19BJ	U23BG	C32BE	P33AK	U35AK	C24AG	P22AE	J34AC
13	E11CP	G12CL	R12CE	F13BP	W30BK	R36BH	K20BE	W15BC	R16AL	K21AH	W29AF	U26AC
14	M31CR	L37CM	X37CF	V19BJ	Y14BM	X32BK	F28BG	Y25BE	X34AJ	F17AJ	Y34AG	R38AD
15	H18CS	D27CP	B27BF	E30BE	T23BM	B20BK	V33BG	T35BE	B21BC	V22AK	T26AH	X11AE
16	S12AB	J13CR	N13CH	M14CF	G36BP	N28BD	E15BM	G16BD	N17BD	E29AL	G38AD	B31AF
17	P37AC	U19CS	A19CK	H23CH	L32CD	A33BM	M25BJ	L24BG	A22BE	M34AM	L11AK	N18AG
18	W27AD	R30AB	Z30CL	S36CJ	D20CE	Z15BP	H35BK	D21BH	Z29BF	H26BC	D31AL	A12AH
19	Y13AE	X14AC	C14CM	P32CK	J28CF	C25BD	S16BL	J17BJ	C34BD	S38BD	J18AM	Z37AJ
20	T19AF	B23AD	K23CP	W20CL	U33CH	K35BE	P24BH	U22BK	K26BH	P11BE	U12BC	C27AK
21	G30AG	N36AE	F36CR	Y28CM	R15CJ	F16CF	W21BJ	R29BL	F38BJ	W31BF	R37BD	K13AL
22	L14AH	A32AF	V32CS	T33CP	X25CK	V24CH	Y17BD	X34BM	V11BK	Y18BG	X27BE	F19AM
23	D23AJ	Z20AG	E20AB	G15CR	B35CL	E21CJ	T22BC	B26BG	E31BL	T12BH	B13BF	V30BC
24	J36AK	C28AH	M28AC	L25CS	N16CM	M17CK	G29BF	N38CD	M18BM	G37BJ	N19BG	E14BD
25	U32AL	K33AJ	H33AD	D35AB	A24CP	H22CL	L34BH	A11CC	H12BP	L27BK	A30BH	M23BE
26	R20AM	F15AK	S15AE	J16AC	Z21CR	S29CM	D26BJ	Z31CF	S37BD	D13BL	Z14BJ	H36BF
27	X28BC	V25AL	P25AF	U24AD	C17CS	P34CP	J38BK	C18CH	P27CE	J19BM	C23BK	S32BG
28	B33BD	E35AM	W35AG	R21AE	K22AB	W26CR	U11BL	K12CJ	W13BB	U30BP	K36BL	P20BH
29	N15BE		Y16AH	X17AF	F29AC	Y38CS	R31CH	F37CK	Y19CH	R14CD	F32BM	W28BJ
30	A25BF		T24AJ	B22AG	V34AD	T11AB	X18CP	V27CL	T30CJ	X23CE	V20BP	Y33BK
31	Z35BG		G21AK		E26AE		B12CR	E13CH		B36CF		T15BL

FIRST LETTER = YOUR PHYSICAL CODE, MIDDLE NUMBER = SENSITIVITY CODE, LAST TWO LETTERS = INTELLECTUAL CODE

1950	JAN	FEB	MAR	APR	MAY	JUN	JUL	AUG	SEP	OCT	NOV	DEC
1	G25BM	N24BK	F24BE	Y22BC	R34AK	F11AH	W18AE	R27AC	F30CS	W23CM	R20CK	K33CF
2	L35BP	A21BL	V21BF	T29BD	X26AL	V31AJ	Y12AF	X13AD	V14AB	Y36CP	X28CL	F15CH
3	D16CD	Z17BM	E17BG	G34BE	B38AM	E18AK	T37AG	B19AE	E23AC	T32CR	B33CM	V25CJ
4	J24CE	C22BP	H22BH	L26BF	N11BC	M12AL	G27AH	N30AF	M36AD	G20CS	N15CP	E35CK
5	U21CF	K29CD	H29BJ	D38BG	A31BD	H37AM	L13AJ	A14AG	H32AE	L28AB	A25CR	M16CL
6	R17CH	F34CE	S34BK	J11BH	Z18BE	S27BC	D19AK	Z23AH	S20AF	D33AC	Z35CS	H24CM
7	X22CJ	V26CF	P26BL	U31BJ	C12BF	P13BD	J30AL	C36AJ	P28AG	J15AD	C16AB	S21CP
8	B29CK	E38CH	W38BM	R18BK	K37BG	W19BE	U14AM	K32AK	W33AH	U25AE	K24AC	P17CR
9	N34CL	M11CJ	Y11BP	X12BL	F27BH	Y30BF	V23BC	F20AL	Y15AJ	R35AF	F21AD	W22CS
10	A26CM	H31CR	T31CD	B37BM	V13BJ	T14BG	X36BD	V28AM	T25AK	X16AG	V17AE	Y29AB
11	Z38CP	S18CL	G18CE	M27BP	E19BK	G23BH	B32BE	E33BC	G35AL	B24AH	E22AF	T34AC
12	C11CR	P12CM	L12CF	A13CD	M30BL	L36BJ	N20BF	M15BD	L16AM	N21AJ	M29AG	G26AD
13	K31CS	W37CP	D37CH	Z19CE	H14BM	D32BK	A28BG	H25BE	D24BC	A17AK	H34AH	L38AE
14	F18AB	Y27CR	J27CJ	C30CF	S23BP	J20BL	Z33BH	S35BF	J21BD	Z22AL	S26AJ	D11AF
15	V12AC	T13CS	U13CK	K14CH	P36CD	U28BM	C15BJ	P16BG	U17BE	C29AM	P38AK	J31AG
16	E37AD	G19AB	R19CL	F23CJ	W32CE	R33BP	K25BK	W24BH	R22BF	K34BC	W11AL	U18AH
17	M27AE	L30AC	X30CM	V36CK	Y20CF	X15CD	F35BL	Y21BJ	X29BG	F26BD	Y31AM	R12AJ
18	H13AF	D14AD	B14CP	E32CL	T28CH	B25CE	V16BM	T17BK	B34BH	V38BE	T18BC	X37AK
19	S19AG	J23AE	N23CR	M20CM	G33CJ	N35CF	E24BP	G22BL	N25BJ	E11BF	G12BD	B27AL
20	P30AH	U36AF	A36CS	H28CP	L15CK	A16CH	M21CD	L29BM	A38BK	M31BG	L37BE	N13AM
21	W14AJ	R32AG	Z32AB	S33CR	D25CL	Z24CJ	H17CE	D34BP	Z11BL	H18BH	D27BF	A19BC
22	Y23AK	X20AH	C20AC	P15CS	J35CM	C21CK	S22CF	J26CD	C31BM	S12BJ	J13BG	Z30BD
23	T36AL	B28AJ	K28AD	W25AB	U16CP	K17CL	P29CH	U38CE	K18BP	P37BK	U19BH	C14BE
24	G32AM	N33AK	F33AE	Y35AC	R24CR	F22CM	W34CJ	R11CF	F12CD	W27BL	R30BJ	K23BF
25	L20BC	A15AL	V15AF	T16AD	X21CS	V29CP	Y26CK	X31CH	V37CE	Y13BM	X14BK	F36BG
26	D28BD	Z25AM	E25AG	G24AE	B17AB	E34CR	T38CL	B18CJ	E27CF	T19BP	B23BL	V32BH
27	J33BE	C35BC	M35AH	L21AF	N22AC	M26CS	G11CM	N12CK	M13CH	G30CD	N36BM	E20BJ
28	U15BF	K16BD	H16AJ	D17AG	A29AD	H38AB	L31CP	A37CL	H19CJ	L14CE	A32BP	M28BK
29	R25BG		S24AK	J22AH	Z34AE	S11AC	D18CR	Z27CM	S30CK	D23CF	Z20CD	H33BL
30	X35BH		P21AL	U29AJ	C26AF	P31AD	J12CS	C13CP	P14CL	J36CH	C28CE	S15BM
31	B16BJ		M17AM		K38AG		U37AB	K19CR		U32CJ		P25BP

1951	JAN	FEB	MAR	APR	MAY	JUN	JUL	AUG	SEP	OCT	NOV	DEC
1	W35CD	R21BM	Z21BG	S29BE	D26AM	Z31AK	H12AG	D13AE	Z14AC	H36CR	D28CM	A15CJ
2	Y16CE	X17BP	C17BH	P34BF	J38BC	C18AL	S37AH	J19AF	C23AD	S32CS	J33CP	Z25CK
3	T24CF	B22CD	K22BJ	W26BG	U11BD	K12AM	P27AJ	U30AG	K36AE	P20AB	U15CR	C35CL
4	G21CH	N29CE	F29BK	Y38BH	R31BE	F37BC	W13AK	R14AH	F32AF	W28AC	R25CS	K16CM
5	L17CJ	A34CF	V34BL	T11BJ	X18BF	V27BD	Y19AL	X23AJ	V20AG	Y33AD	X35AB	F24CP
6	D22CK	Z26CH	E26BM	B35BJ	B12BG	E13BE	T30AM	B36AK	E28AH	T15AE	B16AC	V21CR
7	J29CL	C38CJ	M38BP	L18BL	N37BH	M19BF	G14BC	N32AL	M33AJ	G25AF	N24AD	E17CS
8	U34CM	K11CK	H11CD	D12BM	A27BJ	H30BG	L23BD	A20AM	H15AK	L35AG	A21AE	M22AB
9	R26CP	F31CL	S31CE	J37BP	Z13BK	S14BH	D36BE	Z28BC	S25AL	D16AH	Z17AF	H29AC
10	X38CD	V18CM	P18CF	U27CD	C19BL	P23BJ	J32BF	C33BD	P35AM	J24AJ	C22AG	S34AD
11	B11CS	E12CP	W12CH	R13CE	K30BM	W36BK	U20BG	K15BE	W16BC	U21AK	K29AH	P26AE
12	N31AB	H37CR	Y37CJ	X19CF	F14BP	Y32BL	R28BH	F25BF	Y24BD	R17AL	F34AJ	W38AF
13	A18AC	H27CS	T27CK	B30CH	V23CD	T20BH	X33BJ	V35BG	T21BE	X22AM	V26AK	Y11AG
14	Z12AD	S13AB	G13CL	N14CJ	E36CE	G28BP	B15BK	E16BH	G17BF	B29BC	E38AL	T31AH
15	C37AE	P19AC	L19CM	A23CK	M32CF	L33CD	N25BL	M24BJ	L22BG	N34BD	M11AM	G18AJ
16	K27AF	W30AD	D30CP	Z36CL	H20CH	D15CE	A35BM	H21BK	D29BH	A26BE	H31BC	L12AK
17	F13AG	Y14AE	J14CR	C32CM	S28CJ	J25CF	Z16BP	S17BL	J34BJ	Z38BF	S18BD	D37AL
18	V19AH	T23AF	U23CS	K20CP	P33CK	U35CH	C24CD	P22BM	U26BK	C11BG	P12BE	J27AM
19	E30AJ	G36AG	R36AB	F28CR	W15CL	R16CJ	K21CE	W29BP	R38BL	K31BH	W37BF	U13BC
20	M14AK	L32AH	X32AC	V33CS	Y25CM	X24CK	F17CF	Y34CD	X11BM	F18BJ	Y27BG	R19BD
21	H23AL	D20AJ	B20AD	E15AB	T35CP	B21CL	V22CH	T26CE	B31BP	V12BK	T13BH	X30BE
22	S36AM	J28AK	N28AE	M25AC	G16CR	N17CM	E29CJ	G38CF	N18CD	E37BL	G19BJ	B14BF
23	P32BC	U33AL	A33AF	H35AD	L24CS	A22CP	M34CK	L11CH	A12CE	M27BM	L30BK	N23BG
24	W20BD	R15AM	Z15AG	S16AE	D21AB	Z29CR	H26CL	D31CJ	Z37CF	H13BP	D14BL	A36BH
25	Y28BE	X25BC	C25AH	P24AF	J17AC	C34CS	S38CM	J18CK	C27CH	S19CD	J23BM	Z32BJ
26	T33BF	B35BD	K35AJ	W21AG	U22AD	K26AB	P11CP	U12CL	K13CJ	P30CE	U36BP	C20BK
27	G15BG	N16BE	F16AK	Y17AH	R29AE	F38AC	W31CR	R37CH	F19CK	W14CF	R32CD	K28BL
28	L25BH	A24BF	V24AL	T22AJ	X34AF	V11AD	Y18CS	X27CP	V30CL	Y23CH	X20CE	F33BM
29	D35BJ		E21AM	G29AK	B26AG	E31AE	T12AB	B13CR	E14CM	T36CJ	B28CF	V15BP
30	J16BK		M17BC	L34AL	N38AH	M18AF	G37AC	N19CS	M23CP	G32CK	N33CD	E25CD
31	U24BL		H22BD		A11AJ		L27AD	A30AB		L20CL		M35CE

1952	JAN	FEB	MAR	APR	MAY	JUN	JUL	AUG	SEP	OCT	NOV	DEC
1	H16CF	D17CD	N22BK	M26BH	G11BE	N12BC	E27AK	G30AH	N36AF	E20AC	G15CS	B35CM
2	S24CH	J22CE	A29BL	H38BJ	L31BF	A37BD	M13AL	L14AJ	A32AG	M28AD	L25AB	N16CP
3	P21CJ	U29CF	Z34BM	S11BK	D18BG	Z27BE	H19AM	D23AK	Z20AH	H33AE	D35AC	A24CR
4	W17CK	R34CH	C26BP	P31BL	J12BH	C13BF	S30BC	J36AL	C28AJ	S15AF	J16AD	Z21CS
5	Y22CL	X26CJ	K38CD	W18BM	U37BJ	K19BG	P14BD	U32AM	K33AK	P25AG	U24AE	C17AB
6	T29CM	B38CK	F11CE	Y12BP	R27BK	F30BH	W23BE	R20BC	F15AL	W35AH	R21AF	K22AC
7	G34CP	N11CL	V31CF	T37CD	X13BL	V14BJ	Y36BF	X28BD	V25AM	Y16AJ	X17AG	F29AD
8	L26CR	A31CH	E18CH	G27CE	B19BM	E23BK	T32BG	B33BE	E35BC	T24AK	B22AH	V34AE
9	D38CH	Z18CP	M12CJ	L13CF	N30BP	M36BL	G28BH	N15BF	M16BD	G21AL	N29AJ	E26AF
10	J11AB	C12CR	H37CK	D19CH	A14CD	H32BJ	L28BJ	A25BG	A25BG	H24BE	L17AM	A34AK	M38AG
11	U31AC	K37CS	S27CL	J30CJ	Z23CE	S20BP	D33BK	Z35BH	S21BF	D22BC	Z26AL	H11AH
12	R18AD	F27AB	P13CM	U14CK	C36CF	P28CD	J15BL	C16BJ	P17BG	J29BD	C38AM	S31AJ
13	X12AE	V13AC	W19CP	R23CL	K32CH	W33CE	U25BM	K24BK	W22BH	U34BE	K11BC	P18AK
14	B37AF	E19AD	T30CR	X35CM	F20CJ	Y15CF	R35BP	F21BL	Y29BJ	R26BF	F31BD	W12AL
15	N27AG	M30AE	T14CS	B32CP	V28CK	T25CH	X16CD	V17BM	T34BK	X38BG	V18BE	Y37AM
16	A13AH	H14AF	G23AB	N20CR	E33CL	G35CJ	B24CE	E22BP	G26BL	B11BH	E12BF	T27BC
17	Z19AJ	S23AG	L36AC	A28C9	M15CM	L16CK	N21CF	M29CD	L38BM	N31BJ	M37BG	G13BD
18	C30AK	P36AH	D32AD	Z33AB	H25CP	D24CL	A17CH	H34CE	D11BP	A18BK	H27BH	L19BE
19	K14AL	M32AJ	J20AE	C15AC	S35CR	J21CH	Z22CJ	S26CF	J31CD	Z12BL	S13BJ	D30BF
20	F23AM	Y20AK	U28AF	K25AD	P16CS	U17CP	C29CK	P36CH	U16CE	C37BM	P19BK	J14BG
21	V36BC	T28AL	R33AG	F35AE	W24AB	R22CR	K34CL	W11CJ	R12CF	K27BP	W30BL	U23BH
22	E32BD	G33AM	X15AH	V16AF	Y21AC	X29CS	F26CM	Y31CK	X37CH	F13CD	Y14BM	R36BJ
23	M20BE	L15BC	B25AJ	E24AG	T17AD	B34AB	V38CP	T16CL	B27CJ	V19CE	T23BP	X32BK
24	H28BF	D25BD	N35AK	M21AH	G22AE	N26AC	E11CR	G12CM	N13CK	E30CF	G36CD	B20BL
25	S33BG	J35BE	A16AL	H17AJ	L29AF	A38AD	M31CS	L37CP	A19CL	M14CH	L32CE	N28BM
26	P15BH	U16BF	Z24AM	S22AK	D34AG	Z11AE	H18AB	D27CR	Z30CM	H23CJ	D20CF	A33BP
27	W25BJ	R24BG	C21BC	P29AL	J26AH	C31AF	S12AC	J13CS	C14CP	S36CK	J28CH	Z15CD
28	Y35BK	X21BH	K17BD	W34AM	U38AJ	K18AG	P37AD	U19AB	K23CR	P32CL	U33CJ	C25CE
29	T16BL	B17BJ	F22BE	Y26BC	R11AK	F12AH	W27AE	R30AC	F36CS	W20CH	R15CK	K35CF
30	G24BM		V29BF	T38BD	X31AL	V37AJ	Y13AF	X14AD	V32AB	Y28CP	X25CL	F16CH
31	L21BP		E34BG		B18AM		T19AG	B23AE		T33CR		V24CJ

(152)

FIRST LETTER = YOUR PHYSICAL CODE, MIDDLE NUMBER = SENSITIVITY CODE, LAST TWO LETTERS = INTELLECTUAL CODE

1953	JAN	FEB	MAR	APR	MAY	JUN	JUL	AUG	SEP	OCT	NOV	DEC
1	E21CK	G29CH	R29BM	F38BK	W31BG	R37BE	K13AM	W14AK	R32AH	K28AE	W25AC	U16CR
2	M17CL	L34CJ	X34BP	V11BL	Y18BH	X27BF	F19BC	Y23AL	X20AJ	F33AF	Y35AO	R24CS
3	H22CM	D26CK	B26CD	E31BM	T12BJ	B13BG	V30BD	T36AM	B28AK	V15AG	T16AE	X21AB
4	S29CP	J38CL	N38CE	M18BP	G37BK	N19BH	E14BE	G32BC	N33AL	E25AH	G24AF	B17AC
5	P34CR	U11CM	A11CF	H12CD	L27BL	A30BJ	M23BF	L20BD	A15AM	H35AJ	L21AG	N22AD
6	W26CS	R31CP	Z31CH	S37CE	D13BP	Z14BK	H36BG	D28BE	Z25BC	H16AK	D17AH	A29AE
7	Y38AB	X18CR	C16CJ	P27CF	J19BP	C23BL	S32BH	J33BF	C35BD	S24AL	J22AJ	Z34AF
8	T11AC	B12CS	K12CK	W13CH	U30CO	K36BM	P20BJ	U15BG	K16BE	P21AM	U29AK	C26AG
9	G31AD	N37AB	F37CL	Y19CJ	R14CE	F32BP	W28BK	R25BH	F24BF	W17BC	R34AL	K38AH
10	L18AE	A27AC	V27CH	T30CK	X23CF	V20CO	Y33BL	X35BJ	V21BG	Y22BD	X26AM	F11AJ
11	D12AF	Z13AD	E13CP	G14CL	B36CH	E28CE	T15BM	B16BK	E17BH	T29BE	B38BC	V31AK
12	J37AG	C19AE	M19CR	L23CH	N32CJ	M33CF	G25BP	N24BL	M22BJ	G34BF	N11BD	E18AL
13	U27AH	K30AF	H30CS	D36CP	A20CK	H15CH	L35CO	A21BM	H29BK	L26BG	A31BE	M12AM
14	R13AJ	F14AG	S14AB	J32CR	Z28CL	S25CJ	D16CE	Z17BP	S34BL	D38BH	Z18BF	H37BC
15	X19AK	V23AH	P23AC	U20CS	C33CM	P35CK	J24CF	C22CD	P26BM	J11BJ	C12BG	S27BD
16	B30AL	E36AJ	W36AD	R28AB	K15CP	W16CL	U21CH	K29CE	W38BP	U31BK	K37BH	P13BE
17	N14AM	M32AK	Y32AE	X33AC	F25CR	Y24CH	R17CJ	F34CF	Y11CD	R18BL	F27BJ	W19BF
18	A23BC	H20AL	T20AF	B15AD	V35CS	T21CP	X22CK	V26CH	T31CE	X12BH	V13BK	Y30BG
19	Z36BD	S28AM	G28AG	N25AE	E16AB	G17CR	B29CL	E38CJ	G18CF	B37BP	E19BL	T14BH
20	C32BE	P33BC	L33AH	A35AF	M24AC	L22CS	N34CM	M11CK	L12CH	N27CO	M30BM	G23BJ
21	K20BF	W15BD	D15AJ	Z16AG	H21AD	D29AB	A26CP	H31CL	D37CJ	A13CE	H14BP	L36BK
22	F28BG	Y25BE	J25AK	C24AH	S17AE	J34AC	Z38CR	S18CM	J27CK	Z19CF	S23CD	D32BL
23	W33BH	T35BF	U35AL	K21AJ	P22AF	U26AD	C11CS	P12CP	U13CL	C30CH	P36CE	J20BM
24	E15BJ	G16BG	R16AM	F17AK	W29AG	R38AE	K31AB	W37CR	R19CM	K14CJ	W32CF	U28BP
25	M25BK	L24BH	K24BC	V22AL	Y34AH	X11AF	F18AC	Y27CS	X30CP	F23CK	Y20CH	R33CD
26	H35BL	D21BJ	B21BD	E29AM	T26AJ	B31AG	V12AD	T13AB	B14CR	V36CL	T28CJ	X15CE
27	S16BM	J17BK	N17BE	M34BG	G38AK	N18AH	E37AE	G19AC	N23CS	E32CM	G33CK	B25CF
28	P24BP	U22BL	A22BF	H26BD	L11AL	A12AJ	M27AF	L30AD	A36AB	M20CP	L15CL	N35CH
29	W21CD		Z29BG	S38BE	D31AM	Z37AK	H13AG	D14AE	Z32AC	H28CR	D25CM	A16CJ
30	Y17CE		C34BH	P11BF	J18BC	C27AL	S19AH	J23AF	C20AD	S33CS	J35CP	Z24CK
31	T22CF		K26BJ		U12BD		P30AJ	U36AG		P15AB		C21CL

1954	JAN	FEB	MAR	APR	MAY	JUN	JUL	AUG	SEP	OCT	NOV	DEC
1	K17CM	W34CK	D34CD	Z11BM	H18BJ	D27BG	A19BD	H23AM	D20AK	A33AG	H35AE	L24AB
2	F22CP	Y26CL	J26CE	C31BP	S12BK	J13BH	Z30BE	S36BC	J28AL	Z15AH	S16AF	D21AC
3	V29CR	T38CM	U38CF	K18CO	P37BL	U19BJ	C14BF	P32BD	U33AM	C25AJ	P24AG	J17AD
4	E34CS	G11CP	R11CH	F12CE	M27BM	R30BK	K23BG	M20BE	R15BC	K35AK	M21AH	U22AE
5	M26AB	L31CR	X31CJ	V37CF	Y13BP	X14BL	F36BH	Y28BF	X25BD	F16AL	Y17AJ	R29AF
6	H38AC	D18CS	B18CK	E27CH	T19CO	B23BM	V32BJ	T33BG	B35BE	V24AM	T22AK	X34AG
7	S11AD	J12AB	N12CL	M13CJ	G30CE	N36BP	E20BK	G15BH	N16BF	E21BC	G29AL	B26AH
8	P31AE	U37AC	A37CM	H19CK	L14CF	A32CD	M28BL	L25BJ	A24BG	M17BD	L34AM	N38AJ
9	W18AF	R27AD	Z27CP	S30CL	D23CH	Z20CE	H33BM	D35BK	Z21BH	H22BE	D26BC	A11AK
10	Y12AG	X13AE	C13CR	P14CM	J36CJ	C28CF	S15BP	J16BL	C17BJ	S29BF	J38BD	Z31AL
11	T37AH	B19AF	K19CS	W23CP	U32CK	K33CH	P25CO	U24BM	K22BK	P34BG	U11BH	C18AM
12	G27AJ	N30AG	F30AB	Y36CR	R20CL	F15CJ	W35CE	R21BP	F29BL	W26BH	R31BF	K12BC
13	L13AK	A14AH	V14AC	T32CS	X28CM	V25CK	Y16CF	X17CO	V34BM	Y38BJ	X18BG	F37BD
14	D19AL	Z23AJ	E23AD	G20AB	B33CP	E35CL	T24CH	B22CE	E26BP	T11BM	B12BH	V27BE
15	J30AM	C36AK	M36AE	L28AC	N15CR	M16CM	G21CJ	N29CF	M38CD	G31BL	N37BJ	E13BF
16	U14BC	K32AL	H32AF	D33AD	A25CS	H24CP	L17CK	A34CH	H11CE	L18BM	A27BK	M19BG
17	R23BD	F20AM	S20AG	J15AE	Z35AB	S21CR	D22CL	Z26CJ	S31CF	D12BP	Z13BL	H30BH
18	X36BE	V28BC	P28AH	U25AF	C16AC	P17CS	J29CM	C38CK	P18CH	J37CO	C19BM	S14BJ
19	B32BF	E33BD	W33AJ	R35AG	K24AD	W22AB	U34CP	K11CL	W12CJ	U27CE	K30BP	P23BK
20	N20BG	M15BE	Y15AK	X16AH	F21AE	Y29AC	R26CR	F31CM	Y37CK	R13CF	F14CD	W36BL
21	A28BH	H25BF	T25AL	B24AJ	V17AF	T34AD	X38CS	V18CP	T27CL	X19CH	V23CE	Y32BM
22	Z33BJ	S35BG	G35AM	N21AK	E22AG	G26AE	B11AB	E12CR	G13CM	B30CJ	E36CF	T20BP
23	C15BK	P16BH	L16BC	A17AL	M29AH	L38AF	N31AC	M37CS	L19CP	N14CK	M32CH	G28CD
24	K25BL	W24BJ	D24BD	Z22AM	H11AJ	D11AG	A18AD	H27AB	D30CR	A23CL	H20CJ	L33CE
25	F35BM	Y21BK	J21BE	C29BC	S26AK	J31AH	Z12AE	S13AC	J14CS	Z36CM	S28CK	D15CF
26	V16BP	T17BL	U17BF	K34BD	P38AL	U18AJ	C37AF	P19AD	U23AB	C32CP	P33CL	J25CH
27	E24CD	G22BM	R22BG	F26BE	W11AM	R12AK	K27AG	W30AE	R36AC	K20CR	W15CM	U35CJ
28	M21CE	L29BP	X29BH	V38BF	Y31BC	X37AL	F13AH	Y14AF	X32AD	F28CS	Y25CP	R16CK
29	H17CF	B34BJ	E11BG	T18BD	B27AM	V19AJ	T23AG	B20AE	V33AB	T35CR	X24CL
30	S22CH		N26BK	M31BH	G12BE	N13BC	E30AK	G36AH	N28AF	E15AC	G16CS	B21CM
31	P29CJ		A38BL		L37BF		M14AL	L32AJ		M25AD		N17CP

1955	JAN	FEB	MAR	APR	MAY	JUN	JUL	AUG	SEP	OCT	NOV	DEC
1	A22CR	H26CM	T26CF	B31CD	V12BL	T13BJ	X30BF	V36BD	T28AM	X15AJ	V16AG	Y21AD
2	Z29CS	S38CP	G38CH	N18CE	E37BM	G19BK	B14BG	E32BE	G33BC	B25AK	E24AH	T17AE
3	C34AB	P11CR	L11CJ	A12CF	M27BP	L30BL	N23BH	M20BF	L15BD	N35AL	M21AJ	G22AF
4	K26AC	W31CS	D31CK	Z37CH	H13CO	D14BM	A36BJ	H28BG	D25BE	A16AM	H17AK	L29AG
5	F38AD	Y18AB	J18CL	C27CJ	S19CE	J23BP	Z32BK	S33BH	J35BF	Z24BC	S22AL	D34AH
6	V11AE	T12AC	U12CM	K13CK	P30CF	U36CO	C20BL	P15BJ	U16BG	C21BD	P29AM	J26AJ
7	E31AF	G37AD	R37CP	F19CL	W14CH	R32CE	K28BM	W25BK	R24BH	K17BE	W34BC	U38AK
8	M18AG	L27AE	X27CR	V30CH	Y23CJ	X20CF	F33BP	Y35BL	X21BJ	F22BF	Y26BD	R11AL
9	H12AH	D13AF	B13CS	E14CP	T36CK	B28CH	V15CO	T16BM	B17BK	V29BG	T38BE	X31AM
10	S37AJ	J19AG	N19AB	M23CR	G32CL	N33CJ	E25CE	G24BP	N22BL	E34BH	G11BF	N12BD
11	P27AK	U30AH	A30AC	H36CS	L20CM	A15CK	M35CF	L21CO	A29BM	M26BK	L31BG	N12BO
12	W13AL	R14AJ	Z14AD	S32AB	D28CP	Z25CL	H16CH	D17CE	Z34BP	H38BK	D18BH	A37BE
13	Y19AM	X23AK	C23AE	P20AC	J33CR	C35CM	S24CJ	J22CF	C26CD	S11BL	J12BJ	Z27BF
14	T30BC	B36AL	K36AF	W28AD	U15CS	K16CP	P21CK	U29CH	K38CE	P31BM	U37BK	C13BG
15	G14BD	N32AM	F32AG	Y33AE	R25AB	F24CR	W17CL	R34CJ	F11CF	W18BP	R27BL	K19BH
16	L23BE	A20BC	V20AH	T15AF	X35AC	V21CS	Y22CM	X26CK	V31CH	Y12CD	X13BM	F30BJ
17	D36BF	Z28BD	E28AJ	G25AG	B16AD	E17AB	T29CP	B38CM	E18CJ	T37CO	B19BP	V14BK
18	J32BG	C33BF	M33AK	L35AH	N24AE	M22AC	G34CL	N11CK	M12CK	G27CF	N30CD	E23BL
19	U20BH	K15BF	H15AL	D16AJ	A21AF	H29AD	L26CE	A31CC	H37CL	L13CH	A14CE	M36BM
20	R28BJ	F25BG	S25AM	J24AK	Z17AG	S34AE	D38AB	Z18CR	S27CM	D19CJ	Z23CF	H32BP
21	X33BK	V35BH	P35BC	U21AL	C22AH	P26AF	J11AC	C12CS	P13CP	J30CK	C36CH	S20CD
22	B15BL	E16BJ	W16BD	R17AM	K29AJ	W38AG	U31AD	K37AB	W19CR	U14CL	K32CJ	P28CE
23	N25BM	M24BK	Y24BE	X22BC	F34AK	Y11AH	R18AE	F27AC	Y30CS	R23CM	F20CK	W33CF
24	A35BP	H21BL	T21BF	B29BD	V26AL	T31AJ	X12AF	V13AD	T14AB	X36CR	V28CL	Y15CH
25	Z16CD	S17BM	G17BG	N34BE	E38AM	G18AK	B37AG	E19AE	G23AC	B32CR	E33CM	T25CJ
26	C24CE	P22BP	L22BH	A26BF	M11BD	L12AL	N27AH	M30AF	L36AD	N20CS	M15CP	L16CL
27	K21CF	W29CD	D29BJ	Z38BG	H31BD	D37AM	A13AJ	H14AG	D32AE	A28AB	H25CR	L16CL
28	F17CH	Y34CE	J34BK	C11BB	S18BE	J27BC	Z19AK	S23AH	J20AF	Z33AC	S35CS	D24CM
29	V22CJ		U26BL	K31BJ	P12BF	U13BD	C30AL	P36AJ	U28AG	C15AD	P16AB	J21CP
30	E29CK		R38BM	F18BK	W37BG	R19BE	K14AM	W32AK	R33AH	K25AE	W24AC	U17CR
31	M34CL		X11BP		Y27BH		F23BC	Y20AL		F35AF		R22CS

FIRST LETTER = YOUR PHYSICAL CODE, MIDDLE NUMBER = SENSITIVITY CODE, LAST TWO LETTERS = INTELLECTUAL CODE

1956	JAN	FEB	MAR	APR	MAY	JUN	JUL	AUG	SEP	OCT	NOV	DEC
1	X29AB	V38CR	W11CK	R12CH	K27CD	W30BM	U23BJ	K20BG	W15BE	U35AM	K21AK	P22AG
2	B34AC	E11CS	Y31CL	X37CJ	F13CE	Y14BP	R36BK	F28BH	Y25BF	R16BC	F17AL	W29AH
3	N26AD	M31AB	T18CM	B27CK	V19CF	T23CD	X32BL	V33BJ	T35BG	X24BD	V22AM	Y34AJ
4	A38AE	H18AC	G12CP	N13CL	E30CH	G36CE	B20BM	E15BK	G16BM	B21BE	E29BC	T26AK
5	Z11AF	S12AD	L37CR	A19CM	M14CJ	L32CF	N28BP	M25BL	.24BJ	N17BF	M34BD	G38AL
6	C31AG	P37AE	D27CS	Z30CP	H23CK	D20CH	A33CD	H35BM	D21BK	A22BG	H26BE	L11AM
7	K18AH	W27AF	J13AB	C14CR	S36CL	J28CJ	Z15CE	S16BP	J17BL	Z29BH	S38BF	D31BC
8	F12AJ	Y13AG	U19AC	K23CS	P32CM	U33CK	C25CF	P24CD	U22BM	C34BJ	P11BG	J18BD
9	V37AK	T19AH	R30AD	F36AB	W20CP	R15CL	K35CH	M21CE	R29BP	K26BK	W31BH	U12BE
10	E27AL	G30AJ	X14AE	V32AC	Y28CR	X25CM	F16CJ	Y17CF	X34CD	F38BL	Y18BJ	R37BF
11	M13AM	L14AK	B23AF	E20AD	T33CS	B35CP	V24CK	T22CH	B26CE	V11BH	T12BK	X27BG
12	H19BC	D23AL	N36AG	M28AE	G15AB	N16CR	E21CL	G29CJ	N38CF	E31BP	G37BL	B13BH
13	S30BD	J36AM	A32AH	H33AF	L25AC	A24CS	M17CM	L34CK	A11CH	M18CD	L27BM	N19BJ
14	P14BE	U32BD	Z20AJ	S15AG	D35AD	Z21AB	H22CP	D26CL	Z31CJ	H12CE	D13BP	A30BK
15	W23BF	R20BD	C28AK	P25AH	J16AE	C17AC	S29CR	J38CM	C18CK	S37CF	J19CD	Z14BL
16	Y36BG	X28BE	K33AL	W35AJ	U24AF	K22AD	P34CS	U11CP	K12CL	P27CH	U30CE	C23BH
17	T32BH	B33BF	F15AM	Y16AK	R21AG	F29AE	M26AB	R31CR	F37CM	M13CJ	R14CF	K36BP
18	G20BJ	N15BG	V25BC	T24AL	X17AH	V34AF	Y38AC	X18CS	V27CP	Y19CK	X23CH	F32CD
19	L28BK	A25BH	E35BD	G21AM	B22AJ	E26AG	T11AD	B12AB	E13CR	T30CL	B36CJ	V20CE
20	D33BL	Z35BJ	M16BE	L17BC	N29AK	M38AH	G31AE	N37AC	M19CS	G14CM	N32CK	E28CF
21	J15BM	C16BK	H24BF	D22BD	A34AL	H11AJ	L18AF	A27AD	H30AB	L23CP	A20CL	M33CH
22	U25BP	K24BL	S21BG	J29BE	Z26AM	S31AK	D12AG	Z13AE	S14AC	D36CR	Z28CM	H15CJ
23	R35CD	F21BM	P17BH	U34BF	C38BC	P18AL	J37AH	C19AF	P23AD	J32CS	C33CP	S25CK
24	X16CE	V17BP	W22BJ	R26BG	K11BD	W12AM	U27AJ	K30AG	W36AE	U20AB	K15CR	P35CL
25	B24CF	E22CD	Y29BK	X38BH	F31BE	Y37BC	R13AK	F14AH	Y32AF	R28AC	F25CS	W16CM
26	N21CH	M29CE	T34BL	B11BJ	V18BF	T27BD	X19AL	V23AJ	T20AG	X33AD	V35AB	Y24CP
27	A17CJ	H34CF	G26BM	N31BK	E12BG	G13BE	B30AM	E36AK	G28AH	B15AE	E16AC	T21CR
28	Z22CK	S26CH	L38BP	A18BL	M37BH	L19BF	N14BC	M32AL	L33AJ	N25AF	M24AD	G17CS
29	C29CL	P38CJ	D11CD	Z12BM	H27BJ	D30BG	A23BD	H20AM	D15AK	A35AG	H21AE	L22AB
30	K34CM		J31CE	O37BP	S13BK	J14BH	Z36BE	S28BC	J25AL	Z16AH	S17AF	D29AC
31	F26CP		U18CR		P19BL		C32BF	P33BD		C24AJ		J34AD

1957	JAN	FEB	MAR	APR	MAY	JUN	JUL	AUG	SEP	OCT	NOV	DEC
1	U26AE	K31AC	H31CM	D37CK	A13CF	H14CD	L36BL	A28BJ	H25BG	L16BD	A17AM	M29AJ
2	R38AF	F18AD	S18CP	J27CL	Z19CH	S23CE	D32BM	Z33BK	S35BH	D24BE	Z22BC	H34AK
3	X11AG	V12AE	P12CR	U13CM	C30CJ	P36CF	J20BP	C15BL	P16BJ	J21BF	C29BD	S26AL
4	B31AH	E37AF	M37CS	R19CP	K14CK	M32CH	U28CD	K25BM	M24BK	U17BG	K34BE	P38AM
5	N18AJ	M27AG	Y27AB	X30CR	F23CL	Y20CJ	R33CE	F35BP	Y21BL	R22BH	F26BF	W11BC
6	A12AK	H13AH	T13AG	B14CS	V16CM	T28CK	X15CF	V16CD	T17BM	X29BJ	V38BG	Y31BD
7	Z37AL	S19AJ	G19AD	N23AB	E32CP	G33CL	B25CH	E24CE	G22BP	B34BK	E11BH	T18BE
8	C27AM	P30AK	L30AE	A36AC	M20CR	L15CM	N35CJ	M21CF	L29CD	N26BL	M31BJ	G12BF
9	K13BC	W14AL	D14AF	Z32AD	H28CS	D25CP	A16CK	H17CH	D34CE	A38BM	H18BK	L37BG
10	F19BD	Y23AM	J23AG	C20AE	S33AB	J35CR	Z24CL	S22CJ	J26CF	Z11BP	S12BL	D27BH
11	V30BE	T36BC	U36AH	K28AF	P15AC	U16CS	C21CM	P29CK	U38CH	C31CD	P37BM	J13BJ
12	E14BF	G32BD	R32AJ	F33AG	W25AD	R24AB	K17CP	W34CL	R11CJ	K18CE	W27BP	U19BK
13	M23BG	L20BE	X20AK	V15AH	Y35AE	X21AC	F22CR	Y26CM	X31CK	F12CF	Y13CD	R30BL
14	H36BH	D28BF	B28AL	E25AJ	T16AF	B17AD	V29CS	T38CP	B18CL	V37CH	T19CE	X14BM
15	S32BJ	J33BG	N33AM	H35AK	L21AH	N22AE	E34AB	L31CS	N12CH	E27CJ	G30CF	B23BP
16	P20BK	U15BH	A15BD	H16AL	L21AH	A29AF	M26AC	L31CS	A37CP	M13CK	L14CH	N36CD
17	W28BL	R25BJ	Z25BD	S24AM	D17AJ	Z34AG	H38AD	D18AB	Z27CR	H19CL	D23CJ	A32CE
18	Y33BM	X35BK	C35BE	P21BC	J22AK	C26AH	S11AE	J12AC	C13CS	S30CM	J36CK	Z20CF
19	T15BP	B16BL	K16BF	W17BD	U29AL	K38AJ	P31AF	U37AD	K19AB	P14CP	U32CL	C28CH
20	G25CD	N24BM	F24BG	Y22BE	R34AM	F11AK	W18AG	R27AE	F30AC	W23CR	R20CM	K33CJ
21	L35CE	A21BP	V21BH	T29BF	X26BC	V31AL	Y12AH	X13AF	V14AD	Y36CS	X28CP	F15CK
22	D16CF	Z17CD	E17BJ	G34BG	B38BD	E18AM	T37AJ	B19AG	E23AE	T32AB	B33CR	V25CL
23	J24CH	C22CE	M22BK	L26BH	N11BC	M12BC	G27AK	N30AH	M36AF	G20AC	N15CS	E35CM
24	U21CJ	K29CF	H29BL	D38BJ	A31BF	H37BD	L13AL	A14AJ	H32AG	L28AD	A25AB	M16CP
25	R17CK	F34CH	S34BM	J11BK	Z18BG	S27BE	D19AM	Z23AK	S20AH	D33AE	Z35AC	H24CR
26	X22CL	V26CJ	P26BP	U31BL	C12BH	P13BF	J30BC	C36AL	P28AJ	J15AF	C16AD	S21CS
27	B29CM	E38CK	W38CD	R18BH	K37BJ	W19BG	U14BD	K32AM	W33AH	U25AG	K24AE	P17AB
28	N34CP	M11CL	Y11CE	X12BP	F27BK	Y30BH	R23BE	F20BC	Y15AL	R35AH	F21AF	W22AC
29	A26CR		T31CF	B37CD	V13BL	T14BJ	X36BF	V28BD	T25AM	X16AJ	V17AG	Y29AD
30	Z38CS		G18CH	U27CF	E19BM	G23BK	B32BG	E33BE	G35BC	B24AK	E22AH	T34AE
31	C11AB		L12CJ		M30BP		N20BH	M15BF		N21AL		G26AF

1958	JAN	FEB	MAR	APR	MAY	JUN	JUL	AUG	SEP	OCT	NOV	DEC
1	L38AG	A18AE	V18CR	T27CM	X19CJ	V23CF	Y32BP	X33BL	V35BJ	Y24BF	X22BD	F34AL
2	D11AH	Z12AF	E12CS	G13CP	B30CK	E36CH	T20CD	B15BM	E16BK	T21BG	B29BE	V26AM
3	J31AJ	C37AG	M37AB	U19CR	N14CL	M32CJ	G28CE	N25BP	M24BL	G17BH	N34BF	E38BC
4	U18AK	K27AH	H27AC	D30CS	A23CM	H20CK	L33CF	A35CD	H21BM	L22BJ	A26BG	M11BD
5	R12AL	F13AJ	S13AD	J14AB	Z36CP	S28CL	D15CH	Z16CE	S17BP	D29BK	Z38BH	H31BE
6	X37AM	V19AN	P19AE	U23AC	C32CR	P33CM	J25CJ	C24CF	P22CD	J34BL	C11BJ	S16BF
7	B27BC	E38AL	W30AF	R36AD	K20CS	W15CP	U35CK	K21CH	W29CE	U26BM	K31BK	P12BG
8	N13BD	M14AM	Y14AG	X32AE	F28AB	Y25CR	R16CL	F17CJ	Y34CF	R38BP	F18BL	W37BH
9	A19BE	H23BC	T23AH	B20AF	V33AC	T35CS	X24CH	V22CK	T26CH	X11CD	V12BM	Y27BJ
10	Z30BF	S36BD	G36AJ	N28AG	E15AD	G16AB	B21CP	E29CL	G38CJ	B31CF	E37BP	T13BK
11	C14BG	P32BE	L32AK	A33AH	M25AE	L24AC	N17CR	M34CM	L11CH	N18CF	M27CD	G19BL
12	K23BH	W20BF	D29AL	Z15AJ	H35AF	D21AD	A22CS	H26CP	D31CL	A12CH	H13CE	L30BM
13	F35BJ	Y28BG	J28AM	C25AK	S16AG	J17AE	Z29AB	S38CR	J18CM	Z37CJ	S19CF	D14BP
14	V32BK	T33BH	U33BC	K35AL	P24AH	U22AF	C34AC	P11CS	U12CP	C27CK	P30CH	J23CD
15	E20BL	G15BJ	R15BD	F16AM	W21AJ	R29AG	K26AD	W31CM	R37CR	K13CL	W14CJ	U36CE
16	M28BM	L25BK	X25BE	V24BG	Y17AK	X34AH	F36AE	Y18AC	X27CS	F19CM	Y23CK	R32CF
17	H33BP	D35BL	B35BF	E21BD	T22AL	B26AJ	V11AF	T12AD	B13AB	V30CL	T36CL	X20CH
18	S15CD	J16BM	N16BG	M17BE	G29AM	N38AK	E31AG	G37AE	N19AC	E14CR	G32CH	B28CJ
19	P25CE	U24BP	A24BH	H22BF	L34BC	A11AL	M18AH	L27AF	A30AD	M23CS	L20CP	N33CK
20	W35CF	R21CD	Z21BJ	S29BG	D26BD	Z31AM	H12AJ	D13AG	Z14AE	H36AB	D28CR	A15CL
21	Y16CH	X17CE	C17BK	P34BH	J38BE	C18AK	S37AL	J19AH	C23AF	S32AC	J33CS	Z25CM
22	T24CJ	B22CF	K22BL	W26BJ	U11BF	K12BD	P27AL	U30AJ	K36AG	P20AD	U15AB	C35CP
23	G21CK	N29CH	F29BM	Y38BK	R31BG	F37BE	M13AM	R14AK	F32AH	M28AE	R25AC	K16CR
24	L17CL	A34CJ	V34BP	T11BL	X18BH	V27BF	Y19BC	X23AL	V20AJ	Y33AF	X35AD	F24CS
25	D22CM	Z26CK	E26CD	G31BM	B12BJ	E13BG	T38BD	B36AM	E28AK	T15AG	B16AE	V21AB
26	J29CP	C38CL	M38CE	L18BP	N37BK	M19BH	G14BF	N32BC	M33AL	G25AH	N24AF	E17AC
27	U34CR	K11CM	H11CF	D12CD	A27BL	H30BJ	L23BF	A20BD	H15AM	L35AJ	A21AG	M22AD
28	R26CS	F31CP	S31CH	J37CE	Z13BM	S14BK	D36BG	Z28BE	S25BC	D16AK	Z17AH	H29AE
29	X38AB		P18CJ	U27CF	C19BP	P23BL	J32BH	C33BF	P35BD	J24AL	C22AJ	S34AF
30	B11AC		W12CK	R13CH	K30CD	W36BM	U20BJ	K15BG	W16BE	U21AM	K29AK	P26AG
31	N31AD		Y37CL		F14CE		R28BK	F25BH		R17BC		W38AH

(154)

FIRST LETTER = YOUR PHYSICAL CODE, MIDDLE NUMBER = SENSITIVITY CODE, LAST TWO LETTERS = INTELLECTUAL CODE

1959	JAN	FEB	MAR	APR	MAY	JUN	JUL	AUG	SEP	OCT	NOV	DEC
1	Y11AJ	X12AG	C12AB	P13CR	J30CL	C36CJ	S20CE	J15BP	C16BL	S21BH	J29BF	Z26BC
2	T31AK	B37AM	K37AC	W19CS	U14CM	K32CK	P28CF	U25CD	K24BM	P17BJ	U34BG	C38BD
3	G18AL	N27AJ	F27AD	Y30AB	R23CP	F20CL	W33CH	R35CE	F21BP	W223K	R26BH	K11BE
4	L12AM	A13AK	V13AE	T14AC	X36CR	V28CM	Y15CJ	X16CF	V17CD	Y29BL	X38BJ	F31BF
5	D37BC	Z19AL	E19AF	G23AD	B32CS	E33CP	T25CK	B24CH	E22CE	T343M	B11BK	V18BG
6	J27BD	C30AM	H30AG	L36AE	N20AB	M15CR	G35CL	N21CJ	M29CF	G26BP	N31BL	E12BH
7	U13BE	K14BC	H14AH	D32AF	A28AC	H25CS	L16CM	A17CK	434CH	L38CD	A18BM	M37BJ
8	R19BF	F23BD	S23AJ	J20AG	Z33AD	S35AB	D24CP	Z22CL	S26CJ	D11CE	Z12BP	H27BK
9	X30BG	V36BE	P36AK	U28AH	C15AE	P16AC	J21CR	C29CM	P38CK	J31CF	C37CD	S13BL
10	B14BH	E32BF	W32AL	R33AJ	K25AF	W24AD	U17CS	K34CP	W11CL	U18CH	K27CE	P19BM
11	N23BJ	M20BG	Y20AM	X15AK	F35AG	Y21AE	R22AB	F26CR	Y31CM	R12CJ	F13CF	H30BP
12	A36BK	H28BH	T28BC	B25AL	V16AH	T17AF	X29AC	V38CS	T18CP	X37CK	V19CH	Y14CD
13	Z32BL	S33BJ	G33BD	N35AM	E24AJ	G22AG	B34AD	E11AB	G12CR	B27CL	E30CJ	I23CE
14	C20BM	P15BK	L15BE	A16BC	M21AK	L29AH	N26AE	M31AC	L37CS	N13CM	M14CK	G36CF
15	K28BP	W25BL	D25BF	Z24BD	H17AL	D34AJ	A38AF	H18AD	D27BB	A19CP	H23CL	L32CH
16	F33CD	Y35BM	J35BG	C21BE	S22AM	J26AK	Z11AG	S12AE	J13AC	Z30CR	S36CM	D20CJ
17	V15CE	T16BP	U16BH	K17BF	P29BC	U38AL	C31AH	P37AF	U19AD	C14CS	P32CP	J28CK
18	E25CF	G24CD	R24BJ	F22BG	W34BD	R11AM	K18AJ	W27AG	R30AE	K23AB	W20CR	U33CL
19	M35CH	L21CE	X21BK	V29BH	Y26BE	X38BC	F12AK	Y13AH	X14AF	F36AC	Y28CS	R15CM
20	H16CJ	D17CF	B17BL	E34BJ	T38BF	B18BD	V37AL	T19AJ	B23AG	V32AD	T33AB	X25CP
21	S24CK	J22CH	N22BM	M26BK	G11BG	N12BE	E27AM	G30AK	N36AH	E20AE	G15AC	B35CR
22	P21CL	U29CJ	A29BP	H38BL	L31BH	A37BF	M13BC	L14AL	A32AJ	M28AF	L25AD	N16CS
23	W17CM	R34CK	Z34CD	S11BM	D18BJ	Z27BG	H19BD	D23AM	Z20AK	H33AG	D35AE	A24AB
24	Y22CP	X26CL	C26CE	P31BP	J12BK	C13BH	S30BE	J36BC	C28AL	S15AH	J16AF	Z21AC
25	T29CR	B38CM	K38CF	W18CD	U37BL	K19BJ	P14BF	U32BD	K33AM	P25AJ	U24AG	C17AD
26	G34CS	N11CP	F11CH	Y12CE	R27BM	F30BK	W23BG	R20BE	F15BC	W35AK	R21AH	K22AE
27	L26AB	A31CR	V31CJ	T37CF	X13BP	V14BL	Y36BH	X28BF	V25BD	Y16AL	X17AJ	F29AF
28	D38AC	Z18CS	E18CK	G27CH	B19CD	E23BM	T32BJ	B33BG	E35BE	T24AM	B22AK	V34AG
29	J11AD		M12CK	L13CJ	N30CE	M36BP	G20BK	N15BH	M16BF	G21BC	N29AL	E26AH
30	U31AE		H37CM	D19CK	A14CF	H32CD	L28BL	A25BJ	H24BG	L17BD	A34AM	H38AJ
31	R18AF		S27CP		Z23CH		D33BM	Z35BK		D22BE		H11AK

1960	JAN	FEB	MAR	APR	MAY	JUN	JUL	AUG	SEP	OCT	NOV	DEC
1	S31AL	J37AJ	A27AE	H30AC	L23CR	A20CM	M33CJ	L35CF	A21CD	M28BL	L26BJ	N11BF
2	P18AM	U27AK	Z13AF	S14AD	D36CS	Z28CP	H15CK	D16CM	Z17CE	H29BM	D38BK	A31BG
3	W12BC	R13AL	C19AG	P23AE	J32AB	C33CR	S25CL	J24CJ	C22CF	S34BP	J11BL	Z18BH
4	Y37BD	X19AM	K30AH	W36AF	U20AC	K15CS	P35CM	U21CK	K29CH	P26CD	U31BM	C12BJ
5	T27BE	B30BC	F14AJ	Y32AG	R28AD	F25AB	W16CP	R17CL	F34CJ	W38CE	R18BP	K37BK
6	G13BF	N14BD	V23AK	T20AH	X33AE	V35AC	Y24CR	X22CM	V26CK	Y11CF	X12CD	F27BL
7	L19BG	A23BE	E36AL	G28AJ	B15AF	E16AD	T21CS	B29CP	E38CL	T31CH	B37CE	V13BM
8	D30BH	Z36BF	M32AM	L33AK	N25AG	M24AE	G17AB	N34CR	M11CM	G18CJ	N27CF	E19BP
9	J14BJ	C32BG	H20BC	D15AL	A35AH	H21AF	L22AC	A26CS	H31CP	L12CK	A13CH	H30CD
10	U23BK	K20BH	S24BD	J25AM	Z16AJ	S17AG	D29AD	Z38AB	S16CR	D37CL	Z19CJ	H14CE
11	R36BL	F28BJ	P33BE	U35BC	C24AK	P22AH	J34AE	C11AC	P12CS	J27CM	C30CK	S23CF
12	X32BM	V33BK	W15BF	R16BD	K21AL	W29AJ	U26AF	K31AD	W37AB	U13CP	K14CL	P36CH
13	B20BP	E15BL	Y25BG	X24BE	F17AM	Y34AK	R38AG	F18AE	Y27AC	R19CR	F23CH	W32CJ
14	N28CD	M25BM	T35BH	B21BF	V22BC	T26AL	X11AH	V13AF	T13AD	X30CS	V36CP	Y20CK
15	A33CE	H35BP	G16BJ	N17BG	E29BD	G38AM	B31AJ	E37AG	G19AE	B14AB	E32CR	T28CL
16	Z15CF	S16CD	L24BK	A22BH	M34BE	L11BC	N18AK	M27AH	L30AF	N23AC	M20CS	G33CM
17	C25CH	P24CE	D21BL	Z29BJ	H26BF	D31BD	A12AL	H13AJ	D14AG	A36AD	H28AB	L15CP
18	K35CJ	W21CF	J17BM	C34BK	S38BG	J18BE	Z37AM	S19AK	J23AH	Z32AE	S33AC	D25CR
19	F16CK	Y17CH	U22BP	K26BL	P11BH	U12BF	C27BC	P30AL	U36AJ	C20AF	P15AD	J35CS
20	V24CL	T22CJ	R29CD	F38BM	W31BJ	R37BG	K13BD	W14AM	R32AK	K28AG	W25AE	U16AB
21	E21CM	G29CK	X34CE	V11BP	Y18BK	X27BH	F19BE	Y23BC	X20AL	F33AH	Y35AF	R24AC
22	M17CP	L34CL	B26CF	E31CD	T12BL	B13BJ	V30BF	T36BD	B28AM	V15AJ	T16AG	X21AD
23	H22CR	D26CM	N38CH	M18CE	L27BP	N19BK	E14BG	L28BE	N33BC	E25AK	G24AH	B17AE
24	S29CS	J38CP	A11CJ	H12CF	L27BP	A30BL	M23BH	L20BF	A15BD	M35AL	L21AJ	N22AF
25	P34AB	U11CR	Z31CK	S37CH	D13CD	Z14BM	H36BJ	D28BG	Z25BE	H16AM	D17AK	A29AG
26	W26AC	R31CS	C18CL	P27CJ	J19CE	C23BP	S32BK	J33BH	C35BF	S24BC	J22AL	Z34AH
27	Y38AD	X18AB	K12CM	W13CK	U30CF	K36CD	P20BL	U15BJ	K16BG	P21BD	U29AM	C26AJ
28	T11AE	B12AC	F37CP	Y19CL	R14CH	F32CD	W28BM	R25BK	F24BH	W17BE	R34BC	K38AK
29	G31AF	N37AD	V27CR	T30CM	X23CJ	V20CL	Y33BP	X35BL	V21BJ	Y22BF	X26BD	F11AL
30	L18AG		E13CS	G14CP	B36CK	E28CH	T15CD	B16BM	E17BK	T29BG	B38BD	V31AM
31	D12AH		M19AB		N32CL		G25CE	N24BP		G34BH		E18BC

1961	JAN	FEB	MAR	APR	MAY	JUN	JUL	AUG	SEP	OCT	NOV	DEC
1	M12BD	L13AM	X13AG	V14AE	Y36AB	X28CR	F15CL	Y16CJ	K17CF	F29BP	Y38BL	R31BH
2	H37BE	D19BC	B19AH	E23AF	T32AC	B33CS	V25CM	T24CK	B22CH	V34CD	T11BM	X18BJ
3	S27BF	J39BD	N30AJ	M36AG	G20AD	N15AB	E35CP	G21CL	N29CJ	E26CE	G31BP	B12BK
4	P13BG	U14BE	A14AK	H32AF	L28AC	A25AC	M16CR	L17CM	A34CK	M38CF	L18CD	N37BL
5	W19BH	R23BF	Z23AL	S20AJ	D33AF	Z35AD	H24CS	D22CP	Z26CL	H11CH	D12CC	A27BM
6	Y30BJ	X36BG	C36AM	P28AK	J15AG	C16AE	S21AB	J29CR	C38CM	S31CJ	J37CF	Z13BP
7	T14BK	B32BH	K32BC	W33AL	U25AH	K24AF	P17AC	U34CS	K11CP	P18CK	U27CH	C19CD
8	G23BL	N28BJ	F20BD	Y15AM	R35AK	F21AG	W22AD	R26AE	F31CL	W12CL	R13CJ	K30CE
9	L36BM	A28BK	V28BE	T25BC	X16AK	V17AH	Y32AE	X38AC	V18CS	Y37CM	X19CK	F14CF
10	D32BP	Z33BL	E33BF	G35BD	B24AL	E22AJ	T34AF	B11AD	E12AB	T27CP	B30CL	V23CH
11	J20CD	C15BM	M15BG	L16BE	N21AM	M29AK	G26AG	N31AE	M37AC	G13CR	N14CM	E36CJ
12	U28CE	K25BP	H25BH	D24BF	A17BC	H34AL	L38AH	A18AF	H27AD	L19CS	A23CP	M32CK
13	R33CF	F35CD	S35BJ	J21BG	Z22BD	S26AM	D11AJ	Z12AD	S13AE	D30AB	Z36CR	H20CL
14	X15CH	V16CE	P16BK	U17BH	C29BE	P38AC	J31AH	C37AH	P19AF	J14AC	C32CS	S28CM
15	B25CJ	E24CF	W24BL	R22BF	K34BF	W11BD	U18AL	K27AJ	W30AG	U23AD	K20AB	P33CP
16	N35CK	M21CH	Y21BM	X29BK	F26BG	Y31BE	R12AM	F13AH	Y14AH	R36AC	F28AC	W15CM
17	A16CL	H17CJ	T17BP	B34BL	V38BH	T18BF	X37BC	V19AL	T23AJ	X32AF	V33AD	Y25CS
18	Z24CM	S22CK	G22CD	N26BM	E11BJ	G12BG	B27BD	E30AM	G36AK	B20AG	E15AE	T35AB
19	C21CP	P29CL	L29CE	A38BP	M31BK	L37BH	N13BB	M14AL	L32AL	N28AH	M25AF	G16AC
20	K17CR	W34CM	D34CF	Z11CD	H18BL	D27BJ	A19BF	H28AM	D20AM	A33AJ	H35AG	L24AD
21	F22CS	Y26CH	J26CH	C31CE	S12BM	J13BK	Z30BG	S36BE	J28CC	Z15AK	S16AH	D21AE
22	V29AB	T38CR	U38CJ	K18CF	P37BP	U19BL	C14BH	P32BF	U33BD	C25AL	P24AJ	J17AF
23	E34AC	G11CS	R11CK	F12CH	W27BR	R30BM	K23BJ	W20BG	R15BE	K35AM	W21AK	U22AG
24	M26AD	L31AB	X31CL	V37CJ	Y13CE	X14BP	F36BK	Y28BH	X25BF	F16BC	Y17AL	R29AH
25	H38AE	D18AD	B18CM	E27CK	T19CF	B23BD	V32BL	T33BJ	B35BG	V26BE	T22AM	X34AJ
26	S11AF	J12AD	N12CP	M13CL	G30CH	N36CE	E20BM	G15BK	N16BH	E21BE	G29BC	B26AK
27	P31AG	U37AE	A37CR	H19CM	L14CJ	A32CF	M28BP	L25BL	A24BJ	M17BF	L34BD	N38AL
28	W18AH	R27AF	Z27CS	S30CP	D23CK	Z20CH	H33CD	D35BM	Z21BK	H22BG	D26BE	A11AM
29	Y12AJ		C13BB	P14CR	J36CL	C28CJ	S15CE	J38BP	C17BL	S29BH	J38BF	Z31BC
30	T37AK		K19AC	W23CS	U32CH	K33CK	P25CF	U24CD	K22BM	P34BJ	U11BG	C18BD
31	G27AL		F30AD		R20CP		W35CH	R21CE		W26BK		K12BE

FIRST LETTER = YOUR PHYSICAL CODE, MIDDLE NUMBER = SENSITIVITY CODE, LAST TWO LETTERS = INTELLECTUAL CODE

1962	JAN	FEB	MAR	APR	MAY	JUN	JUL	AUG	SEP	OCT	NOV	DEC
1	F37BF	Y19BD	J19AJ	C23AG	S32AD	J33AB	Z25CP	S24CL	J22CJ	Z34CE	S11BP	D18BK
2	V27BG	T30BE	U30AK	K36AH	P20AE	U15AC	C35CR	P21CM	U29CK	C26CF	P31CD	J12BL
3	E13BH	G14BF	R14AL	F32AJ	W28AF	R25AD	K16CS	W17CP	R34CL	K38CH	W18CE	U37BM
4	M19BJ	L23BG	X23AM	V20AK	Y33AG	X35AE	F24AB	Y22CR	X26CM	F11CJ	Y12CF	R27BP
5	H30BK	D36BH	B36BC	E28AL	T15AH	B16AF	V21AC	T29CS	B38CP	V31CK	T37CH	X13CD
6	S14BL	J32BJ	N32BD	M33AM	G25AJ	N24AG	E17AD	G34AB	N11CR	E18CL	G27CJ	B19CE
7	P23BM	U20BK	A20BE	H15BN	L35AK	A21AH	M22AE	L26AC	A31CS	M12CH	L13CK	N30CF
8	W36BP	R28BL	Z28BF	S25BD	D16AL	Z17AJ	H29AF	D38AD	Z18AB	H37CP	D19CL	A14CH
9	Y32CD	X33BM	C33BG	P35BE	J24AM	C22AK	S34AG	J11AE	C12AC	S27CR	J30CM	Z23CJ
10	T20CE	B15BP	K15BH	W16BF	U21BC	K29AL	P26AH	U31AF	K37AD	P13CS	U14CP	C36CK
11	G28CF	N25CD	F25BJ	Y24BG	R17BD	F34AM	W38AJ	R18AG	F27AE	W19AB	R23CR	K32CL
12	L33CH	A35CE	V35BK	T21BH	X22BE	V26BC	Y11AK	X12AH	V13AF	Y30AC	X36CS	F20CM
13	D15CJ	Z16CF	E16BL	G17BJ	B29BF	E38BD	T31AL	B37AJ	E19AG	T14AD	B32AB	V28CP
14	J25CK	C24CH	H24BM	L22BK	N34BG	M11BE	G18AM	N27AK	M30AH	G23AE	N20AC	E33CR
15	U35CL	K21CJ	H21BP	D29BL	A26BH	H31BF	L12BC	A13AL	H14AF	L36AF	A28AD	M15CS
16	R16CM	F17CK	S17CD	J34BM	Z38BJ	S18BG	D37BD	Z19AM	S23AK	D32AG	Z33AE	H25AB
17	X24CP	V22CL	P22CE	U26BP	C11BK	P12BH	J27BE	C30BC	P36AL	J20AH	C15AF	S35AC
18	B21CR	E29CM	W29CF	R38CD	K31BL	W37BJ	U13BF	K14BD	W32AM	U28AJ	K25AG	P16AD
19	N17CS	M34CP	Y34CH	X11CE	F18BM	Y27BK	R19BG	F23BE	Y20BC	R33AK	F35AH	W24AE
20	A22AB	H26CR	T26CJ	B31CF	V12BP	T13BL	X30BH	V36BF	T28BD	X15AL	V16AJ	Y21AF
21	Z29AC	S38CS	G38CK	N18CH	E37CD	G19BM	B14BJ	E32BG	G33BE	B25AM	E24AK	T17AG
22	C34AD	P11AB	L11CL	A12CJ	M27CE	L30BP	N23BK	M28BJ	L15BF	N35BC	M21AL	G22AH
23	K26AE	W31AC	D31CM	Z37CK	H13CF	D14CD	A36BL	H28BJ	D25BG	A16BD	H17AM	L29AJ
24	F38AF	Y18AD	J16CP	C27CL	S19CH	J23CE	Z32BM	S33BK	J35BH	Z24BE	S28BC	D34AK
25	V11AG	T12AE	U12CR	K13CM	P30CJ	U36CF	C20BP	P15BL	U16BJ	C21BF	P29BD	J26AL
26	E31AH	G37AF	R37CS	F19CP	W14CK	R32CH	K28CD	W25BM	R24BK	K17BG	W34BE	U38AM
27	M18AJ	L27AG	X27AB	V30CR	Y23CL	X20CJ	F33CE	Y35BP	X21BL	F22BH	Y26BF	R11BC
28	H12AK	D13AM	B13AC	E14CS	T36CM	B28CK	V15CF	T16CD	B17BM	V29BJ	T38BG	X31BD
29	S37AL		N19AD	M23AB	G32CP	N33CL	E25CH	G24CE	N22BP	E34BK	G11BH	B18BE
30	P27AM		A30AE	H36AC	L20CR	A15CM	M35CJ	L21CF	A29CD	M26BL	L31BJ	N12BF
31	W13BC		Z14AF		D28CS		H16CK	D17CH		H38BM		A37BG

1963	JAN	FEB	MAR	APR	MAY	JUN	JUL	AUG	SEP	OCT	NOV	DEC
1	Z27BH	S30BF	G30AL	N36AJ	E20AF	G15AD	B35CS	E21CP	G29CL	B26CH	E31CE	T12BH
2	C13BJ	P14BG	L14AM	A32AK	M28AG	L25AE	N16AB	M17CR	L34CM	N38CJ	M18CF	G37BP
3	K19BK	W23BH	D23BC	Z20AL	H33AH	D35AF	A24AC	H22CS	D26CP	A11CK	H12CH	L27CD
4	F30BL	Y35BJ	J36BD	C28AM	S15AJ	J16AG	Z21AD	S29AB	J38CR	Z31CL	S37CJ	D13CE
5	V14BM	T32BK	U32BE	K33BC	P25AK	U24AH	C17AE	P34AC	U11CS	C18CM	P27CK	J19CF
6	E23BP	G20BL	R20BF	F15BD	W35AL	R21AJ	K22AF	W26AD	R31AB	K12CP	W13CL	U30CH
7	M36CD	L28BM	X28BG	V25BE	Y16AM	X17AK	F29AG	Y38AE	X18AC	F37CR	Y19CM	R14CJ
8	H32CE	D33BP	B33BH	E35BF	T24BD	B22AL	V34AH	T11AF	B12AD	V27CS	T30CP	X23CK
9	S20CF	J15CD	N15BJ	M16BG	G21BD	N29AM	E26AK	G31AG	N37AE	E13AB	G14CR	B36CL
10	P28CH	U25CE	A25BK	H24BH	L17BE	A34BC	M38AK	L18AH	A27AF	M19AC	L23CS	N32CM
11	W33CJ	R35CF	Z35BL	S21BJ	D22BF	Z26BD	H11AL	D12AJ	Z13AG	H30AD	D36AB	A20CP
12	Y15CK	X16CH	C16BM	P17BK	J29BG	C38BE	S31AM	J37AK	C19AH	S14AE	J32AC	Z28CR
13	T25CL	B24CJ	K24BP	W22BL	U34BH	K11BF	P18BC	U27AL	K30AJ	P23AF	U20AD	C33CS
14	G35CM	N21CK	F21CD	Y29BM	R26BJ	F31BG	W12BD	R13AM	F14AL	W36AG	R28AE	K15AB
15	L16CP	A17CL	V17CE	T34BP	X38BK	V18BG	Y37BE	X19BC	V23AL	Y32AH	X33AF	F25AC
16	D24CR	Z22CH	E22CF	G26CD	B11BL	E12BJ	T27BF	B30BD	E36AM	T20AJ	B15AG	V35AD
17	J21CS	C29CP	M29CH	L38CE	N31BM	M37BK	G13BG	N14BE	M32BC	G28AK	N25AH	E16AE
18	U17AB	K34CR	H34CJ	D11CF	A18BP	H27BL	L19BH	A23BF	H20BD	L33AL	A35AJ	M24AF
19	R22AC	F26CS	S26CK	J31CH	Z12CD	S13BM	D30BJ	Z36BG	S28BE	D15AM	Z16AK	H21AG
20	X29AD	V38AB	P38CL	U16CJ	C37CE	P19BP	J14BK	C32BH	P33BF	J25BC	C24AL	S17AH
21	B34AE	E11AC	W11CM	R12CK	K27CF	W30CD	U23BL	K20BJ	W15BH	U35BD	K21AM	P22AJ
22	N26AF	M31AD	Y31CP	X37CL	F13CH	Y14CE	R36BM	F28BK	Y25BH	R16BE	F17BC	W29AK
23	A38AG	H18AE	T18CR	B27CM	V19CJ	T23CF	X32BP	V33BL	T35BJ	X24BF	V22BD	Y34AL
24	Z11AH	S12AF	G12CS	N13CP	E30CK	G36CH	B20CD	E15BM	G16BK	B21BG	E29BE	T26AM
25	C31AJ	P37AG	L37CH	A19CR	M14CL	L32CJ	N28CE	M25BP	L24BL	N17BH	M34BF	G38BC
26	K18AK	W27AH	D27AC	Z30CS	H23CM	D20CK	A33CF	H35CD	D21BM	A22BJ	H26BG	L11BD
27	F12AL	Y13AJ	J13AD	C14AB	S36CP	J28CL	Z15CH	S16CE	J17BP	Z29BK	S38BH	D31BE
28	V37AM	T19AK	U19AE	K23AC	P32CR	U33CM	C25CJ	P24CF	U22CD	C34BL	P11BJ	J18BF
29	E27BC		R30AF	F36AD	W20CS	R15CP	K35CK	W21CH	R29CE	K26BM	W31BK	U12BG
30	M13BD		X14AG	V32AE	Y28AB	X25CR	F16CL	Y17CJ	X34CF	F38BP	Y18BL	R37BH
31	H19BE		B23AH		T33AC		V24CM	T22CK		V11CD		X27BJ

1964	JAN	FEB	MAR	APR	MAY	JUN	JUL	AUG	SEP	OCT	NOV	DEC
1	B13BK	E14BH	Y23BD	X20AH	F33AJ	Y35AG	R24AD	F22AB	Y26CR	R11CL	F12CJ	W27CE
2	N19BL	M23BJ	T36BE	B28BC	V15AK	T16AH	X21AE	V29AC	T38CS	X31CH	V37CK	Y13CF
3	A30BM	H36BK	G32BF	N33BD	E25AL	G24AJ	B17AF	E34AD	G11AB	B18CP	E27CL	T19CH
4	Z14BP	S32BL	L28BG	A15BE	M35AM	L21AK	N22AG	M26AE	L31AC	N12CR	M13CM	G30CJ
5	C23CD	P20BM	D28BH	Z25BF	H16BD	D17AL	A29AH	H38AF	D18AD	A37CS	H19CP	L14CK
6	K36CE	W28BP	J33BJ	C35BG	S24BD	J22AM	Z34AJ	S11AG	J12AE	Z27AB	S30CR	D23CL
7	F32CF	Y33CD	U15BK	K16BH	P21BE	U29BC	C26AK	P31AG	U37AF	C13AC	P14CS	J36CM
8	V20CH	T15CE	R25BL	F24BJ	W17BF	R34BH	K38AL	W18AJ	R27AG	K19AD	W23AB	U32CP
9	E28CJ	G25CF	X35BM	V21BK	Y22BG	X26BE	F11AM	Y12AK	X13AH	F30AE	Y36AC	R20CR
10	M33CK	L35CH	B16BP	E17BL	T29BH	B38BF	V31AL	T37AJ	B19AG	V14AF	T32AD	X28CS
11	H15CL	D16CJ	N24CD	M22BM	G34BG	N11BE	E18BD	G27AM	N30AK	E23AG	G20AE	B33AB
12	S25CM	J24CK	A21CE	H29BP	L26BK	A31BC	M12BE	L13BC	A14AL	M36AH	L28AF	N15AC
13	P35CP	U21CL	Z17CF	S34CD	D38BL	Z18BJ	H37BF	D19BD	Z23AM	H32AJ	D33AG	A25AD
14	W16CR	R17CM	C22CH	P26CE	J11BM	C12BK	S27BG	J18BE	C36BC	S20AK	J15AH	Z35AE
15	Y24CS	X22CP	K29CJ	W38CF	U31BP	K37BL	P13BH	U14BF	K32BD	P28AL	U25AJ	C16AF
16	T21AB	B29CR	F34CK	Y11CH	R18CE	F27BM	W19BJ	R23BG	F20BE	W33AM	R35AK	K24AG
17	G17AC	N34CS	V26CL	T31CJ	X12CE	V13BF	Y30BK	X36BE	V28BF	Y15BC	X16AL	F21AH
18	L22AD	A26AB	E36CM	G16CK	B37CF	E19CD	T14BL	B32BJ	E33BG	T25BD	B24AM	V17AJ
19	D29AE	Z38AC	M11CP	L12CL	N27CH	M30CE	G23BM	N20BK	M15BH	G35BE	N21BC	E22AK
20	J34AF	C11AD	H31CR	D37CH	A13CJ	H14CF	L36BP	A28BL	H25BJ	L16BF	A17BD	M29AL
21	U26AG	K31AE	S18CS	J27CP	Z19CK	S23CH	D32BO	Z33BM	S35BK	D24BG	Z22BE	H34AM
22	R38AH	F18AF	P12AB	U13CR	C30CL	P36CJ	J20CE	C15BP	P16BL	J21BH	C29BF	S26BC
23	X11AJ	V12AG	W37AC	R19CS	K14CM	W32CK	U26CF	K25CD	W24BM	U17BJ	K34BG	P38BD
24	B31AK	E37AH	Y27AD	X30AB	F23CP	Y20CL	R33CH	F35CD	Y21BP	R22BK	F26BH	W11BE
25	N18AL	M27AJ	T13AE	B14AC	V36CR	T28CM	X15CJ	V16CF	T17CD	X29BL	V38BJ	Y31BF
26	A12AM	H13AK	G19AF	N23AD	E32CS	G33CP	B25CK	E24CH	G22CE	B34BM	E11BK	T18BG
27	Z37BC	S19AL	L30AG	A36AE	M20AB	L15CR	N35CL	M21CJ	L29CF	N26BP	M31BL	G12BH
28	C27BD	P30AM	D14AH	Z32AF	H28AC	D25CS	A16CM	H17CK	D34CH	A38BJ	H18BM	L37BJ
29	K13BE	W14BC	J23AJ	C20AG	S33AD	J35AB	Z24CP	S22CL	J26CJ	Z11CE	S12BP	D27BK
30	F19BF		U36AK	K28AH	P15AE	U16AC	C21CR	P29CM	U38CK	C31CF	P37CD	J13BL
31	V30BG		R32AL		W25AF		K17CS	W34CP		K18CH		U19BM

(156)

FIRST LETTER = YOUR PHYSICAL CODE, MIDDLE NUMBER = SENSITIVITY CODE, LAST TWO LETTERS = INTELLECTUAL CODE

1965	JAN	FEB	MAR	APR	MAY	JUN	JUL	AUG	SEP	OCT	NOV	DEC
1	R30BP	F36BL	S36BF	J28BD	Z15AL	S16AJ	D21AF	Z29AD	S38AB	D31CP	Z37CL	H13CH
2	X14CD	V32BH	P32BG	U33BE	C25AM	P24AK	J17AG	C34AE	P11AC	J18CR	C27CM	S19CJ
3	B23CE	E20BP	W20BH	R15BF	K35BC	W21AL	U22AH	K26AF	W31AD	U12CS	K13CP	P30CK
4	N36CF	M28CD	Y28BJ	X25BG	F16BD	Y17AM	R29AJ	F38AG	Y18AE	R37AB	F19CR	W14CL
5	A32CH	H33CE	T33BK	B35BH	V24BE	T22BC	X34AK	V11AH	T12AF	X27AC	V30CS	Y23CM
6	Z20CJ	S15CF	G15BL	N16BJ	E21BF	G29BD	B26AL	E31AJ	G37AG	B13AD	E14AB	T36CP
7	C28CK	P25CH	L25BM	A24BK	M17BG	L34BE	N38AM	M18AK	L27AH	N19AE	M23AC	G32CR
8	K33CL	W35CJ	D35BP	Z21BL	H22BH	D26BF	A11BG	H12AL	D13AJ	A30AF	H36AD	L20CS
9	F15CM	Y16CK	J16CD	C17BM	S29BJ	J38BG	Z31BD	S37AM	J19AK	Z14AG	S32AE	D28AB
10	V25CP	T24CL	U24CE	K22BP	P34BK	U11BH	C18BE	P27BC	U30AL	C23AH	P20AF	J33AC
11	E35CR	G21CM	R21CF	F29CD	W26BK	R31BG	K12BF	W13BD	R14AM	K36AJ	W28AG	U15AD
12	M16CS	L17CP	X17CH	V34CE	Y38BM	X18BK	F37BG	Y19BE	X23BC	F32AK	Y33AH	R25AE
13	H24AB	D22CR	B22CJ	E26CF	T11BP	B12BL	V27BH	T30BF	B36BD	V20AL	T15AJ	X35AF
14	S21AC	J29CS	N29CK	M38CH	G31CD	N37BM	E13BJ	G14BG	N32BE	E28AM	G25AK	B16AG
15	P17AD	U34AB	A34CL	H11CJ	L18CE	A27BP	M19BK	L23BH	A20BF	M33BC	L35AL	N24AH
16	W22AE	R26AC	Z26CM	S31CK	D12CF	Z13CD	H30BL	D36BJ	Z28BG	H15BD	D16AM	A21AJ
17	Y29AF	X38AD	C38CP	P18CL	J37CH	C14CE	S14BM	J32BK	C33BH	S25BE	J24BC	Z17AK
18	T34AG	B11AE	K11CR	W12CM	U27CJ	K30CF	P23BP	U20BL	K15BJ	P35BF	U21BD	C22AL
19	G26AH	N31AF	F31CS	Y37CP	R13CK	F14CH	W36CD	R28BM	F25BK	W16BG	R17BE	K29AM
20	L38AJ	A18AG	V18AB	T27CR	X19CL	V23CJ	Y32CE	X33BP	V35BL	Y24BH	X22BF	F34BC
21	D11AK	Z12AH	E12AC	G13CS	B30CR	E36CK	T20CF	B15CD	E16BM	T21BJ	B29BG	V26BD
22	J31AL	C37AJ	M37AD	L19AB	N14CP	M32CL	G28CH	N25CE	M24BP	G17BK	N34BH	E33BE
23	U18AM	K27AK	H27AE	D30AC	A23CR	H20CM	L33CJ	A35CF	H21CD	L22BL	A24BJ	M11BF
24	R12BC	F13AL	S13AF	J14AD	Z36CS	S28CP	D15CK	Z16CH	S17CE	D29BM	Z38BK	H31BG
25	X37BD	V19AM	P19AG	U23AE	C32AB	P33CR	J25CL	C24CJ	P22CF	J34BP	C11BL	S18BH
26	B27BE	E30BC	W30AH	R36AF	K20AD	W15CS	U35CM	K21CK	W29CH	U26CD	K31BM	P12BJ
27	N13BF	M14BD	Y14AJ	X32AG	F28AD	Y25AB	R16CP	F17CL	Y34CJ	R38CE	F18BP	M37BK
28	A19BG	H23BE	T23AK	B20AH	V33AE	T35AC	X24CR	V22CM	T26CK	X11CF	V12CD	Y27BL
29	Z30BH		G36AL	N28AJ	E15AF	G16AD	B21CS	E29CP	G38CL	B31CH	E37CE	T13BM
30	C14BJ		L32AM	A33AK	M25AG	L24AE	N17AB	M34CR	L11CM	N18CJ	M27CF	G19BP
31	K23BK		D20BC		H35AH		A22AC	H26CS		A12CK		L30CD

1966	JAN	FEB	MAR	APR	MAY	JUN	JUL	AUG	SEP	OCT	NOV	DEC
1	D14CE	Z32BP	E32BH	G33BF	B25BC	E24AL	T17AH	B34AF	E11AD	T18CS	B27CP	V19CK
2	J23CF	C20CD	M20BJ	L15BG	N35BD	M21AM	G22AJ	N26AG	M31AE	G12AB	N13CR	E30CL
3	U36CH	K28CE	H28BK	D25BH	A16BF	H17BC	L29AK	A38AH	H18AF	L37AC	A19CS	M14CM
4	R32CJ	F33CF	S33BL	J35BJ	Z24BF	S22BD	D34AL	Z11AJ	S12AG	D27AD	Z30AB	H23CP
5	X20CK	V15CH	P15BM	U16BK	C29BE	P29BE	J26AM	C31AK	P37AH	J13AE	C14AC	S36CR
6	B28CL	E25CJ	W25BP	R24BL	K17BH	W34BF	U38BC	K18AL	W27AJ	U19AF	K23AD	P32CS
7	N33CM	M35CK	Y35CD	X21BM	F22BJ	Y26BG	R11BD	F12AM	Y13AK	R30AG	F36AE	W20AB
8	A15CP	H16CL	T16CE	B17BP	V29BK	T38BH	X31BE	V14BF	T19AL	X14AH	V32AF	Y28AC
9	Z25CR	S24CM	G24CF	N22CD	E34BL	G11BJ	B18BF	E27BD	G30AM	B23AJ	E20AG	T33AD
10	C35CS	P21CP	L21CH	Z34CF	H38BP	L31BK	N12BG	M13BE	L14BC	N36AK	M28AH	G15AE
11	K16AB	W17CR	D17CJ	Z34CF	H38BP	O18BL	A37BH	H19BF	D23BD	A32AL	H33AJ	L25AF
12	F24AC	Y22CS	J22CK	C26CH	S11CD	J12BM	Z27BJ	S30BG	J36BE	Z20AM	S15AK	D35AG
13	V21AD	T29AB	U29CL	K38CJ	P31CE	U37BP	C13BK	P14BH	U32BF	C28BC	P25AL	J16AH
14	E17AE	G34AC	R34CM	F11CK	W18CF	R27CD	K19BL	W23BJ	R20BG	K33BD	W35AM	U24AJ
15	M22AF	L26AD	X26CP	V31CL	Y12CH	X13CE	F30BM	Y36BK	X28BH	F15BE	Y16BC	R21AK
16	H29AG	D38AE	B38CR	E18CM	T37CJ	B19CF	V14BP	T32BL	B33BJ	V25BF	T24BD	X17AL
17	S34AH	J11AF	N11CS	M12CP	G27CK	N30CH	E23CD	G20BM	N15BK	E35BG	G21BE	B22AM
18	P26AJ	U31AG	A31AB	H37CR	L13CL	A14CJ	M36CE	L28BP	A25BL	M16BH	L17BF	N29BC
19	W38AK	R18AH	Z18AC	S27CS	D19CM	Z23CK	H32CF	D33CD	Z35BM	H24BJ	D22BG	A34BD
20	Y11AL	X12AJ	C12AD	P13AB	J30CP	C36CL	S20CH	J15CE	C16BP	S21BK	J29BH	Z26BE
21	T31AM	B37AK	K37AE	W19AC	U14CR	K32CM	P28CJ	U25CF	K24CD	P17BL	U34BJ	C38BF
22	G18BC	N27AL	F27AF	Y30AD	R23CS	F20CP	W33CK	R35CH	F21CE	W22BM	R26BK	K11BG
23	L12BD	A13AM	V13AG	T14AE	X36AB	V28CR	Y15CL	X16CJ	V17CF	Y29BP	X38BL	F31BH
24	D37BE	Z19BC	E19AH	G23AF	B32AC	E33CS	T25CM	E22CH	T34CD	B11BM	V18BJ	
25	J27BF	C30BD	M30AJ	L36AG	N20AD	M15AB	G35CP	N21CL	M29CJ	G26CE	N31BP	E12BK
26	U13BG	K14BE	H14AK	D32AH	A24AG	H25AC	L16CR	A17CM	H34CK	L38CF	A18CD	M37BL
27	R19BH	F23BF	S23AL	J20AJ	Z33AF	S35AD	D24CS	Z22CP	S26CL	D11CH	Z12CE	H27BM
28	X30BJ	V36BG	P36AM	U28AK	C15AG	P16AE	J21AB	C29CR	P38CM	J31CJ	C37CF	S13BP
29	B14BK		W32BC	R33AL	K25AH	W24AF	U17AC	K34CS	W11CP	U18CK	K27CH	P19CD
30	N23BL		Y20BD	X15AM	F35AJ	Y21AG	R22AD	F26AB	Y31CP	R12CL	F13CJ	W30CE
31	A36BM		T28BE		V16AK		X29AE	V38AC		X37CM		Y14CF

1967	JAN	FEB	MAR	APR	MAY	JUN	JUL	AUG	SEP	OCT	NOV	DEC
1	T23CH	B20CE	K20BK	W15BH	U35BE	K21BC	P22AK	U26AH	K31AF	P12AC	U13CS	C30CM
2	G36CJ	N28CF	F28BL	Y25BJ	R16BF	F17BD	M29AL	R38AJ	F18AG	W37AD	R19AB	K14CP
3	L32CK	A33CH	V33BM	T35BK	X24BG	V22BE	Y34AM	X11AK	V12AH	Y27AE	X30AC	F23CR
4	D20CL	Z15CJ	E15BP	G16BL	B21BH	E29BF	T26BC	B31AL	E37AJ	T13AF	B14AD	V36CS
5	J28CM	C25CK	M25CD	L24BM	W34BG	M34BG	N18AM	W27AK	M27AK	G19AG	N23AE	E32AB
6	U33CP	K35CL	H35CE	D21BP	A22BK	H26BH	L11BE	A12BC	H13AL	L30AH	A36AF	M20AC
7	R15CR	F16CM	S16CF	J17CD	Z29BL	S38BJ	D31BF	Z37BD	S19AM	D14AJ	Z32AG	H28AD
8	X25CS	V24CP	P24CH	U22CE	C34BM	P11BK	J18BG	C27BE	P30BC	J23AK	C20AH	S33AE
9	B35AB	E21CR	H21CJ	R29CF	K36BP	W31BL	U12BH	K13BF	W14BD	U36AL	K28AJ	P15AF
10	N16AC	M17CS	Y17CK	X34CH	F38CD	Y18BM	R37BJ	F19BG	Y23BE	R32AM	F33AK	W25AG
11	A24AD	H22AB	T22CL	B26CJ	V11CE	T12BH	X27BK	V30BH	T36BF	X20BC	V15AL	Y35AH
12	Z21AE	S29AC	G29CM	N38CK	E31CF	G37CD	B13BL	E14BJ	G32BG	B28BD	E25AM	T16AJ
13	C17AF	P34AD	L34CP	A11CL	M18CH	L27CE	N19BM	M23BK	L20BH	N33BE	M35BC	G24AK
14	K22AG	W26AE	D26CR	Z31CM	H12CJ	D13CD	A38BP	H36BF	D28BD	A15BF	H16BD	L21AL
15	F29AH	Y38AF	J38CS	C18CP	S37CK	J19CH	Z14CD	S32BM	J33BK	Z25BG	S24BE	D17AM
16	V34AJ	T11AG	U11AB	K12CR	P27CL	U30CH	C23CE	P20BP	U15BL	C35BH	P21BF	J22BC
17	E26AK	G31AH	R31AC	F37CS	M13CM	R14CK	K36CF	W28CD	R25BM	K16BJ	W17BG	U29BD
18	M38AL	L18AJ	X18AD	V27AB	Y19CP	X23CE	F32CH	Y33CK	X35BP	F24BK	Y22BH	R34BE
19	H11AM	D12AK	B12AE	E13AC	T30CR	B36CM	V20CJ	T15CF	B16CD	V21BL	T29BJ	X26BF
20	S31BC	J37AL	N37AF	M19AD	G14CS	N32CP	E28CK	G25CH	N24CE	E17BM	G34BK	B38BG
21	P18BD	U27AM	A27AG	H30AE	L23AB	A20CR	M33CL	L35CJ	A21CF	M22BP	L26BL	N11BH
22	W12BE	R13BC	Z13AH	S14AF	D36AC	Z28CS	H15CM	D16CK	Z17CH	H29CD	D38BM	A31BJ
23	Y37BF	X19BD	C19AJ	P23AG	J32AD	C33AB	S25CP	J24CL	C22CJ	S34CE	J11BP	Z18BK
24	T27BG	B30BE	K30AK	W36AH	U20AE	K15AC	P35CR	U21CM	K29CK	P26CF	U31CD	C12BL
25	G13BH	N14BF	F14AL	Y32AJ	R28AF	Y25AD	W16CS	R17CP	Y34CL	W30CH	R18CE	K37BM
26	L19BJ	A23BG	V23AM	T20AK	X33AG	V35AE	Y24AB	X22CR	V26CM	Y11CJ	X12CF	F27BP
27	D30BK	Z36BH	E36BC	G28AL	B15AH	E16AF	T21AC	B29CS	E38CP	T31CK	B37CH	V13CD
28	J14BL	C32BJ	M32BD	H20BE	D15BC	M24AG	G17AD	N34AB	M11CR	G18CL	N27CJ	E19CE
29	U23BM		H20BE	D15BC	A35AK	H21AH	L22AE	A26AC	H31CS	L12CM	A13CK	H30CF
30	R36BP		S28BF	J25BD	Z16AL	S17AJ	D29AF	Z38AD	S18AB	D37CP	Z19CL	H14CH
31	X32CD		P33BG		C24AM		J34AG	C11AE		J27CR		S23CJ

FIRST LETTER = YOUR PHYSICAL CODE, MIDDLE NUMBER = SENSITIVITY CODE, LAST TWO LETTERS = INTELLECTUAL CODE

1968	JAN	FEB	MAR	APR	MAY	JUN	JUL	AUG	SEP	OCT	NOV	DEC
1	P36CK	U28CH	Z33BP	S35BL	D24BH	Z22BF	H34BC	D11AL	Z12AJ	H27AF	D30AD	A23CS
2	W32CL	R33CJ	C15CD	P16BM	J21BJ	C29BG	S26BD	J31AM	C37AK	S13AG	J14AE	Z36AB
3	Y20CM	X15CK	K25CE	W24BP	U17BK	K34BH	P38BE	U18BC	K27AL	P19AH	U23AF	C32AC
4	T28CP	B25CL	F35CF	Y21CD	R22BL	F26BJ	W11BF	R12BD	F13AM	W30AJ	R36AG	K20AD
5	G33CR	N35CM	V16CH	T17CE	X29BM	V38BK	Y31BG	X37BE	V19BC	Y14AK	X32AH	F28AE
6	L15CS	A16CP	E24CJ	G22CF	B34BP	E11BL	T18BH	B27BF	E30BD	T23AL	B20AJ	V33AF
7	D25AB	Z24CR	M21CK	L29CH	N26CD	M31BM	G12BJ	N13BG	M14BE	G36AM	N28AK	E15AG
8	J35AC	C21CS	H17CL	D34CJ	A38CE	H18BP	L37BK	A19BH	H23BF	L32BC	A33AL	M25AH
9	U16AD	K17AB	S22CM	J26CK	Z11CF	S12CD	D27BL	Z30BJ	S36BG	D20BD	Z15AM	H35AJ
10	R24AE	F22AC	P29CP	U38CL	C31CH	P37CE	J13BM	C14BK	P32BH	J28BE	C25BC	S16AK
11	X21AF	V29AD	W34CR	R11CM	K16CJ	W27CF	U19BP	K23BL	W20BJ	U33BF	K35BD	P24AL
12	B17AG	E34AE	Y26CS	X31CP	F12CK	Y13CH	R30CD	F36BM	Y28BK	R15BG	F16BE	W21AM
13	N22AH	M26AF	T38AB	B18CR	V37CL	T19CJ	X14CE	V32BP	T33BL	X25BH	V24BF	Y17BC
14	A29AJ	H38AG	G11AC	N12CS	E27CM	G30CK	B23CF	E20CD	G15BM	B35BJ	E21BG	T22BD
15	Z34AK	S11AH	L31AD	A37AB	M13CP	L14CL	N36CH	M28CE	L25BP	N16BK	M17BH	G29BE
16	C26AL	P31AJ	D18AE	Z27AC	H19CR	D23CM	A32CJ	H33CF	D35CD	A24BL	H22BJ	L34BF
17	K38AM	H18AK	J12AF	C13AD	S30CS	J36CP	Z20CK	S15CH	J16CE	Z21BM	S29BK	D26BG
18	F11BC	Y12AL	U37AG	K19AE	P14BB	U32CR	C28CL	P25CJ	U24CF	C17BP	P36BC	J38BH
19	V31BD	T37AM	R27AH	F30AF	W23AC	R20CS	K33CM	W35CK	R21CH	K22CD	W26BM	U11BJ
20	E18BE	G27BC	X13AJ	V14AG	Y36AD	X28AB	F15CP	Y16CL	X17CJ	F29CE	Y38BP	R31BK
21	M12BF	L13BD	B19AK	E23AH	T32AE	B33AC	V25CR	T24CM	B22CK	V34CF	T11CD	X18BL
22	H37BG	D19BE	N30AL	M36AJ	G20AF	N15AD	E35CS	G21CP	N29CL	E26CH	G31CE	B12BM
23	S27BH	J30BF	A14AM	H32AK	L28AG	A25AE	M16AB	L17CR	A34CM	M38CJ	L18CF	N37BP
24	P13BJ	U14BG	Z23BC	S20AL	D33AH	Z35AF	H24AC	D22CS	Z26CP	H11CK	D12CH	A27CD
25	W19BK	R23BH	C36BD	P28AM	J15AJ	C16AG	S21AD	J29AB	C38CR	S31CL	J37CJ	Z13CE
26	Y30BL	X36BJ	K32BE	W33BC	U25AK	K24AH	P17AE	U34AC	K11CS	P18CH	U27CK	C19CF
27	T14BM	B32BK	F20BF	Y15BD	R35AL	F21AJ	W22AF	R26AD	F31AB	W12CP	R13CL	K30CH
28	G23BP	N20BL	V28BG	T25BE	X16AM	V17AK	Y29AG	X38AE	V18AC	Y37CR	X19CM	F14CJ
29	L36CD	A28BM	E33BH	G35BF	B24BC	E22AL	T34AH	B11AF	E12AD	T27CS	B30CP	V23CK
30	D32CE		M15BJ	L16BG	N21BD	M29AM	G26AJ	N31AG	M37AE	G13AB	N14CR	E36CL
31	J20CF		H25BK		A17BE		L38AK	A18AH		L19AC		M32CM

1969	JAN	FEB	MAR	APR	MAY	JUN	JUL	AUG	SEP	OCT	NOV	DEC
1	H20CP	D15CL	B15CE	E16BP	T21BK	B29BH	V26BE	T31BC	B37AL	V13AH	T14AF	X36AC
2	S28CR	J25CM	N25CF	M24CD	G17BL	N34BJ	E38BF	G18BD	N27AM	E19AJ	G23AG	B32AD
3	P33CS	U35CP	A35CH	H21CE	L22BM	A26BK	M11BG	L12BE	A13BC	M30AK	L36AH	N20AE
4	W15AB	R15CR	Z16CJ	S17CF	D29BP	Z38BL	H31BH	D37BF	Z19BD	H14AL	D32AJ	A28AF
5	Y25AC	X24CS	C24CK	P22CH	J34CD	C11BM	S18BJ	J27BG	C30BE	S23AM	J20AK	Z33AG
6	T35AD	B21AB	K21CL	W29CJ	U26CE	K31BP	P12BK	U13BH	K14BF	P36BC	U28AL	C15AH
7	G16AE	N17AC	F17CM	Y34CK	R38CF	F18CD	W37BL	R19BJ	F23BG	W32BD	R33AM	K25AJ
8	L24AF	A22AD	V22CP	T26CL	X11CH	V12CE	Y27BM	X30BK	V36BH	Y20BE	X15BC	F35AK
9	D21AG	Z29AE	E29CR	G38CM	B31CJ	E37CF	T13BP	B14BL	E32BJ	T28BF	B25BD	V16AL
10	J17AH	C34AF	M34CS	L11CP	N18CK	M27CH	G19CD	N23BM	M20BK	G33BG	N35BE	E24AM
11	U22AJ	K26AG	H26AB	D31CR	A12CL	H13CJ	L30CE	A36BP	H28BL	L15BH	A16BF	M21BC
12	R29AK	F38AH	S38AC	J18CS	Z37CM	S19CK	D14CF	Z32CD	S33BM	D25BJ	Z24BG	H17BD
13	X34AL	V11AJ	P11AD	U12AB	C27CP	P30CL	J23CH	C20CE	P15BP	J35BK	C21BH	S22BE
14	B26AM	E31AK	W31AE	R37AC	K13CR	W14CM	U36CJ	K28CF	W25CD	U16BL	K17BJ	P29BF
15	N38BC	M18AL	Y18AF	X27AD	F19CS	Y23CP	R32CK	F33CH	Y35CE	R24BM	F22BK	W34BG
16	A11BD	H12AM	T12AG	B13AE	V30AB	T36CR	X20CL	V15CJ	T16CF	X21BP	V29BL	Y26BH
17	Z31BE	S37BC	G37AH	N19AF	E14AC	G32CS	B28CM	E25CK	G24CH	B17CD	E34BM	T38BJ
18	C18BF	P27BD	L27AJ	A30AG	M23AD	L20AB	N33CP	M35CL	L21CJ	N22CE	M26BP	G11BK
19	K12BG	W13BE	D13AK	Z14AH	H36AE	D28AC	A15CR	H16CM	D17CK	A29CF	H38CD	L31BL
20	F37BH	Y19BF	J19AL	C23AJ	S32AF	J33AD	Z25CS	S24CP	J22CL	Z34CH	S11CE	D18BM
21	V27BJ	T30BG	U30AM	K36AK	P20AG	U15AE	C35AB	P21CR	U29CM	C26CJ	P31CF	J12BP
22	E13BK	G14BH	R14BC	F32AL	W28AH	R25AF	K16AC	W17CS	R34CP	K38CK	W18CH	U37CD
23	M19BL	L23BJ	X23BD	V20AM	Y33AJ	X35AG	F24AD	Y22AB	X26CR	F11CL	Y12CJ	R27CE
24	H30BM	D36BK	B36BE	E28BC	T15AK	B16AH	V21AE	T29AC	B38CS	V31CM	T37CK	X13CF
25	S14BP	J32BL	N32BF	M33BD	G25AL	N24AJ	E17AF	G34AD	N11AB	E18CP	G27CL	B19CH
26	P23CD	U20BM	A20BG	H15BE	L35AM	A21AK	M22AG	L26AE	A31AC	M12CR	L13CM	N30CJ
27	W36CE	R28BP	Z28BH	S25BF	D16BC	Z17AL	H29AH	D38AF	Z18AD	H37CS	D19CP	A14CK
28	Y32CF	X33CD	C33BJ	P35BG	J24BD	C22AM	S34AJ	J11AG	C12AE	S27AB	J30CR	Z23CL
29	T20CH		K15BK	W16BH	U21BE	K29BC	P26AK	U31AH	K37AF	P13AC	U14CS	C36CM
30	G28CJ		F25BL	Y24BJ	R17BF	F34BD	W38AL	R18AJ	F27AG	W19AD	R23AB	K32CP
31	L33CK		V35BM		X22BG		Y11AM	X12AK		Y30AE		F20CR

1970	JAN	FEB	MAR	APR	MAY	JUN	JUL	AUG	SEP	OCT	NOV	DEC
1	V28CS	T25CP	U25CH	K24CE	P17BM	U34BK	C38BG	P18BE	U27BC	C19AK	P23AH	J32AE
2	E33AB	G35CM	R35CJ	F21CF	W22BP	R26BL	K11BH	W12BF	R13BD	K30AL	W36AJ	U20AF
3	M15AC	L16CS	X16CK	V17CH	Y29CD	X38BM	F31BJ	Y37BG	X19BE	F14AM	W32AK	R28AG
4	H25AD	D24AB	B24CL	E22CJ	T34CE	B11BP	V18BK	T27BH	B38BF	V23BC	T20AL	X33AH
5	S35AE	J21AC	N21CM	M29CK	G26CF	N31CD	E12BL	G13BJ	N14BG	E36BD	G28AM	B15AJ
6	P16AF	U17AD	A17CP	H34CL	L38CH	A18CE	M37BM	L19BK	A23BH	M32BE	L33BC	N25AK
7	W24AG	R22AE	Z22CR	S26CM	D11CJ	Z12CF	H27BP	D30BL	Z36BJ	H20BF	D15BD	A35AL
8	Y21AH	X29AF	C29CS	P38CP	J31CK	C37CH	S13CD	J14BM	C32BK	S28BG	J25BE	Z16AM
9	T17AJ	B34AG	K34AB	W11CR	U18CL	K27CJ	P19CE	U23BP	K20BL	P33BH	U35BF	C24BC
10	G22AK	N26AH	F26AC	Y31CS	R12CM	F13CK	W30CF	R36CD	F28BM	W15BJ	R16BG	K21BD
11	L29AL	A38AD	V38AD	T18AB	X37CP	V19CL	Y14CH	X32CE	V33BP	Y25BK	X24BH	F17BE
12	D34AM	Z11AK	E11AE	G12AC	B27CR	E30CH	T23CJ	B20CF	E15CD	T35BL	B21BJ	V22BF
13	J26BC	C31AL	M31AF	L37AD	N13CS	M14CP	G36CK	N28CH	M25CE	G16BM	N17BK	E29BG
14	U38BD	K18AM	H18AG	D27AE	A19AB	H23CR	L32CL	A33CJ	H35CF	L24BP	A22BL	M34BH
15	R11BE	F12BC	S12AH	J13AF	Z30AC	S36CS	D20CM	Z15CK	S16CH	D21CD	Z29BM	H26BJ
16	X31BF	V37BD	P37AJ	U19AG	C14AD	P32AB	J28CP	C25CL	P24CJ	J17CE	C34BP	S38BK
17	B18BG	E27BE	W27AK	R30AH	K23AE	W20AC	U33CR	K35CM	W21CK	U22CF	K26CD	P11BL
18	N12BH	M13BF	Y13AL	X14AJ	F36AF	Y28AD	R15CS	F16CP	Y17CL	R29CH	F38CE	W31BM
19	A37BJ	H19BG	T19AM	B23AK	V32AG	T33AE	X25AB	V24CR	T22CM	X34CJ	V11CF	Y18BP
20	Z27BK	S30BH	G30BC	N36AL	E20AH	G15AF	B35AC	E21CS	G29CP	B26CK	E31CH	T12CD
21	C13BL	P14BJ	L14BD	A32AM	M28AJ	L25AG	N16AD	M17AB	L34CR	N38CL	M18CJ	G37CE
22	K19BM	W23BK	D23BE	Z20BC	H33AK	D35AH	A24AE	H22AC	D26CS	A11CM	H12CK	L27CF
23	F30BP	Y36BL	J36BF	C28BD	S15AL	J16AJ	Z21AF	S29AD	J38AB	Z31CR	S37CL	D13CH
24	V14CD	T32BM	U32BG	K33BE	P25AM	U24AK	C17AG	P34AE	U11AC	C18CR	P27CH	J19CJ
25	E23CE	G22BP	Z28BJ	F15BF	W35BC	Z11AL	K22AH	W26AF	Z13AD	K12CS	W13CP	U30CK
26	M36CF	L28CD	X28BJ	V25BG	Y16BD	X17AM	F29AJ	Y38AG	X18AE	F37AB	Y19CR	R14CL
27	H32CH	D33CE	B33BK	E35BH	T24BE	B28BC	V34AK	T11AH	B12AF	V27AC	T30CS	X23CM
28	S20CJ	J15CF	N15BL	M16BJ	G21BF	N29BD	E26AL	G31AJ	N37AG	E13AD	G14AB	B36CP
29	P28CK		A25BM	H24BK	L17BG	A34BE	M38AM	L18AK	A27AH	M19AE	L23AC	N32CR
30	W33CL		Z35BP	S21BL	D22BH	Z26BF	H11BC	D12AL	Z13AJ	H30AF	D36AD	A20CS
31	Y15CM		C16CD		J29BJ		S31BD	J37AM		S14AG		Z28AB

(158)

FIRST LETTER = YOUR PHYSICAL CODE, MIDDLE NUMBER = SENSITIVITY CODE, LAST TWO LETTERS = INTELLECTUAL CODE

1971	JAN	FEB	MAR	APR	MAY	JUN	JUL	AUG	SEP	OCT	NOV	DEC
1	C33AC	P35CS	L35CK	A21CH	M22CD	L26BM	N11BJ	M12BG	L13BE	N30AM	M36AK	G20AG
2	K15AD	W16AB	D16CL	Z17CJ	H29CE	D38BP	A31BK	H37BH	D19BF	A14BC	H32AL	L28AH
3	F25AE	Y24AC	J24CM	C22CK	S34CF	J11CD	Z18BL	S27BJ	J30BG	Z23BD	S20AM	D33AJ
4	V35AF	T21AD	U21CP	K29CL	P26CH	U31CE	C12BH	P13BK	U14BH	C36BE	P28BC	J15AK
5	E16AG	G17AE	R17CR	F34CM	W38CJ	R18CF	K37BP	W19BL	R23BJ	K32BF	W33BD	U25AL
6	M24AH	L22AF	X22CS	V26CP	Y11CK	X12CH	F27CD	Y30BM	X36BK	F20BG	Y15BE	R35AM
7	H21AJ	D29AG	B29AB	E38CR	T31CL	B37CJ	V13CE	T14BP	B32BL	V28BH	T25BF	X16BC
8	S17AK	J34AM	N34AC	M11CS	G18CM	N27CK	E19CF	G23CD	N20BM	E33BJ	G35BG	B24BD
9	P22AL	U26AJ	A26AD	H31AB	L12CP	A13CL	M30CH	L36CE	A28BP	M15BK	L16BH	N21BE
10	W29AM	R38AK	Z38AE	S18AC	D37CR	Z19CM	H14CJ	D32CF	Z33CD	H25BL	D24BJ	A17BF
11	Y34BC	X11AL	C11AF	P12AD	J27CS	C30CP	S23CK	J20CH	C15CE	S35BM	J21BK	Z22BG
12	T26BD	B31AM	K31AG	W37AE	U13AB	K14CR	P36CL	U28CJ	K25CF	P16BP	U17BL	C29BH
13	G38BE	N18BC	F18AH	Y27AF	R19AC	F23CS	M32CM	R33CK	F35CH	M24CD	R22BM	K34BJ
14	L11BF	A12BD	V12AJ	T13AG	X30AD	V36AB	Y20CP	X15CL	V16CJ	Y21CE	X29BP	F26BK
15	D31BG	Z37BE	E37AK	G19AH	B14AE	E32AC	T28CR	B25CM	E24CK	T17CF	B34CD	V38BL
16	J18BH	C27BF	M27AL	L30AJ	N23AF	M20AD	G33CS	N35CP	M21CL	G22CH	N26CE	E11BM
17	U12BJ	K13BG	H13AM	D14AK	A36AG	H28AE	L15AB	A16CR	H17CM	L29CJ	A38CF	M31BP
18	R37BK	F19BH	S19BC	J23AL	Z32AH	S33AF	D25AC	Z24CS	S22CP	D34CK	Z11CH	H18CD
19	X27BL	V30BJ	P30BD	U36AM	C20AJ	P15AG	J35AD	C21AB	P29CR	J26CL	C31CJ	S12CE
20	B13BM	E14BK	W14BE	R32BC	K28AK	W25AH	U16AE	K17AC	W34CS	U38CM	K18CK	P37CF
21	N19BP	M23BL	Y23BF	X20BD	F33AL	Y35AJ	R24AF	F22AD	Y26AB	R11CP	F12CL	M27CH
22	A30CD	H36BM	T36BG	B28BE	V15AM	T16AK	X21AG	V29AE	T38AC	X31CR	V37CH	Y13CJ
23	Z14CE	S32BP	G32BH	N33BF	E25BC	G24AL	B17AH	E34AF	G11AD	B18CS	E27CP	T19CK
24	C23CF	P20CD	L28BJ	A15BG	M35BD	L21AM	N22AJ	M26AG	L31AE	N12AB	M13CR	G30CL
25	K36CH	W28CE	D28BK	Z25BH	H16BE	D17BC	A29AK	H38AH	D18AF	A37AC	H19CS	L14CM
26	F32CJ	Y33CF	J33BL	C35BJ	S24BF	J22BD	Z34AL	S11AJ	J12AG	Z27AD	S30AB	D23CP
27	V20CK	T15CH	U15BM	K16BK	P21BG	U29BE	C26AM	P31AK	U37AH	C13AE	P14AC	J36CR
28	E28CL	G25CJ	R25BP	F24BL	W17BH	R34BF	K38BC	W18AL	R27AJ	K19AF	W23AD	U32CS
29	M33CM		X35CD	V21BM	Y22BJ	X26BG	F11BD	Y12AM	X13CM	F30AG	Y36AE	R20AB
30	H15CP		B16CE	E17BP	T29BK	B38BH	V31BE	T37BC	B19AL	V14AH	T32AF	X28AC
31	S25CR		N24CF		G34BL		E18BF	G27BD		E23AJ		B33AD

1972	JAN	FEB	MAR	APR	MAY	JUN	JUL	AUG	SEP	OCT	NOV	DEC
1	N15AE	M16AC	T24CP	B22CL	V34CH	T11CE	X18BM	V27BK	T30BH	X23BE	V20BC	Y33AK
2	A25AF	H24AD	G21CR	N29CM	E26CJ	G31CF	B12BP	E13BL	G14BJ	B36BF	E28BD	T15AL
3	Z35AG	S21AE	L17CS	A34CP	M38CK	L18CH	N37CD	M19BM	L23BK	N32BG	M33BE	G25AM
4	C16AH	P17AF	D22AB	Z26CR	H11CL	D12CJ	A27CE	H30BP	D36BL	A20BH	H15BF	L35BC
5	K24AJ	W22AG	J29AC	C38CS	S31CM	J37CK	Z13CF	S14CD	J32BM	Z28BJ	S25BG	D16BD
6	F21AK	Y29AM	U34AD	K11AB	P18CP	U27CL	C19CH	P23CE	J20BP	C33BK	P35BH	J24BE
7	V17AL	T34AJ	R26AE	F31AC	W12CR	R13CM	K30CJ	W36CF	R28CD	K15BL	W16BJ	U21BF
8	E22AM	G26AK	X38AF	V18AD	Y37CS	X19CP	F14CK	Y32CH	X33CE	F25BM	Y24BK	R17BG
9	M29BC	L38AL	B11AG	E12AE	T27AB	B30CR	V23CL	T20CJ	B15CF	V35BP	T21BL	X22BH
10	H34BD	D11AM	N31AH	M37AF	G13AC	N14CS	E36CM	G26CK	N25CH	E16CJ	G17BM	B29BJ
11	S26BE	J31BC	A18AJ	H27AG	L19AD	A23AB	M32CP	L33CL	A35CJ	M24CE	L22BP	N34BK
12	P38BF	U18BD	Z12AK	S13AH	D30AE	Z36AC	H20CR	D15CM	Z16CK	H21CF	D29CD	A26BL
13	W11BG	R12BE	C37AL	P19AJ	J14AF	C32AD	S28CS	J25CP	C24CL	S17CH	J34CE	Z38BM
14	Y31BH	X37BF	K27AM	W30AK	U23AG	K20AE	P33AB	U35CR	K21CM	P22CJ	U26CF	C11BP
15	T18BJ	B27BG	F13BC	Y14AL	R36AH	F28AF	W15AC	R16CS	F17CP	W29CK	R38CH	K31CD
16	G12BK	N13BH	V19BD	T23AM	X32AJ	V33AG	Y25AD	X24AB	V22CR	Y34CL	X11CJ	F18CE
17	L37BL	A19BJ	E30BE	G36BC	B20AK	E15AH	T35AE	B21AC	E29CS	T26CM	B31CK	V12CF
18	D27BM	Z30BK	M14BF	L32BD	N28AL	M25AJ	G16AF	N17AD	M34AB	G38CP	N18CL	E37CH
19	J13BP	C14BL	H23BG	D20BE	A33AM	H35AK	L24AG	A22AE	H25AC	L11CR	A12CM	M27CJ
20	U19CD	K23BM	S36BH	J28BF	Z15BC	S16AL	D21AH	Z29AF	S38AD	D31CS	Z37CP	H13CK
21	R30CE	F36BP	P32BJ	U33BG	C25BD	P24AM	J17AJ	C34AG	P11AE	J18AB	C27CR	S19CL
22	X14CF	V32CD	W20BK	R15BH	K35BE	W21BC	U22AK	K26AH	W31AF	U12AC	K13CS	P30CM
23	B23CH	E20CE	Y28BL	X25BJ	F16BF	Y17BD	R29AL	F38AJ	Y18AG	R37AD	F19AB	W14CP
24	N36CJ	M28CF	T33BM	B35BK	T22BG	L27BE	X34AM	V11AK	T12AH	X27AE	V30AC	Y23CR
25	A32CK	H33CH	G15BP	N16BL	E21BH	G29BF	B26BC	E31AL	G37AJ	B13AF	E14AD	T36CS
26	Z20CL	S15CJ	L25CD	A24BM	D24BJ	L34BG	N38BD	M18AH	N25AK	N19AG	M23AE	G32AB
27	C28CM	P25CK	D35CL	Z21BP	H22BK	D26BH	A11BE	H12BC	D13AL	A30AH	H36AF	L20AC
28	K33CP	W35CL	J16CF	C17CD	Z38BL	J38BJ	S37BF	S37BD	J19AM	Z14AJ	S32AG	D28AD
29	F15CR	Y16CM	U24CH	K22CE	P34BM	U11BK	C18BG	P27BE	U30BC	C23AK	P20AH	U15AF
30	V25CS		R21CJ	F29CF	W26BP	R31BL	K12BH	W13BF	R14BD	K36AL	W28AJ	
31	E35AB		X17CK		Y38CD		F37BJ	Y19BG		F32AM		R25AG

1973	JAN	FEB	MAR	APR	MAY	JUN	JUL	AUG	SEP	OCT	NOV	DEC
1	X35AH	V21AF	P21CS	U29CP	C26CK	P31CH	J12CD	C13BM	P14BK	J36BG	C28BE	S15AM
2	B16AJ	E17AG	W17AB	R34CR	K38CL	W18CJ	U37CE	K19BP	W23BL	U32BH	K33BF	P25BC
3	N24AK	M22AM	Y22AC	X26CS	F11CM	Y12CK	R27CF	F30CD	Y36BM	R20BJ	F15BG	W35BD
4	A21AL	H29AJ	T29AD	B38AB	V31CP	T37CL	X13CH	V14CE	T32BP	X28BK	V25BH	Y16BE
5	Z17AM	S34AK	G34AE	N11AC	E23CR	G27CM	B19CJ	E23CF	G20CD	B33BL	E35BJ	T24BF
6	C22BC	P26AL	L26AF	A31AD	M12CS	L13CP	N30CH	M36CH	L28CE	N15BM	M16BK	G21BG
7	K29BD	W38AM	D38AG	Z18AE	H37CR	D19CR	A14CL	H32CJ	D33CF	A25BP	H24BL	L17BH
8	F34BE	Y11BC	J11AH	C12AF	S27AC	J30CS	Z23CM	S20CL	J15CH	Z35CD	S21BM	D22BJ
9	V26BF	T31BD	U31AJ	K37AG	P13AD	U14AB	C36CP	P28CM	U25CJ	C16CE	P17BP	J29BK
10	E38BG	G18BE	R18AK	F27AH	W19AE	R23AC	K32CR	W33CL	R35CK	K24CF	W22CD	U34BL
11	M11BH	L12BF	X12AL	V13AJ	Y30AF	X36AD	F20CS	Y15CM	X16CL	F21CH	Y29CE	R26BM
12	H31BJ	D37BG	B37AM	E19AK	T14AG	B32AE	Z28AB	T25CR	B24CM	Z22CJ	T34CF	X38BP
13	S18BK	J27BH	N27BC	M30AL	G23AH	N20AF	E33AC	G35CS	N21CP	E22CK	G26CH	B11CD
14	P12BL	U13BJ	A13BD	H14AM	L36AJ	A28AB	M15AD	L16CR	A17CM	M29CL	L38CJ	N31CE
15	W37BM	R19BK	Z19BE	S23BC	D32AK	Z33AH	H25AE	D24CC	Z22CS	H34CM	D11CK	A18CF
16	Y27BP	X30BL	C30BF	P36BD	J20AL	C15AJ	S35AF	J21CD	C29AB	S26CP	J31CL	Z12CH
17	T13CD	B14BM	K14BG	W32BE	U28AM	K25AK	P16AG	U17CE	K34AC	P38CR	U18CM	C37CJ
18	G19CE	N23BP	F23BH	Y20BF	R33BC	F35AL	W24AH	R22AF	F26AD	W11CS	R12CP	K27CK
19	L30CF	A36CD	V36BJ	T28BG	X15BD	V16AM	Y21AJ	X29AG	V38AE	Y31AB	X37CR	F13CL
20	D14CH	Z32CE	E32BK	G33BH	B25BE	E24BC	T17AK	B34AH	E11AF	T18AC	B27CS	V19CM
21	J23CJ	C20CF	N20BL	L15BJ	N35BF	M21BD	G22AL	N26AJ	M31AG	G12AD	N13AB	E30CP
22	U36CK	K28CH	H28BM	D25BK	A16BG	H17BE	L29AM	A38AK	H18AH	L37AE	A19AC	H14CR
23	R32CL	F33CJ	S33BP	J35BL	Z14BH	S22BF	D34BC	Z11AL	S12AJ	D27AF	Z30AD	H23CS
24	X20CM	V15CK	P15CD	U16BM	C21BJ	P29BG	J25BD	C31AM	P37AK	J13AG	C14AE	S36AB
25	B28CP	E25CL	W25CE	K17BK	W34BJ	U38BE	K18BC	U38BC	U19AH	U19AH	K23AF	P32AC
26	N33CR	M35CM	Y35CF	X21CD	F22BL	Y26BJ	R11BF	F12BD	Y13AM	R30AJ	F36AG	W20AD
27	A15CS	H16CP	T16CH	B17CE	V29BM	T38BK	S30BG	V37BF	T19BG	X14AK	B23AH	Y28AE
28	Z25AB	S24CR	G24CJ	L21CK	E34BP	G11BL	N12BH	B18BH	E27BF	N36AL	M28AJ	T33AF
29	C35AC		L21CK	A29CH	M26CD	L31BM	N12BG	M13BG	L14BE	A29AM	M28AK	G15AG
30	K16AD		D17CL	Z34CJ	H38CE	D18BP	A37BK	H19BJ	D23BF	L14BE	H33AL	L25AH
31	F24AE		J22CH		S11CF		Z27BL	S30BJ		Z20BD		D35AJ

(159)

FIRST LETTER = YOUR PHYSICAL CODE, MIDDLE NUMBER = SENSITIVITY CODE, LAST TWO LETTERS = INTELLECTUAL CODE

1974	JAN	FEB	MAR	APR	MAY	JUN	JUL	AUG	SEP	OCT	NOV	DEC
1	J16AK	C17AH	M17AC	L34CS	N38CM	M18CK	G37CF	N19CD	M23BM	G32BJ	N33BG	E25BD
2	U24AL	K22AJ	H22AD	D26AB	A11CP	H12CL	L27CH	A30CE	H36BP	L20BK	A15BH	M35BE
3	R21AM	F29AK	S29AE	J38AC	Z31CR	S37CM	D13CJ	Z14CF	S32CD	D28BL	Z25BJ	H16BF
4	X17BC	V34AL	P34AF	U11AD	C18CS	P27CP	J19CK	C23CH	P20CE	J33BM	C35BK	S24BG
5	B22BD	E26AM	W26AG	R31AE	K12AB	W13CR	U30CL	K36CJ	W28CF	U15BP	K16BL	P21BH
6	N29BE	M38BC	Y38AH	X18AF	F37AC	Y19CS	R14CM	F32CK	Y33CH	R25CD	F24BM	W17BJ
7	A34BF	H11BD	T11AJ	B12AG	V27AD	T30AB	X23CP	V20CL	T15CJ	X35CR	V21BP	Y22BK
8	Z26BG	S31BE	G31AK	N37AH	E13AE	G14AC	B36CR	E28CM	G25CK	B16CF	E17CD	T29BL
9	C38BH	P18BF	L18AL	A27AJ	M19AF	L23AQ	N32CS	M33CP	L35CL	N24CH	M22CE	G34BM
10	K11BJ	W12BG	D12AM	Z13AK	H30AG	D36AE	A20AB	H15CM	D16CM	A21CJ	H29CF	L26BP
11	F31BK	Y37BH	J37BC	C19AL	S14AH	J32AF	Z28AC	S25CS	J24CP	Z17CK	S34CH	D38CD
12	V18BL	T27BJ	U27BD	K30AM	P23AJ	U20AG	C33AD	P35AB	J21CR	C22CL	P26CJ	J11CE
13	E12BM	G13BC	R13BE	F14BC	W36AK	R28AH	K15AE	W16AC	J17CS	K29CM	W38CK	U31CF
14	M37BP	L19BL	X19BF	V23BD	Y32AL	X33AJ	F25AF	Y24AD	X22AB	F34CP	Y11CL	R18CH
15	H27CD	D30BM	B30BG	E36BE	T20AM	B15AK	V35AG	T21AE	B29AC	V26CR	T31CH	X12CJ
16	S13CE	J14BP	N14BH	M32BF	G28BC	N25AL	E16AH	G17AF	N34AD	E38CS	G18CP	B37CK
17	P19CF	U23CD	A23BJ	H20BG	L33BD	A35AM	M24AJ	L22AG	A26AE	M11AB	L12CR	N27CL
18	W30CH	R36CE	Z36BK	S28BH	D15BE	Z16BC	H21AK	D29AH	Z38AF	H31AC	D37CS	A13CH
19	Y14CJ	X32CF	C32BL	P33BJ	J25BF	C24BD	S17AL	J34AJ	C11AG	S18AD	J27AB	Z19CP
20	T23CK	B28CH	K20BM	W15BK	U35BG	K21BE	P22AM	U26AK	K31AH	P12AE	U13AC	C30CR
21	G36CL	N28CJ	F28BP	Y25BL	R16BH	F17BF	W29BC	R38AL	F18AJ	W37AF	R19AD	K14CS
22	L32CM	A33CK	V33CD	T35BM	X24BJ	V22BG	Y34BD	X11AM	V12AK	Y27AG	X30AE	F23AB
23	D20CP	Z15CL	E15CE	G16BP	B21BK	E29BH	T26BE	B31BC	E37AL	T13AH	B14AF	V36AC
24	J28CR	C25CM	M25CF	L24CO	N17BL	M34BJ	G38BF	N18BD	M27AM	G19AJ	N23AG	E32AD
25	U33CS	K35CP	H35CH	D21CE	A22BM	H26BK	L11BG	A12BE	H13BC	L30AK	A36AH	M20AE
26	R15AB	F16CR	S16CJ	J17CF	Z29BP	S38BL	D31BH	Z37BF	S19BD	D14AL	Z32AJ	H28AF
27	X25AC	V24CS	P24CK	U22CH	C34CD	P11BM	J18BJ	C27BG	P30BE	J23AM	C20AK	S33AG
28	B35AD	E21AB	W21CL	R29CJ	K26CE	W31BP	U12BK	K13BH	W14BF	U36BC	K28AL	P15AH
29	N16AE		Y17CM	X34CK	F38CF	Y16CO	R37BL	F19BJ	Y23BG	R32BD	F33AM	W25AJ
30	A24AF		T22CP	B26CL	V11CH	T12CE	X27BM	V30BK	T36BH	X20BE	V15BC	Y35AK
31	Z21AG		G29CR		E31CJ		B13BP	E14BL		B28BF		T16AL

1975	JAN	FEB	MAR	APR	MAY	JUN	JUL	AUG	SEP	OCT	NOV	DEC
1	G24AM	N22AK	F22AE	Y26AC	R11CR	F12CM	W27CJ	R30CF	F36CD	W20BL	R15BJ	K35BF
2	L21BC	A29AL	V29AF	T38AD	X31CS	V37CP	Y13CK	X14CH	V32CE	Y28BM	X25BK	F16BG
3	D17BD	Z34AM	E34AG	G11AE	B18AB	E27CR	T19CL	B23CJ	E20CF	T33BP	B35BL	V24BH
4	J22BE	C25BC	M26AH	L31AF	N12AC	M13CS	G30CM	N36CK	M28CM	G15CD	N16BM	E21BJ
5	U29BF	K38BD	H38AJ	D18AG	A37AD	H19AB	L14CP	A32CL	H33CJ	L25CE	A24BP	M17BK
6	R34BG	F11BE	S11AK	J12AH	Z27AE	S30AC	D23CR	Z20CM	S15CK	D35CF	Z21CD	H22BL
7	X26BH	V31BF	P31AL	U37AJ	C13AF	P14AD	J36CS	C28CP	P25CL	J16CH	C17CE	S29BM
8	B38BJ	E18BG	W18AM	R27AK	K19AG	W23AE	U32AB	K33CR	W35CM	U24CJ	K22CF	P34BP
9	N11BK	M12BH	Y12BC	X13AL	F30AH	Y36AF	R20AC	F15CS	Y16CP	R21CK	F29CH	W26CD
10	A31BL	H37BJ	T37BD	B19AM	V14AJ	T32AG	X28AD	V25AB	T24CR	X17CL	V34CJ	Y38CE
11	Z18BM	S27BK	G27BE	N38BC	E23AK	G20AH	B33AE	E35AC	G21CS	B22CM	E26CK	T11CF
12	C12BP	P13BL	L13BF	A14BD	M36AL	L28AJ	N15AF	M16AD	L17AB	N29CP	M38CL	G31CH
13	K37CD	W19BM	D19BG	Z23BE	H32AM	D33AK	A25AG	H24AF	D22AC	A34CR	H11CM	L18CJ
14	F27CE	Y30BP	J30BH	C36BF	S20BC	J15AL	Z35AH	S21AF	J29AD	Z26CS	S31CP	D12CK
15	V13CF	T14CD	U14BJ	K32BG	P28BD	U25AM	C16AJ	P17AG	U34AE	C38AB	P18CR	J37CL
16	E19CH	G23CE	R23BK	F20BH	W33BE	R35BC	K24AK	W22AH	R26AF	K11AC	W12CS	U27CM
17	M30CJ	L36CF	X36BL	V28BJ	Y15BF	X16BD	F21AL	Y29AJ	X38AG	F31AD	Y37AB	R13CP
18	H14CK	D32CH	B32BM	E33BK	T25BG	B24BE	V17AM	T34AK	B11AH	V18AE	T27AC	X19CR
19	S23CL	J20CJ	N20BP	M15BL	G35BH	N21BF	E22BC	G26AL	N31AJ	E12AF	G13AD	B30CS
20	P36CM	U28CK	A28CD	H25BM	L16BJ	A17BG	M29BD	L38AM	A18AK	M37AG	L19AD	N14AB
21	W32CP	R33CL	Z33CE	S35BP	O24BK	Z22BH	H34BC	D11BC	Z12AL	H27AH	D30AF	A23AC
22	Y20CR	X15CM	C15CF	P16CD	J21BL	C29BJ	S26BF	J31BD	C37AM	S13AJ	J14AG	Z36AD
23	T28CS	B25CP	K25CH	W24CE	U17BM	K34BK	P38BG	U18BE	K27BC	P19AK	U23AH	C32AE
24	G33AB	N35CR	F35CJ	Y21CF	R22BP	F26BL	W11BH	R12BF	F13BD	W30AL	R36AJ	K20AF
25	L15AC	A16CS	V16CK	T17CH	X29CO	V38BM	Y31BJ	X37BG	V19BE	Y14AM	X32AK	F28AG
26	D25AD	Z24AB	E24CL	G22CJ	B34CE	E11BP	T18BB	B27BH	E30BF	T23BC	B20AL	V33AH
27	J35AE	C21AC	M21CM	L29CK	N26CF	M31CD	G12BL	N13BJ	M14BG	G36BD	N28AM	E15AJ
28	U16AF	K17AD	H17CP	D34CL	A38CH	H18CE	L37BM	A19BK	H23BH	L32BE	A33BC	M25AK
29	R24AG		S22CR	J26CN	Z11CJ	S12CF	O27BP	Z30BL	S36BJ	D20BF	Z15BD	H35AL
30	X21AH		P29CS	U38CP	C31CK	P37CH	J13CD	C14BM	P32BK	J28BG	C25BE	S16AM
31	B17AJ		W34AB		K18CL		U19CE	K23BP		U33BH		P24BC

1976	JAN	FEB	MAR	APR	MAY	JUN	JUL	AUG	SEP	OCT	NOV	DEC
1	W21BD	R29AM	C34AH	P11AF	J18AC	C27CS	S19CM	J23CK	C20CH	S33CD	J35BM	Z24BJ
2	Y17BE	X34BE	K26AJ	W31AG	U12AD	K13AB	P30CP	U36CL	K28CJ	P15CE	U16BP	C21BK
3	T22BF	B26BD	F38AK	Y18AH	R37AE	F19AC	W14CR	R32CM	F33CK	W25CF	R24CO	K17BL
4	G29BG	N38BE	V11AL	T12AJ	X27AF	V30AD	Y23CS	X20CP	V15CL	Y35CH	X21CE	F22BM
5	L34BH	A11BF	E31AM	G37AK	B13AG	E14AE	T36AB	B26CR	E25CM	T16CJ	B17CF	V29BP
6	O26BJ	Z31BG	M18BC	L27AL	N19AH	M23AF	G32AC	N33CS	M35CP	G24CK	N22CH	E34CD
7	J38BM	C18BH	H12BD	D13AM	A30AJ	H36AG	L20AD	A15AB	H16CR	L21CL	A29CJ	M26CE
8	U11BL	K12BJ	S37BE	J19BC	Z14AK	S32AH	D28AE	Z25AC	S24CS	D17CM	Z34CH	H38CF
9	R31BM	F37BM	P27BF	U30BD	C23AL	P20AJ	J33AF	C35AD	P21AB	J22CP	C26CL	S11CH
10	X18BP	V27BL	W13BG	R14BE	K36AM	W28AK	U15AG	K16AE	W17AC	U29CR	K38CM	P31CJ
11	B12CD	E13BM	Y19BH	X23BF	F32BC	Y33AL	R25AH	F24AF	Y22AD	R34CS	F11CP	W18CK
12	N37CE	M19BP	T30BG	B36BG	V20BD	T15AM	X35AJ	V21AG	T29AE	X26AB	V31CR	Y12CL
13	A27CF	H30CD	G14BK	N32BH	E28BE	G25BC	B16AK	E17AH	G34AF	B38AC	E18CS	T37CM
14	Z13CH	S14CE	L23BL	A20BJ	M33BF	L35BD	N24AL	M22AJ	L26AG	N11AD	M12AB	L13CR
15	C19CJ	P23CF	D36BM	Z28BK	H15BG	D16BE	A21AM	H29AK	D38AH	A31AE	H37AC	C13CR
16	K30CK	W36CH	J32BP	C33BL	S25BH	J24BF	Z17BC	C22BD	S34AL	J11AJ	Z18AF	D19CS
17	F14CL	Y32CJ	U20CD	K15BM	P35BJ	U21BG	C22BD	P26AM	U31AK	C12AG	P13AE	J30AB
18	V23CH	T20CK	R28CE	F25BP	W16BK	R17BH	K29BE	W38BC	R18AL	K37AH	W19AF	U14AC
19	E36CP	G28CL	X33CF	W35CD	Y24BL	X22BJ	F34BF	Y11BD	X12AM	F27AJ	Y30AG	R23AD
20	M32CR	L33CM	B15CH	E16CE	T21BM	B29BK	V26BG	T31BE	B37BC	V13AK	T14AH	X36AE
21	H20CS	D15CP	N25CJ	M24CF	G17BP	N34BL	E38BH	G18BF	N27BD	E19AL	G23AJ	B32AF
22	S28AB	J25CR	A35CK	H21CH	L22CD	A26BM	M11BJ	L12BG	A13BE	M30AM	L36AK	N20AG
23	P33AC	U35CS	Z16CL	S17CJ	D29CE	Z38BP	U16BB	D37BH	Z19BF	U18AD	O32AL	A28AH
24	W15AD	R16AB	C24CM	P22CK	J34CF	C11CD	S18BL	J27BJ	C30BG	S23BD	J20AM	Z33AJ
25	Y25AE	X24CD	K21CP	W29CL	U26CH	K31CE	P12BM	U13BK	K14BH	P36BE	U28BC	C15AK
26	T35AF	B21AD	F17CR	Y34CM	R38CJ	F18CF	W37BP	R19BL	F23BJ	W32BF	R33BD	K25AL
27	G16AG	N17AE	V22CS	T26CP	X11CK	V12CH	Y27CD	X30BM	V36BK	Y20BG	X15BE	F35AM
28	L24AH	A22AF	E29AB	G38CR	B31CL	E37CJ	T13CE	B14BP	E32BD	T28BH	B25BF	V16BG
29	D21AJ	Z29AG	M34AC	L11CS	N18CH	M27CK	G19CF	N23CD	M28BM	G33BJ	N35BG	E24BD
30	J17AK		H26AD	D31AB	A12CP	H13CL	L30CH	A36CE	H28BP	L15BK	A16BH	M21BE
31	U22AL		S38AE		Z37CR		D14CJ	Z32CF		D25BL		H17BF

(160)

FIRST LETTER = YOUR PHYSICAL CODE, MIDDLE NUMBER = SENSITIVITY CODE, LAST TWO LETTERS = INTELLECTUAL CODE

1977	JAN	FEB	MAR	APR	MAY	JUN	JUL	AUG	SEP	OCT	NOV	DEC
1	S22BG	J26BE	N26AK	M31AH	G12AE	N13AC	E30CR	G36CM	N28CK	E15CF	G16CD	B21BL
2	P29BH	U38BF	A38AL	H18AJ	L37AF	A19AD	M14CS	L32CP	A33CL	M25CH	L24CE	N17BM
3	W34BJ	R11BG	Z11AM	S12AK	D27AG	Z30AE	H23AB	D20CR	Z15CM	H35CJ	D21CF	A22BP
4	Y26BK	X31BH	C31BC	P37AL	J13AH	C14AF	S36AC	J28CS	C25CP	S16CK	J17CH	Z29CD
5	T38BL	B18BJ	K18BD	W27AM	U19AJ	K23AG	P32AD	U33AB	K35CR	P24CL	U22CJ	C34CE
6	G11BM	N12BK	F12BE	Y13BC	R30AK	F36AH	W20AE	R15AC	F16CS	W21CM	R29CK	K26CF
7	L31BP	A37BL	V37BF	T19BD	X14AL	V32AJ	Y28AF	X25AD	V24AB	Y17CP	X34CL	F38CH
8	D18CD	Z27BM	E27BG	G30BE	B23AM	E20AK	T33AG	B35AE	E21AC	T22CR	B26CH	V11CJ
9	J12CE	C13BP	M13BH	L14BF	N36BC	M28AL	G15AH	N16AF	M17AD	G29CS	N38CP	E31CK
10	U37CF	K19CD	H19BJ	D23BG	A32BD	H33AM	L25AJ	A24AG	H22AE	L34AB	A11CH	M18CL
11	R27CH	F30CE	S30BK	J36BH	Z20BE	S15BC	D35AK	Z21AH	S29AF	D26AC	Z31CS	H12CM
12	X13CJ	V14CF	P14BL	U32BJ	C28BF	P25BD	J16AL	C17AJ	P34AG	J38AD	C18AB	S37CP
13	B19CK	E23CH	W23BM	R20BK	K33BG	W35BE	U24AM	K22AK	W26AH	U11AE	K12AC	P27CR
14	N30CL	M36CJ	Y36BP	X28BL	F15BH	Y16BF	R18BC	F29AL	V34AM	Y38AJ	F37AD	M13CS
15	A14CM	H32CK	T32CD	B33BM	V25BJ	T24BG	X17BD	V34AM	T11AK	X18AG	V27AE	Y19AB
16	Z23CP	S20CL	G20CE	N15BP	E35BK	G21BH	B22BE	E26BC	G31AL	B12AH	E13AF	T30AC
17	C36CR	P28CM	L28CF	A25CD	M16BL	L17BJ	N29BF	M38BD	L18AM	N37AJ	M19AG	G14AD
18	K32CS	W33CP	D33CH	Z35CE	H24BM	D22BK	A34BG	H11BE	D12BC	A27AK	H30AH	L23AE
19	F20AB	Y15CR	J15CJ	C16CF	S21BP	J29BL	Z26BH	S31BF	J37BD	Z13AL	S14AJ	D36AF
20	V28AC	T25CS	U25CK	K24CH	P17CD	U34BM	C38BJ	P18BG	U27BE	C19AM	P23AK	J32AG
21	E33AD	G35AB	R35CL	F21CJ	W22CE	R26BP	K11BK	W12BH	R13BF	K30BC	W36AL	U20AH
22	M15AE	L16AC	X16CM	V17CK	Y29CF	X38CD	F31BL	Y37BJ	X19BG	F14BD	Y32AM	R28AJ
23	H25AF	D24AD	B24CP	E22CL	T34CH	B11CE	V18BM	T27BK	B30BH	V23BE	T20BC	X33AK
24	S35AG	J21AE	N21CR	M29CM	G26CJ	N31CF	E12BP	G13BE	N14BJ	E36BF	G28BD	B15AL
25	P16AH	U17AF	A17CS	H34CP	L38CK	A18CH	M37CD	L19BM	A23BK	M32BG	L33BE	N25AM
26	W24AJ	R22AG	Z22AB	S26CR	D11CL	Z12CJ	H27CE	D30BP	Z36BL	H20BH	D15BF	A35BC
27	Y21AK	X29AH	C29AC	P38CS	J31CM	C37CK	S13CF	J14CD	C32BM	S28BJ	J25BG	Z16BD
28	T17AL	B34AJ	K34AD	W11AB	U16CF	K27CL	P19CH	U23CE	K20BP	P33BK	U35BH	C24BE
29	G22AM		F26AE	Y31AC	R12CR	F13CM	W30CJ	R36CF	F28CD	W15BL	R16BJ	K21BF
30	L29BC		V38AF	T18AD	X37CS	V19CP	Y14CK	X32CH	V33CE	Y25BM	X24BK	F17BG
31	D34BD		E11AG		B27AB		T23CL	B20CJ		T35BP		V22BH

1978	JAN	FEB	MAR	APR	MAY	JUN	JUL	AUG	SEP	OCT	NOV	DEC
1	E29BJ	G38BG	R38AM	F18AK	W37AG	R19AE	K14AB	W32CR	R33CM	K25CJ	W24CF	U17BP
2	M34BK	L11BH	X11BC	V12AL	Y27AH	X30AF	F23AC	Y20CS	X15CP	F35CK	Y21CH	R22CD
3	H26BL	D31BJ	B31BD	E37AM	T13AJ	B14AG	V36AD	T28AB	B25CR	V16CL	T17CJ	X29CE
4	S38BM	J18BK	N18BE	M27BC	G19AK	N23AH	E32AE	G33AC	N35CS	E24CM	G22CK	B34CF
5	P11BP	U12BL	A12BF	H13BD	L30AL	A36AJ	M20AF	L15AD	A16AB	M21CP	L29CL	N26CH
6	W31CD	R37BM	Z37BG	S19BE	D14AM	Z32AK	H28AG	D25AE	Z24AC	H17CR	D34CM	A38CJ
7	Y18CE	X27BP	C27BH	P30BF	J23BC	C20AL	S33AH	J35AF	C21AD	S22CS	J26CP	Z11CK
8	T12CF	B13CD	K13BJ	W14BG	U36BD	K28AM	P15AJ	U16AG	K17AE	P29AB	U38CR	C31CL
9	G37CH	N19CE	F19BK	Y23BH	R32BE	F33BC	W25AK	R24AH	F22AF	W34AC	R11CS	K18CM
10	L27CJ	A30CF	V38BL	T36BJ	X20BF	V15BD	Y35AL	X21AJ	V29AG	Y26AD	X31AB	F12CP
11	D13CK	Z14CH	E14BM	G32BK	B28BG	E25BE	T16AM	B17AK	E34AH	T38AE	B18AC	V37CR
12	J19CL	C23CJ	H23BP	L20BL	N33BH	M35BF	G24BC	N22AL	M26AJ	G11AF	N12AD	E27CS
13	U30CM	K36CK	H36CD	D28BM	A15BJ	H16BG	L21BD	A29AM	H38AK	L31AG	A37AE	M13AB
14	R14CP	F32CL	S32CE	J33BP	Z25BK	S24BH	D17BE	Z34BC	S11AL	D18AH	Z27AF	H19AC
15	X23CR	V20CM	P20CF	U15CD	C35BL	P21BJ	J22BF	C26BD	P31AM	J12AJ	C13AG	S30AD
16	B36CS	E28CP	W28CH	R25CE	K16BM	W17BK	U29BG	K38BE	W18BC	U37AK	K19AH	P14AE
17	N32AB	M33CR	Y33CJ	X35CF	F24BP	Y22BL	R34BH	F11BF	Y12BD	R27AL	F30AJ	W23AF
18	A20AC	H15CS	T15CK	B16CH	V21CD	T29BM	X26BJ	V31BG	T37BE	X13AM	V14AK	Y36AG
19	Z28AD	S25AB	G25CL	N24CJ	E17CE	G34BP	B38BK	E18BH	G27BF	B19BC	E23AL	T32AH
20	C33AE	P35AC	L35CM	A21CK	H22CF	L26CD	N11BL	H12BJ	L13BG	N30BD	H36AM	G20AJ
21	K15AF	W16AD	D16CP	Z17CL	H29CH	D38CF	A31BM	H37BK	D19BH	A14BE	H32BC	L28AK
22	F25AG	Y24AE	J24CR	C22CM	S34CJ	J11CF	Z18BP	S27BL	J30BJ	Z23BF	S20BD	D33AL
23	V35AH	T21AF	U21CS	K29CP	P26CK	U31CH	C12CD	P13BM	U14BK	C36BG	P28BE	J15AM
24	E16AJ	G17AG	R17AB	F34CR	W38CL	R18CJ	K37CE	W19BP	R23BL	K32BH	W33BF	U25BC
25	M24AK	L22AH	X22AC	V26CS	Y11CM	X12CK	F27CF	Y30CD	X36BM	F20BJ	Y15BG	R35BD
26	H21AL	D29AJ	B29AD	E38AB	T31CP	B37CL	V13CH	T14CE	B32BP	V28BK	T25BH	X16BE
27	S17AM	J34AK	N34AE	M11AC	G18CR	N27CM	E19CJ	G23CF	N20CD	E33BL	G35BJ	B24BF
28	P22BC	U26AL	A26AF	M31AD	L12CS	A13CP	M30CK	L36CH	A28CE	M15BM	L16BK	N21BG
29	W29BD		Z38AG	S18AE	D37AB	Z19CR	H14CL	D32CJ	Z33CD	H25BP	D24BL	A17BH
30	Y34BE		C11AH	P12AF	J27AC	C30CS	S23CH	J20CK	C15CH	S35CD	J21BM	Z22BJ
31	T26BF		K31AJ		U13AD		P36CP	U28CL		P16CE		C29BK

1979	JAN	FEB	MAR	APR	MAY	JUN	JUL	AUG	SEP	OCT	NOV	DEC
1	K34BL	W11BJ	D11BD	Z12AM	H27AJ	D30AG	A23AD	H20AB	D15CR	A35CL	H21CJ	L22CD
2	F26BM	Y31BK	J31BK	C37BC	S13AK	J14AH	Z36AE	S28AC	J25CS	Z16CM	S17CK	D29CF
3	V38BP	T18BL	U18BF	K27BD	P19AL	U23AJ	C32AF	P33AD	U35AB	C24CP	P22CL	J34CH
4	E11CD	G12BM	R12BG	F13BE	W30AM	R36AK	K20AG	W15AE	R16AC	K21CR	W29CM	U26CJ
5	M31CE	L37BP	X37BH	V19BF	Y14BC	X32AL	F28AH	Y25AF	X24AD	F17CS	Y34CP	R38CK
6	H18CF	D27CD	B27BJ	E30BG	T23BD	B20AM	V33AJ	T35AG	B21AE	V22AB	T26CR	X11CL
7	S12CH	J13CE	N13BK	M14BH	G36BE	N28BC	E15AK	G16AH	N17AF	E29AC	G38CS	B31CM
8	P37CJ	U19CF	A19BL	H23BJ	L32BF	A33BD	M25AL	L24AJ	A22AG	M34AD	L11AB	N18CP
9	W27CK	R30CH	Z30BM	S36BK	D20BG	Z15BE	H35AM	D21AK	Z29AH	H26AE	D31AC	A12CR
10	Y13CL	X14CJ	C14BP	P32BL	J28BH	C25BF	S16BC	J17AL	C34AJ	S38AF	J18AD	Z37CS
11	T19CM	B23CK	K23CD	W20BM	U33BJ	K35BG	P24BD	U22AM	K26AK	P11AG	U12AE	C27AB
12	G30CP	N36CL	F36CE	Y28BP	R15BK	F16BH	W21BE	R29BC	F38AL	W31AH	R37AF	K13AC
13	L14CR	A32CM	V32CF	T33CD	X25BL	V24BJ	Y17BF	X34BD	V11AM	Y18AJ	X27AG	F19AD
14	D23CS	Z20CP	E20CH	G15CE	B35BM	E21BK	T22BG	B26BE	E31BC	T12AK	B13AH	V30AE
15	J36AB	C28CR	M28CJ	L25CF	N16BP	M17BL	G29BH	N38BF	M18BD	G37AL	N19AJ	E14AF
16	U32AC	K33CS	H33CK	D35CH	A24CD	H22BM	L34BJ	A11BG	H12BE	L27AM	A30AK	M23AG
17	R20AD	F15AB	S15CL	J16CJ	Z21CE	S29BP	D26BK	Z31BH	S37BF	D13BC	Z14AL	H36AH
18	X28AE	V25AC	P25CM	U24CK	C17CF	P34CF	C18BJ	P27BG	J19BD	C23AM	S32AJ	P20AK
19	B33AF	E35AD	W35CP	R21CL	K22CE	W26CE	U11BM	K12BK	W13BH	U30BE	K36BC	W28AL
20	N15BG	M16AF	Y16CR	X17CM	F29CJ	Y30CF	R31BP	F37BL	Y19BJ	R14BF	F32BD	Y33AM
21	A25AH	H24AF	T24CS	B22CP	V34CK	T11CH	X18CD	V27BM	T30BK	X23BG	V20BE	Y33AM
22	Z35AJ	S21AG	G21AB	N29CR	E26CL	G31CJ	B12CE	E13BP	G14BG	B36BH	E28BF	T15BC
23	C16AK	P17AH	L17AC	A34CS	M38CH	L18CF	N37CD	M19CG	L23BM	N32BJ	M33BG	G25BD
24	K24AL	W22AJ	D22AD	Z26AB	H11CP	D12CL	A27CH	H30CE	D36BP	A20BK	H15BH	L35BE
25	F21AM	Y29AK	J29AE	C38AC	S31CR	J37CH	Z13CJ	S14CF	J32CD	Z28BL	S25BJ	D16BF
26	V17BC	T34AL	U34AF	K11AD	P18CS	U27CP	C19CK	P23CH	U20CE	C33BM	P35BK	J24BG
27	E22BD	G26AM	R26AG	F31AE	W12CM	R13CR	K30CL	W36CJ	R28CF	K15BP	W16BL	U21BH
28	M29BE	L38BC	X38AH	V18AF	Y37CK	X19CS	F14CM	Y32CK	X33CH	F25CD	Y24BM	R17BJ
29	H34BF		B11AJ	E12AG	T27AD	B30AB	V23CP	T20CL	B15CJ	V35CE	T21BP	X22BK
30	S26BG		N31AK	M37AH	G13AE	N14AC	E36CR	G28CH	N25CK	E16CF	G17CD	B29BL
31	P38BH		A18AL		L19AF		M32CS	L33CP		M24CH		N34BM

FIRST LETTER = YOUR PHYSICAL CODE, MIDDLE NUMBER = SENSITIVITY CODE, LAST TWO LETTERS = INTELLECTUAL CODE

1980	JAN	FEB	MAR	APR	MAY	JUN	JUL	AUG	SEP	OCT	NOV	DEC
1	A26BP	H31BL	G18BG	N27BE	E19AM	G23AK	B32AG	E33AE	G35AC	B24CR	E22CM	T34CJ
2	Z38CD	S18BM	L12BH	A13BF	M30BC	L36AL	N20AH	M15AF	L16AD	N21CS	M29CP	G26CK
3	C11CE	P12BP	D37BJ	Z19BG	H14BD	D32AM	A28AJ	H25AG	D24AE	A17AB	H34CR	L38CL
4	K31CF	W37CD	J27BK	C30BH	S23BE	J20BC	Z33AK	S35AH	J21AF	Z22AC	S26CS	D11CM
5	F18CH	Y27CE	U13BL	K14BJ	P36BF	U28BD	C15AL	P16AJ	U17AG	C29AD	P38AB	J31CP
6	V12CJ	T13CF	R19BM	F23BK	W32BG	R33BE	K25AM	W24AK	R22AH	K34AE	W11AC	U18CR
7	E37CK	G19CH	X30BP	V36BL	Y20BH	X15BF	F35BC	Y21AL	X29AJ	F26AF	Y31AD	R12CS
8	M27CL	L30CJ	B14CD	E32BM	T28BJ	B25BG	V16BD	T17AM	B34AK	V38AG	T18AE	X37AB
9	H13CM	D14CK	N23CE	M20BP	G33BK	N35BH	E24BE	G22BC	N26AL	E11AH	G12AF	B27AC
10	S19CP	J23CL	A36CF	H28CD	L15BL	A16BJ	M21BF	L29BD	A38AM	M31AJ	L37AG	N13AD
11	P30CR	U36CM	Z32CH	S33CE	D25BM	Z24BK	H17BG	D34BE	Z11BC	H18AK	D27AH	A19AE
12	W14CS	R32CP	C20CJ	P15CF	J35BP	C21BL	S22BH	J26BF	C31BD	S12AL	J13AJ	Z30AF
13	Y23AB	X20CR	K28CK	W25CH	U16CD	K17BM	P29BJ	U38BG	K18BE	P37AM	U19AK	C14AG
14	T36AC	B28CS	F33CL	Y35CJ	R24CE	F22BP	W34BK	R11BH	F12BF	W27BC	R30AL	K23AH
15	G32AD	N33AB	V15CM	T16CK	X21CF	V29CD	Y26BL	X31BJ	V37BG	Y13BD	X14AM	F36AJ
16	L20AE	A15AC	E25CP	G24CL	B17CH	E34CE	T38BM	B18BK	E27BH	T19BE	B23BC	V32AK
17	D28AF	Z25AD	M35CR	L21CM	N22CJ	M26CF	G11BP	N12BL	M13BJ	G30BF	N36BD	E20AL
18	J33AG	C35AE	H16CS	D17CP	A29CK	H38CH	L31CD	A37BM	H19BK	L14BG	A32BE	M28AM
19	U15AH	K16AF	S24AB	J22CR	Z34CL	S11CJ	D18CE	Z27BP	S30BL	D23BH	Z20BF	H33BC
20	R25AJ	F24AG	P21AC	U29CS	C26CM	P31CK	J12CF	C13CD	P14BM	J36BJ	C28BG	S15BD
21	X35AK	V21AH	W17AD	R34AB	K38CP	W18CL	U37CH	K19CE	W23BP	U32BK	K33BH	P25BE
22	B16AL	E17AJ	Y22AE	X26AC	F11CR	Y12CH	R27CJ	F30CF	Y36CD	R20BL	F15BJ	W35BF
23	N24AM	M22AK	T29AF	B38AD	V31CS	T37CP	X13CK	V14CH	T32CE	X28BM	V25BK	Y16BG
24	A21BC	H29AL	G34AG	N11AE	E18AB	G27CR	B19CL	E23CJ	G20CF	B33BP	E35BL	T24BH
25	Z17BD	S34AM	L26AH	A31AF	M12AC	L13CS	N30CM	M36CK	L28CH	N15CD	M16BM	G21BJ
26	C22BE	P26BC	D38AJ	Z18AG	H37AD	D19AB	A14CP	H32CL	D33CJ	A25CE	H24BP	L17BK
27	K29BF	W38BD	J11AK	C12AH	S27AE	J30AC	Z23CR	S20CM	J15CK	Z35CF	S21CD	D22BL
28	F34BG	Y11BE	U31AL	K37AJ	P13AF	U14AD	C36CS	P28CP	U25CL	C16CH	P17CE	J29BM
29	V26BH	T31BF	R18AM	F27AK	W19AG	R23AE	K32AB	W33CR	R35CM	K24CJ	W22CF	U34BP
30	E38BJ		X12BC	V13AL	Y30AH	X36AF	F20AC	Y15CS	X16CP	F21CK	Y29CH	R26CD
31	M11BK		B37BD		T14AJ		V28AD	T25AB		V17CL		X38CE

1981	JAN	FEB	MAR	APR	MAY	JUN	JUL	AUG	SEP	OCT	NOV	DEC
1	B11CF	E12CD	W12BJ	R13BG	K30BD	W36AM	U20AJ	K15AG	W16AE	U21AB	K29CR	P26CL
2	N31CH	M37CE	Y37BK	X19BH	F14BE	Y32BC	R28AK	F25AH	Y24AF	R17AC	F34CS	W38CM
3	A18CJ	H27CF	T27BL	B30BJ	V23BF	T20BD	X33AL	V35AJ	T21AG	X22AD	V26AB	Y11CP
4	Z12CK	S13CH	G13BM	N14BK	E36BG	G28BE	B15AM	E16AK	G17AH	B29AE	E38AC	T31CR
5	C37CL	P19CJ	L19BP	A23BL	M32BH	L33BF	N25BC	M24AL	L22AJ	N34AF	M11AD	G18CS
6	K27CM	W30CK	D30CD	Z36BM	H20BJ	D15BG	A35BD	H21AM	D29AK	A26AG	H31AE	L12AB
7	F13CP	Y14CL	J14CE	C32BP	S28BK	J25BH	Z16BE	C24BF	J34AL	Z38AH	C11AJ	D37AC
8	V19CR	T23CM	U23CF	K20CD	P33BL	U35BJ	C24BF	P22BD	U26AM	C11AJ	P12AG	J27AD
9	E30CS	G36CP	R36CH	F28CE	W15BM	R16BK	K21BG	W29BE	R38BC	K31AK	W37AH	U13AE
10	M14AB	L32CR	X32CJ	V33CF	Y25BP	X24BL	F17BH	Y34BF	X11BD	F18AL	Y27AJ	R19AF
11	H23AC	D20CS	B20CK	E15CH	T35CD	B21BM	V22BJ	T26BG	B31BE	V12AM	T13AK	X30AG
12	S36AD	J28AB	N28CL	M25CJ	G16CE	N17BP	E29BK	G38BH	N18BF	E37BC	G19AL	B14AH
13	P32AE	U33AC	A33CM	H35CK	L24CF	A22CD	M34BL	L11BJ	A12BG	M27BD	L30AM	N23AJ
14	W20AF	R15AD	Z15CP	S16CL	D21CH	Z29CE	H26BM	D31BK	Z37BH	H13BE	D14BC	A36AK
15	Y28AG	X25AE	C25CR	P24CM	J17CJ	C34CF	S38BP	J18BL	C27BJ	S19BF	J23BD	Z32AL
16	T33AH	B35AF	K35CS	W21CP	U22CK	K26CH	P11CD	U12BM	K13BK	P30BG	U36BE	C20AM
17	G15AJ	N16AG	F16AB	Y17CR	R29CL	F38CJ	W31CE	R37BP	F19BL	W14BH	R32BF	K28BC
18	L25AK	A24AH	V24AC	T22CS	X34CM	V11CK	Y18CF	X27CD	V30BM	Y23BJ	X20BG	F33BD
19	D35AL	Z21AJ	E21AD	G29AB	B26CP	E31CL	T12CH	B13CE	E14BP	T36BK	B28BH	V15BE
20	J16AM	C17AK	M17AE	L34AC	N38CM	M18CH	G37CJ	N19CF	M23CD	G32BL	N33BJ	E25BF
21	U24BC	K22AL	H22AF	D26AD	A11CS	H12CP	L27CK	A30CH	H36CE	L20BM	A15BK	M35BG
22	R21BD	F29AM	S29AG	J38AE	Z31AB	S37CR	D13CL	Z14CJ	S32CF	D28BP	Z25BL	H16BH
23	X17BE	V34BC	P34AH	U11AF	C18AC	P27CS	J19CH	C23CK	P20CH	J13CD	C35BM	S24BJ
24	B22BF	E26BD	W26AJ	R31AG	K12AD	W13AB	U30CP	K36CL	W28CJ	U15CE	K16BP	P21BK
25	N29BG	M38BE	Y38AK	X18AH	F37AE	Y19AC	R14CR	F32CM	Y33CK	R25CF	F24CD	W17BL
26	A34BH	H11BF	T11AL	B12AJ	V27AF	T30AD	X23CS	V20CP	T15CL	X35CH	V21CE	Y22BM
27	Z26BJ	S31BG	G31AM	N37AK	E13AG	G14AE	B36AB	E28CR	G25CM	B16CJ	E17CF	T29BP
28	C38BK	P18BH	L18BC	A27AL	M19AH	L23AF	N32AC	M33CS	L35CP	N24CK	M22CH	G34CD
29	K11BL		D12BD	Z13AM	H30AJ	D36AG	A20AD	H15AB	D16CR	A21CL	H29CJ	L26CE
30	F31BM		J37BE	C19BC	S14AK	J32AH	Z28AE	S25AC	J24CS	Z17CM	S34CK	D38CF
31	V18BP		U27BF		P23AL		C33AF	P35AD		C22CP		J11CH

1982	JAN	FEB	MAR	APR	MAY	JUN	JUL	AUG	SEP	OCT	NOV	DEC
1	U31CJ	K37CF	H37BL	D19BJ	A14BF	H32BD	L28AL	A25AJ	H24AG	L17AD	A34AB	M38CP
2	R18CK	F27CH	S27BM	J30BK	Z23BG	S20BE	D33AM	Z35AK	S21AH	D22AE	Z26AC	H11CR
3	X12CL	V13CJ	P13BP	U14BL	C36BH	P28BF	J15BC	C16AL	P17AJ	J29AF	C38AD	S31CS
4	B37CM	E19CK	W19CD	R23BM	K32BJ	W33BG	U25BD	K24AM	W22AK	U34AG	K11AE	P18AB
5	N27CP	M30CL	Y30CE	X36BP	F20BK	Y15BH	R35BE	F21BC	Y29AL	R26AH	F31AF	W12AC
6	A13CR	H14CH	T14CF	B32CD	V28BL	T25BJ	X16BF	V17BD	T34AM	X38AJ	V18AG	Y37AD
7	Z19CS	S23CP	G23CH	N20CE	E33BM	G35BH	B24BE	E22BF	G26BC	B11AK	E12AH	T27AE
8	C30AB	P36CR	L36CJ	A28CF	M15BP	L16BL	N21BH	M29BF	L38BD	N31AL	M37AJ	G13AF
9	K14AC	W32CS	D32CK	Z33CH	H25CD	D24BM	A17BF	H34BG	D11BE	A18AM	H27AK	L19AG
10	F23AD	Y20AB	J20CL	C15CJ	S35CE	J21BP	Z22BK	S26BH	J31BF	Z12BC	S13AL	D30AH
11	V36AE	T28AC	U28CM	K25CK	P16CF	U17BD	C29BJ	P38BG	U18BE	C37BC	P19AM	J14AJ
12	E32AF	G33AD	R33CP	F35CL	W24CH	R22CE	K34BM	W11BK	R12BH	K27BE	W30BC	U23AK
13	M20AG	L15AE	X15CR	V16CM	Y21CJ	X29CF	F26BP	Y31BL	X37BJ	F13BF	Y14BD	R36AL
14	H28AH	D25AF	B25CS	E24CP	T17CK	B34CH	V38BD	T18BM	B27BK	V19BG	T23BE	X32AM
15	S33AJ	J35AG	N35AB	M21CR	G22CL	N26CJ	E11CE	G12BP	N13BL	E30BH	G36BF	B20BC
16	P15AK	U16AH	A16AC	H17CS	L29CM	A38CK	M31CF	L37CD	A19BM	M14BJ	L32BG	N28BD
17	W25AL	R24AJ	Z24AD	S22AB	D34CP	Z11CL	H18CH	D27CE	Z30BP	H23BK	D20BH	A33BE
18	Y35AM	X21AK	C21AE	P29AC	J26CR	C31CH	S12CJ	J13CF	C14CD	S36BL	J28BJ	Z15BF
19	T16BC	B17AL	K17AF	W34AD	U38CS	K18CP	P37CF	U19CH	K23CE	P32BM	U33BJ	C25BG
20	G24BD	N22AM	F22AG	Y26AE	R11AB	F12CR	W27CS	R30CJ	F36CF	W20BP	R15BL	K35BH
21	L21BE	A29BC	V29AH	T38AF	X31AC	V37CS	Y13CM	X14CK	V32CH	Y28CD	X25BM	F16BJ
22	D17BF	Z34BD	E34AJ	G11AG	B18AD	E27AB	T19CP	B23CL	E20CJ	T33CE	B35BP	V24BK
23	J22BG	C26BE	M26AK	L31AH	N12AE	M13AC	G30CR	N36CM	M28CK	G15CF	N16CD	E21BL
24	U29BH	K38BF	H38AL	D18AJ	A37AF	H19AD	L14CS	A32CP	H33CL	L25CH	A24CE	M17BM
25	R34BJ	F11BG	S11AM	J12AK	Z27AG	S30AE	D23AB	Z20CR	S15CM	D35CJ	Z21CF	H22BP
26	X26BK	V31BH	P31BC	U37AL	C13AH	P14AF	J36AC	C28CS	P25CP	J16CK	C17CH	S29CD
27	B38BL	E18BJ	W18BD	R27AM	K19AJ	W23AG	U32AD	K33AB	W35CR	U24CL	K22CJ	P34CE
28	N11BM	M12BK	Y12BE	X13BC	F30AK	Y36AH	R20AE	F15AC	Y16CS	R21CM	F29CK	W26CF
29	A31BP		T37BF	B19BD	V14AL	T32AJ	X28AF	V25AD	T24AB	X17CP	V34CL	Y38CH
30	Z18CD		G27BG	N30BE	E23AM	G20AK	B33AG	E35AE	G21AC	B22CR	E26CM	T11CJ
31	C12CE		L13BH		M36BC		N15AH	M16AF		N29CS		G31CK

FIRST LETTER = YOUR PHYSICAL CODE, MIDDLE NUMBER = SENSITIVITY CODE, LAST TWO LETTERS = INTELLECTUAL CODE

1983	JAN	FEB	MAR	APR	MAY	JUN	JUL	AUG	SEP	OCT	NOV	DEC
1	L18CL	A27CJ	V27BP	T30BL	X23BH	V20BF	Y33BC	X35AL	V21AJ	Y22AF	X26AD	F11CS
2	D12CM	Z13CK	E13CD	G14BM	B36BJ	E28BG	T15BD	B16AM	E17AK	T29AG	B38AE	V31AB
3	J37CP	C19CL	M19CE	L23BP	N32BK	M33BH	G25BE	N24BC	M22AL	G34AH	N11AF	E18AC
4	U27CR	K30CM	H30CF	D36CD	A20BL	H15BJ	L35BF	A21BD	H29AM	L26AJ	A31AG	M12AD
5	R13CS	F14CP	S14CH	J32CE	Z28BM	S25BK	D16BG	Z17BE	S34BC	D38AK	Z18AH	H37AE
6	X19AB	V23CR	P23CJ	U20CF	C33BP	P35BL	J24BH	C22BF	P26BD	J11AL	C12AJ	S27AF
7	B30AC	E36CS	W36CK	R28CH	K15CD	W16BM	U21BJ	K29BG	W38BE	U31AM	K37AK	P13AG
8	N14AD	M32AB	Y32CL	X33CJ	F25CE	Y24BP	R17BK	F34BH	Y11BF	R18BC	F27AL	W19AH
9	A23AE	H20AC	T20CM	B15CK	V35CF	T21CD	X22BL	V26BJ	T31BG	X12BD	V13AM	Y30AJ
10	Z36AF	S28AD	G28CP	N25CL	E16CH	G17CE	B29BM	E38BK	G18BH	B37BE	E19BC	T14AK
11	C32AG	P33AE	L33CR	A35CM	M24CJ	L22CF	N34BP	M11BL	L12BJ	N27BF	M30BD	G23AL
12	K20AH	W15AF	D15CS	Z16CP	H21CK	D29CH	A26CD	H31BM	D37BK	A13BG	H14BE	L36AM
13	F28AJ	Y25AG	J25AB	C24CR	S17CL	J34CJ	Z38CE	S18BP	J27BL	Z19BH	S23BF	D32BC
14	V33AK	T35AM	U35AC	K21CS	P22CM	U26CK	C11CF	P12CD	U13BM	C30BJ	P36BG	J20BD
15	E15AL	G16AJ	R16AD	F17AB	W29CP	R38CL	K31CH	W37CE	R19BP	K14BK	W32BH	U28BE
16	M25AM	L24AK	X24AE	V22AC	Y34CR	X11CH	F18CJ	Y27CF	X30CD	F23BL	Y20BJ	R33BF
17	H35BC	D21AL	B21AF	E29AD	T26CS	B31CP	V12CK	T13CH	B14CE	V36BM	T28BK	X15BG
18	S16BD	J17AM	N17AG	M34AE	G38AB	N18CR	E37CL	G19CJ	N23CF	E32BP	G33BL	B25BH
19	P24BE	U22BC	A22AH	H26AF	L11AC	A12CS	M27CM	L30CK	A36CH	M20CD	L15BM	N35BJ
20	W21BF	R29BD	Z29AJ	S38AG	D31AD	Z37AB	H13CP	D14CL	Z32CJ	H28CE	D25BP	A16BK
21	Y17BG	X34BE	G34AK	P11AH	J18AE	C27AC	S19CR	J23CH	C20CK	S33CF	J35CD	Z24BL
22	T22BH	B26BF	K26AL	W31AJ	U12AF	K13AD	P30CS	U36CP	K28CL	P15CH	U16CE	C21BM
23	G29BJ	N38BG	F38AM	Y18AK	R37AG	F19AE	W14AB	R32CR	F33CM	W25CJ	R24CF	K17BP
24	L34BK	A11BH	V11BC	T12AL	X27AH	V30AF	Y23AC	X20CS	V15CP	Y35CK	X21CH	F22CD
25	D26BL	Z31BJ	E31BD	G37AM	B13AJ	E14AG	T36AD	B28AB	E25CR	T16CL	B17CJ	V29CE
26	J38BM	C18BK	M18BE	L27BC	N19AK	M23AH	G32AE	N33AC	M35CS	G24CM	N22CK	E34CF
27	U11BP	K12BL	H12BF	D13BD	A30AL	H36AJ	L20AF	A15AQ	H16AB	L21CP	A29CL	M26CH
28	R31CD	F37BM	S37BG	J19BE	Z14AM	S32AK	D28AG	Z25AE	S24AC	D17CR	Z34CM	H38CJ
29	X18CE		P27BH	U30BF	C23BC	P20AL	J33AH	C35AF	P21AD	J22CS	C26CP	S11CK
30	B12CF		W13BJ	R14BG	K36BD	W28AM	U15AJ	K16AG	W17AE	U29AB	K38CR	P31CL
31	N37CH		Y19BK		F32BE		R25AK	F24AH		R34AC		W18CM

1984	JAN	FEB	MAR	APR	MAY	JUN	JUL	AUG	SEP	OCT	NOV	DEC
1	Y12CP	X13CL	K19CF	W23CD	U32BL	K33BJ	P25BF	U24BD	K22AM	P34AJ	U11AG	C18AD
2	T37CR	B19CM	F30CH	Y36CE	R20BM	F15BK	W35BG	R21BE	F29BC	W26AK	R31AH	K12AE
3	G27CS	N30CP	V14CJ	T32CF	X28BP	V25BL	Y16BH	X17BF	V34BD	Y38AL	X18AJ	F37AF
4	L13AB	A14CR	E23CK	G20CH	B33CD	E35BM	T24BJ	B22BG	E26BE	T11AM	B12AK	V27AG
5	D19AC	Z23CS	M36CL	L28CJ	N15CE	M16BP	G21BK	N29BH	M38BF	G31BC	N37AL	E13AH
6	J30AD	C36AB	H32CM	D33CK	A25CF	H24CD	L17BL	A34BJ	H11BG	L18BD	A27AM	M19AJ
7	U14AE	K32AC	S20CP	J15CL	Z35CH	S21CE	D22BM	Z26BK	S31BH	D12BE	Z13BC	H30AK
8	R23AF	F20AD	P28CR	U25CH	C16CJ	P17CF	J29BP	C38BL	P18BJ	J37BF	C19BD	S14AL
9	X36AG	V28AE	W33CS	R35CP	K24CK	W22CH	U34CD	K11BM	W12BK	U27BG	K30BE	P23AM
10	B32AH	E33AF	Y15AB	X16CR	F21CL	Y29CJ	R26CE	F31BP	Y37BL	R13BH	F14BF	H36BC
11	N20AJ	M15AG	T25AC	B24CS	V17CM	T34CK	X38CF	V18CD	T27BM	X19BJ	V23BG	Y32BD
12	A28AK	H25AH	G35AD	N21AB	E22CP	G26CL	B11CH	E12CE	G13BP	B30BK	E36BH	T20BE
13	Z33AL	S35AJ	L16AE	A17AC	M29CR	L38CH	N31CJ	M37CF	L19CD	N14BL	M32BJ	G28BF
14	C15AM	P16AK	D24AF	Z22AD	H34CS	D11CP	A18CK	H27CH	D30CE	A23BM	H20BK	L33BG
15	K25BC	W24AL	J21AG	C29AE	S26AB	J31CR	Z12CL	S13CJ	J14CF	Z36BP	S28BL	D15BH
16	F35BD	Y21AM	U17AH	K34AF	P38AC	U18CS	C37CM	P19CK	U23CH	C32CD	P33BM	J25BJ
17	V16BE	T17BC	R22AJ	F26AG	W11AD	R12AB	K27CP	W30CL	R36CJ	K20CE	W15BP	U35BK
18	E24BF	G22BD	X29AK	V38AH	Y31AE	X37AC	F13CR	Y14CH	X32CK	F28CF	Y25CD	R16BL
19	M21BG	L29BE	B34AL	E11AJ	T18AF	B27AD	V19CS	T23CP	B20CL	V33CH	T35CE	X24BM
20	H17BH	D34BF	N26AM	M31AK	G12AG	N13AE	E30AB	G36CR	N28CM	E15CJ	G16CF	B21BP
21	S22BJ	J26BG	A38BC	H18AL	L37AH	A19AF	M14AC	L32CS	A33CP	M25CK	L24CH	N17CD
22	P29BK	U38BH	Z11BD	S12AM	D27AJ	Z30AG	H23AD	D20AB	Z15CR	H35CL	D21CJ	A22CE
23	W34BL	R11BJ	C31BE	P37BC	J13AK	C14AH	S36AE	J28AC	C25CS	S16CH	J17CK	Z29CF
24	Y26BM	X31BK	K18BF	W27BD	U19AL	K23AJ	P32AF	U33AD	K35AB	P24CP	U22CL	C34CH
25	T38BP	B18BL	F12BG	Y13BE	R30AM	F36AK	W20AG	R15AE	F16AC	W21CR	R29CM	K26CJ
26	G11CD	N12BM	V37BH	T19BF	X14BC	V32AL	Y28AH	X25AF	V24AD	Y17CS	X34CP	F38CK
27	L31CE	A37BP	E27BJ	G30BG	B23BD	E20AM	T33AJ	B35AG	E21AE	T22AB	B26CR	V11CL
28	D18CF	Z27CD	M13BK	L14BH	N36BE	M28BC	G15AK	N16AH	M17AF	G29AC	N38CS	E31CM
29	J12CH	C13CE	H19BL	D23BJ	A32BF	H33BD	L25AL	A24AJ	I22AG	L34AD	A11AB	M18CP
30	U37CJ		S30BM	J36BK	Z20BG	S15BE	D35AM	Z21AK	S29AH	D26AE	Z31AC	H12CR
31	R27CK		P14BP		C28BH		J16BC	C17AL		J38AF		S37CS

1985	JAN	FEB	MAR	APR	MAY	JUN	JUL	AUG	SEP	OCT	NOV	DEC
1	P27AB	U30CR	A30CJ	H36CF	L20BP	A15BL	M35BH	L21BF	A29BD	M26AL	L31AJ	N12AF
2	M13AC	R14CS	Z14CK	S32CH	D28CD	Z25BM	H16BJ	D17BG	Z34BE	H38AM	D18AK	A37AG
3	Y19AD	X23AB	C23CL	P20CJ	J33CE	C35BP	S24BK	J22BH	C26BF	S11BD	J12AL	Z27AH
4	T30AE	B36AC	K36CM	W28CK	U15CF	K16CD	P21BL	U29BJ	K38BG	P31BD	U37AM	C13AJ
5	G14AF	N32AD	F32CP	Y33CL	R25CH	F24CE	W17BM	R34BK	F11BH	W18BE	R27BC	K19AK
6	L23AG	A20AE	V20CR	T15CM	X35CJ	V21CF	Y22BP	X26BL	V31BJ	Y12BF	X13BD	F30AL
7	D36AH	Z28AF	E28CS	G25CP	B16CK	E17CH	T29CD	B38BM	E18BK	T37BG	B19BE	V14AM
8	J32AJ	C33AG	M33AB	L35CR	N24CL	M22CM	G34CE	N11BP	M12BL	G27BH	N30BF	E23BC
9	U20AK	K15AH	H15AC	D16CS	A21CM	H29CK	L26CF	A31BD	H37BM	L13BJ	A14BG	M36BD
10	R28AL	F25AJ	S25AD	J24AB	Z17CP	S34CL	D38CH	Z18CE	S27BP	D19BK	Z23BH	H32BC
11	X33AM	V35AK	P35AE	U21AC	C22CR	P26CM	J11CJ	C12CF	P13CD	J30BL	C36BJ	S20BF
12	B15BC	E16AL	W16AF	R17AD	K29CS	W38CK	U31CH	K37CE	W19CE	U14BM	K32BK	P28BG
13	N25BD	M24AM	Y24AG	X22AE	F34AB	Y11CR	R18CL	F27CJ	Y30CF	R23BP	F20BL	W33BH
14	A35BE	H21BC	T21AH	B29AF	V26AC	T31CS	X12CM	V13CK	T14CH	X36CD	V28BM	Y15BJ
15	Z16BF	S17BD	G17AJ	N34AG	E38AD	G18AB	B37CP	E19CL	G23CJ	B32CE	E33BP	T25BK
16	C24BG	P22BE	L22AK	A26AH	M11AE	L12AC	N27CR	M30CM	L36CK	N20CF	M15CD	G35BL
17	K21BH	W29BF	D29AL	Z38AJ	H31AF	D37AD	A13CS	H14CP	D32CL	A28CH	H25CE	L16BM
18	F17BJ	Y34BG	J34AM	C11AK	S18AG	J27AE	Z19AB	S23CR	J20CM	Z33CJ	S35CF	D24BP
19	V22BK	T26BH	U26BC	K31AL	P12AH	U13AF	C30AC	P36CS	U28CP	C15CK	P16CH	J21CD
20	E29BL	G38BJ	R38BD	F18AM	W37AJ	R19AG	K14AD	W32AB	R33CR	K25CL	W24CJ	U17CE
21	M34BM	L11BK	X11BE	V12BC	Y27AK	X30AH	F23AE	Y20AC	X15CS	F35CH	Y21CK	R22CF
22	H26BP	D31BL	B31BF	E37BD	T13AL	B14AJ	V36AF	T28AD	B25AB	V16CP	T17CL	X29CH
23	S38CD	J18BM	N18BG	M27BE	G19AM	N23AK	E32AG	G33AE	N35AC	E24CR	G22CM	B34CJ
24	P11CE	U12BP	A12BH	H13BF	L30BC	A36AL	H20AH	L15AF	A16AD	H21CS	L29CP	N26CK
25	W31CF	R37CD	Z37BJ	S19BG	D14BD	Z32AM	H28AJ	D25AG	Z24AE	H17AB	D34CR	A38CL
26	Y18CH	X27CF	C27BK	P30BH	J23BE	C20BC	S33AK	J35AM	C21AF	S22AC	J26CS	Z11CM
27	T12CJ	B13CF	K13BL	W14BJ	U36BF	K28BD	P15AL	U16AJ	K17AG	P29AD	U38AB	C31CP
28	G37CK	N19CH	F19BM	Y23BK	R32BG	F33BE	W25AM	R24AK	F22AH	W34AE	R11AC	K18CR
29	L27CL		V30BP	T36BL	X20BH	V15BF	Y35BC	X21AL	V29AJ	Y26AF	X31AD	F12CS
30	D13CM		E14CD	G32BM	B28BJ	E25BG	T16BD	B17AM	E34AK	T38AG	B18AE	V37AB
31	J19CP		M23CE		N33BK		G24BE	N22BC		G11AH		E27AC

THE COMPLETE BOOK OF BIORHYTHM LIFE CYCLES

FIRST LETTER = YOUR PHYSICAL CODE, MIDDLE NUMBER = SENSITIVITY CODE, LAST TWO LETTERS = INTELLECTUAL CODE

1986	JAN	FEB	MAR	APR	MAY	JUN	JUL	AUG	SEP	OCT	NOV	DEC
1	M13AD	L14AB	X14CL	V32CJ	Y28CE	X25BP	F16BK	Y17BH	X34BF	F38BC	Y18AL	R37AH
2	H19AE	D23AC	B23CM	E20CK	T33CF	B35CD	V24BL	T22BJ	B26BG	V11BD	T12AM	X27AJ
3	S30AF	J36AD	N36CP	M28CL	G15CH	N16CE	E21BM	G29BK	N38BH	E31BE	G37BC	B13AK
4	P14AG	U32AE	A32CR	H33CM	L25CJ	A24CF	M17BP	L34BL	A11BJ	M18BF	L27BD	N19AL
5	W23AH	R20AF	Z20CS	S15CP	D35CK	Z21CH	H22CD	D26BM	Z31BK	H12BG	D13BE	A30AM
6	Y36AJ	X28AG	C28AB	P25CR	J16CL	C17CJ	S29CE	J38BP	C18BL	S37BH	J19BF	Z14BC
7	T32AK	B33AH	K33AC	W35CS	U24CM	K22CK	P34CF	U11CD	K12BM	P27BJ	U30BG	C23BD
8	G20AL	N15AJ	F15AD	Y16AB	R21CP	F29CL	W26CH	R31CE	F37BP	W13BK	R14BH	K36BE
9	L28AM	A25AK	V25AE	T24AC	X17CR	V34CM	Y36CJ	X18CF	V27CD	Y19BL	X23BJ	F32BF
10	D33BC	Z35AL	E35AF	G21AD	B22CS	E26CP	T11CK	B12CH	E13CE	T30BM	B36BK	V20BG
11	J15BD	C16AM	M16AG	L17AE	N29AB	M38CR	G31CL	N37CJ	M19CF	G14BP	N32BL	E28BH
12	U25BE	K24BC	H24AH	D22AF	A34AC	H11CS	L18CM	A27CK	H30CH	L23CD	A20BM	M33BJ
13	R35BF	F21BD	S21AJ	J29AG	Z26AD	S31AB	D12CP	Z13CL	S14CJ	D36CE	Z28BP	H15BK
14	X16BG	V17BE	P17AK	U34AH	C38AE	P18AC	J37CR	C19CM	P23CK	J32CF	C33CD	S25BL
15	B24BH	E22BF	W22AL	R26AJ	K11AF	M12AD	U27CS	K30CP	W35CL	U20CH	K15CE	P35BM
16	N21BJ	M29BG	Y29AM	X38AK	F31AG	Y37AE	R13AB	F14CR	Y32CM	R28CJ	F25CF	W16BP
17	A17BK	H34BH	T34BC	B11AL	V18AH	T27AF	X19AC	V23CS	T20CP	X33CK	V35CH	Y24CD
18	Z22BL	S26BJ	G26BD	N31AM	E12AJ	G13AG	B30AD	E36AB	G28CR	B15CL	E16CJ	T21CE
19	C29BM	P38BK	L38BE	A18BC	M37AK	L19AH	N14AE	M32AC	L33CS	N25CM	M24CK	G17CF
20	K34BP	W11BL	D11BF	Z12BD	H27AL	D30AJ	A23AF	H20AD	D15AB	A35CP	H21CL	L22CH
21	F26CD	Y31BM	J31BG	C37BE	S13AM	J14AK	Z36AG	S28AE	J25AC	Z16CR	S17CH	D29CJ
22	V38CE	T18BP	U18BH	K27BF	P19BC	U23AL	C32AH	P33AF	U35AD	C24CS	P22CP	J34CK
23	E11CF	G12CD	R12BJ	F13BG	W30BD	R36AM	K20AJ	W15AG	R16AE	K21AB	W29CR	U26CL
24	M31CH	L37CE	X37BK	V19BH	Y14BE	X32BC	F28AK	Y25AH	X24AF	F17AC	Y34CS	R38CM
25	H18CJ	D27CF	B27BL	E30BJ	T23BF	B20BD	V33AL	T35AJ	B21AG	V22AD	T26AB	X11CP
26	S12CK	J13CH	N13BM	M14BK	G36BG	N28BE	E15AM	G16AK	N17AH	E29AE	G38AC	B31CR
27	P37CL	U19CJ	A19BP	H23BL	L32BH	A33BF	M25BC	L24AL	A22AJ	M34AF	L11AD	N18CS
28	W27CM	R30CK	Z30CD	S36BM	D20BJ	Z15BG	H35BD	D21AM	Z29AK	H26AG	D31AE	A12AB
29	Y13CP		C14CE	P32BP	J28BK	C25BH	S16BE	J17BC	C34AL	S38AH	J18AF	Z37AC
30	T19CR		K23CF	W20CD	U33BL	K35BJ	P24BF	U22BD	K25AM	P11AJ	U12AG	C27AD
31	G30CS		F36CH		R15BM		W21BG	R29BE		W31AK		K13AE

1987	JAN	FEB	MAR	APR	MAY	JUN	JUL	AUG	SEP	OCT	NOV	DEC
1	F19AF	Y23AD	J23CP	C20CL	S33CH	J35CE	Z24BH	S22BK	J26BH	Z11BE	S12BC	D27AK
2	V30AG	T36AE	U36CR	K28CM	P15CJ	U16CF	C21BP	P29BL	J38BJ	C31BF	P37BD	J13AL
3	E14AH	G32AF	R32CS	F33CP	W25CK	R24CH	K17CD	W34BM	R11BK	K18BG	W27BE	U19AM
4	M23AJ	L28AG	X20AB	V15CR	Y35CL	X21CJ	F22CE	Y26BP	X31BL	F12BH	Y13BF	R30BC
5	H36AK	D28AH	B28CS	E25CS	T16CM	B17CK	V29CF	T38CD	B18BM	V37BJ	T19BG	X14BD
6	S32AL	J33AJ	N33AD	M35AB	G24CP	N22CL	E34CH	G11CE	N12BP	E27BK	G30BH	B23BE
7	P20AM	U15AK	A15AE	H16AC	L21CR	A29CM	M26CJ	L31CF	A37CD	M13BL	L14BJ	N36BF
8	W28BC	R25AL	Z25AF	S24AD	D17CS	Z34CP	H38CK	D18CH	Z27CE	H19BM	D23BK	A32BG
9	Y33BD	X35AM	C35AV	P21AE	J22AB	C26CR	S11CL	J12CJ	C13CF	S30BP	J36BL	Z20BH
10	T15BE	B16BC	K16AH	W17AF	U29AC	K38CS	P31CH	U37CK	K19CH	P14CD	U32BM	C28BJ
11	G25BF	N24BD	F24AJ	Y22AG	R34AD	F11AB	W18CP	R27CL	F30CJ	W23CE	R20BP	K33BK
12	L35BG	A21BE	V21AK	T29AH	X26AE	V31AC	Y12CR	X13CM	V14CK	Y36CF	X28CD	F15BL
13	D16BH	Z17BF	E17AL	G34AJ	B38AF	E18AD	T37CS	B19CP	E23CL	T32CH	B33CE	V25BM
14	J24BJ	C22BG	M22AM	L26AK	N11AG	M12AE	G27AB	N30CR	M36CM	G20CJ	N15CF	E35BP
15	U21BK	K29BH	H29BC	D38AL	A31AH	H37AF	L13AC	A14CS	H32CP	L28CK	A25CH	M16CD
16	R17BL	F34BJ	S34BD	J11AM	Z18AJ	S27AG	D19AD	Z23AB	S20CR	D33CL	Z35CJ	H24CE
17	X22BM	V26BK	P26BE	U31BC	C12AK	P13AH	J30AE	C36AC	P28CS	J15CM	C16CK	S21CF
18	B29BP	E38BL	W38BF	R18BD	K37AL	W19AJ	U14AF	K32AD	W33AB	U25CP	K24CL	P17CH
19	N34CD	M11BM	Y11BG	X12BE	F27AM	Y30AK	R23AG	F20AE	Y15AC	R35CR	F21CM	W22CJ
20	A26CE	H31BP	T31BH	B37BF	V13BC	T14AL	X36AH	V28AF	T25AD	X16CS	V17CP	Y29CK
21	Z38CF	S18CD	G18BJ	N27BG	E19BD	G23AM	B32AJ	E33AG	G35AE	B24AB	E22CR	T34CL
22	C11CH	P12CE	L12BK	A13BH	M30BE	L36BC	N20AK	M15AH	L16AF	N21AC	M29CS	G26CM
23	K31CJ	W37CF	D37BL	Z19BJ	H14BF	D32BD	A28AL	H25AJ	D24AG	A17AD	H34AB	L38CP
24	F18CK	Y27CH	C20BM	C30BK	S23BG	J20BE	Z33AM	S35AK	J21AH	Z22AG	S26AC	D11CR
25	V12CL	T13CJ	U13BP	K14BL	P36BH	U28BF	C15BC	P16AL	U17AJ	C29AF	P38AD	J31CS
26	E37CM	G19CK	R19CD	F23BM	W32BJ	R33BG	K25BD	W24AM	R22AK	K34AG	W11AE	U18AB
27	M27CP	L30CL	X30CE	V36BP	Y20BK	X15BJ	F35BE	Y21BC	X29AL	F26AH	Y31AF	R12AC
28	H13CR	D14CM	B14CF	E32CD	T28BL	B25BJ	V16BF	T17BD	B34AM	V38AJ	T18AG	X37AD
29	S19CS		N23CH	M20CE	G33BM	N35BK	E24BG	G28BF	N26BC	E11AK	G12AH	B27AE
30	P30AB		A36CJ	H28CF	L15BP	A16BL	M21BH	L29BF	A38BD	M31AL	L37AJ	N13AF
31	W14AC		Z32CK		D25CD		H17BJ	D34BG		H18AM		A19AG

1988	JAN	FEB	MAR	APR	MAY	JUN	JUL	AUG	SEP	OCT	NOV	DEC
1	Z30AH	S36AF	L32AB	A33CR	M25CL	L24CJ	N17CE	M34BP	L11BL	N18BH	M27BF	G19BC
2	C14AJ	P32AG	D20AC	Z15CS	H35CM	D21CK	A22CF	H26CD	D31BM	A12BJ	H13BG	L30BD
3	K23AK	W20AH	J28AD	C25AB	S16CP	J17CL	Z29CH	S38CE	J18BP	Z37BK	S19BH	D14BE
4	F36AL	Y28AJ	U33AE	K35AC	P24CR	U22CM	C34CJ	P11CF	U12CD	C27BL	P30BJ	J23BF
5	V32AM	T33AK	R15AF	F16AD	W21CS	R29CP	K26CK	W31CH	R37CE	K13BM	W14BK	U36BG
6	E20BC	G15AL	X25AG	V24AE	Y17AB	X34CR	F38CL	Y18CJ	X27CF	F19BP	Y23BL	R32BH
7	M28BD	L25AM	B35AH	E21AF	T22AC	B26CS	V11CH	T12CK	B13CH	V30CD	T36BM	X20BJ
8	H33BE	D35BC	N16AJ	M17AG	G20AD	N38AB	E31CP	G37CL	N19CJ	E14CE	G32BP	B28BK
9	S15BF	J16BD	A24AK	H22AH	L34AE	A11AC	M18CM	L27CH	A30CK	M23CF	L20CD	N33BL
10	P25BG	U24BE	Z21AL	S29AJ	D26AF	Z31AD	H12CS	D13CP	Z14CL	H36CH	D28CE	A15BM
11	W35BH	R21BF	C17AM	P34AK	J38AG	C18AB	S37AB	J19CR	C23CM	S32CJ	J33CF	Z25BP
12	Y16BJ	X17BG	K22BC	W26AL	U11AH	K12AF	P27AC	U30CS	K35CP	P20CK	U15CH	C35CD
13	T24BK	B22BH	F29BD	Y38AM	R31AJ	F37AG	W13AD	R14AB	F32CR	W28CL	R25CJ	K16CE
14	G21BL	N29BJ	V34BE	T11BC	X18AK	V27AH	Y19AE	X23AC	V20CS	Y33CM	X35CK	F24CF
15	L17BM	A34BK	E26BF	G31BD	B12AL	E13AJ	T30AF	B36AD	E28AB	T15CP	B16CL	V21CH
16	D22BP	Z26BL	M38BG	L18BE	N37AM	M19AK	G14AB	N32AE	M33AC	G25CR	N24CM	E17CJ
17	J29CD	C38BM	H11BH	D12BF	A27BC	H30AL	L23AH	A20AF	H15AD	L35CS	A21CP	M22CC
18	U34CE	K11BP	S31BJ	J37BG	Z13BD	S14AM	D36AJ	Z28AG	S25AE	D16AB	Z17CR	H29CL
19	R26CF	F31CD	P18BK	U27BH	C19BE	P23BC	J32AJ	C33AH	P35AF	J24AC	C22CS	S34CM
20	X38CH	V18CE	W12BL	R13BJ	K30BF	W36BD	U20AL	K15AJ	W16AG	U21AD	K29AB	P26CP
21	B11CJ	E12CF	Y37BM	X19BK	F14BG	Y32BE	R28AM	F25AK	Y24AH	R17AE	F34AC	W38CR
22	N31CK	M37CH	T27BP	B30BL	V23BH	T20BF	X33BC	V35AL	T21AJ	X22AF	V26AD	Y11CS
23	A18CL	H27CJ	G13CD	N14BM	E36BJ	G28BG	B15BD	E16AM	G17AK	B29AG	E38AE	T31AB
24	Z12CM	S13CK	L19CE	A23BP	M32BK	L33BH	N25BE	M24BC	L22AL	N34AH	M11AF	G18AC
25	C37CP	P19CL	D30CF	Z36CD	H20BL	D15BJ	A35BF	H21BD	D29AM	A26AJ	H31AG	L12AD
26	K27CR	W30CM	J14CH	C32CE	S28BP	J25BG	Z16BG	S17BE	J36AK	Z38AK	S18AH	D37AE
27	F13CS	Y14CP	U23CJ	K20CF	P33BP	U35BL	C24BH	P22BF	U26BD	C11AL	P12AJ	J27AF
28	V19AB	T23CR	R36CK	F28CH	W15CD	R16BM	K21BK	W29BG	R38BE	K31AM	W37AK	U13AG
29	E30AC	G36CS	X32CL	V33CJ	Y25CK	X24BP	F17BK	Y34BH	X11BF	F18BC	Y27AL	R19AH
30	M14AD		B28CM	E15CK	T35CF	B21CD	V22BL	T26BJ	B31BG	V12BD	T13AM	X30AJ
31	H23AE		N28CP		G16CH		E29BM	G38BK		E37BE		B14AK

FIRST LETTER = YOUR PHYSICAL CODE, MIDDLE NUMBER = SENSITIVITY CODE, LAST TWO LETTERS = INTELLECTUAL CODE

1989	JAN	FEB	MAR	APR	MAY	JUN	JUL	AUG	SEP	OCT	NOV	DEC
1	N23AL	M20AJ	Y20AD	X15AB	F35CP	Y21CL	R22CH	F26CE	Y31BP	R12BK	F13BH	M30BE
2	A36AM	H28AK	T28AE	B25AC	V16CR	T17CN	X29CJ	V38CF	T18CD	X37BL	V19BJ	Y14BF
3	Z32BC	S33AL	G33AF	N35AD	E24CS	G22CP	B34CK	E11CH	G12CE	B27BM	E30BK	T23BG
4	C20BD	P15AM	L15AG	A16AE	M21AB	L29CR	N26CL	M31CJ	L37CF	N13BP	M14BL	G36BH
5	K28BE	W25BC	D25AH	Z24AF	H17AC	D34CS	A38CM	H16CK	D27CH	A19CD	H23BH	L32BJ
6	F33BF	Y35BD	J35AJ	C21AG	S22AD	J26AB	Z11CP	S12CL	J13CJ	Z30CE	S36BP	D20BK
7	V15BG	T16BE	U16AK	K17AH	P29AE	U38AC	C31CR	P37CH	U19CK	C14CF	P32CD	J28BL
8	E25BH	G24BF	R24AL	F22AJ	W34AF	R11AD	K18CS	W27CP	R30CL	K23CH	W20CE	U33BM
9	M35BJ	L21BG	X21AM	V29AK	Y26AG	X31AE	F12AB	Y13CR	X14CM	F36CJ	Y28CF	R15BP
10	H16BK	D17BH	B17BC	E34AL	T38AH	B18AF	V37AC	T19CS	B23CP	V32CK	T33CH	X25CO
11	S24BL	J22BJ	N22BD	M26AM	G11AJ	N12AG	E27AD	G30AB	N36CR	E20CL	G15CJ	B35CE
12	P21BM	U29BK	A29BE	H38BC	L31AK	A37AH	M13AE	L14AC	A32CS	M28CM	L25CK	N16CF
13	W17BP	R34BL	Z34BF	S11BD	D18AL	Z27AJ	H19AF	D23AD	Z20AB	H33CP	D35CL	A24CH
14	Y22CD	X26BM	C26BG	P31BE	J12AM	C13AK	S30AG	J36AE	C28AC	S15CR	J16CM	Z21CJ
15	T29CE	B38BP	K38BH	W18BF	U37BC	K19AL	P14AH	U32AF	K33AD	P25CS	U24CP	C17CK
16	G34CF	N11CD	F11BJ	Y12BG	R27BD	F30AM	W23AJ	R20AG	F15AE	W35AB	R21CR	K22CL
17	L26CH	A31CE	V31BK	T37BH	X13BE	V14BC	Y36AK	X28AH	V25AF	Y16AC	X17CS	F29CM
18	D38CJ	Z18CF	E18BL	G27BJ	B19BF	E23BD	T32AL	B33AJ	E35AG	T24AD	B22AB	V34CP
19	J11CK	C12CH	M12BM	L13BK	N30BG	M36BE	G20AM	N15AK	M16AH	G21AE	N29AC	E26CR
20	U31CL	K37CJ	H37BP	D19BL	A14BH	H32BF	L28BC	A25AL	H24AJ	L17AF	A34AD	M38CS
21	R18CM	F27CK	S27CD	J30BM	Z23BJ	S20BG	D33BD	Z35AM	S21AK	D22AG	Z26AE	H11AB
22	X12CP	V13CL	P13CE	U14BP	C36BK	P28BH	J15BE	C16BC	P17AL	J29AH	C38AF	S31AC
23	B37CR	E19CM	W19CF	R23CD	K32BL	W33BJ	U25BF	K24BD	W22AM	U34AJ	K11AG	P18AD
24	N27CS	M30CP	Y30CH	X36CE	F20BM	Y15BK	R35BG	F21BE	Y29BC	R26AK	F31AH	W12AE
25	A13AB	H14CR	T14CJ	B32CF	V28BP	T25BL	X16BH	V17BF	T34BD	X38AL	V18AJ	Y37AF
26	Z19AC	S23CS	G23CK	N20CH	E33CD	G35BM	B24BJ	E22BG	G26BE	B11AM	E12AK	T27AG
27	C30AD	P36AB	L36CL	A28CJ	M15CE	L16BP	N21BK	M29BH	L38BF	N31BC	M37AL	G13AH
28	K14AE	W32AC	D32CM	Z33CK	H25CF	D24CO	A17BL	H34BJ	D11BG	A18BD	H27AM	L19AJ
29	F23AF		J20CP	C15CL	S35CH	J21CE	Z22BM	S26BK	J31BH	Z12BE	S13BC	D30AK
30	V36AG		U28CR	K25CM	P16CJ	U17CF	C29BP	P38BL	U18BJ	C37BF	P19BD	J14AL
31	E32AH		R33CS		W24CK		K34CD	W11BM		K27BG		U23AM

1990	JAN	FEB	MAR	APR	MAY	JUN	JUL	AUG	SEP	OCT	NOV	DEC
1	R36BC	F28AL	S28AF	J25AD	Z16CS	S17CP	D29CK	Z38CH	S18CE	D37BM	Z19BK	H14BG
2	X32BD	V33AM	P33AG	U35AE	C24AB	P22CR	J34CL	C11CJ	P12CF	J27BP	C30BL	S23BH
3	B20BE	E15BC	W15AH	R16AF	K21AC	W29CS	U26CH	K31CK	W37CH	U13CD	K14BH	P36BJ
4	N28BF	M25BD	Y25AJ	X24AG	F17AD	Y34AB	R38CP	F18CL	Y27CJ	R19CE	F23BP	M32BK
5	A33BG	H35BE	T35AK	B21AH	V22AE	T26AC	X11CR	V12CM	T13CK	X30CF	V36CD	Y20BL
6	Z15BH	S16BF	G16AL	N17AJ	E29AF	G38AD	B31CS	E37CP	G19CL	B14CH	E32CE	T28BM
7	C25BJ	P24BG	L24AM	A22AK	M34AG	L11AE	N18AB	M27CR	L30CM	N23CJ	M20CF	G33BP
8	K35BK	N21BH	D21BC	Z29AL	H26AH	D31AF	A12AC	H13CS	D14CP	A36CK	H28CH	L15CD
9	F16BL	Y17BJ	J17BD	C34AM	S38AJ	J18AG	Z37AD	S19AB	J23CR	Z32CL	S33CJ	D25CE
10	V24BM	T22BK	U22BE	K26BC	P11AK	U12AH	C27AE	P30AC	U36CS	C20CM	P15CK	J35CF
11	E21BP	G29BL	R29BF	F38BD	W31AL	R37AJ	K13AF	W14AD	R32AB	K28CP	W25CL	U16CH
12	M17CD	L34BM	X34BG	V11BE	Y18AM	X27AK	F19AG	Y23AE	X20AC	F33CR	Y35CM	R24CJ
13	H22CE	D26BP	B26BH	E31BF	T12BC	B13AL	V30AH	T36AF	B28AD	V15CS	T16CP	X21CK
14	S29CF	J38CD	N38BJ	M18BG	G37BD	N19AM	E14AJ	G32AG	N33AE	E25AB	G24CR	B17CL
15	P34CH	U11CE	A11BK	H12BH	L27BE	A30BC	M23AK	L20AH	A15AF	M35AC	L21CS	N22CM
16	W26CJ	R31CF	Z31BL	S37BJ	D13BF	Z14BD	H36AL	D28AJ	Z25AG	H16AD	D17AB	A29CP
17	Y38CK	X18CH	C18BM	P27BK	J19BG	C23BE	S32AM	J33AK	C35AH	S24AE	J22AC	Z34CR
18	T11CL	B12CJ	K12BP	W13BL	U30BH	K36BF	P20BC	U15AL	K16AJ	P21AF	U29AD	C26CS
19	G31CM	N37CK	F37CD	Y19BM	R14BJ	F32BG	W28BD	R25AM	F24AK	W17AG	R34AE	K38AB
20	L18CP	A27CL	V27CE	T30BP	X23BK	V20BH	Y33BE	X35BC	V21AL	Y22AH	X26AF	F11AC
21	D12CR	Z13CM	E13CF	G14CD	B36BL	E28BJ	T15BF	B16BD	E17AM	T29AJ	B38AG	V31AD
22	J37CS	C19CR	M19CH	L23CE	N32BH	M33BG	G25BG	N24BE	M22BC	G34AK	N11AH	E18AE
23	U27AB	K30CR	H30CJ	D36CF	A20BP	H15BL	L35BH	A21BF	H29BD	L26AL	A31AJ	M12AF
24	R13AC	F14CS	S14CK	J32CH	Z28CD	S25BM	D16BJ	Z17BG	S34BE	D38AM	Z18AK	H37AG
25	X19AD	V23AB	P23CL	U20CJ	C33CE	P35BP	J24BK	C22BH	P26BF	J11BC	C12AL	S27AH
26	B30AE	E36AC	W36CM	R28CK	K15CF	W16CO	U21BL	K29BJ	W38BG	U31BD	K37AM	P13AJ
27	N14AF	M32AD	Y32CP	X33CL	F25CH	Y24CE	R17BM	F34BK	Y11BH	R18BE	F27BC	W19AK
28	A23AG	H20AE	T20CR	B15CM	V35CJ	T21CF	X22BP	V26BL	T31BJ	X12BF	V13BD	Y30AL
29	Z36AH		G28CS	N25CP	E16CK	G17CN	B29CD	E38BM	G18BK	B37BG	E19BE	T14AM
30	C32AJ		L33AB	A35CR	M24CL	L22CJ	N34CE	M11BP	L12BL	N27BH	M30BF	G23BC
31	K20AK		D15AC		H21CM		A26CF	H31CD		A13BJ		L36BD

1991	JAN	FEB	MAR	APR	MAY	JUN	JUL	AUG	SEP	OCT	NOV	DEC
1	D32BE	Z33BC	E33AH	G35AF	B24AC	E22CS	T34CM	B11CK	E12CH	T27CD	B30BM	V23BJ
2	J20BF	C15BD	M15AJ	L16AF	N21AD	M29AB	G26CP	N31CL	M37CJ	G13CE	N14BP	E36BK
3	U28BG	K25BE	H25AK	D24AH	A17AE	H34AC	L38CR	A18CM	H27CK	L19CF	A23CD	M32BL
4	R33BH	F35BF	S35AL	J21AJ	Z22AF	S26AD	D11CS	Z12CP	S13CL	D30CH	Z36CE	H20BM
5	X15BJ	V16BG	P16AM	U17AK	C29AG	P38AE	J31AB	C37CR	P19CM	J14CJ	C32CF	S28BP
6	B25BK	E24BH	W24BC	R22AL	K34AH	W11AF	U18AC	K27CS	W30CP	U23CK	K20CH	P33CD
7	N35BL	M21BJ	Y21BD	X29AM	F26AJ	Y31AG	R12AD	F13AB	Y14CR	R36CL	F28CJ	M15CE
8	A16BM	H17BK	T17BE	B34BC	V38AK	T18AH	X37AE	V19AC	T23CS	X32CH	V33CK	Y25CF
9	Z24BP	S22BL	G22BF	N26BD	E11AL	G12AJ	B27AF	E30AD	G36AB	B20CP	E15CL	T35CH
10	C21CD	P29BM	L29BG	A38BE	M31AM	L37AK	N13AG	M14AE	L32AC	N28CR	M25CM	G16CJ
11	K17CE	W34BP	D34BH	Z11BF	H18AC	D27AL	A19AH	H23AF	D20AD	A33CS	H35CP	L24CK
12	F22CF	Y26CD	J26BJ	C31BG	S12BD	J13AM	Z30AJ	S36AG	J28AE	Z15AB	S16CR	D21CL
13	V29CH	T38CE	U38BK	K18BH	P37BE	U19BC	C14AK	P32AH	U33AF	C25AC	P24CS	J17CM
14	E34CJ	G11CF	R11BL	F12BJ	W27BF	R30BD	K23AL	W20AH	R15AF	K35AD	W21AB	U22CP
15	M26CK	L31CH	X31BM	V37BK	Y13BG	X14BE	F36AH	Y28AK	X25AH	F16AC	Y17AC	R29CR
16	H38CL	D18CJ	B18BF	E27BL	T19BH	B23BF	V28BC	T33AL	B35AJ	V24AF	T22AD	X34CS
17	S11CM	J12CK	N12CD	M13BM	G30BJ	N36BG	E20BD	G15AM	N16AK	E21AG	G29AE	B26AB
18	P31CP	U37CL	A37CE	H19BP	L14BK	A32BH	M28BE	L25BC	A24AL	M17AH	L34AF	N38AC
19	W18CR	R27CM	Z27CF	S30CD	D23BL	Z20BJ	H33BF	D35BD	Z21AM	H22AJ	D26AG	A11AD
20	Y12CS	X13CP	C13CH	P14CE	J36BM	C28BK	S15BG	J16BE	C17BC	S29AK	J38AH	Z31AE
21	T37AB	B19CR	K19CJ	W23CF	F25CH	K33BL	P25BH	U24BF	K22BD	P34AL	U11AJ	C18AF
22	G27AC	N30CS	F30CK	Y36CH	R20CD	F15BM	W35BG	R21BE	F29BC	W26AM	R31AK	K12AG
23	L13AD	A14AB	V14CL	T32CJ	X26CE	V25BP	Y16BK	X17BH	V34BF	Y38BC	X18AL	F37AH
24	D19AE	Z23AC	E23CM	G20CK	B33CF	E35CO	T24BL	B22BJ	E26BG	T11BD	B12AM	V27AJ
25	J30AF	C36AD	M36CP	L28CL	N15CH	M16CE	G21BP	N29BH	M38BF	G31BE	N37BC	E13AK
26	U14AG	K32AE	H32CR	D33CM	A25CJ	H24CF	L17BM	A34BL	H11BJ	L18BF	A27BD	M19AL
27	R23AH	F20AF	S20CS	J15CP	Z35CK	S21CH	D22CD	Z26BM	S31BK	D12BG	Z13BE	H30AM
28	X36AJ	V28AG	P29AB	U25CR	C16CL	P17CJ	J29CE	C38BP	P18BL	J37BH	C19BF	S14BC
29	B32AK		W33AC	R35CS	K24CH	W22CK	U34CF	K11CD	W12BM	U27BJ	K30BG	P23BD
30	N20AL		Y15AD	X16AB	F21CP	Y29CL	R26CH	F31CE	Y37BP	R13BK	F14BH	W36BE
31	A28AH		T25AE		V17CR		X38CJ	V18CF		X19BL		Y32BF

FIRST LETTER = YOUR PHYSICAL CODE, MIDDLE NUMBER = SENSITIVITY CODE, LAST TWO LETTERS = INTELLECTUAL CODE

1992	JAN	FEB	MAR	APR	MAY	JUN	JUL	AUG	SEP	OCT	NOV	DEC
1	T20BG	B15BE	F25AL	Y24AJ	R17AF	F34AD	W38CS	R18CP	F27CL	W19CH	R23CE	K32BM
2	G28BH	N25BF	V35AM	T21AK	X22AG	V26AE	Y11AB	X12CR	V13CM	Y30CJ	X36CF	F20BP
3	L33BJ	A35BG	E16BC	G17AL	B29AH	E38AF	T31AC	B37CS	E19CP	T14CK	B32CH	V28CD
4	D15BK	Z16BM	M24BD	L22AM	N34AJ	M11AG	G18AD	N27AB	M30CR	G23CL	L36CM	E33CE
5	J25BL	C24BJ	H21BE	D29BC	A26AK	H31AH	L12AE	A13AC	H14CS	L36CM	A28CK	M15CF
6	U35BM	K21BK	S17BF	J34BD	Z38AL	S18AJ	D37AF	Z19AD	S23AB	D32CP	Z33CL	H25CH
7	R16BP	F17BL	P22BG	U26BE	C11AM	P12AK	J27AG	C30AE	P36AC	J20CR	C15CM	S35CJ
8	X24CO	V22BM	W29BH	R38BF	K31BC	W37AL	U13AH	K14AF	W32AD	U28CS	K25CP	P16CK
9	B21CE	E29BP	Y34BJ	X11BG	F18BD	Y27AM	R19AJ	F23AG	Y20AE	R33AB	F35CR	H24CL
10	N17CF	M34CD	T26BK	B31BH	V12BE	T13BC	X30AK	V36AH	T28AF	X15AC	V16CS	Y21CM
11	A22CH	H26CE	G38BL	N18BJ	E37BF	G19BD	B14AL	E32AJ	G33AG	B25AD	E24AB	T17CP
12	Z29CJ	S38CF	L11BM	A12BK	M27BG	L30BE	N23AH	M20AK	L15AH	N35AE	M21AC	G22CR
13	C34CK	P11CH	D31BP	Z37BL	H13BH	D14BF	A36BC	H28AL	D25AJ	A16AF	H17AD	L29CS
14	K26CL	W31CJ	J18CD	C27BM	S19BJ	J23BG	Z32BD	S33AH	J35AK	Z24AG	S22AE	D34AB
15	F38CM	Y18CK	U12CE	K13BP	P30BK	U36BH	C20BE	P15BC	U16AL	C21AH	P29AF	J26AC
16	V11CP	T12CL	R37CF	F19CD	W14BL	R32BJ	K28BF	W25BD	R24AM	K17AJ	W34AG	U38AD
17	E31CR	G37CM	X27CH	V30CE	Y23BM	X20BK	F33BG	Y35BE	X21BC	F22AK	Y26AH	R11AE
18	M18CS	L27CP	B13CJ	E14CF	T36BP	B28BL	V15BH	T16BF	B17BD	V29AL	T38AJ	X31AF
19	H12AB	D13CR	N19CK	M23CH	G32CD	N33BM	E25BJ	G24BG	N22BE	E34AM	G11AK	B18AG
20	S37AC	J19CS	A30CL	H36CJ	L20CE	A15BP	M35BK	L21BH	A29BF	M26BC	L31AL	N12AH
21	P27AD	U30AB	Z14CM	S32CK	D28CF	Z25CD	H16BL	D17BJ	Z34BG	H38BD	D18AM	A37AJ
22	W13AE	R14AC	C23CP	P20CL	J33CH	C35CE	S24BM	J22BK	C26BH	S11BE	J12BC	Z27AK
23	Y19AF	X23AD	K36CR	W28CM	U15CJ	K16CF	P21BP	U29BL	K38BJ	P31BF	U37BD	C13AL
24	T30AG	B36AE	F32CS	Y33CP	R25CK	F24CH	W17CD	R34BM	F11BK	W18BG	R27BE	K19AM
25	G14AH	N32AF	V20AB	T15CR	X35CL	V21CJ	Y22CE	X26BP	V31BL	Y12BH	X13BF	F30BC
26	L23AJ	A20AG	E28AC	G25CS	B16CM	E17CK	T29CF	B38CD	E18BM	T37BJ	B19BG	V14BD
27	D36AK	Z28AH	M33AD	L35AB	N24CP	M22CL	G34CH	N11CE	M12BP	G27BK	N30BH	E23BE
28	J32AL	C33AJ	H15AE	D16AC	A21CR	H29CM	L26CJ	A31CF	H37CD	L13BL	A14BJ	M36BF
29	U20AM	K15AK	S24AF	J24AD	Z17CS	S34CP	D38CK	Z18CH	S27CE	D19BM	Z23BK	H32BG
30	R28BC		P35AG	U21AE	C22AB	P26CR	J11CL	C12CJ	P13CF	J30BP	C36BL	S20BH
31	X33BD		W16AH		K29AC		U31CH	K37CK		U14CD		P28BJ

1993	JAN	FEB	MAR	APR	MAY	JUN	JUL	AUG	SEP	OCT	NOV	DEC
1	W33BK	R35BH	Z35BC	S21AL	D22AH	Z26AF	H11AC	D12CS	Z13CP	H30CK	D36CH	A20CD
2	Y15BL	X16BJ	C16BD	P17AM	J29AJ	C38AG	S31AD	J37AB	C19CR	S14CL	J32CJ	Z28CE
3	T25BM	B24BK	K24BE	W22BO	U34AK	K11AH	P18AE	U27AC	K30CS	P23CH	U20CK	C33CF
4	G35BP	N21BL	F21BF	Y29BD	R26AL	F31AJ	M12AF	R13AD	F14AB	M36CP	R28CL	K15CH
5	L16CD	A17BM	V17BG	T34BE	X38AM	V18AK	Y37AG	X19AE	V23AC	Y32CR	X33CH	F25CJ
6	D24CE	Z22BP	E22BH	G26BF	B11BC	E12AL	T27AH	B30AF	E36AD	T20CS	B15CP	V35CK
7	J21CF	C29CD	M29BJ	L38BG	N31BD	M37AM	G13AJ	N14AG	M32AE	G28AB	N25CR	E16CL
8	U17CH	K34CE	H34BK	D11BH	A18BE	H27BC	L19AK	A23AH	H20AF	L33AC	A35CS	M24CH
9	R22CJ	F26CF	S26BL	J31BJ	Z12BF	S13BD	D30AL	Z36AJ	S28AG	D15AD	Z16AB	H21CP
10	X29CK	V38CH	P38BM	U18BK	C37BG	P19BE	J14AM	C32AK	P33AH	J25AE	C24AC	S17CR
11	B34CL	E11CJ	W11BP	R12BL	K27BH	W30BF	U23BC	K20AL	W15AJ	U35AF	K21AD	P22CS
12	N26CM	M31CK	Y31CD	X37BM	F13BJ	Y14BG	R36BD	F28AM	Y25AK	R16AG	F17AE	W29AB
13	A38CP	H18CL	T18CE	B27BP	V19BK	T23BH	X32BE	V33BC	T35AL	X24AH	V22AF	Y34AC
14	Z11CR	S12CM	G12CF	N13CD	E30BL	G36BJ	B20BF	E15BD	G16AM	B21AJ	E29AG	T26AD
15	C31CS	P37CP	L37CH	A19CE	M14BM	L32BF	N28BG	M25BE	L24BC	N17AK	M34AH	G38AE
16	K18AB	W27CR	D27CJ	Z30CF	H23BP	D20BL	A33BH	H35BF	D21BD	A22AL	H26AJ	L11AF
17	F12AC	Y13CS	J13CK	C14CH	S36CD	J28BM	Z15BJ	S16BG	J17BE	Z29AM	S38AK	D31AG
18	V37AD	T19AB	U19CL	K23CJ	P32CE	U33BP	C25BK	P24BH	U22BF	C34BC	P11AL	J18AH
19	E27AE	G30AC	R30CM	F36CK	W20CF	R15CD	K35BL	W21BJ	R29BG	K26BD	W31AM	U12AJ
20	M13AF	L14AD	X14CP	V32CL	Y28CH	X25CD	F16BM	Y17BK	X34BH	F38BE	Y18BC	R37AK
21	H19AG	D23AE	B23CR	E20CM	T33CJ	B35CF	V24BP	T28BL	B26BJ	V11BF	T12BD	X27AL
22	S30AH	J36AF	N36CS	M28CP	G15CK	N16CH	E21CD	G29BM	N38BK	E31BG	G37BE	B13AM
23	P14AJ	U32AG	A32AB	H33CR	L25CL	A24CJ	M17CE	L34BP	A11BL	M18BH	L27BF	N19BC
24	W23AK	R20AH	Z20AC	S15CS	D35CM	Z21CK	H22CF	D26CD	Z31BM	H12BJ	D13BG	A30BD
25	Y36AL	X28AJ	C28AD	P25AB	J16CP	C17CL	S29CH	J38CE	C18BP	S37BK	J19BH	Z14BE
26	T32AM	B33AK	K33AE	W35AC	U24CR	K22CH	P34CJ	U11CF	K12CD	P27BL	U30BJ	C23BF
27	G20BC	N15AL	F15AF	Y16AD	R21CS	F29CP	W26CK	R31CH	F37CE	W13BM	R14BK	K36BG
28	L28BD	A25AM	V25AG	T17AE	X17AB	V34CR	Y38CL	X18CJ	V27CF	Y19BP	X23BL	F32BH
29	D33BE		E35AH	G21AF	B22AC	E26CS	T11CM	B12CH	E13CH	T30CD	B36BM	V20BJ
30	J15BF		M16AJ	L17AG	N29AD	M38AB	G31CP	N37CL	M19CJ	G14CE	N32BP	E28BK
31	U25BG		H24AK		A34AE		L18CR	A27CM		L23CF		M33BL

1994	JAN	FEB	MAR	APR	MAY	JUN	JUL	AUG	SEP	OCT	NOV	DEC
1	H15BM	D16BK	B16BE	E17BG	T29AK	B38AH	V31AE	T37AC	B19CS	V14CM	T32CK	X28CF
2	S25BP	J24BL	N24BF	M22BD	G34AL	N11AJ	E18AF	G27AD	N30AB	E23CP	G20CL	B33CH
3	P35CD	U21BM	A21BG	H29BE	L26AM	A31AK	M12AG	L13AE	A11AC	M36CR	L28CM	N15CJ
4	W16CE	R17BP	Z17BH	S34BF	D38BC	Z18AL	H37AH	D19AF	Z23AD	H32CS	D33CP	A25CK
5	Y24CF	X22CD	C22BJ	P26BG	J11BD	C12AM	S27AJ	J30AG	C36AE	S20AB	J15CR	Z35CL
6	T21CH	B29CE	K29BK	W38BH	U31BE	K37BC	P13AK	U14AH	K32AF	P28AC	U25CS	C16CM
7	G17CJ	N34CF	F34BL	Y11BJ	R18BF	F27BD	W19AL	R23AJ	F20AG	W33AD	R35AB	K24CP
8	L22CK	A26CH	V26BM	T31BK	X12BG	V13BE	Y30AM	X36AH	V28AH	Y15AE	X16AC	F21CR
9	D29CL	Z38CJ	E38BP	B37BG	N37BH	E19BF	T14BC	B32AL	E33AJ	T25AF	B24AD	V17CS
10	J34CM	C11CK	M11CD	L12BH	N27BJ	M30BG	G23BD	N20AM	M15AK	G35AG	N21AE	E22AB
11	U26CP	K31CL	H31CE	D37BP	A13BK	H14BH	L36BC	A28AL	H25AJ	L16AH	A17AF	M29AC
12	R38CR	F18CM	S18CF	J27CD	Z19BL	S23BJ	D32BF	Z33BD	S35AM	D24AJ	Z22AG	H34AD
13	X11CS	V12CP	P12CH	U13CE	C30BM	P36BK	J20BG	C15BE	P16BC	J21AK	C29AH	S26AE
14	B31AB	E37CR	W37CJ	R19CF	K14BP	W32BH	U28BF	K25BD	W24BD	U17AL	K34AJ	P38AF
15	N18AC	M27CS	Y27CK	X30CH	F23CD	Y20BM	R33BJ	F35BG	Y21BE	R22AM	F26AK	W11AG
16	A12AD	H13AB	T13CL	B14CJ	V36CE	T28BP	X15BK	V16BF	T17BF	X29CC	V38AL	Y31AH
17	Z37AE	S19AC	G19CM	N23CK	E32CF	G33CD	B25BL	E24BJ	G22BG	B34BD	E11AM	T18AJ
18	C27AF	P30AD	L30CP	A36CL	M20CH	L15CC	N35BM	M21BK	L29BH	N26BE	M31BC	G12AK
19	K13AG	W14AE	D14CR	Z32CM	H28CJ	D25CF	A16BP	H17BL	D34BJ	A38BF	H18BD	L37AL
20	F19AH	Y23AF	J23CS	C20CP	S33CK	J35CH	Z24CD	S22BP	J26BK	Z11BG	S12BE	D27AM
21	V30AJ	T35AG	U36AB	K28CR	P15CL	U16CJ	C21CE	P29BP	J38BL	J36BP	C31BH	C31BC
22	E14AK	G32AH	R32AC	F33CS	W25CH	R24CK	K17CF	W34CD	R11BM	K18BJ	W27BG	U19BD
23	M23AL	L20AJ	X20AD	V15AB	Y35CP	X21CL	F22CH	Y26CE	X31BF	F12BH	Y13BH	R30BE
24	H36AM	D28AK	B28AE	E25AC	T16CR	B17CH	V29CJ	T38CF	B18CD	V37BL	T19BJ	X14BF
25	S32BC	J33AL	N33AF	M35AD	G24CS	N22CP	E34CK	G11CH	N12CE	E27BM	G30BG	B23BG
26	P20BD	U15AM	A15AG	H16AE	L21AB	A29CR	M26CL	L31CJ	A37CF	M13BP	L14BL	N36BH
27	W28BE	R25BD	Z25AH	D17AC	Z34CS	H38CD	D18CK	Z27CH		H19CD	D23BM	A32BJ
28	Y33BF	X35BD	C35AJ	P21AG	J22AD	C26AB	S11CP	J12CL	C13CJ	S30CE	J36BP	Z20BK
29	T15BG		K16AK	H17AH	U29AE	K38AC	P31CH	U37CK	K19CK	P14CF	U32CD	C28BL
30	G25BH		F24AH	Y22AJ	R34AF	F11AD	M18CS	R27CP	F30CL	M23CH	R20CE	K33BM
31	L35BJ		V21AM		X26AG		Y12AB	X13CR		Y36CJ		F15BP

FIRST LETTER = YOUR PHYSICAL CODE, MIDDLE NUMBER = SENSITIVITY CODE, LAST TWO LETTERS = INTELLECTUAL CODE

1995	JAN	FEB	MAR	APR	MAY	JUN	JUL	AUG	SEP	OCT	NOV	DEC
1	V25CD	T24BM	U24BG	K22BE	P34AM	U11AK	C18AG	P27AE	U30AC	C23CR	P20CM	J33CJ
2	E35CE	G21BP	R21BH	F29BF	W26BC	R31AL	K12AH	W13AF	R14AD	K36CS	W28CP	U15CK
3	M16CF	L17CD	X17BJ	V34BG	Y38BD	X18AM	F37AJ	Y19AG	X23AE	F32AB	Y33CR	R25CL
4	H24CH	D22CE	B22BK	E26BH	T11BE	B12BC	V27AK	T30AH	B36AF	V20AC	T15CS	X35CM
5	S21CJ	J29CF	N29BL	M38BJ	G31BF	N37BD	E13AL	G14AJ	N32AG	E28AD	G25AB	B16CP
6	P17CK	U34CM	A34BM	H11BK	L18BG	A27BE	M19AM	L23AK	A20AH	M33AE	L35AK	N24CR
7	W22CL	R26CJ	Z26BP	S31BL	D12BH	Z13BF	H30BC	D36AL	Z28AJ	H15AF	D16AD	A21CS
8	Y29CM	X38CK	C38CD	P18BM	J37BJ	C19BG	S14BD	J32AM	C33AK	S25AG	J24AE	Z17AB
9	T34CP	B11CL	K11CE	W12BP	U27BK	K30BH	P23BE	U20BC	K15AL	P35AH	U21AF	C22AC
10	G26CR	N31CH	F31CF	Y37CD	R13BL	F14BJ	W36BF	R28BD	F25AM	W16AJ	R17AG	K29AD
11	L38CS	A18CP	V18CH	T27CE	X19BM	V23BK	Y32BG	X33BE	V35BC	Y24AK	X22AH	F34AE
12	D11AB	Z12CR	E12CJ	G13CF	B30BP	E36BL	T20BH	B15BF	E16BD	T21AL	B29AJ	V26AF
13	J31AC	C37CS	M37CK	L19CH	N14CD	M32BH	G28BJ	N25BG	M24BE	G17AM	N34AK	E38AG
14	U18AD	K27AB	H27CL	D30CJ	A23CE	H20BP	L33BK	A35BH	H21BF	L22BC	A26AL	M11AH
15	R12AE	F13AC	S13CM	J14CK	Z36CF	S28CD	D15BL	Z16BJ	S17BG	D29BD	Z38AM	H31AJ
16	X37AF	V19AD	P19CP	U23CL	C32CH	P33CE	J25BM	C24BK	P22BH	J34BE	C11BC	S18AK
17	B27AG	E30AE	W30CR	R36CM	K20CJ	W15CF	U35BP	K21BL	W29BJ	U26BF	K31BD	P12AL
18	N13AH	M14AF	Y14CS	X32CP	F28CK	Y25CH	R16CD	F17BM	Y34BK	R38BG	F18BE	W37AM
19	A19AJ	H23AG	T23AB	B20CR	V33CL	T35CJ	X24CE	V22BP	T26BL	X11BH	V12BF	Y27BC
20	Z30AK	S36AH	G36AC	N28CS	E15CM	G16CK	B21CF	E29CD	G38BM	B31BJ	E37BG	T13BD
21	C14AL	P32AJ	L32AD	A33AB	M25CP	L24CL	N17CH	M34CE	L11BP	N18BK	M27BH	G19BE
22	K23AM	W20AK	D28AE	Z15AC	H35CR	D21CM	A22CJ	H26CF	D31CD	A12BL	H13BJ	L30BF
23	F36BC	Y28AL	J28AF	C25AD	S16CS	J17CP	Z29CK	S38CH	J18CE	Z37BM	S19BK	D14BG
24	V32BD	T33AM	U33AG	K35AE	P24AB	U22CR	C34CL	P11CJ	U12CF	C27BP	P30BL	J23BH
25	E20BE	G15BC	R15AH	F16AF	W21AC	R29CS	K26CM	M31CK	R37CH	K13CD	W14BM	U36BJ
26	M28BF	L25BD	X25AJ	V24AG	Y17AD	X34AB	F38CP	Y18CL	X27CJ	F19CE	Y23BP	R32BK
27	H33BG	D35BE	B35AK	E21AH	T22AE	B26AC	V11CR	T12CM	B13CK	V30CF	T36CD	X20BL
28	S15BH	J16BF	N16AL	M17AJ	G29AF	N38AD	E31CS	G37CP	N19CL	E14CH	G32CE	B28BM
29	P25BJ		A24AM	H22AK	L34AG	A11AE	M18AB	L27CR	A30CM	M23CJ	L20CF	N33BP
30	W35BK		Z21BC	S29AL	D26AH	Z31AF	H12AC	D13CS	Z14CP	H36CK	D28CH	A15CD
31	Y16BL		C17BD		J38AJ		S37AD	J19AB		S32CL		Z25CE

1996	JAN	FEB	MAR	APR	MAY	JUN	JUL	AUG	SEP	OCT	NOV	DEC
1	C35CF	P21CD	D17BK	Z34BH	H38BE	D18BC	A37AK	H19AH	D23AF	A32AC	H33CS	L25CM
2	K16CH	W17CE	J22BL	C26BJ	S11BF	J12BD	Z27AL	S30AJ	J36AG	Z20AD	S15AB	D35CP
3	F24CJ	Y22CF	U29BM	K38BK	P31BG	U37BE	C13AM	P14AK	U32AH	C28AE	P25AC	J16CR
4	V21CK	T29CH	R34BP	F11BL	W18BH	R27BF	K19BC	W23AL	R20AJ	K33AF	W35AD	U24CS
5	E17CL	G34CJ	X26CD	V31BM	Y12BJ	X13BG	F30BD	Y36AM	X28AK	F15AG	Y16AE	R21AB
6	M22CM	L26CK	B38CE	E18BP	T37BK	B19BH	V14BE	T32BG	B33AL	V25AH	T24AF	X17AC
7	H29CP	D38CL	N11CF	M12CD	G27BL	N30BJ	E23BF	G20BD	N15AM	E35AJ	G21AG	B22AD
8	S34CR	J11CM	A31CH	H37CE	L13BM	A14BK	M36BG	L28BE	A25BC	M16AK	L17AH	N29AE
9	P26CS	U31CP	Z18CP	S27CF	D19BP	Z23BL	H32BH	D33BF	Z35BD	H24AL	D22AJ	A34AF
10	M38AB	R18CR	C12CK	P13CH	J30CD	C36BM	S20BJ	J15BG	C16BE	S21AM	J29AK	Z26AG
11	Y11AC	X12CS	K37CL	W19CJ	U14CE	K32BP	P28BK	U35BH	K24BF	P17BC	U34AL	C38AH
12	T31AD	B37AB	F27CM	Y30CK	R23CF	F20CD	W33BL	R35BJ	F21BG	W22BD	R26AM	K11AJ
13	G18AE	N27AC	V13CP	T14CL	X36CH	V28CE	Y15BM	X16BK	V17BH	Y29BE	X38BC	F31AK
14	L12AF	A13AD	E19CR	G23CH	B32CJ	E33CF	T25BP	B24BL	E22BJ	T34BF	B11BD	V18AL
15	D37AG	Z19AE	M30CS	L36CP	N20CK	M15CH	G35CD	N21BM	M29BK	G26BG	N31BE	E12AM
16	J27AH	C30AF	H14AB	D32CR	A28CL	H25CJ	L16CE	A17BP	H34BL	L38BH	A13BF	M37BC
17	U13AJ	K14AG	S23AC	J20CS	Z33CM	S35CK	D24CF	Z22CD	S26BM	D11BJ	Z12BG	H27BD
18	R19AK	F23AH	P36AD	U28AB	C15CP	P16CL	J21CH	C29CE	P38BP	J31BK	C37BH	S13BE
19	X30AL	W36AJ	W32AE	R33AC	K25CR	W24CH	U17CJ	K34CF	W11CD	U18BL	K27BJ	P19BF
20	B14AM	E32AK	Y20AF	X15AD	F35CS	Y21CP	R22CK	F26CH	Y31CE	R12BM	F13BK	W30BG
21	N23BC	M20AL	T28AG	B25AE	V16AB	T17CR	X29CL	V38CJ	T18CF	X37BP	V19BL	Y14BH
22	A36BD	H28AM	G33AH	N35AF	E24AC	G22CS	B34CM	E11CK	G12CH	B27CD	E30BM	T23BJ
23	Z32BE	S33BC	L15AJ	A16AG	M21AD	L29AB	N26CP	M31CL	L37CJ	N13CE	M14BP	G36BK
24	C20BF	P15BD	D25AK	Z24AH	H17AE	D34AC	A38CR	H18CM	D27CK	A19CF	H23CD	L32BL
25	K28BG	N25BE	J35AL	C21AJ	S22AF	J26AD	Z11CS	S12CP	J13CL	Z30CH	S36CE	D20BM
26	F33BH	Y35BF	U16AM	K17AK	P29AG	U38AE	C31AB	P37CR	U19CM	C14CJ	P32CF	J28BP
27	V15BJ	T16BG	R24BC	P22AL	W34AH	R11AF	K18AC	W27CS	R30CP	K23CK	W20CH	U33CD
28	E25BK	G24BH	X21BD	V29AM	Y26AJ	X31AG	F12AD	Y13AB	X14CR	F36CL	Y28CJ	R15CE
29	M35BL	L21BJ	B17BE	E34BC	T38AK	B18AH	V37AE	T19AC	B23CS	V32CH	T33CK	X25CF
30	H16BM		N22BF	M26BD	G11AL	N12AJ	E27AF	G30AD	N36AB	E20CP	G15CL	B35CH
31	S24BP		A29BG		L31AM		M13AG	L14AE		M28CR		N16CJ

1997	JAN	FEB	MAR	APR	MAY	JUN	JUL	AUG	SEP	OCT	NOV	DEC
1	A24CK	H22CM	T22BM	B26BK	V11BG	T12BE	X27AM	V30AK	T36AH	X20AE	V15AC	Y35CR
2	Z21CL	S29CJ	G29BP	N38BL	E31BH	G37BF	B13BC	E14AL	G32AJ	B28AF	E25AD	T16CS
3	C17CM	P34CK	L34CD	A11BM	M18BJ	L27BG	N19BD	M23AM	L20AK	N33AG	M35AE	G24AB
4	K22CP	W26CL	D26CE	Z31BP	H12BG	D13BH	A30BE	H36BC	D28AL	A15AH	H16AF	L21AC
5	F29CR	Y38CM	J38CF	C18CD	S37BL	J19BJ	Z14BF	S32BD	J33AM	Z25AJ	S24AG	D17AD
6	W34CS	T11CP	U11CH	K12CE	P27BM	U30BK	C23BG	P20BE	U15BC	C35AK	P21AH	J22AE
7	E26AB	G31CR	R31CJ	F37CF	W13BP	R14BL	K36BH	W28BF	R25BD	K16AL	W17AJ	U29AF
8	M38AC	L18CS	X18CK	V27CH	Y19CJ	X23BM	F32BJ	Y33BG	X35BE	F24AM	Y22AK	R34AG
9	H11AD	D12AB	B12CL	E13CJ	T30CE	B36BP	V20BK	T15BH	B16BF	V21AK	T29AL	X26AH
10	S31AE	J37AC	N37CH	M19CK	G14CF	N32CD	E28BL	G25BJ	N24BG	E17BD	G34AM	B38AJ
11	P18AF	U27AD	A27CP	H30CL	L23CH	A20CE	M33BH	L35BK	A21BH	M22BE	L26BC	N11AK
12	W12AG	R13AE	Z13CR	S14CM	D36CJ	Z28CF	H15BP	D16BL	Z17BJ	H29BE	D38BD	A31AL
13	Y37AH	X19AF	C19CS	P23CP	J32CK	C33CH	S25CD	J24BM	C22BH	S34BG	J11BE	Z18AM
14	T27AJ	B30AG	K30AB	W36CR	U20CL	K15CJ	P35CE	U21BF	K29BL	P26BH	U31BF	C12BC
15	G13AK	N14AH	F14AC	Y32CS	R28CH	F25CF	W16CF	R17CD	F34BH	W38BJ	R18BG	K37BD
16	L19AL	A23AJ	V23AD	T20AB	X33CP	V35CL	Y24CH	X22CE	V26BP	Y11BK	X12BH	F27BE
17	D30AM	Z36AK	E36AE	G28AC	B15CR	E16CM	T21CJ	B29CF	E38CD	T31BL	B37BJ	V13BF
18	J14BC	C32AL	M32AF	L33AD	N25CS	M24CP	G17CH	N34CH	M11CE	G18BM	N27BK	E19BG
19	U23BD	K20AM	H20AG	D15AE	A35AB	H21CR	L22CL	A26CJ	H31CF	L12BP	A13BL	M30BH
20	R36BE	F28BC	S28AH	J25AF	Z16AC	S17CS	D29CM	Z38CH	S18CL	D37CD	Z19BM	H14BJ
21	X32BF	V33BD	P33AJ	U35AG	C24AD	P22AB	J34CP	C11CL	P12CJ	J27CE	C30BP	S23BK
22	B20BG	E15BE	W15AK	R16AH	K21AE	W29AC	U26CR	K31CH	W37CL	U13CF	K14CD	P36BL
23	N28BH	M25BF	Y25AL	X24AJ	F17AF	Y34AD	R38CS	F18CP	Y27CL	R19CH	F23CE	W32BM
24	A33BJ	H35BG	T35AM	B21AK	V22AG	T26AE	X11AB	V12CR	T13CM	X30CJ	V36CF	Y20BP
25	Z15BK	S16BH	G16BC	N17AL	E29AH	G38AF	B31AC	E37CS	G19CP	B14CK	E32CH	T28CD
26	C25BL	P24BJ	L24BD	A22AM	M34AJ	L11AG	N18AD	M27AB	L30CR	N23CL	M20CJ	G33CE
27	K35BM	W21BK	D21BE	Z29BG	H26AK	D31AH	A12AE	H13AC	D14CS	A36CH	H28CK	L15CF
28	F16BP	Y17BL	J17BF	C34BD	S38AL	J18AJ	Z37AF	S19AD	J23AB	Z32CP	S33CL	D25CH
29	V24CD		U22BG	K26BE	P11AM	U12AJ	C27AG	P30AE	U36AC	C20CR	P15CM	J35CJ
30	E21CE		R29BH	F38BF	W31BC	R37AL	K13AH	W14AF	R32AD	K28CS	W25CP	U16CK
31	M17CF		X34BJ		Y18BD		F19AJ	Y23AG		F33AB		R24CL

FIRST LETTER = YOUR PHYSICAL CODE, MIDDLE NUMBER = SENSITIVITY CODE, LAST TWO LETTERS = INTELLECTUAL CODE

1998	JAN	FEB	MAR	APR	MAY	JUN	JUL	AUG	SEP	OCT	NOV	DEC
1	X21CM	V29CK	P29CD	U38BM	C31BJ	P37BG	J13BD	C14AM	P32AK	J28AG	C25AE	S16AB
2	B17CP	E34CL	W34CE	R11BP	K18BK	W27BH	U19BE	K23BC	W20AL	U33AH	K35AF	P24AC
3	N22CR	M26CM	Y26CF	X31CD	F12BL	Y13BJ	R30BF	F36BD	Y28AM	R15AJ	F16AG	W21AD
4	A29CS	H38CP	T36CH	B18CE	V37BM	T19BK	X14BG	V32BE	T33BC	X25AK	V24AH	Y17AE
5	Z34AB	S11CR	G11CJ	N12CF	E27BP	G30BL	B23BH	E20BF	G15BD	B35AL	E21AJ	T22AF
6	C26AC	P31CS	L31CK	A37CH	M13CD	L14BM	N36BJ	M28BG	L25BE	N16AM	M17AK	G29AG
7	K38AD	W18AB	D18CL	Z27CJ	H19CE	D23BP	A32BK	H33BH	D35BF	A24BC	H22AL	L34AH
8	F11AE	Y12AC	J12CH	C13CK	S30CF	J36CD	Z20BL	S15BJ	J16BG	Z21BD	S29AM	D26AJ
9	V31AF	T37AD	U37CP	K19CL	P14CH	U32CE	C28BM	P25BK	U24BH	C17BE	P34BC	J38AK
10	E18AG	G27AE	R27CR	F30CH	W23CJ	R20CF	K33BP	W35BL	R21BJ	K22BF	W26BD	U11AL
11	M12AH	L13AF	X13CS	V14CP	Y36CK	X28CH	F15CD	Y16BM	X17BK	F29BG	Y38BE	R31AM
12	H37AJ	D19AG	B19AB	E23CF	T32CL	B33CJ	V25CE	T24BP	B22BL	V34BH	T11BF	X18BC
13	S27AK	J30AH	N30AC	M36CS	G20CM	N15CK	E35CF	G21CD	N29BM	E26BJ	G31BG	B12BD
14	P13AL	U14AJ	A14AD	H32AB	L28CP	A25CL	M16CH	L17CE	A34BP	M38BK	L18BH	N37BE
15	W19AM	R23AK	Z23AE	S20AC	D33CR	Z35CM	H24CJ	D22CF	Z25CD	H11BL	D12BJ	A27BF
16	Y30BC	X36AL	C36AF	P28AD	J15CS	C16CP	S21CK	J29CH	C38CE	S31BM	J37BK	Z13BG
17	T14BD	B32AM	K32AG	W33AE	U25AB	K24CR	P17CL	U34CJ	K11CF	P18BP	U27BL	C19BH
18	G23BE	N20BC	F20AH	Y15AF	R35AC	F21CS	W22CM	R26CK	F31CH	W12CD	R13BM	K30BJ
19	L36BF	A28BD	V28AJ	T25AG	X16AD	V17AB	Y29CP	X38CL	V18CJ	Y37CE	X19BP	F14BK
20	D32BG	Z33BE	E33AK	G35AH	B24AE	E22AC	T34CR	B11CM	E12CK	T27CF	B30CD	V23BL
21	J20BH	C15BF	M15AL	L16AJ	N21AF	M29AD	G26CS	N31CP	M37CL	G13CH	N14CE	E36BM
22	U28BJ	K25BG	H25AM	D24AK	A17AG	H34AE	L38AB	A18CR	H27CM	L19CJ	A23CF	M32BP
23	R33BK	F35BM	S35BC	J21AL	Z22AH	S26AF	D11AC	Z12CS	S13CP	D30CK	Z36CH	H20CD
24	X15BL	V16BJ	P16BD	U17AM	C29AJ	P38AG	J31AD	C37AB	P19CR	J14CL	C32CJ	S28CE
25	B25BM	E24BK	H24BE	R22BC	K34AK	H11AH	U18AE	K27AC	W30CS	U23CH	K20CK	P33CF
26	N35BP	M21BL	Y21BF	X29BD	F26AL	Y31AJ	R12AF	F13AD	Y14AB	R36CP	F28CL	W15CH
27	A16CD	H17BM	T17BG	B34BE	V38AM	T18AK	X37AG	V19AE	T23AC	X32CS	V33CM	Y25CJ
28	Z24CE	S22BP	G22BH	N26BF	E11BC	G12AL	B27AH	E30AF	G36AD	B20CS	E15CP	T35CK
29	C21CF		L29BJ	A38BG	M31BD	L37AM	N13AJ	M14AG	L32AE	N28AB	M25CR	G16CL
30	K17CH		D34BK	Z11BH	H18BE	D27BC	A19AK	H23AH	D20AF	A33AC	H35CS	L24CM
31	F22CJ		J26BL		S12BF		Z30AL	S36AJ		Z15AD		D21CP

1999	JAN	FEB	MAR	APR	MAY	JUN	JUL	AUG	SEP	OCT	NOV	DEC
1	J17CR	C34CM	H34CF	L11CD	N18BL	M27BJ	G19BF	N23BD	M20AM	G33AJ	N35AG	E24AD
2	U22CS	K26CP	H26CH	D31CE	A12BM	H13BK	L30BG	A36BE	H28BC	L15AK	A16AH	M21AE
3	R29AB	F38CR	S38CJ	J18CF	Z37BP	S19BL	D14BH	Z32BF	S33BD	D25AL	Z24AJ	H17AF
4	X34AC	V11CS	P11CK	U12CH	C27CD	P30BM	J23BJ	C20BG	P15BE	J35AM	C21AK	S22AG
5	B26AD	E31AB	W31CL	R37CJ	K13CE	W14BP	U36BK	K28BH	W25BF	U16BC	K17AL	P29AH
6	N38AE	M18AC	Y18CM	X27CK	F19CF	Y23CD	R32BL	F33BJ	Y35BG	R24BD	F22AM	W34AJ
7	A11AF	H12AD	T12CP	B13CL	V30CH	T36CE	X20BM	V15BK	T16BH	X21BE	V29BC	Y26AK
8	Z31AG	S37AE	G37CR	N19CM	E14CJ	G32CF	B28BP	E25BL	G24BJ	B17BF	E34BD	T38AL
9	C18AH	P27AF	L27CS	A30CP	H23CK	L20CH	N33CD	M35BM	L21BK	N22BG	M26BE	G11AM
10	K12AJ	H13AG	D13AB	Z14CR	H36CL	D28CJ	A15CE	H16BP	D17BL	A29BH	H38BF	L31BC
11	F37AK	Y19AH	J19AC	C23CS	S32CM	J33CK	Z25CF	S24CD	J22BM	Z34BJ	S11BG	D18BD
12	V27AL	T30AJ	U30AD	K36AB	P20CP	U15CL	C35CH	P21CE	U29BP	C26BK	P31BH	J12BE
13	E13AM	G14AK	R14AE	F32AC	W28CR	R25CM	K16CJ	W17CF	R34CD	K38BL	W18BJ	U37BF
14	M19BC	L23AL	X23AF	V20AD	Y33CS	X35CP	F24CK	Y22CH	X25CE	F11BM	Y12BK	R27BG
15	H30BD	D36AM	B36AG	E28AE	T15AB	B16CR	V21CL	T29CJ	B38CF	V31BP	T37BL	X13BH
16	S14BE	J32BC	N32AH	M33AF	G25AC	N24CS	E17CM	G34CK	N11CH	E18CD	G27BM	B19BJ
17	P23BF	U20BD	A20AJ	H15AG	L35AD	A21AB	M22CP	L26CL	A31CJ	M12CE	L13BP	N30BK
18	W36BG	R28BE	Z28AK	S25AH	D16AE	Z17AC	H29CR	D38CM	Z18CK	H37CF	D19CD	A14BL
19	Y32BH	X33BF	C33AL	P35AJ	J24AF	C22AD	S34CS	J11CP	C12CL	S27CH	J30CE	Z23BM
20	T20BJ	B15BG	K15AM	W16AK	U21AG	K29AE	P26AB	U31CR	K37CM	P13CJ	U14CF	C36BP
21	G28BK	N25BH	F25BC	Y24AL	R17AH	F34AF	W38AC	R18CS	F27CP	W19CK	R23CH	K32CD
22	L33BL	A35BJ	V35BD	D21AM	X22AJ	V26AG	Y11AD	X12AB	V13CR	Y30CL	X36CJ	F20CE
23	D15BM	Z16BK	E16BE	G17BC	B29AK	E38AH	T31AE	B37AC	E19CS	T14CM	B32CK	V28CF
24	J25BP	C24BL	M24BF	L22BD	N34AL	M11AJ	G18AF	N27AD	M30AB	G23CP	N20CL	E33CH
25	U35CD	K21BM	S17BH	J34BF	Z38BC	S18AL	D37AH	Z19AF	S23AD	D32CS	Z33CP	H25CK
26	R16CF	F17BP	P22BJ	U26BG	C11BD	P12AM	J27AJ	C30AG	P36AE	J20AB	C15CR	S35CL
27	X24CF	V22CD										
28	B21CH	E29CE	W29BK	R38BH	K31BE	W37BC	U13AK	K14AH	W32AF	U28AC	K25CS	P16CM
29	N17CJ		Y34BL	X11BJ	F18BF	Y27BD	R19AL	F23AJ	Y20AG	R33AD	F35AB	W24CP
30	A22CK		T26BM	B31BK	V12BG	T13BE	X30AM	V36AK	T28AH	X15AE	V16AC	Y21CR
31	Z29CL		G38BP		E37BH		B14BC	E32AL		B25AF		T17CS

2000	JAN	FEB	MAR	APR	MAY	JUN	JUL	AUG	SEP	OCT	NOV	DEC
1	G22AB	N26CR	S39CK	T18CH	X37CD	V19BM	Y14BJ	X32BG	V33BE	Y25AM	X24AK	F17AG
2	L29AC	A38CS	E11CL	G12CJ	B27CE	E30BP	T23BK	B20BH	E15BF	T35BC	B21AL	V22AH
3	D34AD	Z11AB	M31CM	L37CK	N13CF	M14CD	G36BL	N26BJ	M25BG	G16BD	N17AM	E29AJ
4	J26AE	C31AC	H18CP	D27CL	A19CH	H23CE	L32BM	A33BK	H35BH	L24BE	A22BC	M34AK
5	U38AF	K18AD	S12CR	J13CM	Z30CJ	S36CF	D20BP	Z15BL	S16BJ	D21BF	Z29BD	H26AL
6	R11AG	F12AE	P37CS	U19CP	C14CK	P32CH	J28CD	C25BM	P24BK	J17BG	C34BE	S38AM
7	X31AH	V37AF	W27AB	R30CP	K23CL	W20CJ	U33CE	K35BP	W21BL	U22BH	K26BF	P11BC
8	B18AJ	E27AG	Y13AC	X14CS	V32CP	Y28CH	T33CL	V24CE	Y17BM	T22BP	V23BG	W31BD
9	N12AK	M13AH	T19AD	B23AB	V32CP	T33CL	X25CH	V24CE	T22BP	X34BK	V11BH	Y18BE
10	A37AL	H19AJ	G30AE	N36AC	E20CR	G15CM	B35CJ	E21CF	G29CD	B26BL	E31BJ	T12BF
11	Z27AM	S30AK	L14AF	A32AD	M28CS	L25CP	N16CK	M17CH	L34CE	N38BM	M18BK	G37BG
12	C13BC	P14AL	D23AG	Z20AE	H33AB	D35CR	A24CL	H22CJ	D26CF	A11BP	H12BL	L27BH
13	K19BD	W23AM	N31CM	C28AF	S15AD	J16CS	Z21CH	S29CK	J38CH	Z31CD	S37BM	D13BJ
14	F30BE	Y36BC	U32AJ	K33AG	P25AD	U24AB	C17CP	P34CL	U11CJ	C18CE	P27BP	J19BK
15	V14BF	T32BD	R20AK	F15AH	W35AE	R21AC	K22CR	W26CH	R31CK	K12CF	W13CD	U30BL
16	E23BG	G20BE	X28AL	V25AJ	Y16AF	X17AD	F29CS	Y38CP	X18CL	F37CH	Y19CE	R14BM
17	M36BH	L28BF	B33AM	E35AK	T24AG	B22AE	V34AB	T11CR	B12CH	V27CJ	T30CF	X23BP
18	H32BJ	D33BG	N15BC	M16AL	G21AH	N29AF	E26AC	G31CS	N37CP	E13CK	G14CH	B36CD
19	S20BK	J15BH	A25BD	H24AM	L17AJ	A34AG	M38AD	L18AB	A27CR	M19CL	L23CJ	N32CE
20	P28BL	U25BJ	Z35BE	S21BC	D22AK	Z26AH	H11AE	D12AC	Z13CS	H30CH	D36CK	A20CF
21	W33BM	R35BK	G16BF	P17BD	J29AL	G38AJ	S31AF	J37AD	G19AB	S14CP	J32CL	Z28CH
22	Y15BP	X16BL	L26BL	W22BE	U34AM	K11AK	P18AG	U27AE	K30AC	P23CR	U20CH	C33CJ
23	T25CD	B24BM	F21BH	Y29BF	R26BC	F31AL	W12AH	R13AF	F14AD	W36CS	R28CP	K15CK
24	G35CE	N21BP	V17BJ	T34BG	X38BD	V18AM	Y37AJ	X19AH	V23AE	Y32AB	X33CR	F25CL
25	L16CF	A17CD	E22BK	G26BH	B11BE	E12BC	T27AK	B30AH	E36AF	T20AC	B15CS	V35CM
26	D24CH	Z22CE	M29BL	L38BJ	N31BG	M37BD	G13AL	N14AJ	M32AG	G28AD	N25AB	E16CP
27	J21CJ	C29CF	H34BM	D11BK	A18BG	H27BE	L19AM	A23AK	H20AH	L33AE	A35AC	M24CR
28	U17CK	K34CH	S26BP	J31BL	Z12BH	S13BF	D38BC	Z36AL	S28AJ	D15AF	Z16AD	H21CS
29	R22CL	F26CJ	P38CD	U18BM	C37BJ	P19BG	J14BD	C32AM	P33AK	J25AG	C24AE	S17AB
30	X29CM		W11CE	R12BP	K27BK	W30BH	U23BE	K20BC	W15AL	U35AH	K21AF	P22AC
31	B34CP		Y31CF		F13BL		R36BF	F28BD		R16AJ		W29AD

FIRST LETTER = YOUR PHYSICAL CODE, MIDDLE NUMBER = SENSITIVITY CODE, LAST TWO LETTERS = INTELLECTUAL CODE

2001	JAN	FEB	MAR	APR	MAY	JUN	JUL	AUG	SEP	OCT	NOV	DEC
1	D21CP	Z29CL	E29CE	G38BP	B31BK	E37BH	T13BE	B14BC	E32AL	T28AH	B25AF	V16AC
2	J17CR	C34CM	M34CF	L11CD	N18BL	M27BJ	G19BF	N23BD	M20AM	G33AJ	N35AG	E24AD
3	U22CS	K26CP	H26CH	D31CE	A12BM	H13BK	L30BG	A36BE	H28BC	L15AK	A16AH	M21AE
4	R29AB	F38CR	S38CJ	J18CF	Z37BP	S19BL	D14BH	Z32BF	S33BD	D25AL	Z24AJ	H17AF
5	X34AC	V11CS	P11CK	U12CH	C27CD	P30BM	J23BJ	C20BG	P15BE	J35AM	C21AK	S22AG
6	B26AD	E31AB	W31CL	R37CJ	K13CE	W14BP	U36BK	K28BH	W25BF	U16BC	K17AL	P29AH
7	N38AE	M18AC	Y18CM	X27CK	F19CF	Y23CD	R32BL	F33BJ	Y35BG	R24BD	F22AM	W34AJ
8	A11AF	H12AD	T12CP	B13CL	V30CH	T36CE	X20BM	V15BL	T16BH	X21BE	V29BC	Y26AK
9	Z31AG	S37AE	G37CR	N19CM	E14CJ	G32CF	B28BP	E25BL	G24BJ	B17BF	E34BD	T38AL
10	C18AH	P27AF	L27CS	A30CP	M23CL	L20CH	N33CD	M35BM	L21BK	N22BG	M26BE	G11AM
11	K12AJ	W13AG	D13AB	Z14CR	H36CL	D28CJ	A15CE	H16BP	D17BL	A29BH	H38BF	L31BC
12	F37AK	Y19AH	J19AC	C23CS	S32CM	J33CK	Z25CF	S24CD	J22BM	Z34BJ	S11BG	D18BD
13	V27AL	T30AJ	U30AD	K36AB	P20CP	U15CL	C35CH	P21CE	U29BP	C26BK	P31BH	J12BE
14	E13AM	G14AK	R14AE	F32AC	W28CR	R25CM	K16CJ	W17CF	R34CD	K38BL	W18BJ	U37BF
15	M19BC	L23AL	X23AF	V20AD	Y33CS	X35CP	F24CK	Y22CH	X26CE	F11BM	Y12BK	R27BG
16	H30BD	O36AM	B36AG	E28AE	T15AB	B16CR	V21CL	T29CJ	B38CF	V31BP	T37BL	X13BH
17	S14BE	J32BC	N32AH	M33AF	G25AC	N24CS	E17CM	G34CK	N11CH	E18CD	G27BM	B19BJ
18	P23BF	U20BD	A20AJ	H15AG	L35AD	A21AB	M22CP	L26CL	A31CJ	M12CE	L13BP	N30BK
19	W36BG	R28BE	Z28AK	S25AH	D16AE	Z17AC	H29CR	D38CM	Z18CK	H37CF	D19CO	A14BL
20	Y32BH	X33BF	C33AL	P35AJ	J24AF	C22AD	S34CS	J11CP	C12CL	S27CH	J30CE	Z23BM
21	T20BJ	B15BG	K15AM	W16AK	U21AG	K29AE	P26AB	U31CR	K37CM	P13CJ	U14CF	C36BP
22	G28BK	N25BH	F25BC	Y24AL	R17AH	F34AF	W38AC	R18CS	F27CP	W19CK	R23CH	K32CD
23	L33BL	A35BJ	V35BD	T21AM	X22AJ	V26AG	Y11AD	X12AB	V13CR	Y30CL	X36CJ	F20CE
24	O15BM	Z16BK	E16BE	G17BC	B29AK	E38AH	T31AE	B37AC	E19CS	T14CM	B32CK	V28CF
25	J25BP	C24BL	M24BF	L22BD	N34AL	M11AJ	G18AF	N27AD	M30AB	G23CP	N20CL	E33CH
26	U35CD	K21BM	H21BG	D29BE	A26AM	H31AK	L12AG	A13AE	H14AC	L36CR	A28CM	M15CJ
27	R16CE	F17BP	S17BH	J34BF	Z38BC	S18AL	D37AH	Z19AF	S23AO	D32CS	Z33CP	H25CH
28	X24CF	V22CD	P22BJ	U26BG	C11BD	P12AM	J27AJ	C30AG	P36AE	J20AB	C15CR	S35CL
29	B21CH		W29BK	R38BH	K31BE	M37BC	U13AK	K14AH	W32AF	U28AC	K25CS	P16CM
30	N17CJ		Y34BL	X11BJ	F18BF	Y27BD	R19AL	F23AJ	Y20AG	R33AD	F35AB	H24CP
31	A22CK		T26BM		V12BG		X30AM	V36AK		X15AE		Y21CR

PHYSICAL TABLES FOR INQUIRY YEARS 1800--1900

PHYSICAL BIRTH CODES ARE LISTED BY COLUMNS--USE TABLE IN PROPER ROW AND COLUMN FOR BIORHYTHM PHYSICAL DATA

YEAR OF INQUIRY	*	A	B	C	D	E	F	G	H	J	K	L	M	N	P	R	S	T	U	V	W	X	Y	Z
1800	*	P11	P12	P13	P14	P15	P16	P17	P18	P19	P20	P21	P22	P23	P24	P25	P26	P27	P28	P29	P30	P31	P32	P33
1801	*	P20	P33	P29	P25	P26	P22	P19	P30	P31	P15	P28	P24	P13	P27	P23	P32	P14	P12	P18	P17	P11	P21	P16
1802	*	P15	P16	P18	P23	P32	P24	P31	P17	P11	P26	P12	P27	P29	P14	P13	P21	P25	P33	P30	P19	P20	P28	P22
1803	*	P26	P22	P30	P13	P21	P27	P11	P19	P20	P32	P33	P14	P18	P25	P29	P28	P23	P16	P17	P31	P15	P12	P24
1804	*	P72	P64	P57	P69	P68	P54	P60	P71	P55	P61	P56	P65	P70	P63	P58	P52	P53	P62	P59	P51	P66	P73	P67
1805	*	P17	P32	P14	P22	P31	P28	P29	P23	P18	P19	P15	P12	P27	P33	P24	P11	P16	P26	P25	P13	P30	P20	P21
1806	*	P19	P21	P25	P24	P11	P12	P18	P13	P30	P31	P26	P33	P14	P16	P27	P20	P22	P32	P23	P29	P17	P15	P28
1807	*	P31	P28	P23	P27	P20	P33	P30	P29	P17	P11	P32	P16	P25	P22	P14	P15	P24	P21	P13	P18	P19	P26	P12
1808	*	P51	P52	P53	P54	P55	P56	P57	P58	P59	P60	P61	P62	P63	P64	P65	P66	P67	P68	P69	P70	P71	P72	P73
1809	*	P13	P11	P16	P28	P18	P15	P14	P24	P25	P29	P19	P26	P33	P32	P12	P30	P21	P31	P22	P27	P23	P17	P20
1810	*	P29	P20	P22	P12	P30	P26	P25	P27	P23	P18	P31	P32	P16	P21	P33	P17	P28	P11	P24	P14	P13	P19	P15
1811	*	P18	P15	P24	P33	P17	P32	P23	P14	P13	P30	P11	P21	P22	P28	P16	P19	P12	P20	P27	P25	P29	P31	P26
1812	*	P70	P66	P67	P56	P59	P61	P53	P65	P69	P57	P60	P68	P64	P52	P62	P71	P73	P55	P54	P63	P58	P51	P72
1813	*	P27	P30	P21	P15	P25	P19	P16	P12	P22	P14	P29	P31	P32	P11	P26	P23	P20	P18	P28	P33	P24	P13	P17
1814	*	P14	P17	P28	P26	P23	P31	P22	P33	P24	P25	P18	P11	P21	P20	P32	P13	P15	P30	P12	P16	P27	P29	P19
1815	*	P25	P19	P12	P32	P13	P11	P24	P16	P27	P23	P30	P20	P28	P15	P21	P29	P26	P17	P33	P22	P14	P18	P31
1816	*	P63	P71	P73	P61	P69	P60	P67	P62	P54	P53	P57	P55	P52	P66	P68	P58	P72	P59	P56	P64	P65	P70	P51
1817	*	P33	P23	P20	P19	P22	P29	P21	P26	P28	P16	P14	P18	P11	P30	P31	P24	P17	P25	P15	P32	P12	P27	P13
1818	*	P16	P13	P15	P31	P24	P18	P28	P32	P12	P22	P25	P30	P20	P17	P11	P27	P19	P23	P26	P21	P33	P14	P29
1819	*	P22	P29	P26	P11	P27	P30	P12	P21	P33	P24	P23	P17	P15	P19	P20	P14	P31	P13	P32	P28	P16	P25	P18
1820	*	P64	P58	P72	P60	P54	P57	P73	P68	P56	P67	P53	P59	P66	P71	P55	P65	P51	P69	P61	P52	P62	P63	P70
1821	*	P32	P24	P17	P29	P28	P14	P20	P31	P15	P21	P16	P25	P30	P23	P18	P12	P13	P22	P19	P11	P26	P33	P27
1822	*	P21	P27	P19	P18	P12	P25	P15	P11	P26	P28	P22	P23	P17	P13	P30	P33	P29	P24	P31	P20	P32	P16	P14
1823	*	P28	P14	P31	P30	P33	P23	P26	P20	P32	P12	P24	P13	P19	P29	P17	P16	P18	P27	P11	P15	P21	P22	P25
1824	*	P52	P65	P51	P57	P56	P53	P72	P55	P61	P73	P67	P69	P71	P58	P59	P62	P70	P54	P60	P66	P68	P64	P63
1825	*	P11	P12	P13	P14	P15	P16	P17	P18	P19	P20	P21	P22	P23	P24	P25	P26	P27	P28	P29	P30	P31	P32	P33
1826	*	P20	P33	P29	P25	P26	P22	P19	P30	P31	P15	P28	P24	P13	P27	P23	P32	P14	P12	P18	P17	P11	P21	P16
1827	*	P15	P16	P18	P23	P32	P24	P31	P17	P11	P26	P12	P27	P29	P14	P13	P21	P25	P33	P30	P19	P20	P28	P22
1828	*	P66	P62	P70	P53	P61	P67	P51	P59	P60	P72	P73	P54	P58	P65	P69	P68	P63	P56	P57	P71	P55	P52	P64
1829	*	P30	P26	P27	P16	P19	P21	P13	P25	P29	P17	P20	P28	P24	P12	P22	P31	P33	P15	P14	P23	P18	P11	P32
1830	*	P17	P32	P14	P22	P31	P28	P29	P23	P18	P19	P15	P12	P27	P33	P24	P11	P16	P26	P25	P13	P30	P20	P21
1831	*	P19	P21	P25	P24	P11	P12	P18	P13	P30	P31	P26	P33	P14	P16	P27	P20	P22	P32	P23	P29	P17	P15	P28
1832	*	P71	P68	P63	P67	P60	P73	P70	P69	P57	P51	P72	P56	P65	P62	P54	P55	P64	P61	P53	P58	P59	P66	P52
1833	*	P23	P31	P33	P21	P29	P20	P27	P22	P14	P13	P17	P15	P12	P26	P28	P18	P32	P19	P16	P24	P25	P30	P11
1834	*	P13	P11	P16	P28	P18	P15	P14	P24	P25	P29	P19	P26	P33	P32	P12	P30	P21	P31	P22	P27	P23	P17	P20
1835	*	P29	P20	P22	P12	P30	P26	P25	P27	P23	P18	P31	P32	P16	P21	P33	P17	P28	P11	P24	P14	P13	P19	P15
1836	*	P58	P55	P64	P73	P57	P72	P63	P54	P53	P70	P51	P61	P62	P68	P56	P59	P52	P60	P67	P65	P69	P71	P66
1837	*	P24	P18	P32	P20	P14	P17	P33	P28	P16	P27	P13	P19	P26	P31	P15	P25	P11	P29	P21	P12	P22	P23	P30
1838	*	P27	P30	P21	P15	P25	P19	P16	P12	P22	P14	P29	P31	P32	P11	P26	P23	P20	P18	P28	P33	P24	P13	P17
1839	*	P14	P17	P28	P26	P23	P31	P22	P33	P24	P25	P18	P11	P21	P20	P32	P13	P15	P30	P12	P16	P27	P29	P19
1840	*	P65	P59	P52	P72	P53	P51	P64	P56	P67	P63	P70	P60	P68	P55	P61	P69	P66	P57	P73	P62	P54	P58	P71
1841	*	P12	P25	P11	P17	P16	P13	P32	P15	P21	P33	P27	P29	P31	P18	P19	P22	P30	P14	P20	P26	P28	P24	P23
1842	*	P33	P23	P20	P19	P22	P29	P21	P26	P28	P16	P14	P18	P11	P30	P31	P24	P17	P25	P15	P32	P12	P27	P13
1843	*	P16	P13	P15	P31	P24	P18	P28	P32	P12	P22	P25	P30	P20	P17	P11	P27	P19	P23	P26	P21	P33	P14	P29
1844	*	P62	P69	P66	P51	P67	P70	P52	P61	P73	P64	P63	P57	P55	P59	P60	P54	P71	P53	P72	P68	P56	P65	P58
1845	*	P26	P22	P30	P13	P21	P27	P11	P19	P20	P32	P33	P14	P18	P25	P29	P28	P23	P16	P17	P31	P15	P12	P24
1846	*	P32	P24	P17	P29	P28	P14	P20	P31	P15	P21	P16	P25	P30	P23	P18	P12	P13	P22	P19	P11	P26	P33	P27
1847	*	P21	P27	P19	P18	P12	P25	P15	P11	P26	P28	P22	P23	P17	P13	P30	P33	P29	P24	P31	P20	P32	P16	P14
1848	*	P68	P54	P71	P70	P73	P63	P66	P60	P72	P52	P64	P53	P59	P69	P57	P56	P58	P67	P51	P55	P61	P62	P65
1849	*	P31	P28	P23	P27	P20	P33	P30	P29	P17	P11	P32	P16	P25	P22	P14	P15	P24	P21	P13	P18	P19	P26	P12
1850	*	P11	P12	P13	P14	P15	P16	P17	P18	P19	P20	P21	P22	P23	P24	P25	P26	P27	P28	P29	P30	P31	P32	P33
1851	*	P20	P33	P29	P25	P26	P22	P19	P30	P31	P15	P28	P24	P13	P27	P23	P32	P14	P12	P18	P17	P11	P21	P16
1852	*	P55	P56	P58	P63	P72	P64	P71	P57	P51	P66	P52	P67	P69	P54	P53	P61	P65	P73	P70	P59	P60	P68	P62
1853	*	P18	P15	P24	P33	P17	P32	P23	P14	P13	P30	P11	P21	P22	P28	P16	P19	P12	P20	P27	P25	P29	P31	P26
1854	*	P30	P26	P27	P16	P19	P21	P13	P25	P29	P17	P20	P28	P24	P12	P22	P31	P33	P15	P14	P23	P18	P11	P32
1855	*	P17	P32	P14	P22	P31	P28	P29	P23	P18	P19	P15	P12	P27	P33	P24	P11	P16	P26	P25	P13	P30	P20	P21
1856	*	P59	P61	P65	P64	P51	P52	P58	P53	P70	P71	P66	P73	P54	P56	P67	P60	P62	P72	P63	P69	P57	P55	P68
1857	*	P25	P19	P12	P32	P13	P11	P24	P16	P27	P23	P30	P20	P28	P15	P21	P29	P26	P17	P33	P22	P14	P18	P31
1858	*	P23	P31	P33	P21	P29	P20	P27	P22	P14	P13	P17	P15	P12	P26	P28	P18	P32	P19	P16	P24	P25	P30	P11
1859	*	P13	P11	P16	P28	P18	P15	P14	P24	P25	P29	P19	P26	P33	P32	P12	P30	P21	P31	P22	P27	P23	P17	P20
1860	*	P69	P60	P62	P52	P70	P66	P65	P67	P63	P58	P71	P72	P56	P61	P73	P57	P68	P51	P64	P54	P53	P59	P55
1861	*	P22	P29	P26	P11	P27	P30	P12	P21	P33	P24	P23	P17	P15	P19	P20	P14	P31	P13	P32	P28	P16	P25	P18
1862	*	P24	P18	P32	P20	P14	P17	P33	P28	P16	P27	P13	P19	P26	P31	P15	P25	P11	P29	P21	P12	P22	P23	P30
1863	*	P27	P30	P21	P15	P25	P19	P16	P12	P22	P14	P29	P31	P32	P11	P26	P23	P20	P18	P28	P33	P24	P13	P17
1864	*	P54	P57	P68	P66	P63	P71	P62	P73	P64	P65	P58	P51	P61	P60	P72	P53	P55	P70	P52	P56	P67	P69	P59
1865	*	P28	P14	P31	P30	P33	P23	P26	P20	P32	P12	P24	P13	P19	P29	P17	P16	P18	P27	P11	P15	P21	P22	P25
1866	*	P12	P25	P11	P17	P16	P13	P32	P15	P21	P33	P27	P29	P31	P18	P19	P22	P30	P14	P20	P26	P28	P24	P23
1867	*	P33	P23	P20	P19	P22	P29	P21	P26	P28	P16	P14	P18	P11	P30	P31	P24	P17	P25	P15	P32	P12	P27	P13
1868	*	P56	P53	P55	P71	P64	P58	P68	P72	P52	P62	P65	P70	P60	P57	P51	P67	P59	P63	P66	P61	P73	P54	P69
1869	*	P15	P16	P18	P23	P32	P24	P31	P17	P11	P26	P12	P27	P29	P14	P13	P21	P25	P33	P30	P19	P20	P28	P22
1870	*	P26	P22	P30	P13	P21	P27	P11	P19	P20	P32	P33	P14	P18	P25	P29	P28	P23	P16	P17	P31	P15	P12	P24
1871	*	P32	P24	P17	P29	P28	P14	P20	P31	P15	P21	P16	P25	P30	P23	P18	P12	P13	P22	P19	P11	P26	P33	P27
1872	*	P61	P67	P59	P58	P52	P65	P55	P51	P66	P68	P62	P63	P57	P53	P70	P73	P69	P64	P71	P60	P72	P56	P54
1873	*	P19	P21	P25	P24	P11	P12	P18	P13	P30	P31	P26	P33	P14	P16	P27	P20	P22	P32	P23	P29	P17	P15	P28
1874	*	P31	P28	P23	P27	P20	P33	P30	P29	P17	P11	P32	P16	P25	P22	P14	P15	P24	P21	P13	P18	P19	P26	P12
1875	*	P11	P12	P13	P14	P15	P16	P17	P18	P19	P20	P21	P22	P23	P24	P25	P26	P27	P28	P29	P30	P31	P32	P33
1876	*	P60	P73	P69	P65	P66	P62	P59	P70	P71	P55	P68	P64	P53	P67	P63	P72	P54	P52	P58	P57	P51	P61	P56
1877	*	P29	P20	P22	P12	P30	P26	P25	P27	P23	P18	P31	P32	P16	P21	P33	P17	P28	P11	P24	P14	P13	P19	P15
1878	*	P18	P15	P24	P33	P17	P32	P23	P14	P13	P30	P11	P21	P22	P28	P16	P19	P12	P20	P27	P25	P29	P31	P26
1879	*	P30	P26	P27	P16	P19	P21	P13	P25	P29	P17	P20	P28	P24	P12	P22	P31	P33	P15	P14	P23	P18	P11	P32
1880	*	P57	P72	P54	P62	P71	P68	P69	P63	P58	P59	P55	P52	P67	P73	P64	P51	P56	P66	P65	P53	P70	P60	P61
1881	*	P14	P17	P28	P26	P23	P31	P22	P33	P24	P25	P18	P11	P21	P20	P32	P13	P15	P30	P12	P16	P27	P29	P19
1882	*	P25	P19	P12	P32	P13	P11	P24	P16	P27	P23	P30	P20	P28	P15	P21	P29	P26	P17	P33	P22	P14	P18	P31
1883	*	P23	P31	P33	P21	P29	P20	P27	P22	P14	P13	P17	P15	P12	P26	P28	P18	P32	P19	P16	P24	P25	P30	P11
1884	*	P53	P51	P56	P68	P58	P55	P54	P64	P65	P69	P59	P66	P73	P72	P52	P70	P61	P71	P62	P67	P63	P57	P60
1885	*	P16	P13	P15	P31	P24	P18	P28	P32	P12	P22	P25	P30	P20	P17	P11	P27	P19	P23	P26	P21	P33	P14	P29
1886	*	P22	P29	P26	P11	P27	P30	P12	P21	P33	P24	P23	P17	P15	P19	P20	P14	P31	P13	P32	P28	P16	P25	P18
1887	*	P24	P18	P32	P20	P14	P17	P33	P28	P16	P27	P13	P19	P26	P31	P15	P25	P11	P29	P21	P12	P22	P23	P30
1888	*	P67	P70	P61	P55	P65	P59	P56	P52	P62	P54	P69	P71	P72	P51	P66	P63	P60	P58	P68	P73	P64	P53	P57
1889	*	P21	P27	P19	P18	P12	P25	P15	P11	P26	P28	P22	P23	P17	P13	P30	P33	P29	P24	P31	P20	P32	P16	P14
1890	*	P28	P14	P31	P30	P33	P23	P26	P20	P32	P12	P24	P13	P19	P29	P17	P16	P18	P27	P11	P15	P21	P22	P25
1891	*	P12	P25	P11	P17	P16	P13	P32	P15	P21	P33	P27	P29	P31	P18	P19	P22	P30	P14	P20	P26	P28	P24	P23
1892	*	P73	P63	P60	P59	P62	P69	P61	P66	P68	P56	P54	P58	P51	P70	P71	P64	P57	P65	P55	P72	P52	P67	P53
1893	*	P20	P33	P29	P25	P26	P22	P19	P30	P31	P15	P28	P24	P13	P27	P23	P32	P14	P12	P18	P17	P11	P21	P16
1894	*	P15	P16	P18	P23	P32	P24	P31	P17	P11	P26	P12	P27	P29	P14	P13	P21	P25	P33	P30	P19	P20	P28	P22
1895	*	P26	P22	P30	P13	P21	P27	P11	P19	P20	P32	P33	P14	P18	P25	P29	P28	P23	P16	P17	P31	P15	P12	P24
1896	*	P72	P64	P57	P69	P68	P54	P60	P71	P55	P61	P56	P65	P70	P63	P58	P52	P53	P62	P59	P51	P66	P73	P67
1897	*	P17	P32	P14	P22	P31	P28	P29	P23	P18	P19	P15	P12	P27	P33	P24	P11	P16	P26	P25	P13	P30	P20	P21
1898	*	P19	P21	P25	P24	P11	P12	P18	P13	P30	P31	P26	P33	P14	P16	P27	P20	P22	P32	P23	P29	P17	P15	P28
1899	*	P31	P28	P23	P27	P20	P33	P30	P29	P17	P11	P32	P16	P25	P22	P14	P15	P24	P21	P13	P18	P19	P26	P12
1900	*	P11	P12	P13	P14	P15	P16	P17	P18	P19	P20	P21	P22	P23	P24	P25	P26	P27	P28	P29	P30	P31	P32	P33

PHYSICAL TABLES FOR INQUIRY YEARS 1901--2001

PHYSICAL BIRTH CODES ARE LISTED BY COLUMNS--USE TABLE IN PROPER ROW AND COLUMN FOR BIORHYTHM PHYSICAL DATA

YEAR OF INQUIRY	*	A	B	C	D	E	F	G	H	J	K	L	M	N	P	R	S	T	U	V	W	X	Y	Z
1901	*	P20	P33	P29	P25	P26	P22	P19	P30	P31	P15	P28	P24	P13	P27	P23	P32	P14	P12	P18	P17	P11	P21	P16
1902	*	P15	P16	P18	P23	P32	P24	P31	P17	P11	P26	P12	P27	P29	P14	P13	P21	P25	P33	P30	P19	P20	P28	P22
1903	*	P26	P22	P30	P13	P21	P27	P11	P19	P20	P32	P33	P14	P18	P25	P29	P28	P23	P16	P17	P31	P15	P12	P24
1904	*	P72	P64	P57	P69	P68	P54	P60	P71	P55	P61	P56	P65	P70	P63	P58	P52	P53	P62	P59	P51	P66	P73	P67
1905	*	P17	P32	P14	P22	P31	P28	P29	P23	P18	P19	P15	P12	P27	P33	P24	P11	P16	P26	P25	P13	P30	P20	P21
1906	*	P19	P21	P25	P24	P11	P12	P18	P13	P30	P31	P26	P33	P14	P16	P27	P20	P22	P32	P23	P29	P17	P15	P28
1907	*	P31	P28	P23	P27	P20	P33	P30	P29	P17	P11	P32	P16	P25	P22	P14	P15	P24	P21	P13	P18	P19	P26	P12
1908	*	P51	P52	P53	P54	P55	P56	P57	P58	P59	P60	P61	P62	P63	P64	P65	P66	P67	P68	P69	P70	P71	P72	P73
1909	*	P13	P11	P16	P28	P18	P15	P14	P24	P25	P29	P19	P26	P33	P32	P12	P30	P21	P31	P22	P27	P23	P17	P20
1910	*	P29	P20	P22	P12	P30	P26	P25	P27	P23	P18	P31	P32	P16	P21	P33	P17	P28	P11	P24	P14	P13	P19	P15
1911	*	P18	P15	P24	P33	P17	P32	P23	P14	P13	P30	P11	P21	P22	P28	P16	P19	P12	P20	P27	P25	P29	P31	P26
1912	*	P70	P66	P67	P56	P59	P61	P53	P65	P69	P57	P60	P68	P64	P52	P62	P71	P73	P55	P54	P63	P58	P51	P72
1913	*	P27	P30	P21	P15	P25	P19	P16	P12	P22	P14	P29	P31	P32	P11	P26	P23	P20	P18	P28	P33	P24	P13	P17
1914	*	P14	P17	P28	P26	P23	P31	P22	P33	P24	P25	P18	P11	P21	P20	P32	P13	P15	P30	P12	P16	P27	P29	P19
1915	*	P25	P19	P12	P32	P13	P11	P24	P16	P27	P23	P30	P20	P28	P15	P21	P29	P26	P17	P33	P22	P14	P18	P31
1916	*	P63	P71	P73	P61	P69	P60	P67	P62	P54	P53	P57	P55	P52	P66	P68	P58	P72	P59	P56	P64	P65	P70	P51
1917	*	P33	P23	P20	P19	P22	P29	P21	P26	P28	P16	P14	P18	P11	P30	P31	P24	P17	P25	P15	P32	P12	P27	P13
1918	*	P16	P13	P15	P31	P24	P18	P28	P32	P12	P22	P25	P30	P20	P17	P11	P27	P19	P23	P26	P21	P33	P14	P29
1919	*	P22	P29	P26	P11	P27	P30	P12	P21	P33	P24	P23	P17	P15	P19	P20	P14	P31	P13	P32	P28	P16	P25	P18
1920	*	P64	P58	P72	P60	P54	P57	P73	P68	P56	P67	P53	P59	P66	P71	P55	P65	P51	P69	P61	P52	P62	P63	P70
1921	*	P32	P24	P17	P29	P28	P14	P20	P31	P15	P21	P16	P25	P30	P23	P18	P12	P13	P22	P19	P11	P26	P33	P27
1922	*	P21	P27	P19	P18	P12	P25	P15	P11	P26	P28	P22	P23	P17	P13	P30	P33	P29	P24	P31	P20	P32	P16	P14
1923	*	P28	P14	P31	P30	P33	P23	P26	P20	P32	P12	P24	P13	P19	P29	P17	P16	P18	P27	P11	P15	P21	P22	P25
1924	*	P52	P65	P51	P57	P56	P53	P72	P55	P61	P73	P67	P69	P71	P58	P59	P62	P70	P54	P60	P66	P68	P64	P63
1925	*	P11	P12	P13	P14	P15	P16	P17	P18	P19	P20	P21	P22	P23	P24	P25	P26	P27	P28	P29	P30	P31	P32	P33
1926	*	P20	P33	P29	P25	P26	P22	P19	P30	P31	P15	P28	P24	P13	P27	P23	P32	P14	P12	P18	P17	P11	P21	P16
1927	*	P15	P16	P18	P23	P32	P24	P31	P17	P11	P26	P12	P27	P29	P14	P13	P21	P25	P33	P30	P19	P20	P28	P22
1928	*	P66	P62	P70	P53	P61	P67	P51	P59	P60	P72	P73	P54	P58	P65	P69	P68	P63	P56	P57	P71	P55	P52	P64
1929	*	P30	P26	P27	P16	P19	P21	P13	P25	P29	P17	P20	P28	P24	P12	P22	P31	P33	P15	P14	P23	P18	P11	P32
1930	*	P17	P32	P14	P22	P31	P28	P29	P23	P18	P19	P15	P12	P27	P33	P24	P11	P16	P26	P25	P13	P30	P20	P21
1931	*	P19	P21	P25	P24	P11	P12	P18	P13	P30	P31	P26	P33	P14	P16	P27	P20	P22	P32	P23	P29	P17	P15	P28
1932	*	P71	P68	P63	P67	P60	P73	P70	P69	P57	P51	P72	P56	P65	P62	P54	P55	P64	P61	P53	P58	P59	P66	P52
1933	*	P23	P31	P33	P21	P29	P20	P27	P22	P14	P13	P17	P15	P12	P26	P28	P18	P32	P19	P16	P24	P25	P30	P11
1934	*	P13	P11	P16	P28	P18	P15	P14	P24	P25	P29	P19	P26	P33	P32	P12	P30	P21	P31	P22	P27	P23	P17	P20
1935	*	P29	P20	P22	P12	P30	P26	P25	P27	P23	P18	P31	P32	P16	P21	P33	P17	P28	P11	P24	P14	P13	P19	P15
1936	*	P58	P55	P64	P73	P57	P72	P63	P54	P53	P70	P51	P61	P62	P68	P56	P59	P52	P60	P67	P65	P69	P71	P66
1937	*	P24	P18	P32	P20	P14	P17	P33	P28	P16	P27	P13	P19	P26	P31	P15	P25	P11	P29	P21	P12	P22	P23	P30
1938	*	P27	P30	P21	P15	P25	P19	P16	P12	P22	P14	P29	P31	P32	P11	P26	P23	P20	P18	P28	P33	P24	P13	P17
1939	*	P14	P17	P28	P26	P23	P31	P22	P33	P24	P25	P18	P11	P21	P20	P32	P13	P15	P30	P12	P16	P27	P29	P19
1940	*	P65	P59	P52	P72	P53	P51	P64	P56	P67	P63	P70	P60	P68	P55	P61	P69	P66	P57	P73	P62	P54	P58	P71
1941	*	P12	P25	P11	P17	P16	P13	P32	P15	P21	P33	P27	P29	P31	P18	P19	P22	P30	P14	P20	P26	P28	P24	P23
1942	*	P33	P23	P20	P19	P22	P29	P21	P26	P28	P16	P14	P18	P11	P30	P31	P24	P17	P25	P15	P32	P12	P27	P13
1943	*	P16	P13	P15	P31	P24	P18	P28	P32	P12	P22	P25	P30	P20	P17	P11	P27	P19	P23	P26	P21	P33	P14	P29
1944	*	P62	P69	P66	P51	P67	P70	P52	P61	P73	P64	P63	P57	P55	P59	P60	P54	P71	P53	P72	P68	P56	P65	P58
1945	*	P26	P22	P30	P13	P21	P27	P11	P19	P20	P32	P33	P14	P18	P25	P29	P28	P23	P16	P17	P31	P15	P12	P24
1946	*	P32	P24	P17	P29	P28	P14	P20	P31	P15	P21	P16	P25	P30	P23	P18	P12	P13	P22	P19	P11	P26	P33	P27
1947	*	P21	P27	P19	P18	P12	P25	P15	P11	P26	P28	P22	P23	P17	P13	P30	P33	P29	P24	P31	P20	P32	P16	P14
1948	*	P68	P54	P71	P70	P73	P63	P66	P60	P72	P52	P64	P53	P59	P69	P57	P56	P58	P67	P51	P55	P61	P62	P65
1949	*	P31	P28	P23	P27	P20	P33	P30	P29	P17	P11	P32	P16	P25	P22	P14	P15	P24	P21	P13	P18	P19	P26	P12
1950	*	P11	P12	P13	P14	P15	P16	P17	P18	P19	P20	P21	P22	P23	P24	P25	P26	P27	P28	P29	P30	P31	P32	P33
1951	*	P20	P33	P29	P25	P26	P22	P19	P30	P31	P15	P28	P24	P13	P27	P23	P32	P14	P12	P18	P17	P11	P21	P16
1952	*	P55	P56	P58	P63	P72	P64	P71	P57	P51	P66	P52	P67	P69	P54	P53	P61	P65	P73	P70	P59	P60	P68	P62
1953	*	P18	P15	P24	P33	P17	P32	P23	P14	P13	P30	P11	P21	P22	P28	P16	P19	P12	P20	P27	P25	P29	P31	P26
1954	*	P30	P26	P27	P16	P19	P21	P13	P25	P29	P17	P20	P28	P24	P12	P22	P31	P33	P15	P14	P23	P18	P11	P32
1955	*	P17	P32	P14	P22	P31	P28	P29	P23	P18	P19	P15	P12	P27	P33	P24	P11	P16	P26	P25	P13	P30	P20	P21
1956	*	P59	P61	P65	P64	P51	P52	P58	P53	P70	P71	P66	P73	P54	P56	P67	P60	P62	P72	P63	P69	P57	P55	P68
1957	*	P25	P19	P12	P32	P13	P11	P24	P16	P27	P23	P30	P20	P28	P15	P21	P29	P26	P17	P33	P22	P14	P18	P31
1958	*	P23	P31	P33	P21	P29	P20	P27	P22	P14	P13	P17	P15	P12	P26	P28	P18	P32	P19	P16	P24	P25	P30	P11
1959	*	P13	P11	P16	P28	P18	P15	P14	P24	P25	P29	P19	P26	P33	P32	P12	P30	P21	P31	P22	P27	P23	P17	P20
1960	*	P69	P60	P62	P52	P70	P66	P65	P67	P63	P58	P71	P72	P56	P61	P73	P57	P68	P51	P64	P54	P53	P59	P55
1961	*	P22	P29	P26	P11	P27	P30	P12	P21	P33	P24	P23	P17	P15	P19	P20	P14	P31	P13	P32	P28	P16	P25	P18
1962	*	P24	P18	P32	P20	P14	P17	P33	P28	P16	P27	P13	P19	P26	P31	P15	P25	P11	P29	P21	P12	P22	P23	P30
1963	*	P27	P30	P21	P15	P25	P19	P16	P12	P22	P14	P29	P31	P32	P11	P26	P23	P20	P18	P28	P33	P24	P13	P17
1964	*	P54	P57	P68	P66	P63	P71	P62	P73	P64	P65	P58	P51	P61	P60	P72	P53	P55	P70	P52	P56	P67	P69	P59
1965	*	P28	P14	P31	P30	P33	P23	P26	P20	P32	P12	P24	P13	P19	P29	P17	P16	P18	P27	P11	P15	P21	P22	P25
1966	*	P12	P25	P11	P17	P16	P13	P32	P15	P21	P33	P27	P29	P31	P18	P19	P22	P30	P14	P20	P26	P28	P24	P23
1967	*	P33	P23	P20	P19	P22	P29	P21	P26	P28	P16	P14	P18	P11	P30	P31	P24	P17	P25	P15	P32	P12	P27	P13
1968	*	P56	P53	P55	P71	P64	P58	P68	P72	P52	P62	P65	P70	P60	P57	P51	P67	P59	P63	P66	P61	P73	P54	P69
1969	*	P15	P16	P18	P23	P32	P24	P31	P17	P11	P26	P12	P27	P29	P14	P13	P21	P25	P33	P30	P19	P20	P28	P22
1970	*	P26	P22	P30	P13	P21	P27	P11	P19	P20	P32	P33	P14	P18	P25	P29	P28	P23	P16	P17	P31	P15	P12	P24
1971	*	P32	P24	P17	P29	P28	P14	P20	P31	P15	P21	P16	P25	P30	P23	P18	P12	P13	P22	P19	P11	P26	P33	P27
1972	*	P61	P67	P59	P58	P52	P65	P55	P51	P66	P68	P62	P63	P57	P53	P70	P73	P69	P64	P71	P60	P72	P56	P54
1973	*	P19	P21	P25	P24	P11	P12	P18	P13	P30	P31	P26	P33	P14	P16	P27	P20	P22	P32	P23	P29	P17	P15	P28
1974	*	P31	P28	P23	P27	P20	P33	P30	P29	P17	P11	P32	P16	P25	P22	P14	P15	P24	P21	P13	P18	P19	P26	P12
1975	*	P11	P12	P13	P14	P15	P16	P17	P18	P19	P20	P21	P22	P23	P24	P25	P26	P27	P28	P29	P30	P31	P32	P33
1976	*	P60	P73	P69	P65	P66	P62	P59	P70	P71	P55	P68	P64	P53	P67	P63	P72	P54	P52	P58	P57	P51	P61	P56
1977	*	P29	P20	P22	P12	P30	P26	P25	P27	P23	P18	P31	P32	P16	P21	P33	P17	P28	P11	P24	P14	P13	P19	P15
1978	*	P18	P15	P24	P33	P17	P32	P23	P14	P13	P30	P11	P21	P22	P28	P16	P19	P12	P20	P27	P25	P29	P31	P26
1979	*	P30	P26	P27	P16	P19	P21	P13	P25	P29	P17	P20	P28	P24	P12	P22	P31	P33	P15	P14	P23	P18	P11	P32
1980	*	P57	P72	P54	P62	P71	P68	P69	P63	P58	P59	P55	P52	P67	P73	P64	P51	P56	P66	P65	P53	P70	P60	P61
1981	*	P14	P17	P28	P26	P23	P31	P22	P33	P24	P25	P18	P11	P21	P20	P32	P13	P15	P30	P12	P16	P27	P29	P19
1982	*	P25	P19	P12	P32	P13	P11	P24	P16	P27	P23	P30	P20	P28	P15	P21	P29	P26	P17	P33	P22	P14	P18	P31
1983	*	P23	P31	P33	P21	P29	P20	P27	P22	P14	P13	P17	P15	P12	P26	P28	P18	P32	P19	P16	P24	P25	P30	P11
1984	*	P53	P51	P56	P68	P58	P55	P54	P64	P65	P69	P59	P66	P73	P72	P52	P70	P61	P71	P62	P67	P63	P57	P60
1985	*	P16	P13	P15	P31	P24	P18	P28	P32	P12	P22	P25	P30	P20	P17	P11	P27	P19	P23	P26	P21	P33	P14	P29
1986	*	P22	P29	P26	P11	P27	P30	P12	P21	P33	P24	P23	P17	P15	P19	P20	P14	P31	P13	P32	P28	P16	P25	P18
1987	*	P24	P18	P32	P20	P14	P17	P33	P28	P16	P27	P13	P19	P26	P31	P15	P25	P11	P29	P21	P12	P22	P23	P30
1988	*	P67	P70	P61	P55	P65	P59	P56	P52	P62	P54	P69	P71	P72	P51	P66	P63	P60	P58	P68	P73	P64	P53	P57
1989	*	P21	P27	P19	P18	P12	P25	P15	P11	P26	P28	P22	P23	P17	P13	P30	P33	P29	P24	P31	P20	P32	P16	P14
1990	*	P28	P14	P31	P30	P33	P23	P26	P20	P32	P12	P24	P13	P19	P29	P17	P16	P18	P27	P11	P15	P21	P22	P25
1991	*	P12	P25	P11	P17	P16	P13	P32	P15	P21	P33	P27	P29	P31	P18	P19	P22	P30	P14	P20	P26	P28	P24	P23
1992	*	P73	P63	P60	P59	P62	P69	P61	P66	P68	P56	P54	P58	P51	P70	P71	P64	P57	P65	P55	P72	P52	P67	P53
1993	*	P20	P33	P29	P25	P26	P22	P19	P30	P31	P15	P28	P24	P13	P27	P23	P32	P14	P12	P18	P17	P11	P21	P16
1994	*	P15	P16	P18	P23	P32	P24	P31	P17	P11	P26	P12	P27	P29	P14	P13	P21	P25	P33	P30	P19	P20	P28	P22
1995	*	P26	P22	P30	P13	P21	P27	P11	P19	P20	P32	P33	P14	P18	P25	P29	P28	P23	P16	P17	P31	P15	P12	P24
1996	*	P72	P64	P57	P69	P68	P54	P60	P71	P55	P61	P56	P65	P70	P63	P58	P52	P53	P62	P59	P51	P66	P73	P67
1997	*	P17	P32	P14	P22	P31	P28	P29	P23	P18	P19	P15	P12	P27	P33	P24	P11	P16	P26	P25	P13	P30	P20	P21
1998	*	P19	P21	P25	P24	P11	P12	P18	P13	P30	P31	P26	P33	P14	P16	P27	P20	P22	P32	P23	P29	P17	P15	P28
1999	*	P31	P28	P23	P27	P20	P33	P30	P29	P17	P11	P32	P16	P25	P22	P14	P15	P24	P21	P13	P18	P19	P26	P12
2000	*	P51	P52	P53	P54	P55	P56	P57	P58	P59	P60	P61	P62	P63	P64	P65	P66	P67	P68	P69	P70	P71	P72	P73
2001	*	P13	P11	P16	P28	P18	P15	P14	P24	P25	P29	P19	P26	P33	P32	P12	P30	P21	P31	P22	P27	P23	P17	P20

```
YEAR   *                        SENSITIVITY TABLES FOR INQUIRY YEARS 1800--1900
 OF    *   SENSITIVITY BIRTH CODES ARE LISTED BY COLUMNS--USE TABLE IN PROPER ROW AND COLUMN FOR BIORHYTHM SENSITIVITY DATA
INQUIRY*   11   12   13   14   15   16   17   18   19   20   21   22   23   24   25   26   27   28   29   30   31   32   33   34   35   36   37   38
-------
1800  *  S29  S38  S18  S27  S32  S33  S35  S26  S12  S14  S25  S16  S13  S15  S20  S17  S31  S23  S24  S37  S34  S30  S36  S21  S28  S19  S11  S22
1801  *  S22  S26  S31  S37  S36  S28  S25  S34  S18  S30  S15  S35  S27  S33  S32  S21  S11  S14  S16  S12  S29  S19  S23  S24  S20  S13  S38  S17
1802  *  S17  S34  S11  S12  S23  S20  S15  S29  S31  S19  S33  S25  S37  S28  S36  S30  S35  S18  S22  S13  S14  S16  S32  S27  S26  S21  S24
1803  *  S21  S29  S38  S18  S14  S32  S33  S22  S11  S13  S28  S15  S12  S20  S23  S16  S26  S19  S25  S31  S17  S27  S30  S35  S36  S37  S34  S24
1804  *  S64  S62  S66  S71  S70  S76  S68  S57  S78  S67  S60  S73  S58  S72  S54  S75  S74  S53  S55  S51  S61  S77  S59  S65  S63  S52  S69  S56
1805  *  S35  S21  S29  S38  S13  S14  S32  S24  S34  S12  S36  S20  S11  S23  S19  S15  S22  S37  S28  S26  S16  S18  S27  S33  S30  S31  S17  S25
1806  *  S25  S24  S22  S26  S27  S30  S36  S16  S29  S18  S23  S32  S38  S14  S13  S33  S17  S12  S20  S30  S36  S31  S37  S28  S24  S20  S13  S31
1807  *  S15  S16  S17  S34  S37  S19  S23  S35  S22  S31  S14  S36  S26  S30  S27  S28  S21  S18  S32  S29  S25  S11  S12  S20  S13  S38  S24  S33
1808  *  S73  S75  S61  S69  S52  S53  S54  S65  S57  S51  S70  S63  S74  S59  S77  S60  S64  S71  S76  S62  S55  S78  S58  S72  S67  S66  S56  S68
1809  *  S20  S15  S16  S17  S31  S37  S19  S33  S24  S26  S13  S30  S22  S27  S18  S36  S35  S38  S14  S21  S28  S34  S11  S23  S12  S29  S25  S32
1810  *  S32  S33  S35  S21  S11  S12  S13  S28  S16  S34  S27  S19  S17  S37  S31  S23  S25  S26  S30  S24  S20  S29  S38  S14  S18  S22  S15  S36
1811  *  S36  S28  S25  S24  S38  S18  S27  S20  S35  S29  S37  S13  S21  S12  S11  S14  S15  S34  S19  S16  S32  S22  S26  S30  S31  S17  S33  S23
1812  *  S63  S60  S55  S56  S66  S71  S77  S72  S65  S62  S67  S64  S58  S78  S73  S69  S53  S75  S76  S57  S74  S59  S51  S61  S68  S54
1813  *  S30  S36  S28  S25  S29  S38  S18  S23  S33  S21  S31  S12  S35  S11  S34  S13  S20  S17  S37  S15  S14  S24  S22  S27  S26  S16  S32  S19
1814  *  S19  S23  S20  S15  S22  S26  S31  S14  S28  S24  S11  S18  S25  S38  S29  S27  S30  S16  S17  S37  S34  S35  S36  S13
1815  *  S13  S14  S32  S33  S17  S34  S11  S30  S20  S16  S38  S31  S15  S26  S22  S37  S36  S24  S18  S28  S19  S35  S21  S12  S29  S25  S23  S27
1816  *  S67  S70  S76  S68  S61  S69  S78  S59  S72  S75  S66  S51  S73  S74  S57  S52  S63  S56  S71  S60  S53  S65  S64  S58  S62  S55  S54  S77
1817  *  S12  S13  S14  S32  S16  S17  S34  S27  S23  S15  S29  S26  S20  S22  S24  S31  S30  S25  S38  S36  S37  S33  S35  S11  S21  S28  S19  S18
1818  *  S18  S27  S30  S36  S35  S21  S29  S37  S14  S33  S22  S34  S32  S17  S16  S11  S19  S15  S26  S23  S12  S28  S25  S38  S24  S20  S13  S31
1819  *  S31  S37  S19  S23  S25  S24  S22  S12  S30  S28  S17  S29  S36  S21  S35  S38  S13  S33  S34  S14  S18  S20  S15  S26  S16  S32  S27  S11
1820  *  S51  S52  S53  S54  S55  S56  S57  S58  S59  S60  S61  S62  S63  S64  S65  S66  S67  S68  S69  S77  S71  S72  S73  S74  S75  S76  S77  S78
1821  *  S26  S31  S37  S19  S28  S25  S24  S11  S27  S36  S16  S21  S30  S35  S33  S29  S12  S32  S17  S13  S38  S23  S20  S22  S15  S14  S18  S34
1822  *  S34  S11  S12  S13  S20  S15  S16  S38  S37  S23  S35  S24  S19  S25  S28  S22  S18  S36  S21  S27  S33  S30  S31  S32
1823  *  S29  S38  S18  S27  S32  S32  S33  S35  S26  S12  S14  S25  S16  S13  S15  S20  S17  S31  S23  S24  S37  S34  S30  S36  S21  S28  S19  S11  S22
1824  *  S62  S66  S71  S77  S76  S68  S74  S58  S70  S55  S75  S67  S73  S72  S61  S51  S54  S56  S52  S69  S59  S63  S64  S60  S53  S78  S57
1825  *  S21  S29  S38  S18  S14  S32  S33  S22  S11  S13  S28  S15  S12  S20  S23  S16  S26  S19  S25  S31  S17  S27  S30  S35  S36  S37  S34  S24
1826  *  S24  S22  S26  S31  S30  S36  S28  S17  S38  S27  S20  S33  S18  S32  S14  S13  S15  S11  S37  S19  S25  S21  S12  S29  S23  S16
1827  *  S16  S17  S34  S11  S19  S23  S20  S21  S26  S37  S32  S28  S31  S36  S30  S25  S29  S27  S33  S38  S24  S12  S13  S15  S14  S18  S22  S35
1828  *  S75  S61  S69  S78  S53  S54  S72  S64  S74  S52  S78  S60  S51  S63  S59  S55  S62  S77  S68  S66  S56  S58  S67  S73  S70  S57  S71  S65
1829  *  S15  S16  S17  S34  S37  S19  S23  S35  S22  S32  S31  S14  S36  S26  S30  S27  S28  S21  S18  S32  S29  S25  S11  S12  S20  S13  S38  S24  S33
1830  *  S33  S35  S21  S29  S12  S13  S14  S32  S25  S17  S11  S30  S33  S24  S19  S37  S20  S23  S34  S31  S38  S18  S32  S27  S26  S16  S28
1831  *  S28  S25  S24  S22  S18  S27  S30  S15  S21  S38  S19  S14  S29  S13  S12  S32  S16  S11  S23  S17  S33  S26  S31  S36  S37  S34  S35  S20
1832  *  S60  S55  S56  S57  S71  S77  S73  S64  S58  S65  S53  S70  S62  S67  S78  S75  S74  S61  S68  S54  S51  S66  S52  S59  S69  S65  S72
1833  *  S36  S28  S25  S24  S38  S18  S27  S20  S35  S29  S37  S13  S21  S12  S11  S14  S15  S34  S19  S16  S32  S22  S26  S30  S31  S17  S33  S23
1834  *  S23  S20  S15  S16  S26  S31  S37  S32  S25  S12  S27  S24  S18  S38  S30  S33  S29  S13  S35  S17  S34  S19  S11  S21  S28  S14
1835  *  S14  S32  S33  S35  S34  S11  S12  S36  S15  S17  S18  S37  S16  S31  S26  S19  S28  S22  S27  S25  S23  S21  S29  S13  S38  S24  S20  S30
1836  *  S70  S76  S68  S65  S69  S78  S63  S73  S61  S71  S52  S75  S51  S74  S53  S60  S57  S77  S55  S54  S62  S64  S60  S53  S72  S59
1837  *  S13  S14  S32  S33  S17  S34  S11  S30  S20  S16  S38  S31  S15  S26  S22  S37  S36  S24  S18  S28  S19  S35  S21  S12  S29  S25  S23  S27
1838  *  S27  S30  S36  S28  S21  S29  S38  S19  S32  S35  S24  S11  S33  S34  S17  S12  S23  S16  S31  S25  S18  S28  S24  S18  S12  S15  S14  S37
1839  *  S37  S19  S23  S20  S24  S22  S26  S13  S36  S25  S34  S38  S28  S29  S21  S18  S14  S35  S11  S32  S27  S15  S16  S31  S17  S33  S30  S12
1840  *  S52  S53  S54  S72  S56  S57  S74  S67  S63  S65  S69  S66  S60  S62  S64  S70  S65  S78  S77  S73  S75  S51  S61  S68  S59  S58
1841  *  S31  S37  S19  S23  S25  S24  S22  S12  S30  S28  S17  S29  S36  S21  S35  S38  S13  S33  S34  S14  S18  S20  S15  S26  S16  S32  S27  S11
1842  *  S11  S12  S13  S14  S15  S16  S17  S18  S19  S20  S21  S22  S23  S24  S25  S26  S27  S28  S29  S30  S31  S32  S33  S34  S35  S36  S37  S38
1843  *  S38  S18  S27  S30  S33  S35  S21  S31  S13  S32  S24  S17  S14  S16  S15  S34  S37  S20  S22  S19  S11  S36  S28  S29  S25  S23  S12  S26
1844  *  S66  S71  S77  S59  S68  S65  S64  S51  S67  S76  S56  S61  S70  S75  S73  S69  S52  S72  S57  S53  S78  S63  S60  S62  S55  S54  S58  S74
1845  *  S29  S38  S18  S27  S32  S32  S33  S35  S26  S12  S14  S25  S16  S13  S15  S20  S17  S31  S23  S24  S37  S34  S30  S36  S21  S28  S19  S11  S22
1846  *  S22  S26  S31  S37  S36  S28  S25  S34  S18  S30  S15  S35  S27  S33  S32  S21  S11  S14  S16  S12  S29  S19  S23  S24  S20  S13  S38  S17
1847  *  S17  S34  S11  S12  S23  S20  S15  S29  S31  S19  S33  S25  S37  S28  S36  S30  S35  S18  S22  S13  S14  S16  S32  S27  S26  S21
1848  *  S61  S69  S78  S58  S54  S72  S73  S62  S51  S53  S68  S55  S52  S60  S63  S50  S57  S78  S56  S67  S75  S70  S71  S77  S74  S64
1849  *  S16  S17  S34  S11  S19  S23  S20  S21  S26  S37  S32  S28  S31  S36  S30  S25  S29  S27  S33  S38  S24  S12  S13  S15  S14  S18  S22  S35
1850  *  S35  S21  S29  S38  S13  S14  S32  S24  S34  S12  S36  S20  S11  S23  S19  S15  S22  S37  S28  S26  S16  S18  S27  S33  S30  S31  S17  S25
1851  *  S25  S24  S22  S26  S27  S30  S36  S16  S29  S18  S23  S32  S38  S14  S13  S33  S17  S12  S20  S34  S35  S31  S37  S28  S19  S11  S21  S15
1852  *  S55  S56  S57  S74  S77  S53  S75  S62  S71  S54  S78  S64  S70  S67  S68  S61  S58  S72  S56  S51  S52  S60  S53  S78  S64  S73
1853  *  S28  S25  S24  S22  S18  S27  S30  S15  S21  S38  S19  S14  S29  S13  S12  S32  S16  S11  S23  S17  S33  S26  S31  S36  S37  S34  S35  S20
1854  *  S20  S15  S16  S17  S31  S37  S19  S33  S24  S26  S13  S30  S22  S27  S18  S36  S35  S38  S14  S21  S28  S34  S11  S23  S12  S29  S25  S32
1855  *  S32  S33  S35  S21  S11  S12  S13  S28  S16  S34  S27  S19  S17  S37  S31  S23  S25  S26  S30  S24  S20  S29  S38  S14  S18  S22  S15  S36
1856  *  S76  S68  S65  S64  S78  S58  S60  S75  S59  S77  S53  S61  S51  S52  S55  S54  S74  S59  S56  S62  S66  S70  S71  S57  S73  S63
1857  *  S14  S32  S33  S35  S34  S11  S12  S36  S15  S17  S18  S37  S16  S31  S26  S19  S28  S22  S27  S23  S21  S29  S13  S38  S24  S20  S30
1858  *  S30  S36  S28  S25  S29  S38  S18  S23  S33  S21  S31  S12  S35  S11  S34  S13  S20  S17  S37  S15  S14  S24  S22  S27  S26  S16  S32  S19
1859  *  S19  S23  S20  S15  S22  S26  S31  S14  S28  S24  S11  S18  S25  S38  S29  S27  S30  S32  S17  S37  S12  S13  S33  S30  S16  S35  S32  S13
1860  *  S53  S54  S72  S73  S57  S74  S51  S70  S60  S56  S78  S71  S55  S66  S62  S77  S76  S64  S58  S68  S59  S75  S61  S52  S69  S65  S63  S67
1861  *  S37  S19  S23  S20  S24  S22  S26  S13  S36  S25  S34  S38  S28  S29  S21  S18  S14  S35  S11  S32  S27  S15  S16  S31  S17  S33  S30  S12
1862  *  S12  S13  S14  S32  S16  S17  S34  S27  S23  S15  S29  S26  S20  S22  S24  S31  S30  S25  S38  S36  S37  S33  S35  S11  S21  S28  S19  S18
1863  *  S18  S27  S30  S36  S35  S21  S29  S37  S14  S33  S22  S34  S32  S17  S16  S11  S19  S15  S26  S23  S12  S28  S25  S38  S24  S20  S13  S31
1864  *  S71  S77  S59  S63  S65  S64  S52  S70  S68  S69  S78  S61  S75  S78  S53  S73  S74  S54  S58  S60  S55  S66  S56  S72  S67  S51
1865  *  S38  S18  S27  S30  S33  S35  S21  S31  S13  S32  S24  S17  S14  S16  S15  S34  S37  S20  S22  S19  S11  S36  S28  S29  S25  S23  S12  S26
1866  *  S26  S31  S37  S19  S28  S25  S24  S11  S27  S36  S16  S21  S30  S35  S33  S29  S12  S32  S17  S13  S38  S23  S20  S22  S15  S14  S18  S34
1867  *  S34  S11  S12  S13  S20  S15  S16  S38  S37  S23  S35  S24  S19  S25  S28  S22  S18  S36  S21  S27  S33  S30  S31  S32  S15  S14  S18  S34
1868  *  S69  S78  S58  S67  S72  S73  S75  S66  S52  S54  S65  S56  S53  S55  S60  S57  S71  S63  S64  S77  S74  S70  S76  S61  S68  S59  S51  S62
1869  *  S17  S34  S11  S12  S23  S20  S15  S29  S31  S19  S33  S25  S37  S28  S36  S30  S35  S18  S22  S13  S14  S16  S32  S27  S26  S21
1870  *  S21  S29  S38  S18  S14  S32  S33  S22  S11  S13  S28  S15  S12  S20  S23  S16  S26  S19  S25  S31  S17  S27  S30  S35  S36  S37  S34  S24
1871  *  S24  S22  S26  S31  S30  S36  S28  S17  S38  S27  S20  S33  S18  S32  S14  S13  S15  S11  S34  S35  S13  S21  S19  S12  S29  S23  S16
1872  *  S56  S57  S74  S51  S59  S63  S60  S61  S66  S77  S72  S68  S71  S76  S70  S65  S69  S67  S73  S78  S64  S52  S53  S55  S54  S58  S62  S75
1873  *  S25  S24  S22  S26  S27  S30  S36  S16  S29  S18  S23  S32  S38  S14  S13  S33  S17  S12  S20  S34  S35  S37  S28  S19  S11  S21  S15
1874  *  S15  S16  S17  S34  S37  S19  S23  S35  S22  S31  S14  S36  S26  S30  S27  S28  S21  S18  S32  S29  S25  S11  S12  S20  S13  S38  S24  S33
1875  *  S33  S35  S21  S29  S12  S13  S14  S32  S25  S17  S11  S30  S33  S24  S19  S37  S20  S23  S34  S31  S38  S18  S22  S15  S26  S16  S28
1876  *  S68  S65  S64  S62  S58  S67  S70  S55  S61  S78  S59  S54  S69  S53  S52  S72  S56  S51  S63  S57  S73  S66  S71  S76  S77  S74  S75  S60
1877  *  S32  S33  S35  S21  S11  S12  S13  S28  S16  S34  S27  S19  S17  S37  S31  S23  S25  S26  S30  S24  S20  S29  S38  S14  S18  S22  S15  S36
1878  *  S36  S28  S25  S24  S38  S18  S27  S20  S35  S29  S37  S13  S21  S12  S11  S14  S15  S34  S19  S16  S32  S22  S26  S30  S31  S17  S33  S23
1879  *  S23  S20  S15  S16  S26  S31  S37  S32  S25  S12  S27  S24  S18  S38  S30  S33  S29  S13  S35  S17  S34  S19  S11  S21  S28  S14
1880  *  S54  S72  S73  S75  S74  S51  S52  S76  S55  S57  S58  S77  S56  S71  S66  S59  S68  S62  S67  S65  S63  S61  S69  S53  S78  S64  S60  S70
1881  *  S19  S23  S20  S15  S22  S26  S31  S14  S28  S24  S11  S18  S25  S38  S29  S27  S30  S16  S17  S37  S34  S35  S36  S13
1882  *  S13  S14  S32  S33  S17  S34  S11  S30  S20  S16  S38  S31  S15  S26  S22  S37  S36  S24  S18  S28  S19  S35  S21  S12  S29  S25  S23  S27
1883  *  S27  S30  S36  S28  S21  S29  S38  S19  S32  S35  S24  S11  S33  S34  S17  S12  S23  S16  S31  S25  S18  S28  S24  S18  S12  S15  S14  S37
1884  *  S77  S59  S63  S60  S64  S62  S66  S53  S76  S65  S74  S78  S68  S69  S61  S58  S54  S75  S51  S72  S67  S55  S56  S71  S57  S73  S70  S52
1885  *  S18  S27  S30  S36  S35  S21  S29  S37  S14  S33  S22  S34  S32  S17  S16  S11  S19  S15  S26  S23  S12  S28  S25  S38  S24  S20  S13  S31
1886  *  S31  S37  S19  S23  S25  S24  S22  S12  S30  S28  S17  S29  S36  S21  S35  S38  S13  S33  S34  S14  S18  S20  S15  S26  S16  S32  S27  S11
1887  *  S11  S12  S13  S14  S15  S16  S17  S18  S19  S20  S21  S22  S23  S24  S25  S26  S27  S28  S29  S30  S31  S32  S33  S34  S35  S36  S37  S38
1888  *  S78  S58  S67  S70  S73  S75  S61  S71  S53  S72  S64  S57  S54  S56  S55  S74  S77  S60  S62  S59  S51  S76  S68  S69  S65  S63  S52  S66
1889  *  S34  S11  S12  S13  S20  S15  S16  S38  S37  S23  S35  S24  S19  S25  S28  S22  S18  S36  S21  S27  S33  S30  S31  S32  S17  S33  S31  S29
1890  *  S29  S38  S18  S27  S32  S32  S33  S35  S26  S12  S14  S25  S16  S13  S15  S20  S17  S31  S23  S24  S37  S34  S30  S36  S21  S28  S19  S11  S22
1891  *  S22  S26  S31  S37  S36  S28  S25  S34  S18  S30  S15  S35  S27  S33  S32  S21  S11  S14  S16  S12  S29  S19  S23  S24  S20  S13  S38  S17
1892  *  S57  S74  S51  S52  S63  S60  S55  S69  S71  S59  S73  S65  S77  S68  S76  S64  S78  S70  S75  S58  S62  S53  S54  S56  S72  S67  S66  S61
1893  *  S24  S22  S26  S31  S30  S36  S28  S17  S38  S27  S20  S33  S18  S32  S14  S13  S15  S11  S37  S19  S25  S21  S12  S29  S23  S16
1894  *  S16  S17  S34  S11  S19  S23  S20  S21  S26  S37  S32  S28  S31  S36  S30  S25  S29  S27  S33  S38  S24  S12  S13  S15  S14  S18  S22  S35
1895  *  S35  S21  S29  S38  S13  S14  S32  S24  S34  S12  S36  S20  S11  S23  S19  S15  S22  S37  S28  S26  S16  S18  S27  S33  S30  S31  S17  S25
1896  *  S65  S64  S62  S66  S67  S70  S76  S56  S69  S58  S63  S72  S78  S54  S53  S73  S57  S52  S60  S74  S75  S71  S77  S68  S59  S51  S61  S55
1897  *  S33  S35  S21  S29  S12  S13  S14  S32  S25  S17  S11  S30  S33  S24  S19  S37  S20  S23  S34  S31  S38  S18  S22  S15  S26  S16  S28
1898  *  S28  S25  S24  S22  S18  S27  S30  S15  S21  S38  S19  S14  S29  S13  S12  S32  S16  S11  S23  S17  S33  S26  S31  S36  S37  S34  S35  S20
1899  *  S20  S15  S16  S17  S31  S37  S19  S33  S24  S26  S13  S30  S22  S27  S18  S36  S35  S38  S14  S21  S28  S34  S11  S23  S12  S29  S25  S32
1900  *  S32  S33  S35  S21  S11  S12  S13  S28  S16  S34  S27  S19  S17  S37  S31  S23  S25  S26  S30  S24  S20  S29  S38  S14  S18  S22  S15  S36
```

SENSITIVITY TABLES FOR INQUIRY YEARS 1901--2001

SENSITIVITY BIRTH CODES ARE LISTED BY COLUMNS--USE TABLE IN PROPER ROW AND COLUMN FOR BIORHYTHM SENSITIVITY DATA

YEAR OF INQUIRY	11	12	13	14	15	16	17	18	19	20	21	22	23	24	25	26	27	28	29	30	31	32	33	34	35	36	37	38
1901	S36	S28	S25	S24	S38	S18	S27	S20	S35	S29	S37	S13	S21	S12	S11	S14	S15	S34	S19	S16	S32	S22	S26	S30	S31	S17	S33	S23
1902	S23	S20	S15	S16	S26	S31	S37	S32	S25	S22	S12	S27	S24	S18	S30	S33	S29	S13	S35	S36	S17	S34	S19	S11	S21	S28	S14	S30
1903	S14	S32	S33	S35	S34	S11	S12	S36	S15	S17	S18	S37	S16	S31	S26	S19	S28	S22	S27	S25	S23	S21	S29	S13	S38	S24	S20	S30
1904	S70	S76	S68	S65	S69	S78	S58	S63	S73	S61	S71	S52	S75	S51	S74	S53	S60	S57	S77	S55	S54	S64	S62	S67	S66	S56	S72	S59
1905	S13	S14	S32	S33	S17	S34	S11	S30	S20	S16	S38	S31	S15	S26	S22	S37	S36	S24	S18	S28	S19	S35	S21	S12	S29	S25	S23	S27
1906	S27	S30	S36	S28	S21	S29	S38	S19	S32	S35	S26	S11	S33	S34	S17	S12	S23	S16	S31	S20	S13	S25	S24	S18	S22	S15	S14	S37
1907	S37	S19	S23	S20	S24	S22	S26	S13	S36	S25	S34	S38	S28	S29	S21	S18	S14	S35	S11	S32	S27	S15	S16	S31	S17	S33	S30	S12
1908	S52	S53	S54	S72	S56	S57	S74	S67	S63	S55	S69	S66	S60	S62	S64	S71	S70	S65	S78	S76	S77	S73	S75	S51	S61	S68	S59	S58
1909	S31	S37	S19	S23	S25	S24	S22	S12	S30	S28	S17	S29	S36	S21	S35	S38	S13	S33	S34	S14	S18	S20	S15	S26	S16	S32	S27	S11
1910	S11	S12	S13	S14	S15	S16	S17	S18	S19	S20	S21	S22	S23	S24	S25	S26	S27	S28	S29	S30	S31	S32	S33	S34	S35	S36	S37	S38
1911	S38	S18	S27	S30	S33	S35	S21	S31	S13	S32	S24	S17	S14	S16	S15	S34	S37	S20	S22	S19	S11	S36	S28	S29	S25	S23	S12	S26
1912	S66	S71	S77	S59	S68	S65	S64	S51	S67	S76	S56	S61	S70	S75	S73	S69	S52	S72	S57	S53	S78	S63	S60	S62	S55	S54	S58	S74
1913	S29	S38	S18	S27	S32	S33	S35	S26	S12	S14	S25	S16	S13	S15	S20	S17	S31	S23	S24	S37	S34	S30	S36	S21	S28	S19	S11	S22
1914	S22	S26	S31	S37	S36	S28	S25	S34	S18	S30	S15	S35	S27	S33	S32	S21	S11	S16	S12	S29	S19	S23	S24	S20	S13	S38	S17	
1915	S17	S34	S11	S12	S23	S20	S15	S29	S31	S19	S33	S25	S37	S28	S36	S24	S38	S30	S35	S18	S22	S13	S14	S16	S32	S27	S26	S21
1916	S61	S69	S78	S58	S54	S72	S73	S62	S51	S53	S68	S55	S52	S56	S63	S56	S65	S71	S57	S67	S70	S75	S52	S60	S53	S78	S77	S64
1917	S16	S17	S34	S11	S19	S23	S20	S21	S26	S37	S32	S28	S31	S36	S30	S25	S29	S27	S33	S38	S24	S12	S13	S15	S14	S18	S22	S35
1918	S35	S21	S29	S38	S13	S14	S32	S24	S34	S12	S36	S20	S11	S23	S19	S15	S22	S37	S28	S16	S27	S33	S30	S31	S17	S25		
1919	S25	S24	S22	S26	S27	S30	S36	S16	S29	S18	S23	S32	S38	S14	S13	S33	S17	S12	S20	S34	S35	S31	S37	S28	S19	S11	S21	S15
1920	S55	S56	S52	S74	S51	S77	S59	S62	S71	S54	S76	S66	S70	S57	S68	S61	S53	S60	S69	S65	S51	S52	S50	S53	S78	S64	S73	
1921	S28	S25	S24	S22	S18	S27	S30	S15	S21	S38	S19	S14	S29	S13	S12	S32	S16	S11	S23	S17	S33	S26	S31	S36	S37	S34	S35	S20
1922	S20	S15	S16	S17	S31	S37	S19	S33	S24	S26	S13	S30	S22	S27	S18	S36	S35	S38	S14	S21	S34	S11	S12	S29	S25	S23	S28	S32
1923	S32	S33	S35	S21	S11	S12	S13	S28	S16	S34	S27	S19	S17	S37	S31	S23	S25	S26	S30	S24	S20	S29	S38	S14	S18	S22	S15	S36
1924	S76	S68	S65	S64	S78	S58	S67	S60	S75	S69	S77	S53	S61	S52	S51	S54	S55	S74	S59	S56	S72	S62	S66	S70	S71	S57	S73	S63
1925	S14	S32	S33	S35	S34	S11	S12	S36	S15	S17	S18	S37	S16	S31	S26	S19	S28	S22	S27	S25	S23	S21	S29	S13	S38	S24	S20	S30
1926	S30	S36	S28	S25	S29	S38	S18	S23	S33	S21	S31	S12	S35	S11	S34	S13	S20	S17	S15	S14	S24	S22	S27	S26	S16	S32	S19	
1927	S19	S23	S20	S15	S22	S26	S31	S14	S28	S24	S11	S18	S25	S38	S29	S27	S32	S21	S12	S33	S30	S16	S17	S37	S34	S35	S36	S13
1928	S53	S54	S72	S57	S74	S51	S70	S60	S64	S58	S71	S55	S66	S78	S68	S59	S75	S61	S52	S72	S69	S65	S77	S73	S63	S56	S65	S67
1929	S37	S19	S23	S20	S24	S22	S26	S13	S36	S25	S34	S38	S28	S29	S21	S18	S14	S35	S11	S32	S27	S15	S16	S31	S17	S33	S30	S12
1930	S12	S13	S14	S32	S16	S17	S34	S27	S23	S15	S29	S26	S20	S22	S24	S31	S30	S33	S36	S37	S33	S34	S30	S35	S11	S21	S28	S18
1931	S18	S27	S30	S36	S35	S21	S29	S37	S14	S33	S22	S34	S32	S12	S16	S11	S19	S15	S26	S23	S12	S28	S25	S38	S24	S20	S13	S31
1932	S71	S77	S59	S63	S65	S64	S62	S52	S70	S68	S57	S73	S74	S54	S58	S60	S55	S56	S56	S72	S67	S51	S75	S78	S53	S76	S72	S51
1933	S38	S18	S27	S30	S33	S35	S21	S31	S13	S32	S24	S17	S14	S16	S15	S34	S37	S20	S22	S19	S11	S36	S28	S29	S25	S23	S12	S26
1934	S26	S31	S37	S13	S20	S25	S24	S11	S27	S36	S16	S21	S30	S35	S33	S29	S12	S32	S17	S13	S38	S23	S20	S15	S14	S18	S34	
1935	S34	S11	S12	S13	S28	S15	S16	S38	S37	S23	S35	S24	S19	S20	S25	S28	S22	S18	S36	S21	S27	S26	S14	S32	S17	S33	S30	S31
1936	S69	S78	S58	S67	S72	S73	S57	S66	S52	S54	S65	S55	S53	S55	S60	S57	S71	S54	S70	S77	S74	S59	S56	S51	S68	S59	S51	S62
1937	S17	S34	S11	S12	S23	S20	S15	S29	S31	S19	S33	S25	S37	S28	S36	S24	S38	S30	S35	S18	S22	S13	S14	S16	S32	S27	S26	S21
1938	S21	S29	S38	S18	S14	S32	S33	S22	S11	S13	S25	S28	S15	S12	S20	S23	S16	S26	S37	S31	S17	S27	S30	S35	S36	S34	S24	
1939	S24	S22	S26	S31	S30	S36	S28	S17	S38	S27	S20	S33	S18	S32	S14	S35	S34	S13	S15	S11	S21	S37	S19	S29	S23	S12	S25	S16
1940	S56	S57	S74	S51	S59	S63	S75	S66	S77	S72	S68	S71	S70	S65	S73	S67	S52	S53	S55	S54	S58	S62	S75					
1941	S25	S24	S22	S26	S27	S30	S36	S16	S29	S18	S23	S32	S38	S14	S13	S33	S17	S12	S20	S34	S35	S31	S37	S28	S19	S11	S21	S15
1942	S15	S16	S17	S34	S37	S19	S23	S35	S22	S31	S14	S36	S26	S13	S21	S20	S27	S28	S21	S18	S32	S25	S11	S12	S20	S13	S38	S24
1943	S33	S35	S21	S29	S12	S13	S14	S25	S17	S11	S30	S23	S34	S19	S37	S20	S24	S31	S36	S22	S15	S38	S18	S32	S27	S26	S16	S28
1944	S68	S65	S64	S62	S58	S67	S70	S55	S66	S78	S59	S53	S69	S53	S52	S72	S56	S51	S63	S57	S73	S76	S77	S74	S75	S71	S60	S61
1945	S32	S33	S35	S21	S11	S12	S13	S28	S16	S34	S27	S19	S17	S37	S31	S23	S25	S26	S30	S24	S20	S29	S38	S14	S18	S22	S15	S36
1946	S36	S28	S25	S24	S38	S18	S27	S20	S35	S29	S37	S13	S21	S12	S11	S14	S15	S34	S19	S16	S32	S22	S26	S30	S31	S17	S33	S23
1947	S23	S20	S15	S16	S26	S31	S37	S32	S25	S22	S12	S27	S24	S18	S38	S30	S33	S29	S13	S35	S36	S17	S34	S19	S11	S21	S28	S14
1948	S54	S72	S73	S75	S74	S51	S52	S76	S55	S57	S58	S77	S56	S71	S66	S59	S69	S53	S67	S63	S61	S65	S60	S78	S64	S60	S70	
1949	S19	S23	S20	S15	S22	S26	S31	S14	S28	S24	S11	S18	S25	S38	S29	S27	S32	S21	S12	S33	S30	S16	S17	S37	S34	S35	S36	S13
1950	S13	S14	S32	S33	S17	S34	S11	S30	S20	S16	S38	S31	S15	S26	S22	S37	S36	S24	S18	S28	S19	S35	S21	S12	S29	S25	S23	S27
1951	S27	S30	S36	S28	S21	S29	S38	S19	S32	S35	S26	S11	S33	S34	S17	S12	S23	S16	S31	S20	S13	S25	S24	S18	S22	S15	S14	S37
1952	S77	S59	S63	S60	S64	S62	S66	S65	S51	S70	S54	S75	S55	S72	S67	S55	S56	S71	S57	S53	S73	S70	S52	S74	S77	S73	S70	S52
1953	S18	S27	S30	S36	S35	S21	S29	S37	S14	S33	S22	S34	S32	S17	S16	S11	S19	S15	S26	S23	S12	S28	S25	S38	S24	S20	S13	S31
1954	S31	S37	S19	S23	S25	S24	S22	S12	S30	S28	S17	S29	S36	S21	S35	S38	S13	S33	S34	S14	S18	S20	S15	S26	S16	S32	S27	S11
1955	S11	S12	S13	S14	S15	S16	S17	S18	S19	S20	S21	S22	S23	S24	S25	S26	S27	S28	S29	S30	S31	S32	S33	S34	S35	S36	S37	S38
1956	S78	S58	S67	S73	S75	S61	S71	S53	S72	S64	S57	S55	S54	S55	S74	S77	S76	S68	S69	S51	S52	S66	S65	S63	S62	S56	S60	S70
1957	S34	S11	S12	S13	S20	S15	S16	S38	S37	S23	S35	S24	S19	S25	S28	S22	S18	S36	S21	S27	S26	S14	S32	S17	S33	S30	S31	S29
1958	S29	S38	S18	S27	S32	S33	S35	S26	S12	S14	S25	S16	S13	S15	S20	S17	S31	S23	S24	S37	S34	S30	S36	S21	S28	S19	S11	S22
1959	S22	S26	S31	S37	S36	S28	S25	S34	S18	S30	S15	S35	S27	S33	S32	S21	S11	S14	S16	S12	S29	S19	S23	S24	S20	S13	S38	S17
1960	S57	S74	S51	S52	S63	S60	S59	S71	S59	S73	S65	S77	S58	S68	S76	S64	S70	S75	S58	S62	S53	S54	S56	S72	S67	S66	S61	
1961	S24	S22	S26	S31	S30	S36	S28	S17	S38	S27	S20	S33	S18	S32	S14	S35	S34	S13	S15	S11	S21	S37	S19	S25	S23	S12	S29	S16
1962	S16	S17	S34	S11	S19	S23	S20	S21	S26	S37	S32	S28	S31	S36	S30	S25	S29	S27	S33	S38	S24	S12	S13	S15	S14	S18	S22	S35
1963	S35	S21	S29	S38	S13	S14	S32	S24	S34	S12	S36	S20	S11	S23	S19	S15	S22	S37	S28	S26	S16	S18	S27	S33	S30	S31	S17	S25
1964	S65	S64	S62	S66	S67	S70	S56	S69	S58	S63	S57	S71	S77	S78	S54	S53	S57	S51	S52	S74	S75	S70	S61	S55	S68	S73	S60	S59
1965	S33	S35	S21	S29	S12	S13	S14	S25	S17	S11	S30	S23	S34	S19	S37	S20	S24	S31	S36	S22	S15	S38	S18	S32	S27	S26	S16	S28
1966	S28	S25	S24	S22	S18	S27	S30	S15	S21	S38	S19	S14	S29	S13	S12	S32	S16	S11	S23	S17	S33	S26	S31	S36	S37	S34	S35	S20
1967	S20	S15	S16	S17	S31	S37	S19	S33	S24	S26	S13	S30	S22	S27	S18	S36	S35	S38	S14	S21	S23	S28	S34	S11	S23	S25	S29	S32
1968	S72	S73	S75	S61	S51	S52	S53	S68	S56	S74	S67	S59	S57	S77	S71	S63	S65	S58	S70	S60	S66	S78	S54	S58	S55	S64	S73	S76
1969	S23	S20	S15	S16	S26	S31	S37	S32	S25	S22	S12	S27	S24	S18	S30	S33	S29	S13	S35	S36	S17	S34	S19	S11	S21	S28	S14	S30
1970	S14	S32	S33	S35	S34	S11	S12	S36	S15	S17	S18	S37	S16	S31	S26	S19	S28	S22	S27	S25	S23	S21	S29	S13	S38	S24	S20	S30
1971	S30	S36	S28	S25	S29	S38	S18	S23	S33	S21	S31	S12	S35	S11	S34	S13	S20	S17	S15	S14	S24	S22	S27	S26	S16	S32	S19	
1972	S59	S63	S60	S55	S62	S66	S71	S54	S68	S64	S51	S58	S65	S78	S69	S67	S72	S61	S52	S73	S70	S56	S57	S74	S75	S76	S53	
1973	S27	S30	S36	S28	S21	S29	S38	S19	S32	S35	S26	S11	S33	S34	S17	S12	S23	S16	S31	S20	S13	S25	S24	S18	S22	S15	S14	S37
1974	S37	S19	S23	S20	S24	S22	S26	S13	S36	S25	S34	S38	S28	S29	S21	S18	S14	S35	S11	S32	S27	S15	S16	S31	S17	S33	S30	S12
1975	S12	S13	S14	S32	S16	S17	S34	S27	S23	S15	S29	S26	S20	S22	S24	S31	S30	S33	S36	S37	S33	S35	S11	S21	S28	S19	S18	
1976	S58	S67	S70	S76	S75	S61	S69	S77	S54	S73	S62	S74	S72	S57	S56	S51	S59	S55	S66	S63	S52	S68	S65	S64	S60	S53	S71	
1977	S11	S12	S13	S14	S15	S16	S17	S18	S19	S20	S21	S22	S23	S24	S25	S26	S27	S28	S29	S30	S31	S32	S33	S34	S35	S36	S37	S38
1978	S38	S18	S27	S30	S33	S35	S21	S31	S13	S32	S24	S17	S14	S16	S15	S34	S37	S20	S22	S19	S11	S36	S28	S29	S25	S23	S12	S26
1979	S26	S31	S37	S19	S23	S20	S15	S22	S11	S27	S18	S25	S33	S30	S13	S21	S12	S35	S17	S37	S15	S14	S24	S22	S27	S26	S16	S32
1980	S74	S51	S52	S53	S60	S55	S56	S78	S77	S63	S75	S64	S59	S65	S68	S62	S58	S76	S61	S67	S66	S54	S72	S57	S73	S70	S71	S69
1981	S22	S26	S31	S37	S36	S28	S25	S34	S18	S30	S15	S35	S27	S33	S32	S21	S11	S14	S16	S12	S29	S19	S23	S24	S20	S13	S38	S17
1982	S17	S34	S11	S12	S23	S20	S15	S29	S31	S19	S33	S25	S37	S28	S36	S24	S38	S30	S35	S18	S22	S13	S14	S16	S32	S27	S26	S21
1983	S21	S29	S38	S18	S14	S32	S11	S33	S24	S13	S28	S15	S12	S20	S23	S16	S26	S37	S31	S17	S27	S30	S35	S36	S34	S24		
1984	S64	S62	S66	S71	S70	S76	S68	S57	S78	S67	S60	S73	S58	S72	S54	S75	S74	S53	S55	S51	S61	S77	S59	S65	S63	S52	S69	S56
1985	S35	S21	S29	S38	S13	S14	S32	S24	S34	S12	S36	S20	S11	S23	S19	S15	S22	S37	S28	S16	S18	S27	S33	S30	S31	S17	S25	
1986	S25	S24	S22	S26	S27	S30	S36	S16	S29	S18	S23	S32	S38	S14	S13	S33	S17	S12	S20	S34	S35	S31	S37	S28	S19	S11	S21	S15
1987	S15	S16	S17	S34	S37	S19	S23	S35	S22	S31	S14	S36	S26	S13	S21	S20	S27	S28	S21	S18	S32	S25	S11	S12	S20	S13	S38	S24
1988	S73	S75	S61	S69	S52	S53	S54	S65	S57	S51	S70	S63	S74	S59	S77	S60	S64	S71	S76	S62	S55	S78	S58	S72	S67	S66	S56	S68
1989	S20	S15	S16	S17	S31	S37	S19	S33	S24	S26	S13	S30	S22	S27	S18	S36	S35	S38	S14	S21	S23	S28	S34	S11	S23	S25	S29	S32
1990	S32	S33	S35	S21	S11	S12	S13	S28	S16	S34	S27	S19	S17	S37	S31	S23	S25	S26	S30	S24	S20	S29	S38	S14	S18	S22	S15	S36
1991	S36	S28	S25	S24	S38	S18	S27	S20	S35	S29	S37	S13	S21	S12	S11	S14	S15	S34	S19	S16	S32	S22	S26	S30	S31	S17	S33	S23
1992	S63	S60	S55	S56	S66	S71	S77	S72	S65	S62	S52	S67	S64	S58	S78	S70	S73	S69	S53	S75	S76	S57	S74	S59	S51	S61	S68	S54
1993	S30	S36	S28	S25	S29	S38	S18	S23	S33	S21	S31	S12	S35	S11	S34	S13	S20	S17	S15	S14	S24	S22	S27	S26	S16	S32	S19	
1994	S19	S23	S20	S15	S22	S26	S31	S14	S28	S24	S11	S18	S25	S38	S29	S27	S32	S21	S12	S33	S30	S16	S17	S37	S34	S35	S36	S13
1995	S13	S14	S32	S33	S17	S34	S11	S30	S20	S16	S38	S31	S15	S26	S22	S37	S36	S24	S18	S28	S19	S35	S21	S12	S29	S25	S23	S27
1996	S67	S70	S76	S68	S61	S69	S78	S59	S72	S75	S66	S51	S73	S74	S57	S52	S63	S56	S71	S60	S53	S65	S64	S58	S62	S55	S54	S77
1997	S12	S13	S14	S32	S16	S17	S34	S27	S23	S15	S29	S26	S20	S22	S24	S31	S30	S35	S33	S37	S33	S35	S11	S21	S28	S19	S18	
1998	S18	S27	S30	S36	S35	S21	S29	S37	S14	S33	S22	S34	S32	S17	S16	S11	S19	S15	S26	S23	S12	S28	S25	S38	S24	S20	S13	S31
1999	S31	S37	S19	S23	S25	S24	S22	S12	S30	S28	S17	S29	S36	S21	S35	S38	S13	S33	S34	S14	S18	S20	S15	S26	S16	S32	S27	S11
2000	S51	S52	S53	S54	S55	S56	S57	S58	S59	S60	S61	S62	S63	S64	S65	S66	S67	S68	S69	S70	S71	S72	S73	S74	S75	S76	S77	S78
2001	S26	S31	S37	S19	S28	S25	S24	S11	S27	S36	S16	S21	S30	S35	S33	S29	S12	S32	S17	S13	S38	S23	S20	S22	S15	S14	S18	S34

INTELLECTUAL TABLES FOR INQUIRY YEARS 1800--1900

INTELLECTUAL BIRTH CODES ARE LISTED BY COLUMNS--USE TABLE IN PROPER ROW AND COLUMN TO FIND BIORHYTHM INTELLECTUAL DATA

YEAR* OF * INQY*	AB	AC	AD	AE	AF	AG	AH	AJ	AK	AL	AM	BC	BD	BE	BF	BG	BH	BJ	BK	BL	BM	BP	CD	CE	CF	CH	CJ	CK	CL	CM	CP	CR	CS
1800	M30	M31	M32	M33	M34	M35	M36	M37	M38	M39	M40	M41	M42	M43	M11	M12	M13	M14	M15	M16	M17	M18	M19	M20	M21	M22	M23	M24	M25	M26	M27	M28	M29
1801	M28	M29	M30	M31	M32	M33	M34	M35	M36	M37	M38	M39	M40	M41	M42	M43	M11	M12	M13	M14	M15	M16	M17	M18	M19	M20	M21	M22	M23	M24	M25	M26	M27
1802	M26	M27	M28	M29	M30	M31	M32	M33	M34	M35	M36	M37	M38	M39	M40	M41	M42	M43	M11	M12	M13	M14	M15	M16	M17	M18	M19	M20	M21	M22	M23	M24	M25
1803	M24	M25	M26	M27	M28	M29	M30	M31	M32	M33	M34	M35	M36	M37	M38	M39	M40	M41	M42	M43	M11	M12	M13	M14	M15	M16	M17	M18	M19	M20	M21	M22	M23
1804	M62	M63	M64	M65	M66	M67	M68	M69	M70	M71	M72	M73	M74	M75	M76	M77	M78	M79	M80	M81	M82	M83	M51	M52	M53	M54	M55	M56	M57	M58	M59	M60	M61
1805	M19	M20	M21	M22	M23	M24	M25	M26	M27	M28	M29	M30	M31	M32	M33	M34	M35	M36	M37	M38	M39	M40	M41	M42	M43	M11	M12	M13	M14	M15	M16	M17	M18
1806	M17	M18	M19	M20	M21	M22	M23	M24	M25	M26	M27	M28	M29	M30	M31	M32	M33	M34	M35	M36	M37	M38	M39	M40	M41	M42	M43	M11	M12	M13	M14	M15	M16
1807	M15	M16	M17	M18	M19	M20	M21	M22	M23	M24	M25	M26	M27	M28	M29	M30	M31	M32	M33	M34	M35	M36	M37	M38	M39	M40	M41	M42	M43	M11	M12	M13	M14
1808	M53	M54	M55	M56	M57	M58	M59	M60	M61	M62	M63	M64	M65	M66	M67	M68	M69	M70	M71	M72	M73	M74	M75	M76	M77	M78	M79	M80	M81	M82	M83	M51	M52
1809	M43	M11	M12	M13	M14	M15	M16	M17	M18	M19	M20	M21	M22	M23	M24	M25	M26	M27	M28	M29	M30	M31	M32	M33	M34	M35	M36	M37	M38	M39	M40	M41	M42
1810	M41	M42	M43	M11	M12	M13	M14	M15	M16	M17	M18	M19	M20	M21	M22	M23	M24	M25	M26	M27	M28	M29	M30	M31	M32	M33	M34	M35	M36	M37	M38	M39	M40
1811	M39	M40	M41	M42	M43	M11	M12	M13	M14	M15	M16	M17	M18	M19	M20	M21	M22	M23	M24	M25	M26	M27	M28	M29	M30	M31	M32	M33	M34	M35	M36	M37	M38
1812	M77	M78	M79	M80	M81	M82	M83	M51	M52	M53	M54	M55	M56	M57	M58	M59	M60	M61	M62	M63	M64	M65	M66	M67	M68	M69	M70	M71	M72	M73	M74	M75	M76
1813	M34	M35	M36	M37	M38	M39	M40	M41	M42	M43	M11	M12	M13	M14	M15	M16	M17	M18	M19	M20	M21	M22	M23	M24	M25	M26	M27	M28	M29	M30	M31	M32	M33
1814	M32	M33	M34	M35	M36	M37	M38	M39	M40	M41	M42	M43	M11	M12	M13	M14	M15	M16	M17	M18	M19	M20	M21	M22	M23	M24	M25	M26	M27	M28	M29	M30	M31
1815	M30	M31	M32	M33	M34	M35	M36	M37	M38	M39	M40	M41	M42	M43	M11	M12	M13	M14	M15	M16	M17	M18	M19	M20	M21	M22	M23	M24	M25	M26	M27	M28	M29
1816	M68	M69	M70	M71	M72	M73	M74	M75	M76	M77	M78	M79	M80	M81	M82	M83	M51	M52	M53	M54	M55	M56	M57	M58	M59	M60	M61	M62	M63	M64	M65	M66	M67
1817	M25	M26	M27	M28	M29	M30	M31	M32	M33	M34	M35	M36	M37	M38	M39	M40	M41	M42	M43	M11	M12	M13	M14	M15	M16	M17	M18	M19	M20	M21	M22	M23	M24
1818	M23	M24	M25	M26	M27	M28	M29	M30	M31	M32	M33	M34	M35	M36	M37	M38	M39	M40	M41	M42	M43	M11	M12	M13	M14	M15	M16	M17	M18	M19	M20	M21	M22
1819	M21	M22	M23	M24	M25	M26	M27	M28	M29	M30	M31	M32	M33	M34	M35	M36	M37	M38	M39	M40	M41	M42	M43	M11	M12	M13	M14	M15	M16	M17	M18	M19	M20
1820	M59	M60	M61	M62	M63	M64	M65	M66	M67	M68	M69	M70	M71	M72	M73	M74	M75	M76	M77	M78	M79	M80	M81	M82	M83	M51	M52	M53	M54	M55	M56	M57	M58
1821	M16	M17	M18	M19	M20	M21	M22	M23	M24	M25	M26	M27	M28	M29	M30	M31	M32	M33	M34	M35	M36	M37	M38	M39	M40	M41	M42	M43	M11	M12	M13	M14	M15
1822	M14	M15	M16	M17	M18	M19	M20	M21	M22	M23	M24	M25	M26	M27	M28	M29	M30	M31	M32	M33	M34	M35	M36	M37	M38	M39	M40	M41	M42	M43	M11	M12	M13
1823	M12	M13	M14	M15	M16	M17	M18	M19	M20	M21	M22	M23	M24	M25	M26	M27	M28	M29	M30	M31	M32	M33	M34	M35	M36	M37	M38	M39	M40	M41	M42	M43	M11
1824	M83	M51	M52	M53	M54	M55	M56	M57	M58	M59	M60	M61	M62	M63	M64	M65	M66	M67	M68	M69	M70	M71	M72	M73	M74	M75	M76	M77	M78	M79	M80	M81	M82
1825	M40	M41	M42	M43	M11	M12	M13	M14	M15	M16	M17	M18	M19	M20	M21	M22	M23	M24	M25	M26	M27	M28	M29	M30	M31	M32	M33	M34	M35	M36	M37	M38	M39
1826	M38	M39	M40	M41	M42	M43	M11	M12	M13	M14	M15	M16	M17	M18	M19	M20	M21	M22	M23	M24	M25	M26	M27	M28	M29	M30	M31	M32	M33	M34	M35	M36	M37
1827	M36	M37	M38	M39	M40	M41	M42	M43	M11	M12	M13	M14	M15	M16	M17	M18	M19	M20	M21	M22	M23	M24	M25	M26	M27	M28	M29	M30	M31	M32	M33	M34	M35
1828	M74	M75	M76	M77	M78	M79	M80	M81	M82	M83	M51	M52	M53	M54	M55	M56	M57	M58	M59	M60	M61	M62	M63	M64	M65	M66	M67	M68	M69	M70	M71	M72	M73
1829	M31	M32	M33	M34	M35	M36	M37	M38	M39	M40	M41	M42	M43	M11	M12	M13	M14	M15	M16	M17	M18	M19	M20	M21	M22	M23	M24	M25	M26	M27	M28	M29	M30
1830	M29	M30	M31	M32	M33	M34	M35	M36	M37	M38	M39	M40	M41	M42	M43	M11	M12	M13	M14	M15	M16	M17	M18	M19	M20	M21	M22	M23	M24	M25	M26	M27	M28
1831	M27	M28	M29	M30	M31	M32	M33	M34	M35	M36	M37	M38	M39	M40	M41	M42	M43	M11	M12	M13	M14	M15	M16	M17	M18	M19	M20	M21	M22	M23	M24	M25	M26
1832	M65	M66	M67	M68	M69	M70	M71	M72	M73	M74	M75	M76	M77	M78	M79	M80	M81	M82	M83	M51	M52	M53	M54	M55	M56	M57	M58	M59	M60	M61	M62	M63	M64
1833	M22	M23	M24	M25	M26	M27	M28	M29	M30	M31	M32	M33	M34	M35	M36	M37	M38	M39	M40	M41	M42	M43	M11	M12	M13	M14	M15	M16	M17	M18	M19	M20	M21
1834	M20	M21	M22	M23	M24	M25	M26	M27	M28	M29	M30	M31	M32	M33	M34	M35	M36	M37	M38	M39	M40	M41	M42	M43	M11	M12	M13	M14	M15	M16	M17	M18	M19
1835	M18	M19	M20	M21	M22	M23	M24	M25	M26	M27	M28	M29	M30	M31	M32	M33	M34	M35	M36	M37	M38	M39	M40	M41	M42	M43	M11	M12	M13	M14	M15	M16	M17
1836	M56	M57	M58	M59	M60	M61	M62	M63	M64	M65	M66	M67	M68	M69	M70	M71	M72	M73	M74	M75	M76	M77	M78	M79	M80	M81	M82	M83	M51	M52	M53	M54	M55
1837	M13	M14	M15	M16	M17	M18	M19	M20	M21	M22	M23	M24	M25	M26	M27	M28	M29	M30	M31	M32	M33	M34	M35	M36	M37	M38	M39	M40	M41	M42	M43	M11	M12
1838	M11	M12	M13	M14	M15	M16	M17	M18	M19	M20	M21	M22	M23	M24	M25	M26	M27	M28	M29	M30	M31	M32	M33	M34	M35	M36	M37	M38	M39	M40	M41	M42	M43
1839	M42	M43	M11	M12	M13	M14	M15	M16	M17	M18	M19	M20	M21	M22	M23	M24	M25	M26	M27	M28	M29	M30	M31	M32	M33	M34	M35	M36	M37	M38	M39	M40	M41
1840	M80	M81	M82	M83	M51	M52	M53	M54	M55	M56	M57	M58	M59	M60	M61	M62	M63	M64	M65	M66	M67	M68	M69	M70	M71	M72	M73	M74	M75	M76	M77	M78	M79
1841	M37	M38	M39	M40	M41	M42	M43	M11	M12	M13	M14	M15	M16	M17	M18	M19	M20	M21	M22	M23	M24	M25	M26	M27	M28	M29	M30	M31	M32	M33	M34	M35	M36
1842	M35	M36	M37	M38	M39	M40	M41	M42	M43	M11	M12	M13	M14	M15	M16	M17	M18	M19	M20	M21	M22	M23	M24	M25	M26	M27	M28	M29	M30	M31	M32	M33	M34
1843	M33	M34	M35	M36	M37	M38	M39	M40	M41	M42	M43	M11	M12	M13	M14	M15	M16	M17	M18	M19	M20	M21	M22	M23	M24	M25	M26	M27	M28	M29	M30	M31	M32
1844	M71	M72	M73	M74	M75	M76	M77	M78	M79	M80	M81	M82	M83	M51	M52	M53	M54	M55	M56	M57	M58	M59	M60	M61	M62	M63	M64	M65	M66	M67	M68	M69	M70
1845	M28	M29	M30	M31	M32	M33	M34	M35	M36	M37	M38	M39	M40	M41	M42	M43	M11	M12	M13	M14	M15	M16	M17	M18	M19	M20	M21	M22	M23	M24	M25	M26	M27
1846	M26	M27	M28	M29	M30	M31	M32	M33	M34	M35	M36	M37	M38	M39	M40	M41	M42	M43	M11	M12	M13	M14	M15	M16	M17	M18	M19	M20	M21	M22	M23	M24	M25
1847	M24	M25	M26	M27	M28	M29	M30	M31	M32	M33	M34	M35	M36	M37	M38	M39	M40	M41	M42	M43	M11	M12	M13	M14	M15	M16	M17	M18	M19	M20	M21	M22	M23
1848	M62	M63	M64	M65	M66	M67	M68	M69	M70	M71	M72	M73	M74	M75	M76	M77	M78	M79	M80	M81	M82	M83	M51	M52	M53	M54	M55	M56	M57	M58	M59	M60	M61
1849	M19	M20	M21	M22	M23	M24	M25	M26	M27	M28	M29	M30	M31	M32	M33	M34	M35	M36	M37	M38	M39	M40	M41	M42	M43	M11	M12	M13	M14	M15	M16	M17	M18
1850	M17	M18	M19	M20	M21	M22	M23	M24	M25	M26	M27	M28	M29	M30	M31	M32	M33	M34	M35	M36	M37	M38	M39	M40	M41	M42	M43	M11	M12	M13	M14	M15	M16
1851	M15	M16	M17	M18	M19	M20	M21	M22	M23	M24	M25	M26	M27	M28	M29	M30	M31	M32	M33	M34	M35	M36	M37	M38	M39	M40	M41	M42	M43	M11	M12	M13	M14
1852	M53	M54	M55	M56	M57	M58	M59	M60	M61	M62	M63	M64	M65	M66	M67	M68	M69	M70	M71	M72	M73	M74	M75	M76	M77	M78	M79	M80	M81	M82	M83	M51	M52
1853	M43	M11	M12	M13	M14	M15	M16	M17	M18	M19	M20	M21	M22	M23	M24	M25	M26	M27	M28	M29	M30	M31	M32	M33	M34	M35	M36	M37	M38	M39	M40	M41	M42
1854	M41	M42	M43	M11	M12	M13	M14	M15	M16	M17	M18	M19	M20	M21	M22	M23	M24	M25	M26	M27	M28	M29	M30	M31	M32	M33	M34	M35	M36	M37	M38	M39	M40
1855	M39	M40	M41	M42	M43	M11	M12	M13	M14	M15	M16	M17	M18	M19	M20	M21	M22	M23	M24	M25	M26	M27	M28	M29	M30	M31	M32	M33	M34	M35	M36	M37	M38
1856	M77	M78	M79	M80	M81	M82	M83	M51	M52	M53	M54	M55	M56	M57	M58	M59	M60	M61	M62	M63	M64	M65	M66	M67	M68	M69	M70	M71	M72	M73	M74	M75	M76
1857	M34	M35	M36	M37	M38	M39	M40	M41	M42	M43	M11	M12	M13	M14	M15	M16	M17	M18	M19	M20	M21	M22	M23	M24	M25	M26	M27	M28	M29	M30	M31	M32	M33
1858	M32	M33	M34	M35	M36	M37	M38	M39	M40	M41	M42	M43	M11	M12	M13	M14	M15	M16	M17	M18	M19	M20	M21	M22	M23	M24	M25	M26	M27	M28	M29	M30	M31
1859	M30	M31	M32	M33	M34	M35	M36	M37	M38	M39	M40	M41	M42	M43	M11	M12	M13	M14	M15	M16	M17	M18	M19	M20	M21	M22	M23	M24	M25	M26	M27	M28	M29
1860	M68	M69	M70	M71	M72	M73	M74	M75	M76	M77	M78	M79	M80	M81	M82	M83	M51	M52	M53	M54	M55	M56	M57	M58	M59	M60	M61	M62	M63	M64	M65	M66	M67
1861	M25	M26	M27	M28	M29	M30	M31	M32	M33	M34	M35	M36	M37	M38	M39	M40	M41	M42	M43	M11	M12	M13	M14	M15	M16	M17	M18	M19	M20	M21	M22	M23	M24
1862	M23	M24	M25	M26	M27	M28	M29	M30	M31	M32	M33	M34	M35	M36	M37	M38	M39	M40	M41	M42	M43	M11	M12	M13	M14	M15	M16	M17	M18	M19	M20	M21	M22
1863	M21	M22	M23	M24	M25	M26	M27	M28	M29	M30	M31	M32	M33	M34	M35	M36	M37	M38	M39	M40	M41	M42	M43	M11	M12	M13	M14	M15	M16	M17	M18	M19	M20
1864	M59	M60	M61	M62	M63	M64	M65	M66	M67	M68	M69	M70	M71	M72	M73	M74	M75	M76	M77	M78	M79	M80	M81	M82	M83	M51	M52	M53	M54	M55	M56	M57	M58
1865	M16	M17	M18	M19	M20	M21	M22	M23	M24	M25	M26	M27	M28	M29	M30	M31	M32	M33	M34	M35	M36	M37	M38	M39	M40	M41	M42	M43	M11	M12	M13	M14	M15
1866	M14	M15	M16	M17	M18	M19	M20	M21	M22	M23	M24	M25	M26	M27	M28	M29	M30	M31	M32	M33	M34	M35	M36	M37	M38	M39	M40	M41	M42	M43	M11	M12	M13
1867	M12	M13	M14	M15	M16	M17	M18	M19	M20	M21	M22	M23	M24	M25	M26	M27	M28	M29	M30	M31	M32	M33	M34	M35	M36	M37	M38	M39	M40	M41	M42	M43	M11
1868	M83	M51	M52	M53	M54	M55	M56	M57	M58	M59	M60	M61	M62	M63	M64	M65	M66	M67	M68	M69	M70	M71	M72	M73	M74	M75	M76	M77	M78	M79	M80	M81	M82
1869	M40	M41	M42	M43	M11	M12	M13	M14	M15	M16	M17	M18	M19	M20	M21	M22	M23	M24	M25	M26	M27	M28	M29	M30	M31	M32	M33	M34	M35	M36	M37	M38	M39
1870	M38	M39	M40	M41	M42	M43	M11	M12	M13	M14	M15	M16	M17	M18	M19	M20	M21	M22	M23	M24	M25	M26	M27	M28	M29	M30	M31	M32	M33	M34	M35	M36	M37
1871	M36	M37	M38	M39	M40	M41	M42	M43	M11	M12	M13	M14	M15	M16	M17	M18	M19	M20	M21	M22	M23	M24	M25	M26	M27	M28	M29	M30	M31	M32	M33	M34	M35
1872	M74	M75	M76	M77	M78	M79	M80	M81	M82	M83	M51	M52	M53	M54	M55	M56	M57	M58	M59	M60	M61	M62	M63	M64	M65	M66	M67	M68	M69	M70	M71	M72	M73
1873	M31	M32	M33	M34	M35	M36	M37	M38	M39	M40	M41	M42	M43	M11	M12	M13	M14	M15	M16	M17	M18	M19	M20	M21	M22	M23	M24	M25	M26	M27	M28	M29	M30
1874	M29	M30	M31	M32	M33	M34	M35	M36	M37	M38	M39	M40	M41	M42	M43	M11	M12	M13	M14	M15	M16	M17	M18	M19	M20	M21	M22	M23	M24	M25	M26	M27	M28
1875	M27	M28	M29	M30	M31	M32	M33	M34	M35	M36	M37	M38	M39	M40	M41	M42	M43	M11	M12	M13	M14	M15	M16	M17	M18	M19	M20	M21	M22	M23	M24	M25	M26
1876	M65	M66	M67	M68	M69	M70	M71	M72	M73	M74	M75	M76	M77	M78	M79	M80	M81	M82	M83	M51	M52	M53	M54	M55	M56	M57	M58	M59	M60	M61	M62	M63	M64
1877	M22	M23	M24	M25	M26	M27	M28	M29	M30	M31	M32	M33	M34	M35	M36	M37	M38	M39	M40	M41	M42	M43	M11	M12	M13	M14	M15	M16	M17	M18	M19	M20	M21
1878	M20	M21	M22	M23	M24	M25	M26	M27	M28	M29	M30	M31	M32	M33	M34	M35	M36	M37	M38	M39	M40	M41	M42	M43	M11	M12	M13	M14	M15	M16	M17	M18	M19
1879	M18	M19	M20	M21	M22	M23	M24	M25	M26	M27	M28	M29	M30	M31	M32	M33	M34	M35	M36	M37	M38	M39	M40	M41	M42	M43	M11	M12	M13	M14	M15	M16	M17
1880	M56	M57	M58	M59	M60	M61	M62	M63	M64	M65	M66	M67	M68	M69	M70	M71	M72	M73	M74	M75	M76	M77	M78	M79	M80	M81	M82	M83	M51	M52	M53	M54	M55
1881	M13	M14	M15	M16	M17	M18	M19	M20	M21	M22	M23	M24	M25	M26	M27	M28	M29	M30	M31	M32	M33	M34	M35	M36	M37	M38	M39	M40	M41	M42	M43	M11	M12
1882	M11	M12	M13	M14	M15	M16	M17	M18	M19	M20	M21	M22	M23	M24	M25	M26	M27	M28	M29	M30	M31	M32	M33	M34	M35	M36	M37	M38	M39	M40	M41	M42	M43
1883	M42	M43	M11	M12	M13	M14	M15	M16	M17	M18	M19	M20	M21	M22	M23	M24	M25	M26	M27	M28	M29	M30	M31	M32	M33	M34	M35	M36	M37	M38	M39	M40	M41
1884	M80	M81	M82	M83	M51	M52	M53	M54	M55	M56	M57	M58	M59	M60	M61	M62	M63	M64	M65	M66	M67	M68	M69	M70	M71	M72	M73	M74	M75	M76	M77	M78	M79
1885	M37	M38	M39	M40	M41	M42	M43	M11	M12	M13	M14	M15	M16	M17	M18	M19	M20	M21	M22	M23	M24	M25	M26	M27	M28	M29	M30	M31	M32	M33	M34	M35	M36
1886	M35	M36	M37	M38	M39	M40	M41	M42	M43	M11	M12	M13	M14	M15	M16	M17	M18	M19	M20	M21	M22	M23	M24	M25	M26	M27	M28	M29	M30	M31	M32	M33	M34
1887	M33	M34	M35	M36	M37	M38	M39	M40	M41	M42	M43	M11	M12	M13	M14	M15	M16	M17	M18	M19	M20	M21	M22	M23	M24	M25	M26	M27	M28	M29	M30	M31	M32
1888	M71	M72	M73	M74	M75	M76	M77	M78	M79	M80	M81	M82	M83	M51	M52	M53	M54	M55	M56	M57	M58	M59	M60	M61	M62	M63	M64	M65	M66	M67	M68	M69	M70
1889	M28	M29	M30	M31	M32	M33	M34	M35	M36	M37	M38	M39	M40	M41	M42	M43	M11	M12	M13	M14	M15	M16	M17	M18	M19	M20	M21	M22	M23	M24	M25	M26	M27
1890	M26	M27	M28	M29	M30	M31	M32	M33	M34	M35	M36	M37	M38	M39	M40	M41	M42	M43	M11	M12	M13	M14	M15	M16	M17	M18	M19	M20	M21	M22	M23	M24	M25
1891	M24	M25	M26	M27	M28	M29	M30	M31	M32	M33	M34	M35	M36	M37	M38	M39	M40	M41	M42	M43	M11	M12	M13	M14	M15	M16	M17	M18	M19	M20	M21	M22	M23
1892	M62	M63	M64	M65	M66	M67	M68	M69	M70	M71	M72	M73	M74	M75	M76	M77	M78	M79	M80	M81	M82	M83	M51	M52	M53	M54	M55	M56	M57	M58	M59	M60	M61
1893	M19	M20	M21	M22	M23	M24	M25	M26	M27	M28	M29	M30	M31	M32	M33	M34	M35	M36	M37	M38	M39	M40	M41	M42	M43	M11	M12	M13	M14	M15	M16	M17	M18
1894	M17	M18	M19	M20	M21	M22	M23	M24	M25	M26	M27	M28	M29	M30	M31	M32	M33	M34	M35	M36	M37	M38	M39	M40	M41	M42	M43	M11	M12	M13	M14	M15	M16
1895	M15	M16	M17	M18	M19	M20	M21	M22	M23	M24	M25	M26	M27	M28	M29	M30	M31	M32	M33	M34	M35	M36	M37	M38	M39	M40	M41	M42	M43	M11	M12	M13	M14
1896	M53	M54	M55	M56	M57	M58	M59	M60	M61	M62	M63	M64	M65	M66	M67	M68	M69	M70	M71	M72	M73	M74	M75	M76	M77	M78	M79	M80	M81	M82	M83	M51	M52
1897	M43	M11	M12	M13	M14	M15	M16	M17	M18	M19	M20	M21	M22	M23	M24	M25	M26	M27	M28	M29	M30	M31	M32	M33	M34	M35	M36	M37	M38	M39	M40	M41	M42
1898	M41	M42	M43	M11	M12	M13	M14	M15	M16	M17	M18	M19	M20	M21	M22	M23	M24	M25	M26	M27	M28	M29	M30	M31	M32	M33	M34	M35	M36	M37	M38	M39	M40
1899	M39	M40	M41	M42	M43	M11	M12	M13	M14	M15	M16	M17	M18	M19	M20	M21	M22	M23	M24	M25	M26	M27	M28	M29	M30	M31	M32	M33	M34	M35	M36	M37	M38
1900	M37	M38	M39	M40	M41	M42	M43	M11	M12	M13	M14	M15	M16	M17	M18	M19	M20	M21	M22	M23	M24	M25	M26	M27	M28	M29	M30	M31	M32	M33	M34	M35	M36

INTELLECTUAL TABLES FOR INQUIRY YEARS 1901--2001

INTELLECTUAL BIRTH CODES ARE LISTED BY COLUMNS--USE TABLE IN PROPER ROW AND COLUMN TO FIND BIORHYTHM INTELLECTUAL DATA

YEAR OF INQY	AB	AC	AD	AE	AF	AG	AH	AJ	AK	AL	AM	BC	BD	BE	BF	BG	BH	BJ	BK	BL	BM	BP	CD	CE	CF	CH	CJ	CK	CL	CM	CP	CR	CS
1901	M35	M36	M37	M38	M39	M40	M41	M42	M43	M11	M12	M13	M14	M15	M16	M17	M18	M19	M20	M21	M22	M23	M24	M25	M26	M27	M28	M29	M30	M31	M32	M33	M34
1902	M33	M34	M35	M36	M37	M38	M39	M40	M41	M42	M43	M11	M12	M13	M14	M15	M16	M17	M18	M19	M20	M21	M22	M23	M24	M25	M26	M27	M28	M29	M30	M31	M32
1903	M31	M32	M33	M34	M35	M36	M37	M38	M39	M40	M41	M42	M43	M11	M12	M13	M14	M15	M16	M17	M18	M19	M20	M21	M22	M23	M24	M25	M26	M27	M28	M29	M30
1904	M69	M70	M71	M72	M73	M74	M75	M76	M77	M78	M79	M80	M81	M82	M83	M51	M52	M53	M54	M55	M56	M57	M58	M59	M60	M61	M62	M63	M64	M65	M66	M67	M68
1905	M26	M27	M28	M29	M30	M31	M32	M33	M34	M35	M36	M37	M38	M39	M40	M41	M42	M43	M11	M12	M13	M14	M15	M16	M17	M18	M19	M20	M21	M22	M23	M24	M25
1906	M24	M25	M26	M27	M28	M29	M30	M31	M32	M33	M34	M35	M36	M37	M38	M39	M40	M41	M42	M43	M11	M12	M13	M14	M15	M16	M17	M18	M19	M20	M21	M22	M23
1907	M22	M23	M24	M25	M26	M27	M28	M29	M30	M31	M32	M33	M34	M35	M36	M37	M38	M39	M40	M41	M42	M43	M11	M12	M13	M14	M15	M16	M17	M18	M19	M20	M21
1908	M60	M61	M62	M63	M64	M65	M66	M67	M68	M69	M70	M71	M72	M73	M74	M75	M76	M77	M78	M79	M80	M81	M82	M83	M51	M52	M53	M54	M55	M56	M57	M58	M59
1909	M17	M18	M19	M20	M21	M22	M23	M24	M25	M26	M27	M28	M29	M30	M31	M32	M33	M34	M35	M36	M37	M38	M39	M40	M41	M42	M43	M11	M12	M13	M14	M15	M16
1910	M15	M16	M17	M18	M19	M20	M21	M22	M23	M24	M25	M26	M27	M28	M29	M30	M31	M32	M33	M34	M35	M36	M37	M38	M39	M40	M41	M42	M43	M11	M12	M13	M14
1911	M13	M14	M15	M16	M17	M18	M19	M20	M21	M22	M23	M24	M25	M26	M27	M28	M29	M30	M31	M32	M33	M34	M35	M36	M37	M38	M39	M40	M41	M42	M43	M11	M12
1912	M51	M52	M53	M54	M55	M56	M57	M58	M59	M60	M61	M62	M63	M64	M65	M66	M67	M68	M69	M70	M71	M72	M73	M74	M75	M76	M77	M78	M79	M80	M81	M82	M83
1913	M41	M42	M43	M11	M12	M13	M14	M15	M16	M17	M18	M19	M20	M21	M22	M23	M24	M25	M26	M27	M28	M29	M30	M31	M32	M33	M34	M35	M36	M37	M38	M39	M40
1914	M39	M40	M41	M42	M43	M11	M12	M13	M14	M15	M16	M17	M18	M19	M20	M21	M22	M23	M24	M25	M26	M27	M28	M29	M30	M31	M32	M33	M34	M35	M36	M37	M38
1915	M37	M38	M39	M40	M41	M42	M43	M11	M12	M13	M14	M15	M16	M17	M18	M19	M20	M21	M22	M23	M24	M25	M26	M27	M28	M29	M30	M31	M32	M33	M34	M35	M36
1916	M75	M76	M77	M78	M79	M80	M81	M82	M83	M51	M52	M53	M54	M55	M56	M57	M58	M59	M60	M61	M62	M63	M64	M65	M66	M67	M68	M69	M70	M71	M72	M73	M74
1917	M32	M33	M34	M35	M36	M37	M38	M39	M40	M41	M42	M43	M11	M12	M13	M14	M15	M16	M17	M18	M19	M20	M21	M22	M23	M24	M25	M26	M27	M28	M29	M30	M31
1918	M30	M31	M32	M33	M34	M35	M36	M37	M38	M39	M40	M41	M42	M43	M11	M12	M13	M14	M15	M16	M17	M18	M19	M20	M21	M22	M23	M24	M25	M26	M27	M28	M29
1919	M28	M29	M30	M31	M32	M33	M34	M35	M36	M37	M38	M39	M40	M41	M42	M43	M11	M12	M13	M14	M15	M16	M17	M18	M19	M20	M21	M22	M23	M24	M25	M26	M27
1920	M66	M67	M68	M69	M70	M71	M72	M73	M74	M75	M76	M77	M78	M79	M80	M81	M82	M83	M51	M52	M53	M54	M55	M56	M57	M58	M59	M60	M61	M62	M63	M64	M65
1921	M23	M24	M25	M26	M27	M28	M29	M30	M31	M32	M33	M34	M35	M36	M37	M38	M39	M40	M41	M42	M43	M11	M12	M13	M14	M15	M16	M17	M18	M19	M20	M21	M22
1922	M21	M22	M23	M24	M25	M26	M27	M28	M29	M30	M31	M32	M33	M34	M35	M36	M37	M38	M39	M40	M41	M42	M43	M11	M12	M13	M14	M15	M16	M17	M18	M19	M20
1923	M19	M20	M21	M22	M23	M24	M25	M26	M27	M28	M29	M30	M31	M32	M33	M34	M35	M36	M37	M38	M39	M40	M41	M42	M43	M11	M12	M13	M14	M15	M16	M17	M18
1924	M57	M58	M59	M60	M61	M62	M63	M64	M65	M66	M67	M68	M69	M70	M71	M72	M73	M74	M75	M76	M77	M78	M79	M80	M81	M82	M83	M51	M52	M53	M54	M55	M56
1925	M14	M15	M16	M17	M18	M19	M20	M21	M22	M23	M24	M25	M26	M27	M28	M29	M30	M31	M32	M33	M34	M35	M36	M37	M38	M39	M40	M41	M42	M43	M11	M12	M13
1926	M12	M13	M14	M15	M16	M17	M18	M19	M20	M21	M22	M23	M24	M25	M26	M27	M28	M29	M30	M31	M32	M33	M34	M35	M36	M37	M38	M39	M40	M41	M42	M43	M11
1927	M43	M11	M12	M13	M14	M15	M16	M17	M18	M19	M20	M21	M22	M23	M24	M25	M26	M27	M28	M29	M30	M31	M32	M33	M34	M35	M36	M37	M38	M39	M40	M41	M42
1928	M81	M82	M83	M51	M52	M53	M54	M55	M56	M57	M58	M59	M60	M61	M62	M63	M64	M65	M66	M67	M68	M69	M70	M71	M72	M73	M74	M75	M76	M77	M78	M79	M80
1929	M38	M39	M40	M41	M42	M43	M11	M12	M13	M14	M15	M16	M17	M18	M19	M20	M21	M22	M23	M24	M25	M26	M27	M28	M29	M30	M31	M32	M33	M34	M35	M36	M37
1930	M36	M37	M38	M39	M40	M41	M42	M43	M11	M12	M13	M14	M15	M16	M17	M18	M19	M20	M21	M22	M23	M24	M25	M26	M27	M28	M29	M30	M31	M32	M33	M34	M35
1931	M34	M35	M36	M37	M38	M39	M40	M41	M42	M43	M11	M12	M13	M14	M15	M16	M17	M18	M19	M20	M21	M22	M23	M24	M25	M26	M27	M28	M29	M30	M31	M32	M33
1932	M72	M73	M74	M75	M76	M77	M78	M79	M80	M81	M82	M83	M51	M52	M53	M54	M55	M56	M57	M58	M59	M60	M61	M62	M63	M64	M65	M66	M67	M68	M69	M70	M71
1933	M29	M30	M31	M32	M33	M34	M35	M36	M37	M38	M39	M40	M41	M42	M43	M11	M12	M13	M14	M15	M16	M17	M18	M19	M20	M21	M22	M23	M24	M25	M26	M27	M28
1934	M27	M28	M29	M30	M31	M32	M33	M34	M35	M36	M37	M38	M39	M40	M41	M42	M43	M11	M12	M13	M14	M15	M16	M17	M18	M19	M20	M21	M22	M23	M24	M25	M26
1935	M25	M26	M27	M28	M29	M30	M31	M32	M33	M34	M35	M36	M37	M38	M39	M40	M41	M42	M43	M11	M12	M13	M14	M15	M16	M17	M18	M19	M20	M21	M22	M23	M24
1936	M63	M64	M65	M66	M67	M68	M69	M70	M71	M72	M73	M74	M75	M76	M77	M78	M79	M80	M81	M82	M83	M51	M52	M53	M54	M55	M56	M57	M58	M59	M60	M61	M62
1937	M20	M21	M22	M23	M24	M25	M26	M27	M28	M29	M30	M31	M32	M33	M34	M35	M36	M37	M38	M39	M40	M41	M42	M43	M11	M12	M13	M14	M15	M16	M17	M18	M19
1938	M18	M19	M20	M21	M22	M23	M24	M25	M26	M27	M28	M29	M30	M31	M32	M33	M34	M35	M36	M37	M38	M39	M40	M41	M42	M43	M11	M12	M13	M14	M15	M16	M17
1939	M16	M17	M18	M19	M20	M21	M22	M23	M24	M25	M26	M27	M28	M29	M30	M31	M32	M33	M34	M35	M36	M37	M38	M39	M40	M41	M42	M43	M11	M12	M13	M14	M15
1940	M54	M55	M56	M57	M58	M59	M60	M61	M62	M63	M64	M65	M66	M67	M68	M69	M70	M71	M72	M73	M74	M75	M76	M77	M78	M79	M80	M81	M82	M83	M51	M52	M53
1941	M11	M12	M13	M14	M15	M16	M17	M18	M19	M20	M21	M22	M23	M24	M25	M26	M27	M28	M29	M30	M31	M32	M33	M34	M35	M36	M37	M38	M39	M40	M41	M42	M43
1942	M42	M43	M11	M12	M13	M14	M15	M16	M17	M18	M19	M20	M21	M22	M23	M24	M25	M26	M27	M28	M29	M30	M31	M32	M33	M34	M35	M36	M37	M38	M39	M40	M41
1943	M40	M41	M42	M43	M11	M12	M13	M14	M15	M16	M17	M18	M19	M20	M21	M22	M23	M24	M25	M26	M27	M28	M29	M30	M31	M32	M33	M34	M35	M36	M37	M38	M39
1944	M78	M79	M80	M81	M82	M83	M51	M52	M53	M54	M55	M56	M57	M58	M59	M60	M61	M62	M63	M64	M65	M66	M67	M68	M69	M70	M71	M72	M73	M74	M75	M76	M77
1945	M35	M36	M37	M38	M39	M40	M41	M42	M43	M11	M12	M13	M14	M15	M16	M17	M18	M19	M20	M21	M22	M23	M24	M25	M26	M27	M28	M29	M30	M31	M32	M33	M34
1946	M33	M34	M35	M36	M37	M38	M39	M40	M41	M42	M43	M11	M12	M13	M14	M15	M16	M17	M18	M19	M20	M21	M22	M23	M24	M25	M26	M27	M28	M29	M30	M31	M32
1947	M31	M32	M33	M34	M35	M36	M37	M38	M39	M40	M41	M42	M43	M11	M12	M13	M14	M15	M16	M17	M18	M19	M20	M21	M22	M23	M24	M25	M26	M27	M28	M29	M30
1948	M69	M70	M71	M72	M73	M74	M75	M76	M77	M78	M79	M80	M81	M82	M83	M51	M52	M53	M54	M55	M56	M57	M58	M59	M60	M61	M62	M63	M64	M65	M66	M67	M68
1949	M26	M27	M28	M29	M30	M31	M32	M33	M34	M35	M36	M37	M38	M39	M40	M41	M42	M43	M11	M12	M13	M14	M15	M16	M17	M18	M19	M20	M21	M22	M23	M24	M25
1950	M24	M25	M26	M27	M28	M29	M30	M31	M32	M33	M34	M35	M36	M37	M38	M39	M40	M41	M42	M43	M11	M12	M13	M14	M15	M16	M17	M18	M19	M20	M21	M22	M23
1951	M22	M23	M24	M25	M26	M27	M28	M29	M30	M31	M32	M33	M34	M35	M36	M37	M38	M39	M40	M41	M42	M43	M11	M12	M13	M14	M15	M16	M17	M18	M19	M20	M21
1952	M60	M61	M62	M63	M64	M65	M66	M67	M68	M69	M70	M71	M72	M73	M74	M75	M76	M77	M78	M79	M80	M81	M82	M83	M51	M52	M53	M54	M55	M56	M57	M58	M59
1953	M17	M18	M19	M20	M21	M22	M23	M24	M25	M26	M27	M28	M29	M30	M31	M32	M33	M34	M35	M36	M37	M38	M39	M40	M41	M42	M43	M11	M12	M13	M14	M15	M16
1954	M15	M16	M17	M18	M19	M20	M21	M22	M23	M24	M25	M26	M27	M28	M29	M30	M31	M32	M33	M34	M35	M36	M37	M38	M39	M40	M41	M42	M43	M11	M12	M13	M14
1955	M13	M14	M15	M16	M17	M18	M19	M20	M21	M22	M23	M24	M25	M26	M27	M28	M29	M30	M31	M32	M33	M34	M35	M36	M37	M38	M39	M40	M41	M42	M43	M11	M12
1956	M51	M52	M53	M54	M55	M56	M57	M58	M59	M60	M61	M62	M63	M64	M65	M66	M67	M68	M69	M70	M71	M72	M73	M74	M75	M76	M77	M78	M79	M80	M81	M82	M83
1957	M41	M42	M43	M11	M12	M13	M14	M15	M16	M17	M18	M19	M20	M21	M22	M23	M24	M25	M26	M27	M28	M29	M30	M31	M32	M33	M34	M35	M36	M37	M38	M39	M40
1958	M39	M40	M41	M42	M43	M11	M12	M13	M14	M15	M16	M17	M18	M19	M20	M21	M22	M23	M24	M25	M26	M27	M28	M29	M30	M31	M32	M33	M34	M35	M36	M37	M38
1959	M37	M38	M39	M40	M41	M42	M43	M11	M12	M13	M14	M15	M16	M17	M18	M19	M20	M21	M22	M23	M24	M25	M26	M27	M28	M29	M30	M31	M32	M33	M34	M35	M36
1960	M75	M76	M77	M78	M79	M80	M81	M82	M83	M51	M52	M53	M54	M55	M56	M57	M58	M59	M60	M61	M62	M63	M64	M65	M66	M67	M68	M69	M70	M71	M72	M73	M74
1961	M32	M33	M34	M35	M36	M37	M38	M39	M40	M41	M42	M43	M11	M12	M13	M14	M15	M16	M17	M18	M19	M20	M21	M22	M23	M24	M25	M26	M27	M28	M29	M30	M31
1962	M30	M31	M32	M33	M34	M35	M36	M37	M38	M39	M40	M41	M42	M43	M11	M12	M13	M14	M15	M16	M17	M18	M19	M20	M21	M22	M23	M24	M25	M26	M27	M28	M29
1963	M28	M29	M30	M31	M32	M33	M34	M35	M36	M37	M38	M39	M40	M41	M42	M43	M11	M12	M13	M14	M15	M16	M17	M18	M19	M20	M21	M22	M23	M24	M25	M26	M27
1964	M66	M67	M68	M69	M70	M71	M72	M73	M74	M75	M76	M77	M78	M79	M80	M81	M82	M83	M51	M52	M53	M54	M55	M56	M57	M58	M59	M60	M61	M62	M63	M64	M65
1965	M23	M24	M25	M26	M27	M28	M29	M30	M31	M32	M33	M34	M35	M36	M37	M38	M39	M40	M41	M42	M43	M11	M12	M13	M14	M15	M16	M17	M18	M19	M20	M21	M22
1966	M21	M22	M23	M24	M25	M26	M27	M28	M29	M30	M31	M32	M33	M34	M35	M36	M37	M38	M39	M40	M41	M42	M43	M11	M12	M13	M14	M15	M16	M17	M18	M19	M20
1967	M19	M20	M21	M22	M23	M24	M25	M26	M27	M28	M29	M30	M31	M32	M33	M34	M35	M36	M37	M38	M39	M40	M41	M42	M43	M11	M12	M13	M14	M15	M16	M17	M18
1968	M57	M58	M59	M60	M61	M62	M63	M64	M65	M66	M67	M68	M69	M70	M71	M72	M73	M74	M75	M76	M77	M78	M79	M80	M81	M82	M83	M51	M52	M53	M54	M55	M56
1969	M14	M15	M16	M17	M18	M19	M20	M21	M22	M23	M24	M25	M26	M27	M28	M29	M30	M31	M32	M33	M34	M35	M36	M37	M38	M39	M40	M41	M42	M43	M11	M12	M13
1970	M12	M13	M14	M15	M16	M17	M18	M19	M20	M21	M22	M23	M24	M25	M26	M27	M28	M29	M30	M31	M32	M33	M34	M35	M36	M37	M38	M39	M40	M41	M42	M43	M11
1971	M43	M11	M12	M13	M14	M15	M16	M17	M18	M19	M20	M21	M22	M23	M24	M25	M26	M27	M28	M29	M30	M31	M32	M33	M34	M35	M36	M37	M38	M39	M40	M41	M42
1972	M81	M82	M83	M51	M52	M53	M54	M55	M56	M57	M58	M59	M60	M61	M62	M63	M64	M65	M66	M67	M68	M69	M70	M71	M72	M73	M74	M75	M76	M77	M78	M79	M80
1973	M38	M39	M40	M41	M42	M43	M11	M12	M13	M14	M15	M16	M17	M18	M19	M20	M21	M22	M23	M24	M25	M26	M27	M28	M29	M30	M31	M32	M33	M34	M35	M36	M37
1974	M36	M37	M38	M39	M40	M41	M42	M43	M11	M12	M13	M14	M15	M16	M17	M18	M19	M20	M21	M22	M23	M24	M25	M26	M27	M28	M29	M30	M31	M32	M33	M34	M35
1975	M34	M35	M36	M37	M38	M39	M40	M41	M42	M43	M11	M12	M13	M14	M15	M16	M17	M18	M19	M20	M21	M22	M23	M24	M25	M26	M27	M28	M29	M30	M31	M32	M33
1976	M72	M73	M74	M75	M76	M77	M78	M79	M80	M81	M82	M83	M51	M52	M53	M54	M55	M56	M57	M58	M59	M60	M61	M62	M63	M64	M65	M66	M67	M68	M69	M70	M71
1977	M29	M30	M31	M32	M33	M34	M35	M36	M37	M38	M39	M40	M41	M42	M43	M11	M12	M13	M14	M15	M16	M17	M18	M19	M20	M21	M22	M23	M24	M25	M26	M27	M28
1978	M27	M28	M29	M30	M31	M32	M33	M34	M35	M36	M37	M38	M39	M40	M41	M42	M43	M11	M12	M13	M14	M15	M16	M17	M18	M19	M20	M21	M22	M23	M24	M25	M26
1979	M25	M26	M27	M28	M29	M30	M31	M32	M33	M34	M35	M36	M37	M38	M39	M40	M41	M42	M43	M11	M12	M13	M14	M15	M16	M17	M18	M19	M20	M21	M22	M23	M24
1980	M63	M64	M65	M66	M67	M68	M69	M70	M71	M72	M73	M74	M75	M76	M77	M78	M79	M80	M81	M82	M83	M51	M52	M53	M54	M55	M56	M57	M58	M59	M60	M61	M62
1981	M20	M21	M22	M23	M24	M25	M26	M27	M28	M29	M30	M31	M32	M33	M34	M35	M36	M37	M38	M39	M40	M41	M42	M43	M11	M12	M13	M14	M15	M16	M17	M18	M19
1982	M18	M19	M20	M21	M22	M23	M24	M25	M26	M27	M28	M29	M30	M31	M32	M33	M34	M35	M36	M37	M38	M39	M40	M41	M42	M43	M11	M12	M13	M14	M15	M16	M17
1983	M16	M17	M18	M19	M20	M21	M22	M23	M24	M25	M26	M27	M28	M29	M30	M31	M32	M33	M34	M35	M36	M37	M38	M39	M40	M41	M42	M43	M11	M12	M13	M14	M15
1984	M54	M55	M56	M57	M58	M59	M60	M61	M62	M63	M64	M65	M66	M67	M68	M69	M70	M71	M72	M73	M74	M75	M76	M77	M78	M79	M80	M81	M82	M83	M51	M52	M53
1985	M11	M12	M13	M14	M15	M16	M17	M18	M19	M20	M21	M22	M23	M24	M25	M26	M27	M28	M29	M30	M31	M32	M33	M34	M35	M36	M37	M38	M39	M40	M41	M42	M43
1986	M42	M43	M11	M12	M13	M14	M15	M16	M17	M18	M19	M20	M21	M22	M23	M24	M25	M26	M27	M28	M29	M30	M31	M32	M33	M34	M35	M36	M37	M38	M39	M40	M41
1987	M40	M41	M42	M43	M11	M12	M13	M14	M15	M16	M17	M18	M19	M20	M21	M22	M23	M24	M25	M26	M27	M28	M29	M30	M31	M32	M33	M34	M35	M36	M37	M38	M39
1988	M78	M79	M80	M81	M82	M83	M51	M52	M53	M54	M55	M56	M57	M58	M59	M60	M61	M62	M63	M64	M65	M66	M67	M68	M69	M70	M71	M72	M73	M74	M75	M76	M77
1989	M35	M36	M37	M38	M39	M40	M41	M42	M43	M11	M12	M13	M14	M15	M16	M17	M18	M19	M20	M21	M22	M23	M24	M25	M26	M27	M28	M29	M30	M31	M32	M33	M34
1990	M33	M34	M35	M36	M37	M38	M39	M40	M41	M42	M43	M11	M12	M13	M14	M15	M16	M17	M18	M19	M20	M21	M22	M23	M24	M25	M26	M27	M28	M29	M30	M31	M32
1991	M31	M32	M33	M34	M35	M36	M37	M38	M39	M40	M41	M42	M43	M11	M12	M13	M14	M15	M16	M17	M18	M19	M20	M21	M22	M23	M24	M25	M26	M27	M28	M29	M30
1992	M69	M70	M71	M72	M73	M74	M75	M76	M77	M78	M79	M80	M81	M82	M83	M51	M52	M53	M54	M55	M56	M57	M58	M59	M60	M61	M62	M63	M64	M65	M66	M67	M68
1993	M26	M27	M28	M29	M30	M31	M32	M33	M34	M35	M36	M37	M38	M39	M40	M41	M42	M43	M11	M12	M13	M14	M15	M16	M17	M18	M19	M20	M21	M22	M23	M24	M25
1994	M24	M25	M26	M27	M28	M29	M30	M31	M32	M33	M34	M35	M36	M37	M38	M39	M40	M41	M42	M43	M11	M12	M13	M14	M15	M16	M17	M18	M19	M20	M21	M22	M23
1995	M22	M23	M24	M25	M26	M27	M28	M29	M30	M31	M32	M33	M34	M35	M36	M37	M38	M39	M40	M41	M42	M43	M11	M12	M13	M14	M15	M16	M17	M18	M19	M20	M21
1996	M60	M61	M62	M63	M64	M65	M66	M67	M68	M69	M70	M71	M72	M73	M74	M75	M76	M77	M78	M79	M80	M81	M82	M83	M51	M52	M53	M54	M55	M56	M57	M58	M59
1997	M17	M18	M19	M20	M21	M22	M23	M24	M25	M26	M27	M28	M29	M30	M31	M32	M33	M34	M35	M36	M37	M38	M39	M40	M41	M42	M43	M11	M12	M13	M14	M15	M16
1998	M15	M16	M17	M18	M19	M20	M21	M22	M23	M24	M25	M26	M27	M28	M29	M30	M31	M32	M33	M34	M35	M36	M37	M38	M39	M40	M41	M42	M43	M11	M12	M13	M14
1999	M13	M14	M15	M16	M17	M18	M19	M20	M21	M22	M23	M24	M25	M26	M27	M28	M29	M30	M31	M32	M33	M34	M35	M36	M37	M38	M39	M40	M41	M42	M43	M11	M12
2000	M51	M52	M53	M54	M55	M56	M57	M58	M59	M60	M61	M62	M63	M64	M65	M66	M67	M68	M69	M70	M71	M72	M73	M74	M75	M76	M77	M78	M79	M80	M81	M82	M83
2001	M41	M42	M43	M11	M12	M13	M14	M15	M16	M17	M18	M19	M20	M21	M22	M23	M24	M25	M26	M27	M28	M29	M30	M31	M32	M33	M34	M35	M36	M37	M38	M39	M40

TABLE P11

	CRITICALS	HIGHS	LOWS
JAN	10 21	16	4 27
FEB	2 13 25	8	19
MAR	8 20 31	3 26	14
APR	12 23	18	6 29
MAY	5 16 28	11	22
JUN	8 20	3 26	14
JUL	1 13 24	19	7 30
AUG	5 16 28	11	22
SEP	8 20	3 26	14
OCT	1 13 24	19	7 30
NOV	5 16 28	11	22
DEC	9 21	4 27	15

TABLE P12

	CRITICALS	HIGHS	LOWS
JAN	8 19 31	14	2 25
FEB	11 23	6	17
MAR	6 18 29	1 24	12
APR	10 21	16	4 27
MAY	3 14 26	9	20
JUN	6 18 29	1 24	12
JUL	11 22	17	5 28
AUG	3 14 26	9	20
SEP	6 18 29	1 24	12
OCT	11 22	17	5 28
NOV	3 14 26	9	20
DEC	7 19 30	2 25	13

TABLE P13

	CRITICALS	HIGHS	LOWS
JAN	12 23	18	6 29
FEB	4 15 27	10	21
MAR	10 22	5 28	16
APR	2 14 25	20	8
MAY	7 18 30	13	1 24
JUN	10 22	5 28	16
JUL	3 15 26	21	9
AUG	7 18 30	13	1 24
SEP	10 22	5 28	16
OCT	3 15 26	21	9
NOV	7 18 30	13	1 24
DEC	11 23	6 29	17

TABLE P14

	CRITICALS	HIGHS	LOWS
JAN	3 14 26	9	20
FEB	6 18	1 24	12
MAR	1 13 24	19	7 30
APR	5 16 28	11	22
MAY	9 21	4 27	15
JUN	1 13 24	19	7 30
JUL	6 17 29	12	23
AUG	9 21	4 27	15
SEP	1 13 24	19	7 30
OCT	6 17 29	12	23
NOV	9 21	4 27	15
DEC	2 14 25	20	8 31

TABLE P15

	CRITICALS	HIGHS	LOWS
JAN	4 16 27	22	10
FEB	8 19	14	2 25
MAR	3 14 26	9	20
APR	6 18 29	1 24	12
MAY	11 22	17	5 28
JUN	3 14 26	9	20
JUL	7 19 30	2 25	13
AUG	11 22	17	5 28
SEP	3 14 26	9	20
OCT	7 19 30	2 25	13
NOV	11 22	17	5 28
DEC	4 15 27	10	21

TABLE P16

	CRITICALS	HIGHS	LOWS
JAN	2 14 25	20	8 31
FEB	6 17	12	23
MAR	1 12 24	7 30	18
APR	4 16 27	22	10
MAY	9 20	15	3 26
JUN	1 12 24	7 30	18
JUL	5 17 28	23	11
AUG	9 20	15	3 26
SEP	1 12 24	7 30	18
OCT	5 17 28	23	11
NOV	9 20	15	3 26
DEC	2 13 25	8 31	19

TABLE P17

	CRITICALS	HIGHS	LOWS
JAN	1 12 24	7 30	18
FEB	4 16 27	22	10
MAR	11 22	17	5 28
APR	3 14 26	9	20
MAY	7 19 30	2 25	13
JUN	11 22	17	5 28
JUL	4 15 27	10	21
AUG	7 19 30	2 25	13
SEP	11 22	17	5 28
OCT	4 15 27	10	21
NOV	7 19 30	2 25	13
DEC	12 23	18	6 29

TABLE P18

	CRITICALS	HIGHS	LOWS
JAN	6 18 29	24	12
FEB	10 21	16	4 27
MAR	5 16 28	11	22
APR	8 20	3 26	14
MAY	1 13 24	19	7 30
JUN	5 16 28	11	22
JUL	9 21	4 27	15
AUG	1 13 24	19	7 30
SEP	5 16 28	11	22
OCT	9 21	4 27	15
NOV	1 13 24	19	7 30
DEC	6 17 29	12	23

TABLE P19

	CRITICALS	HIGHS	LOWS
JAN	4 15 27	10	21
FEB	7 19	2 25	13
MAR	2 14 25	20	8 31
APR	6 17 29	12	23
MAY	10 22	5 28	16
JUN	2 14 25	20	8
JUL	7 18 30	13	1 24
AUG	10 22	5 28	16
SEP	2 14 25	20	8
OCT	7 18 30	13	1 24
NOV	10 22	5 28	16
DEC	3 15 26	21	9

TABLE P20

	CRITICALS	HIGHS	LOWS
JAN	1 13 24	19	7 30
FEB	5 16 28	11	22
MAR	11 23	6 29	17
APR	3 15 26	21	9
MAY	8 19 31	14	2 25
JUN	11 23	6 29	17
JUL	4 16 27	22	10
AUG	8 19 31	14	2 25
SEP	11 23	6 29	17
OCT	4 16 27	22	10
NOV	8 19	14	2 25
DEC	1 12 24	7 30	18

TABLE P21

	CRITICALS	HIGHS	LOWS
JAN	2 13 25	8 31	19
FEB	5 17 28	23	11
MAR	11 23	18	6 29
APR	4 15 27	10	21
MAY	8 20 31	3 26	14
JUN	12 23	18	6 29
JUL	5 16 28	11	22
AUG	8 20 31	3 26	14
SEP	12 23	18	6 29
OCT	5 16 28	11	22
NOV	8 20	3 26	14
DEC	1 13 24	19	7 30

TABLE P22

	CRITICALS	HIGHS	LOWS
JAN	5 17 28	23	11
FEB	9 20	15	3 26
MAR	4 15 27	10	21
APR	7 19 30	2 25	13
MAY	12 23	18	6 29
JUN	4 15 27	10	21
JUL	8 20 31	3 26	14
AUG	12 23	18	6 29
SEP	4 15 27	10	21
OCT	8 20 31	3 26	14
NOV	12 23	18	6 29
DEC	5 16 28	11	22

TABLE P23

	CRITICALS	HIGHS	LOWS
JAN	9 20	15	3 26
FEB	1 12 24	7	18
MAR	7 19 30	2 25	13
APR	11 22	17	5 28
MAY	4 15 27	10	21
JUN	7 19 30	2 25	13
JUL	12 23	18	6 29
AUG	4 15 27	10	21
SEP	7 19 30	2 25	13
OCT	12 23	18	6 29
NOV	4 15 27	10	21
DEC	8 20 31	3 26	14

TABLE P24

	CRITICALS	HIGHS	LOWS
JAN	8 20 31	2 26	14
FEB	12 23	18	6
MAR	7 18 30	13	1 24
APR	10 22	5 28	16
MAY	3 15 26	21	9
JUN	7 18 30	13	1 24
JUL	11 23	6 29	17
AUG	3 15 26	21	9
SEP	7 18 30	13	1 24
OCT	11 23	6 29	17
NOV	3 15 26	21	9
DEC	8 19 31	14	2 25

TABLE P25

	CRITICALS	HIGHS	LOWS
JAN	6 17 29	12	23
FEB	9 21	4 27	15
MAR	4 16 27	10	21
APR	8 19	14	2 25
MAY	1 12 24	7 30	18
JUN	4 16 27	10	21
JUL	9 21	4 27	15
AUG	1 12 24	7 30	18
SEP	4 16 27	10	21
OCT	9 21	4 27	15
NOV	1 12 24	7 30	18
DEC	5 17 28	23	11

TABLE P26

	CRITICALS	HIGHS	LOWS
JAN	7 19 30	1 25	13
FEB	11 22	17	5 28
MAR	6 17 29	12	23
APR	9 21	4 27	15
MAY	2 14 25	20	8 31
JUN	6 17 29	12	23
JUL	10 22	5 28	16
AUG	2 14 25	20	8 31
SEP	6 17 29	12	23
OCT	10 22	5 28	16
NOV	2 14 25	20	8
DEC	7 18 30	13	1 24

TABLE P27

	CRITICALS	HIGHS	LOWS
JAN	11 23	5 29	17
FEB	3 15 26	21	9
MAR	10 21	16	4 27
APR	2 13 25	8	19
MAY	6 18 29	1 24	12
JUN	10 21	16	4 27
JUL	3 14 26	9	20
AUG	6 18 29	1 24	12
SEP	10 21	16	4 27
OCT	3 14 26	9	20
NOV	6 18 29	1 24	12
DEC	11 22	17	5 28

TABLE P28

	CRITICALS	HIGHS	LOWS
JAN	5 16 28	11	22
FEB	8 20	3 26	14
MAR	3 15 26	21	9
APR	7 18 30	13	1 24
MAY	11 23	6 29	17
JUN	3 15 26	21	9
JUL	8 19 31	14	2 25
AUG	11 23	6 29	17
SEP	3 15 26	21	9
OCT	8 19 31	14	2 25
NOV	11 23	6 29	17
DEC	4 16 27	22	10

TABLE P29

	CRITICALS	HIGHS	LOWS
JAN	3 15 26	21	9
FEB	7 18	13	1 24
MAR	2 13 25	8 31	19
APR	5 17 28	23	11
MAY	10 21	16	4 27
JUN	2 13 25	8	19
JUL	6 18 29	1 24	12
AUG	10 21	16	4 27
SEP	2 13 25	8	19
OCT	6 18 29	1 24	12
NOV	10 21	16	4 27
DEC	3 14 26	9	20

TABLE P30

	CRITICALS	HIGHS	LOWS
JAN	9 21	3 27	15
FEB	1 13 24	19	7
MAR	8 19 31	14	2 25
APR	11 23	6 29	17
MAY	4 16 27	22	10
JUN	8 19	14	2 25
JUL	1 12 24	7 30	18
AUG	4 16 27	22	10
SEP	8 19	14	2 25
OCT	1 12 24	7 30	18
NOV	4 16 27	22	10
DEC	9 20	15	3 26

TABLE P31

	CRITICALS	HIGHS	LOWS
JAN	7 18 30	13	1 24
FEB	10 22	5 28	16
MAR	5 17 28	23	11
APR	9 20	15	3 26
MAY	2 13 25	8 31	19
JUN	5 17 28	23	11
JUL	10 21	16	4 27
AUG	2 13 25	8 31	19
SEP	5 17 28	23	11
OCT	10 21	16	4 27
NOV	2 13 25	8	19
DEC	6 18 29	1 24	12

TABLE P32

	CRITICALS	HIGHS	LOWS
JAN	10 22	4 28	16
FEB	2 14 25	20	8
MAR	9 20	15	3 26
APR	1 12 24	7 30	18
MAY	5 17 28	23	11
JUN	9 20	15	3 26
JUL	2 13 25	8 31	19
AUG	5 17 28	23	11
SEP	9 20	15	3 26
OCT	2 13 25	8 31	19
NOV	5 17 28	23	11
DEC	10 21	16	4 27

TABLE P33

	CRITICALS	HIGHS	LOWS
JAN	11 22	17	5 28
FEB	3 14 26	9	20
MAR	9 21	4 27	15
APR	1 13 24	19	7 30
MAY	6 17 29	12	23
JUN	9 21	4 27	15
JUL	2 14 25	20	8 31
AUG	6 17 29	12	23
SEP	9 21	4 27	15
OCT	2 14 25	20	8 31
NOV	6 17 29	12	23
DEC	10 22	5 28	16

TABLE P51

	CRITICALS	HIGHS	LOWS
JAN	10 21	16	4 27
FEB	2 13 25	8	19
MAR	7 19 30	2 25	13
APR	11 22	17	5 28
MAY	4 15 27	10	21
JUN	7 19 30	2 25	13
JUL	12 23	18	6 29
AUG	4 15 27	10	21
SEP	7 19 30	2 25	13
OCT	12 23	18	6 29
NOV	4 15 27	10	21
DEC	8 20 31	3 26	14

TABLE P52

	CRITICALS	HIGHS	LOWS
JAN	8 19 31	14	2 25
FEB	11 23	6 29	17
MAR	5 17 28	23	11
APR	9 20	15	3 26
MAY	2 13 25	8 31	19
JUN	5 17 28	23	11
JUL	10 21	16	4 27
AUG	2 13 25	8 31	19
SEP	5 17 28	23	11
OCT	10 21	16	4 27
NOV	2 13 25	8	19
DEC	6 18 29	1 24	12

TABLE P53

	CRITICALS	HIGHS	LOWS
JAN	12 23	18	6 29
FEB	4 15 27	10	21
MAR	9 21	4 27	15
APR	1 13 24	19	7 30
MAY	6 17 29	12	23
JUN	9 21	4 27	15
JUL	2 14 25	20	8 31
AUG	6 17 29	12	23
SEP	9 21	4 27	15
OCT	2 14 25	20	8 31
NOV	6 17 29	12	23
DEC	10 22	5 28	16

TABLE P54

	CRITICALS	HIGHS	LOWS
JAN	3 14 26	9	20
FEB	6 18 29	1 24	12
MAR	12 23	18	6 29
APR	4 15 27	10	21
MAY	8 20 31	3 26	14
JUN	12 23	18	6 29
JUL	5 16 28	11	22
AUG	8 20 31	3 26	14
SEP	12 23	18	6 29
OCT	5 16 28	11	22
NOV	8 20	3 26	14
DEC	1 13 24	19	7 30

TABLE P55

	CRITICALS	HIGHS	LOWS
JAN	4 16 27	22	10
FEB	8 19	14	2 25
MAR	2 13 25	8 31	19
APR	5 17 28	23	11
MAY	10 21	16	4 27
JUN	2 13 25	8	19
JUL	6 18 29	1 24	12
AUG	10 21	16	4 27
SEP	2 13 25	8	19
OCT	6 18 29	1 24	12
NOV	10 21	16	4 27
DEC	3 14 26	9	20

TABLE P56

	CRITICALS	HIGHS	LOWS
JAN	2 14 25	20	8 31
FEB	6 17 29	12	23
MAR	11 23	6 29	17
APR	3 15 26	21	9
MAY	8 19 31	14	2 25
JUN	11 23	6 29	17
JUL	4 16 27	22	10
AUG	8 19 31	14	2 25
SEP	11 23	6 29	17
OCT	4 16 27	22	10
NOV	8 19	14	2 25
DEC	1 12 24	7 30	18

TABLE P57

	CRITICALS	HIGHS	LOWS
JAN	1 12 24	7 30	18
FEB	4 16 27	22	10
MAR	9 21	4 27	15
APR	2 13 25	8	19
MAY	6 18 29	1 24	12
JUN	10 21	16	4 27
JUL	3 14 26	9	20
AUG	6 18 29	1 24	12
SEP	10 21	16	4 27
OCT	3 14 26	9	20
NOV	6 18 29	1 24	12
DEC	11 22	17	5 28

TABLE P58

	CRITICALS	HIGHS	LOWS
JAN	6 18 29	1 24	12
FEB	10 21	16	4 27
MAR	4 15 27	10	21
APR	7 19 30	2 25	13
MAY	12 23	18	6 29
JUN	4 15 27	10	21
JUL	8 20 31	3 26	14
AUG	12 23	18	6 29
SEP	4 15 27	10	21
OCT	8 20 31	3 26	14
NOV	12 23	18	6 29
DEC	5 16 28	11	22

TABLE P59

	CRITICALS	HIGHS	LOWS
JAN	4 15 27	10	21
FEB	7 19	2 25	13
MAR	1 13 24	19	7 30
APR	5 16 28	11	22
MAY	9 21	4 27	15
JUN	1 13 24	19	7 30
JUL	6 17 29	12	23
AUG	9 21	4 27	15
SEP	1 13 24	19	7 30
OCT	6 17 29	12	23
NOV	9 21	4 27	15
DEC	2 14 25	20	8 31

TABLE P60

	CRITICALS	HIGHS	LOWS
JAN	1 13 24	19	7 30
FEB	5 16 28	11	22
MAR	10 22	5 28	16
APR	2 14 25	20	8
MAY	7 18 30	13	1 24
JUN	10 22	5 28	16
JUL	3 15 26	21	9
AUG	7 18 30	13	1 24
SEP	10 22	5 28	16
OCT	3 15 26	21	9
NOV	7 18 30	13	1 24
DEC	11 23	6 29	17

TABLE P61

	CRITICALS	HIGHS	LOWS
JAN	2 13 25	8 31	19
FEB	5 17 28	23	11
MAR	11 22	17	5 28
APR	3 14 26	9	20
MAY	7 19 30	2 25	13
JUN	11 22	17	5 28
JUL	4 15 27	10	21
AUG	7 19 30	2 25	13
SEP	11 22	17	5 28
OCT	4 15 27	10	21
NOV	7 19 30	2 25	13
DEC	12 23	18	6 29

TABLE P62

	CRITICALS	HIGHS	LOWS
JAN	5 17 28	23	11
FEB	9 20	15	3 26
MAR	3 14 26	9	20
APR	6 18 29	1 24	12
MAY	11 22	17	5 28
JUN	3 14 26	9	20
JUL	7 19 30	2 25	13
AUG	11 22	17	5 28
SEP	3 14 26	9	20
OCT	7 19 30	2 25	13
NOV	11 22	17	5 28
DEC	4 15 27	10	21

TABLE P63

	CRITICALS	HIGHS	LOWS
JAN	9 20	15	3 26
FEB	1 12 24	7	18
MAR	6 18 29	1 24	12
APR	10 21	16	4 27
MAY	3 14 26	9	20
JUN	6 18 29	1 24	12
JUL	11 22	17	5 28
AUG	3 14 26	9	20
SEP	6 18 29	1 24	12
OCT	11 22	17	5 28
NOV	3 14 26	9	20
DEC	7 19 30	2 25	13

TABLE P64

	CRITICALS	HIGHS	LOWS
JAN	8 20 31	2 26	14
FEB	12 23	18	6 29
MAR	6 17 29	12	23
APR	9 21	4 27	15
MAY	2 14 25	20	8 31
JUN	6 17 29	12	23
JUL	10 22	5 28	16
AUG	2 14 25	20	8 31
SEP	6 17 29	12	23
OCT	10 22	5 28	16
NOV	2 14 25	20	8 31
DEC	7 18 30	13	1 24

TABLE P65

	CRITICALS	HIGHS	LOWS
JAN	6 17 29	12	23
FEB	9 21	4 27	15
MAR	3 15 26	21	9
APR	7 18 30	13	1 24
MAY	11 23	6 29	17
JUN	3 15 26	21	9
JUL	8 19 31	14	2 25
AUG	11 23	6 29	17
SEP	3 15 26	21	9
OCT	8 19 31	14	2 25
NOV	11 23	6 29	17
DEC	4 16 27	22	10

TABLE P66

	CRITICALS	HIGHS	LOWS
JAN	7 19 30	1 25	13
FEB	11 22	17	5 28
MAR	5 16 28	11	22
APR	8 20	3 26	14
MAY	1 13 24	19	7 30
JUN	5 16 28	11	22
JUL	9 21	4 27	15
AUG	1 13 24	19	7 30
SEP	5 16 28	11	22
OCT	9 21	4 27	15
NOV	1 13 24	19	7 30
DEC	6 17 29	12	23

TABLE P67

	CRITICALS	HIGHS	LOWS
JAN	11 23	6 29	17
FEB	3 15 26	21	9
MAR	9 20	15	3 26
APR	1 12 24	7 30	18
MAY	5 17 28	23	11
JUN	9 20	15	3 26
JUL	2 13 25	8 31	19
AUG	5 17 28	23	11
SEP	9 20	15	3 26
OCT	2 13 25	8 31	19
NOV	5 17 28	23	11
DEC	10 21	16	4 27

TABLE P68

	CRITICALS	HIGHS	LOWS
JAN	5 16 28	11	22
FEB	8 20	3 26	14
MAR	2 14 25	20	8 31
APR	6 17 29	12	23
MAY	10 22	5 28	16
JUN	2 14 25	20	8
JUL	7 18 30	13	1 24
AUG	10 22	5 28	16
SEP	2 14 25	20	8
OCT	7 18 30	13	1 24
NOV	10 22	5 28	16
DEC	3 15 26	21	9

TABLE P69

	CRITICALS	HIGHS	LOWS
JAN	3 15 26	21	9
FEB	7 18	13	1 24
MAR	1 12 24	7 30	18
APR	4 16 27	22	10
MAY	9 20	15	3 26
JUN	1 12 24	7 30	18
JUL	5 17 28	23	11
AUG	9 20	15	3 26
SEP	1 12 24	7 30	18
OCT	5 17 28	23	11
NOV	9 20	15	3 26
DEC	2 13 25	8 31	19

TABLE P70

	CRITICALS	HIGHS	LOWS
JAN	9 21	3 27	15
FEB	1 13 24	19	7
MAR	7 18 30	13	1 24
APR	10 22	5 28	16
MAY	3 15 26	21	9
JUN	7 18 30	13	1 24
JUL	11 23	6 29	17
AUG	3 15 26	21	9
SEP	7 18 30	13	1 24
OCT	11 23	6 29	17
NOV	3 15 26	21	9
DEC	8 19 31	14	2 25

TABLE P71

	CRITICALS	HIGHS	LOWS
JAN	7 18 30	13	1 24
FEB	10 22	5 28	16
MAR	4 16 27	22	10
APR	8 19	14	2 25
MAY	1 12 24	7 30	18
JUN	4 16 27	22	10
JUL	9 20	15	3 26
AUG	1 12 24	7 30	18
SEP	4 16 27	22	10
OCT	9 20	15	3 26
NOV	1 12 24	7 30	18
DEC	5 17 28	23	11

TABLE P72

	CRITICALS	HIGHS	LOWS
JAN	10 22	4 28	16
FEB	2 14 26	20	8
MAR	8 19 31	14	2 25
APR	11 23	6 29	17
MAY	4 16 27	22	10
JUN	8 19	14	2 25
JUL	1 12 24	7 30	18
AUG	4 16 27	22	10
SEP	8 19	14	2 25
OCT	1 12 24	7 30	18
NOV	4 16 27	22	10
DEC	9 20	15	3 26

TABLE P73

	CRITICALS	HIGHS	LOWS
JAN	11 22	17	5 28
FEB	3 14 26	9	20
MAR	8 20 31	3 26	14
APR	12 23	18	6 29
MAY	5 16 28	11	22
JUN	8 20	3 26	14
JUL	1 13 24	19	7 30
AUG	5 16 28	11	22
SEP	8 20	3 26	14
OCT	1 13 24	19	7 30
NOV	5 16 28	11	22
DEC	9 21	4 27	15

TABLE S11

	CRITICALS	HIGHS	LOWS
JAN	6 20	13	27
FEB	3 17	10	24
MAR	3 17 31	10	24
APR	14 28	7	21
MAY	12 26	5	19
JUN	9 23	2 30	16
JUL	7 21	28	14
AUG	4 18	25	11
SEP	1 15 29	22	8
OCT	13 27	20	6
NOV	10 24	17	3
DEC	8 22	15	1 29

TABLE S12

	CRITICALS	HIGHS	LOWS
JAN	9 23	16	2 30
FEB	6 20	13	27
MAR	6 20	13	27
APR	3 17	10	24
MAY	1 15 29	8	22
JUN	12 26	5	19
JUL	10 24	3 31	17
AUG	7 21	28	14
SEP	4 18	25	11
OCT	2 16 30	23	9
NOV	13 27	20	6
DEC	11 25	18	4

TABLE S13

	CRITICALS	HIGHS	LOWS
JAN	12 26	19	5
FEB	9 23	16	2
MAR	9 23	16	2 30
APR	6 20	13	27
MAY	4 18	11	25
JUN	1 15 29	8	22
JUL	13 27	6	20
AUG	10 24	3 31	17
SEP	7 21	28	14
OCT	5 19	26	12
NOV	2 16 30	23	9
DEC	14 28	21	7

TABLE S14

	CRITICALS	HIGHS	LOWS
JAN	1 15 29	22	8
FEB	12 26	19	5
MAR	12 26	19	5
APR	9 23	16	2 30
MAY	7 21	14	28
JUN	4 18	11	25
JUL	2 16 30	9	23
AUG	13 27	6	20
SEP	10 24	3	17
OCT	8 22	1 29	15
NOV	5 19	26	12
DEC	3 17 31	24	10

TABLE S15

	CRITICALS	HIGHS	LOWS
JAN	8 22	1 29	15
FEB	5 19	26	12
MAR	5 19	26	12
APR	2 16 30	23	9
MAY	14 28	21	7
JUN	11 25	18	4
JUL	9 23	16	2 30
AUG	6 20	13	27
SEP	3 17	10	24
OCT	1 15 29	8	22
NOV	12 26	5	19
DEC	10 24	3 31	17

TABLE S16

	CRITICALS	HIGHS	LOWS
JAN	11 25	4	18
FEB	8 22	1	15
MAR	8 22	1 29	15
APR	5 19	26	12
MAY	3 17 31	24	10
JUN	14 28	21	7
JUL	12 26	19	5
AUG	9 23	16	2 30
SEP	6 20	13	27
OCT	4 18	11	25
NOV	1 15 29	8	22
DEC	13 27	6	20

TABLE S17

	CRITICALS	HIGHS	LOWS
JAN	14 28	7	21
FEB	11 25	4	18
MAR	11 25	4	18
APR	8 22	1 29	15
MAY	6 20	27	13
JUN	3 17	24	10
JUL	1 15 29	22	8
AUG	12 26	19	5
SEP	9 23	16	2 30
OCT	7 21	14	28
NOV	4 18	11	25
DEC	2 16 30	9	23

TABLE S18

	CRITICALS	HIGHS	LOWS
JAN	8 22	15	1 29
FEB	5 19	12	26
MAR	5 19	12	26
APR	2 16 30	9	23
MAY	14 28	7	21
JUN	11 25	4	18
JUL	9 23	2 30	16
AUG	6 20	27	13
SEP	3 17	24	10
OCT	1 15 29	22	8
NOV	12 26	19	5
DEC	10 24	17	3 31

TABLE S19

	CRITICALS	HIGHS	LOWS
JAN	13 27	20	6
FEB	10 24	17	3
MAR	10 24	17	3 31
APR	7 21	14	28
MAY	5 19	12	26
JUN	2 16 30	9	23
JUL	14 28	7	21
AUG	11 25	4	18
SEP	8 22	1 29	15
OCT	6 20	27	13
NOV	3 17	24	10
DEC	1 15 29	22	8

TABLE S20

	CRITICALS	HIGHS	LOWS
JAN	5 19	26	12
FEB	2 16	23	9
MAR	2 16 30	23	9
APR	13 27	20	6
MAY	11 25	18	4
JUN	8 22	15	1 29
JUL	6 20	13	27
AUG	3 17 31	10	24
SEP	14 28	7	21
OCT	12 26	5	19
NOV	9 23	2 30	16
DEC	7 21	28	14

TABLE S21

	CRITICALS	HIGHS	LOWS
JAN	13 27	6	20
FEB	10 24	3	17
MAR	10 24	3 31	17
APR	7 21	28	14
MAY	5 19	26	12
JUN	2 16 30	23	9
JUL	14 28	21	7
AUG	11 25	18	4
SEP	8 22	15	1 29
OCT	6 20	13	27
NOV	3 17	10	24
DEC	1 15 29	8	22

TABLE S22

	CRITICALS	HIGHS	LOWS
JAN	1 15 29	8	22
FEB	12 26	5	19
MAR	12 26	5	19
APR	9 23	2 30	16
MAY	7 21	28	14
JUN	4 18	25	11
JUL	2 16 30	23	9
AUG	13 27	20	6
SEP	10 24	17	3
OCT	8 22	15	1 29
NOV	5 19	12	26
DEC	3 17 31	10	24

TABLE S23

	CRITICALS	HIGHS	LOWS
JAN	2 16 30	23	9
FEB	13 27	20	6
MAR	13 27	20	6
APR	10 24	17	3
MAY	8 22	15	1 29
JUN	5 19	12	26
JUL	3 17 31	10	24
AUG	14 28	7	21
SEP	11 25	4	18
OCT	9 23	2 30	16
NOV	6 20	27	13
DEC	4 18	25	11

TABLE S24

	CRITICALS	HIGHS	LOWS
JAN	12 26	5	19
FEB	9 23	2	16
MAR	9 23	2 30	16
APR	6 20	27	13
MAY	4 18	25	11
JUN	1 15 29	22	8
JUL	13 27	20	6
AUG	10 24	17	3 31
SEP	7 21	14	28
OCT	5 19	12	26
NOV	2 16 30	9	23
DEC	14 28	7	21

TABLE S25

	CRITICALS	HIGHS	LOWS
JAN	9 23	2 30	16
FEB	6 20	27	13
MAR	6 20	27	13
APR	3 17	24	10
MAY	1 15 29	22	8
JUN	12 26	19	5
JUL	10 24	17	3 31
AUG	7 21	14	28
SEP	4 18	11	25
OCT	2 16 30	9	23
NOV	13 27	6	20
DEC	11 25	4	18

TABLE S26

	CRITICALS	HIGHS	LOWS
JAN	4 18	11	25
FEB	1 15	8	22
MAR	1 15 29	8	22
APR	12 26	5	19
MAY	10 24	3 31	17
JUN	7 21	28	14
JUL	5 19	26	12
AUG	2 16 30	23	9
SEP	13 27	20	6
OCT	11 25	18	4
NOV	8 22	15	1 29
DEC	6 20	13	27

TABLE S27

	CRITICALS	HIGHS	LOWS
JAN	11 25	18	4
FEB	8 22	15	1
MAR	8 22	15	1 29
APR	5 19	12	26
MAY	3 17 31	10	24
JUN	14 28	7	21
JUL	12 26	5	19
AUG	9 23	2 30	16
SEP	6 20	27	13
OCT	4 18	25	11
NOV	1 15 29	22	8
DEC	13 27	20	6

TABLE S28

	CRITICALS	HIGHS	LOWS
JAN	6 20	27	13
FEB	3 17	24	10
MAR	3 17 31	24	10
APR	14 28	21	7
MAY	12 26	19	5
JUN	9 23	16	2 30
JUL	7 21	14	28
AUG	4 18	11	25
SEP	1 15 29	8	22
OCT	13 27	6	20
NOV	10 24	3	17
DEC	8 22	1 29	15

TABLE S29

	CRITICALS	HIGHS	LOWS
JAN	2 16 30	9	23
FEB	13 27	6	20
MAR	13 27	6	20.
APR	10 24	3	17
MAY	8 22	1 29	15
JUN	5 19	26	12
JUL	3 17 31	24	10
AUG	14 28	21	7
SEP	11 25	18	4
OCT	9 23	16	2 30
NOV	6 20	13	27
DEC	4 18	11	25

TABLE S30

	CRITICALS	HIGHS	LOWS
JAN	14 28	21	7
FEB	11 25	18	4
MAR	11 25	18	4
APR	8 22	15	1 29
MAY	6 20	13	27
JUN	3 17	10	24
JUL	1 15 29	8	22
AUG	12 26	5	19
SEP	9 23	2 30	16
OCT	7 21	28	14
NOV	4 18	25	11
DEC	2 16 30	23	9

TABLE S31

	CRITICALS	HIGHS	LOWS
JAN	7 21	14	28
FEB	4 18	11	25
MAR	4 18	11	25
APR	1 15 29	8	22
MAY	13 27	6	20
JUN	10 24	3	17
JUL	8 22	1 29	15
AUG	5 19	26	12
SEP	2 16 30	23	9
OCT	14 28	21	7
NOV	11 25	18	4
DEC	9 23	16	2 30

TABLE S32

	CRITICALS	HIGHS	LOWS
JAN	4 18	25	11
FEB	1 15	22	8
MAR	1 15 29	22	8
APR	12 26	19	5
MAY	10 24	17	3 31
JUN	7 21	14	28
JUL	5 19	12	26
AUG	2 16 30	9	23
SEP	14 28	7	21
OCT	11 25	4	18
NOV	8 22	1 29	15
DEC	6 20	27	13

TABLE S33

	CRITICALS	HIGHS	LOWS
JAN	7 21	28	14
FEB	4 18	25	11
MAR	4 18	25	11
APR	1 15 29	22	8
MAY	13 27	20	6
JUN	10 24	17	3
JUL	8 22	15	1 29
AUG	5 19	12	26
SEP	2 16 30	9	23
OCT	14 28	7	21
NOV	11 25	4	18
DEC	9 23	2 30	16

TABLE S34

	CRITICALS	HIGHS	LOWS
JAN	3 17 31	10	24
FEB	14 28	7	21
MAR	14 28	7	21
APR	11 25	4	18
MAY	9 23	2 30	16
JUN	6 20	27	13
JUL	4 18	25	11
AUG	1 15 29	22	8
SEP	12 26	19	5
OCT	10 24	17	3 31
NOV	7 21	14	28
DEC	5 19	12	26

TABLE S35

	CRITICALS	HIGHS	LOWS
JAN	10 24	3 31	17
FEB	7 21	28	14
MAR	7 21	28	14
APR	4 18	25	11
MAY	2 16 30	23	9
JUN	13 27	20	6
JUL	11 25	18	4
AUG	8 22	15	1 29
SEP	5 19	12	26
OCT	3 17 31	10	24
NOV	14 28	7	21
DEC	12 26	5	19

TABLE S36

	CRITICALS	HIGHS	LOWS
JAN	3 17 31	24	10
FEB	14 28	21	7
MAR	14 28	21	7
APR	11 25	18	4
MAY	9 23	16	2 30
JUN	6 20	13	27
JUL	4 18	11	25
AUG	1 15 29	8	22
SEP	12 26	5	19
OCT	10 24	3 31	17
NOV	7 21	28	14
DEC	5 19	26	12

TABLE S37

	CRITICALS	HIGHS	LOWS
JAN	10 24	17	3 31
FEB	7 21	14	28
MAR	7 21	14	28
APR	4 18	11	25
MAY	2 16 30	9	23
JUN	13 27	6	20
JUL	11 25	4	18
AUG	8 22	1 29	15
SEP	5 19	26	12
OCT	3 17 31	24	10
NOV	14 28	21	7
DEC	12 26	19	5

TABLE S38

	CRITICALS	HIGHS	LOWS
JAN	5 19	12	26
FEB	2 16	9	23
MAR	2 16 30	9	23
APR	13 27	6	20
MAY	11 25	4	18
JUN	8 22	1 29	15
JUL	6 20	27	13
AUG	3 17 31	24	10
SEP	14 28	21	7
OCT	12 26	19	5
NOV	9 23	16	2 30
DEC	7 21	14	28

TABLE S51

	CRITICALS	HIGHS	LOWS
JAN	6 20	13	27
FEB	3 17	10	24
MAR	2 16 30	9	23
APR	13 27	6	20
MAY	11 25	4	18
JUN	8 22	1 29	15
JUL	6 20	27	13
AUG	3 17 31	24	10
SEP	14 28	21	7
OCT	12 26	19	5
NOV	9 23	16	2 30
DEC	7 21	14	28

TABLE S52

	CRITICALS	HIGHS	LOWS
JAN	9 23	16	2 30
FEB	6 20	13	27
MAR	5 19	12	26
APR	2 16 30	9	23
MAY	14 28	7	21
JUN	11 25	4	18
JUL	9 23	2 30	16
AUG	6 20	27	13
SEP	3 17	24	10
OCT	1 15 29	22	8
NOV	12 26	19	5
DEC	10 24	17	3 31

TABLE S53

	CRITICALS	HIGHS	LOWS
JAN	12 26	19	5
FEB	9 23	16	2
MAR	8 22	15	1 29
APR	5 19	12	26
MAY	3 17 31	10	24
JUN	14 28	7	21
JUL	12 26	5	19
AUG	9 23	2 30	16
SEP	6 20	27	13
OCT	4 18	25	11
NOV	1 15 29	22	8
DEC	13 27	20	6

TABLE S54

	CRITICALS	HIGHS	LOWS
JAN	1 15 29	22	8
FEB	12 26	19	5
MAR	11 25	18	4
APR	8 22	15	1 29
MAY	6 20	13	27
JUN	3 17	10	24
JUL	1 15 29	8	22
AUG	12 26	5	19
SEP	9 23	2 30	16
OCT	7 21	28	14
NOV	4 18	25	11
DEC	2 16 30	23	9

TABLE S55

	CRITICALS	HIGHS	LOWS
JAN	8 22	1 29	15
FEB	5 19	26	12
MAR	4 18	25	11
APR	1 15 29	22	8
MAY	13 27	20	6
JUN	10 24	17	3
JUL	8 22	15	1 29
AUG	5 19	12	26
SEP	2 16 30	9	23
OCT	7 21	28	14
NOV	11 25	4	18
DEC	9 23	2 30	16

TABLE S56

	CRITICALS	HIGHS	LOWS
JAN	11 25	4	18
FEB	8 22	1 29	15
MAR	7 21	28	14
APR	4 18	25	11
MAY	2 16 30	23	9
JUN	13 27	20	6
JUL	11 25	18	4
AUG	8 22	15	1 29
SEP	5 19	12	26
OCT	3 17 31	10	24
NOV	14 28	7	21
DEC	12 26	5	19

TABLE S57

	CRITICALS	HIGHS	LOWS
JAN	14 28	7	21
FEB	11 25	4	18
MAR	10 24	3 31	17
APR	7 21	28	14
MAY	5 19	26	12
JUN	2 16 30	23	9
JUL	14 28	21	7
AUG	11 25	18	4
SEP	8 22	15	1 29
OCT	6 20	13	27
NOV	3 17	10	24
DEC	1 15 29	8	22

TABLE S58

	CRITICALS	HIGHS	LOWS
JAN	8 22	15	1 29
FEB	5 19	12	26
MAR	4 18	11	25
APR	1 15 29	8	22
MAY	13 27	6	20
JUN	10 24	3	17
JUL	8 22	1 29	15
AUG	5 19	26	12
SEP	2 16 30	23	9
OCT	14 28	21	7
NOV	11 25	18	4
DEC	9 23	16	2 30

TABLE S59

	CRITICALS	HIGHS	LOWS
JAN	13 27	20	6
FEB	10 24	17	3
MAR	9 23	16	2 30
APR	6 20	13	27
MAY	4 18	11	25
JUN	1 15 29	8	22
JUL	13 27	6	20
AUG	10 24	3 31	17
SEP	7 21	28	14
OCT	5 19	26	12
NOV	2 16 30	23	9
DEC	14 28	21	7

TABLE S60

	CRITICALS	HIGHS	LOWS
JAN	5 19	26	12
FEB	2 16	23	9
MAR	1 15 29	22	8
APR	12 26	19	5
MAY	10 24	17	3 31
JUN	7 21	14	28
JUL	5 19	12	26
AUG	2 16 30	9	23
SEP	13 27	6	20
OCT	11 25	4	18
NOV	8 22	1 29	15
DEC	6 20	27	13

TABLE S61

	CRITICALS	HIGHS	LOWS
JAN	13 27	6	20
FEB	10 24	3	17
MAR	9 23	2 30	16
APR	6 20	27	13
MAY	4 18	25	11
JUN	1 15 29	22	8
JUL	13 27	20	6
AUG	10 24	17	3 31
SEP	7 21	14	28
OCT	5 19	12	26
NOV	2 16 30	9	23
DEC	14 28	7	21

TABLE S62

	CRITICALS	HIGHS	LOWS
JAN	1 15 29	8	22
FEB	12 26	5	19
MAR	11 25	4	18
APR	8 22	1 29	15
MAY	6 20	27	13
JUN	3 17	24	10
JUL	1 15 29	22	8
AUG	12 26	19	5
SEP	9 23	16	2 30
OCT	7 21	14	28
NOV	4 18	11	25
DEC	2 16 30	9	23

TABLE S63

	CRITICALS	HIGHS	LOWS
JAN	2 16 30	23	9
FEB	13 27	20	6
MAR	12 26	19	5
APR	9 23	16	2 30
MAY	7 21	14	28
JUN	4 18	11	25
JUL	2 16 30	9	23
AUG	13 27	6	20
SEP	10 24	3	17
OCT	8 22	1 29	15
NOV	5 19	26	12
DEC	3 17 31	24	10

TABLE S64

	CRITICALS	HIGHS	LOWS
JAN	12 26	5	19
FEB	9 23	2	16
MAR	8 22	1 29	15
APR	5 19	26	12
MAY	3 17 31	24	10
JUN	14 28	21	7
JUL	12 26	19	5
AUG	9 23	16	2 30
SEP	6 20	13	27
OCT	4 18	11	25
NOV	1 15 29	8	22
DEC	13 27	6	20

TABLE S65

	CRITICALS	HIGHS	LOWS
JAN	9 23	2 30	16
FEB	6 20	27	13
MAR	5 19	26	12
APR	2 16 30	23	9
MAY	14 28	21	7
JUN	11 25	18	4
JUL	9 23	16	2 30
AUG	6 20	13	27
SEP	3 17	10	24
OCT	1 15 29	8	22
NOV	12 26	5	19
DEC	10 24	3 31	17

TABLE S66

	CRITICALS	HIGHS	LOWS
JAN	4 18	11	25
FEB	1 15 29	8	22
MAR	14 28	7	21
APR	11 25	4	18
MAY	9 23	2 30	16
JUN	6 20	27	13
JUL	4 18	25	11
AUG	1 15 29	22	8
SEP	12 26	19	5
OCT	10 24	17	3 31
NOV	7 21	14	28
DEC	5 19	12	26

TABLE S67

	CRITICALS	HIGHS	LOWS
JAN	11 25	18	4
FEB	8 22	15	1 29
MAR	7 21	14	28
APR	4 18	11	25
MAY	2 16 30	9	23
JUN	13 27	6	20
JUL	11 25	4	18
AUG	8 22	1 29	15
SEP	5 19	26	12
OCT	3 17 31	24	10
NOV	14 28	21	7
DEC	12 26	19	5

TABLE S68

	CRITICALS	HIGHS	LOWS
JAN	6 20	27	13
FEB	3 17	24	10
MAR	2 16 30	23	9
APR	13 27	20	6
MAY	11 25	18	4
JUN	8 22	15	1 29
JUL	6 20	13	27
AUG	3 17 31	10	24
SEP	14 28	7	21
OCT	12 26	5	19
NOV	9 23	2 30	16
DEC	7 21	28	14

TABLE S69

	CRITICALS	HIGHS	LOWS
JAN	2 16 30	9	23
FEB	13 27	6	20
MAR	12 26	5	19
APR	9 23	2 30	16
MAY	7 21	28	14
JUN	4 18	25	11
JUL	2 16 30	23	9
AUG	13 27	20	6
SEP	10 24	17	3
OCT	8 22	15	1 29
NOV	5 19	12	26
DEC	3 17 31	10	24

TABLE S70

	CRITICALS	HIGHS	LOWS
JAN	14 28	21	7
FEB	11 25	18	4
MAR	10 24	17	3 31
APR	7 21	14	28
MAY	5 19	12	26
JUN	2 16 30	9	23
JUL	14 28	7	21
AUG	11 25	4	18
SEP	8 22	1 29	15
OCT	6 20	27	13
NOV	3 17	24	10
DEC	1 15 29	22	8

TABLE S71

	CRITICALS	HIGHS	LOWS
JAN	7 21	14	28
FEB	4 18	11	25
MAR	3 17 31	10	24
APR	14 28	7	21
MAY	12 26	5	19
JUN	9 23	2 30	16
JUL	7 21	28	14
AUG	4 18	25	11
SEP	1 15 29	22	8
OCT	13 27	20	6
NOV	10 24	17	3
DEC	8 22	15	1 29

TABLE S72

	CRITICALS	HIGHS	LOWS
JAN	4 18	25	11
FEB	1 15 29	22	8
MAR	14 28	21	7
APR	11 25	18	4
MAY	9 23	16	2 30
JUN	6 20	13	27
JUL	4 18	11	25
AUG	1 15 29	8	22
SEP	12 26	5	19
OCT	10 24	3 31	17
NOV	7 21	28	14
DEC	5 19	26	12

TABLE S73

	CRITICALS	HIGHS	LOWS
JAN	7 21	28	14
FEB	4 18	25	11
MAR	3 17 31	24	10
APR	14 28	21	7
MAY	12 26	19	5
JUN	9 23	16	2 30
JUL	7 21	14	28
AUG	4 18	11	25
SEP	1 15 29	8	22
OCT	13 27	6	20
NOV	10 24	3	17
DEC	8 22	1 29	15

TABLE S74

	CRITICALS	HIGHS	LOWS
JAN	3 17 31	10	24
FEB	14 28	7	21
MAR	13 27	6	20
APR	10 24	3	17
MAY	8 22	1 29	15
JUN	5 19	26	12
JUL	3 17 31	24	10
AUG	14 28	21	7
SEP	11 25	18	4
OCT	9 23	16	2 30
NOV	6 20	13	27
DEC	4 18	11	25

TABLE S75

	CRITICALS	HIGHS	LOWS
JAN	10 24	3 31	17
FEB	7 21	28	14
MAR	6 20	27	13
APR	3 17	24	10
MAY	1 15 29	22	8
JUN	12 26	19	5
JUL	10 24	17	3 31
AUG	7 21	14	28
SEP	4 18	11	25
OCT	2 16 30	9	23
NOV	13 27	6	20
DEC	11 25	4	18

TABLE S76

	CRITICALS	HIGHS	LOWS
JAN	3 17 31	24	10
FEB	14 28	21	7
MAR	13 27	20	6
APR	10 24	17	3
MAY	8 22	15	1 29
JUN	5 19	12	26
JUL	3 17 31	10	24
AUG	14 28	7	21
SEP	11 25	4	18
OCT	9 23	2 30	16
NOV	6 20	27	13
DEC	4 18	25	11

TABLE S77

	CRITICALS	HIGHS	LOWS
JAN	10 24	17	3 31
FEB	7 21	14	28
MAR	6 20	13	27
APR	3 17	10	24
MAY	1 15 29	8	22
JUN	12 26	5	19
JUL	10 24	3 31	17
AUG	7 21	28	14
SEP	4 18	25	11
OCT	2 16 30	23	9
NOV	13 27	20	6
DEC	11 25	18	4

TABLE S78

	CRITICALS	HIGHS	LOWS
JAN	5 19	12	26
FEB	2 16	9	23
MAR	1 15 29	8	22
APR	12 26	5	19
MAY	10 24	3 31	17
JUN	7 21	28	14
JUL	5 19	26	12
AUG	2 16 30	23	9
SEP	13 27	20	6
OCT	11 25	18	4
NOV	8 22	15	1 29
DEC	6 20	13	27

TABLE M11

	CRITICALS	HIGHS	LOWS
JAN	1 17	9	25
FEB	3 19	11	27
MAR	8 24	16	
APR	10 26	18	1
MAY	13 29	21	4
JUN	15	23	6
JUL	1 18	26	9
AUG	3 20	28	11
SEP	5 22	30	13
OCT	8 25		16
NOV	10 27	2	18
DEC	13 30	5	21

TABLE M12

	CRITICALS	HIGHS	LOWS
JAN	2 18	10	26
FEB	4 20	12	28
MAR	9 25	17	
APR	11 27	19	2
MAY	14 30	22	5
JUN	16	24	7
JUL	2 19	27	10
AUG	4 21	29	12
SEP	6 23		14
OCT	9 26	1	17
NOV	11 28	3	19
DEC	14 31	6	22

TABLE M13

	CRITICALS	HIGHS	LOWS
JAN	3 19	11	27
FEB	5 21	13	
MAR	10 26	18	1
APR	12 28	20	3
MAY	15 31	23	6
JUN	17	25	8
JUL	3 20	28	11
AUG	4 21	30	13
SEP	7 24		15
OCT	10 27	2	18
NOV	12 29	4	20
DEC	15	7	23

TABLE M14

	CRITICALS	HIGHS	LOWS
JAN	4 20	12	28
FEB	6 22	14	
MAR	11 27	19	2
APR	13 29	21	4
MAY	16	24	7
JUN	1 18	26	9
JUL	4 21	29	12
AUG	6 23	31	14
SEP	8 25		16
OCT	11 28	3	19
NOV	13 30	5	21
DEC	16	8	24

TABLE M15

	CRITICALS	HIGHS	LOWS
JAN	5 21	13	29
FEB	7 23	15	
MAR	12 28	20	3
APR	14 30	22	5
MAY	17	25	8
JUN	2 19	27	10
JUL	5 22	30	13
AUG	7 24		15
SEP	9 26	1	17
OCT	12 29	4	20
NOV	14	6	22
DEC	1 17	9	25

TABLE M16

	CRITICALS	HIGHS	LOWS
JAN	6 22	14	30
FEB	8 24	16	
MAR	13 29	21	4
APR	15	23	6
MAY	1 18	26	9
JUN	3 20	28	11
JUL	6 23	31	14
AUG	8 25		16
SEP	10 27	2	18
OCT	13 30	5	21
NOV	15	7	23
DEC	2 18	10	26

TABLE M17

	CRITICALS	HIGHS	LOWS
JAN	7 23	15	31
FEB	9 25	17	
MAR	14 30	22	5
APR	16	24	7
MAY	2 19	27	10
JUN	4 21	29	12
JUL	7 24		15
AUG	9 26	1	17
SEP	11 28	3	19
OCT	14 31	6	22
NOV	16	8	24
DEC	3 19	11	27

TABLE M18

	CRITICALS	HIGHS	LOWS
JAN	8 24	16	
FEB	10 26	18	1
MAR	15 31	23	6
APR	17	25	8
MAY	3 20	28	11
JUN	5 22	30	13
JUL	8 25		16
AUG	10 27	2	18
SEP	12 29	4	20
OCT	15	7	23
NOV	1 17	9	25
DEC	4 20	12	28

TABLE M19

	CRITICALS	HIGHS	LOWS
JAN	9 25	17	1
FEB	11 27	19	2
MAR	16	24	7
APR	1 18	26	9
MAY	4 21	29	12
JUN	6 23		14
JUL	9 26	1	17
AUG	11 28	3	19
SEP	13 30	5	21
OCT	16	8	24
NOV	2 18	10	26
DEC	5 21	13	29

TABLE M20

	CRITICALS	HIGHS	LOWS
JAN	10 26	18	2
FEB	12 28	20	3
MAR	17	25	8
APR	2 19	27	10
MAY	5 22	30	13
JUN	7 24		15
JUL	10 27	2	18
AUG	12 29	4	20
SEP	14	6	22
OCT	1 17	9	25
NOV	3 19	11	27
DEC	6 22	14	30

TABLE M21

	CRITICALS	HIGHS	LOWS
JAN	11 27	19	3
FEB	13	21	4
MAR	1 18	26	9
APR	3 20	28	11
MAY	6 23	31	14
JUN	8 25		16
JUL	11 28	3	19
AUG	13 30	5	21
SEP	15	7	23
OCT	2 18	10	26
NOV	4 20	12	28
DEC	7 23	15	31

TABLE M22

	CRITICALS	HIGHS	LOWS
JAN	12 28	20	4
FEB	14	22	5
MAR	2 19	27	10
APR	4 21	29	12
MAY	7 24		15
JUN	9 26	1	17
JUL	12 29	4	20
AUG	14 31	6	22
SEP	16	8	24
OCT	3 19	11	27
NOV	5 21	13	29
DEC	8 24		

TABLE M23

	CRITICALS	HIGHS	LOWS
JAN	13 29	21	5
FEB	15	23	6
MAR	3 20	28	11
APR	5 22	30	13
MAY	8 25		16
JUN	10 27	2	18
JUL	13 30	5	21
AUG	15	7	23
SEP	1 17	9	25
OCT	4 20	12	28
NOV	6 22	14	30
DEC	9 25	17	

TABLE M24

	CRITICALS	HIGHS	LOWS
JAN	14 30	22	6
FEB	16	24	7
MAR	4 21	29	12
APR	6 23		14
MAY	9 26	1	17
JUN	11 28	3	19
JUL	14 31	6	22
AUG	16	8	24
SEP	2 18	10	26
OCT	5 21	13	29
NOV	7 23	15	
DEC	10 26	18	1

TABLE M25

	CRITICALS	HIGHS	LOWS
JAN	15 31	23	7
FEB	17	25	8
MAR	5 22	30	13
APR	7 24		15
MAY	10 27	2	18
JUN	12 29	4	20
JUL	15	7	23
AUG	1 17	9	25
SEP	3 19	11	27
OCT	6 22	14	30
NOV	8 24	16	
DEC	11 27	19	2

TABLE M26

	CRITICALS	HIGHS	LOWS
JAN	16	24	8
FEB	1 18	26	9
MAR	6 23	31	14
APR	8 25		16
MAY	11 28	3	19
JUN	13 30	5	21
JUL	16	8	24
AUG	2 18	10	26
SEP	4 20	12	28
OCT	7 23	15	31
NOV	9 25	17	
DEC	12 28	20	3

TABLE M27

	CRITICALS	HIGHS	LOWS
JAN	17	25	9
FEB	2 19	27	10
MAR	7 24		15
APR	9 26	1	17
MAY	12 29	4	20
JUN	14	6	22
JUL	1 17	9	25
AUG	3 19	11	27
SEP	5 21	13	29
OCT	8 24	16	
NOV	10 26	18	1
DEC	13 29	21	4

TABLE M28

	CRITICALS	HIGHS	LOWS
JAN	1 18	26	9
FEB	3 20	28	11
MAR	8 25		16
APR	10 27	2	18
MAY	13 30	5	21
JUN	15	7	23
JUL	2 18	10	26
AUG	4 20	12	28
SEP	6 22	14	30
OCT	9 25	17	
NOV	11 27	19	2
DEC	14 30	22	5

TABLE M29

	CRITICALS	HIGHS	LOWS
JAN	2 19	27	10
FEB	4 21		12
MAR	9 26	1	17
APR	11 28	3	19
MAY	14 31	6	22
JUN	16	8	24
JUL	3 19	11	27
AUG	5 21	13	29
SEP	7 23	15	
OCT	10 26	18	1
NOV	12 28	20	3
DEC	15 31	23	6

TABLE M30

	CRITICALS	HIGHS	LOWS
JAN	3 20	28	11
FEB	5 22		13
MAR	10 27	2	18
APR	12 29	4	20
MAY	15	7	23
JUN	1 17	9	25
JUL	4 20	12	28
AUG	6 22	14	30
SEP	8 24	16	
OCT	11 27	19	2
NOV	13 29	21	4
DEC	16	24	7

TABLE M31

	CRITICALS	HIGHS	LOWS
JAN	4 21	29	12
FEB	6 23		14
MAR	11 28	3	19
APR	13 30	5	21
MAY	16	8	24
JUN	2 18	10	26
JUL	5 21	13	29
AUG	7 23	15	31
SEP	9 25	17	
OCT	12 28	20	3
NOV	14 30	22	5
DEC	17	25	8

TABLE M32

	CRITICALS	HIGHS	LOWS
JAN	5 22	30	13
FEB	7 24		15
MAR	12 29	4	20
APR	14	6	22
MAY	1 17	9	25
JUN	3 19	11	27
JUL	6 22	14	30
AUG	8 24	16	
SEP	10 26	18	1
OCT	13 29	21	4
NOV	15	23	6
DEC	1 18	26	9

TABLE M33

	CRITICALS	HIGHS	LOWS
JAN	6 23	31	14
FEB	8 25		16
MAR	13 30	5	21
APR	15	7	23
MAY	2 18	10	26
JUN	4 20	12	28
JUL	7 23	15	31
AUG	9 25	17	
SEP	11 27	19	2
OCT	14 30	22	5
NOV	16	24	7
DEC	2 19	27	10

TABLE M34

	CRITICALS	HIGHS	LOWS
JAN	7 24		15
FEB	9 26	1	17
MAR	14 31	6	22
APR	16	8	24
MAY	3 19	11	27
JUN	5 21	13	29
JUL	8 24	16	
AUG	10 26	18	1
SEP	12 28	20	3
OCT	15 31	23	6
NOV	17	25	8
DEC	3 20	28	11

TABLE M35

	CRITICALS	HIGHS	LOWS
JAN	8 25		16
FEB	10 27	2	18
MAR	15	7	23
APR	1 17	9	25
MAY	4 20	12	28
JUN	6 22	14	30
JUL	9 25	17	
AUG	11 27	19	2
SEP	13 29	21	4
OCT	16	24	7
NOV	1 18	26	9
DEC	4 21	29	12

TABLE M36

	CRITICALS	HIGHS	LOWS
JAN	9 26	1	17
FEB	11 28	3	19
MAR	16	8	24
APR	2 18	10	26
MAY	5 21	13	29
JUN	7 23	15	
JUL	11 27	18	1
AUG	12 28	20	3
SEP	14 30	22	5
OCT	17	25	8
NOV	2 19	27	10
DEC	5 22	30	13

TABLE M37

	CRITICALS	HIGHS	LOWS
JAN	10 27	2	18
FEB	12	4	20
MAR	1 17	9	25
APR	3 19	11	27
MAY	6 22	14	30
JUN	8 24	16	
JUL	11 27	19	2
AUG	13 29	21	4
SEP	15	23	6
OCT	1 18	26	9
NOV	3 20	28	11
DEC	6 23	31	14

TABLE M38

	CRITICALS	HIGHS	LOWS
JAN	11 28	3	19
FEB	13	5	21
MAR	2 18	10	26
APR	4 20	12	28
MAY	7 23	15	31
JUN	9 25	17	
JUL	12 28	20	3
AUG	14 30	22	5
SEP	16	24	7
OCT	2 19	27	10
NOV	4 21	29	12
DEC	7 24		15

TABLE M39

	CRITICALS	HIGHS	LOWS
JAN	12 29	4	20
FEB	14	6	22
MAR	3 19	11	27
APR	5 21	13	29
MAY	8 24	16	
JUN	10 26	18	1
JUL	13 29	21	4
AUG	15 31	23	6
SEP	17	25	8
OCT	3 20	28	11
NOV	5 22	30	13
DEC	8 25		16

TABLE M40

	CRITICALS	HIGHS	LOWS
JAN	13 30	5	21
FEB	15	7	23
MAR	4 20	12	28
APR	6 22	14	30
MAY	9 25	17	
JUN	11 27	19	2
JUL	14 30	22	5
AUG	16	24	7
SEP	1 18	26	9
OCT	4 21	29	12
NOV	6 23		14
DEC	9 26	1	17

TABLE M41

	CRITICALS	HIGHS	LOWS
JAN	14 31	6	22
FEB	16	8	24
MAR	5 21	13	29
APR	7 23	15	
MAY	10 26	18	1
JUN	12 28	20	3
JUL	15 31	23	6
AUG	17	25	8
SEP	2 19	27	10
OCT	5 22	30	13
NOV	7 24		15
DEC	10 27	2	18

TABLE M42

	CRITICALS	HIGHS	LOWS
JAN	15	7	23
FEB	1 17	9	25
MAR	6 22	14	30
APR	8 24	16	
MAY	11 27	19	2
JUN	13 29	21	4
JUL	16	24	7
AUG	1 18	26	9
SEP	3 20	28	11
OCT	6 23	31	14
NOV	8 25		16
DEC	11 28	3	19

TABLE M43

	CRITICALS	HIGHS	LOWS
JAN	16	8	24
FEB	2 18	10	26
MAR	7 23	15	31
APR	9 25	17	
MAY	12 28	20	3
JUN	14 30	22	5
JUL	17	25	8
AUG	2 19	27	10
SEP	4 21	29	12
OCT	7 24		15
NOV	9 26	1	17
DEC	12 29	4	20

TABLE M51

	CRITICALS	HIGHS	LOWS
JAN	1 17	9	25
FEB	3 19	11	27
MAR	7 23	15	31
APR	9 25	17	
MAY	12 28	20	3
JUN	14 30	22	5
JUL	17	25	8
AUG	2 19	27	10
SEP	4 21	29	12
OCT	7 24		15
NOV	9 26	1	17
DEC	12 29	4	20

TABLE M52

	CRITICALS	HIGHS	LOWS
JAN	2 18	10	26
FEB	4 20	12	28
MAR	8 24	16	
APR	10 26	18	1
MAY	13 29	21	4
JUN	15	23	6
JUL	1 18	26	9
AUG	3 20	28	11
SEP	5 22	30	13
OCT	8 25		16
NOV	10 27	2	18
DEC	13 30	5	21

TABLE M53

	CRITICALS	HIGHS	LOWS
JAN	3 19	11	27
FEB	5 21	13	29
MAR	9 25	17	
APR	11 27	19	2
MAY	14 30	22	5
JUN	16	24	7
JUL	1 18	27	10
AUG	4 21	29	12
SEP	6 23		14
OCT	9 26	1	17
NOV	11 28	3	19
DEC	14 31	6	22

TABLE M54

	CRITICALS	HIGHS	LOWS
JAN	4 20	12	28
FEB	6 22	14	
MAR	10 26	18	1
APR	12 28	20	3
MAY	15 31	23	6
JUN	17	25	8
JUL	3 20	28	11
AUG	5 22	30	13
SEP	7 24		15
OCT	10 27	2	18
NOV	12 29	4	20
DEC	15	7	23

TABLE M55

	CRITICALS	HIGHS	LOWS
JAN	5 21	13	29
FEB	7 23	15	
MAR	11 27	19	2
APR	13 29	21	4
MAY	16	24	7
JUN	1 18	26	9
JUL	4 21	29	12
AUG	6 23	31	14
SEP	8 25		16
OCT	11 28	3	19
NOV	13 30	5	21
DEC	16	8	24

TABLE M56

	CRITICALS	HIGHS	LOWS
JAN	6 22	14	30
FEB	8 24	16	
MAR	12 28	20	3
APR	14 30	22	5
MAY	17	25	8
JUN	2 19	27	10
JUL	5 22	30	13
AUG	7 24		15
SEP	9 26	1	17
OCT	12 29	4	20
NOV	14	6	22
DEC	1 17	9	25

TABLE M57

	CRITICALS	HIGHS	LOWS
JAN	7 23	15	31
FEB	9 25	17	
MAR	13 29	21	4
APR	15	23	6
MAY	1 18	26	9
JUN	3 20	28	11
JUL	6 23	31	14
AUG	8 25		16
SEP	10 27	2	18
OCT	13 30	5	21
NOV	15	7	23
DEC	2 18	10	26

TABLE M58

	CRITICALS	HIGHS	LOWS
JAN	8 24	16	
FEB	10 26	18	1
MAR	14 30	22	5
APR	16	24	7
MAY	2 19	27	10
JUN	4 21	29	12
JUL	7 24		15
AUG	9 26	1	17
SEP	11 28	3	19
OCT	14 31	6	22
NOV	16	8	24
DEC	3 19	11	27

TABLE M59

	CRITICALS	HIGHS	LOWS
JAN	9 25	17	1
FEB	11 27	19	2
MAR	15 31	23	6
APR	17	25	8
MAY	3 20	28	11
JUN	5 22	30	13
JUL	8 25		16
AUG	10 27	2	18
SEP	12 29	4	20
OCT	15	7	23
NOV	1 17	9	25
DEC	4 20	12	28

TABLE M60

	CRITICALS	HIGHS	LOWS
JAN	10 26	18	2
FEB	12 28	20	3
MAR	16	24	7
APR	2 18	26	9
MAY	4 21	29	12
JUN	6 23		14
JUL	9 26	1	17
AUG	11 28	3	19
SEP	13 30	5	21
OCT	16	8	24
NOV	2 18	10	26
DEC	5 21	13	29

TABLE M61

	CRITICALS	HIGHS	LOWS
JAN	11 27	19	3
FEB	13 29	21	4
MAR	17	25	8
APR	2 19	27	10
MAY	5 22	30	13
JUN	7 24		15
JUL	10 27	2	18
AUG	12 29	4	20
SEP	14	6	22
OCT	1 17	9	25
NOV	3 19	11	27
DEC	6 22	14	30

TABLE M62

	CRITICALS	HIGHS	LOWS
JAN	12 28	20	4
FEB	14	22	5
MAR	1 18	26	9
APR	3 20	28	11
MAY	6 23	31	14
JUN	8 25		16
JUL	11 28	3	19
AUG	13 30	5	21
SEP	15	7	23
OCT	2 18	10	26
NOV	4 20	12	28
DEC	7 23	15	31

TABLE M63

	CRITICALS	HIGHS	LOWS
JAN	13 29	21	5
FEB	15	23	6
MAR	2 19	27	10
APR	4 21	29	12
MAY	7 24		15
JUN	9 26	1	17
JUL	12 29	4	20
AUG	14 31	6	22
SEP	16	8	24
OCT	3 19	11	27
NOV	5 21	13	29
DEC	8 24	16	

TABLE M64

	CRITICALS	HIGHS	LOWS
JAN	14 30	22	6
FEB	16	24	7
MAR	3 20	28	11
APR	5 22	30	13
MAY	8 25		16
JUN	10 27	2	18
JUL	13 30	5	21
AUG	15	7	23
SEP	1 18	9	25
OCT	4 20	12	28
NOV	6 22	14	30
DEC	9 25	17	

TABLE M65

	CRITICALS	HIGHS	LOWS
JAN	15 31	23	7
FEB	17	25	8
MAR	4 21	29	12
APR	6 23		14
MAY	9 26	1	17
JUN	11 28	3	19
JUL	14 31	6	22
AUG	16	8	24
SEP	2 18	10	26
OCT	5 21	13	29
NOV	7 23	15	
DEC	10 26	18	1

TABLE M66

	CRITICALS	HIGHS	LOWS
JAN	16	24	8
FEB	1 18	26	9
MAR	5 22	30	13
APR	7 24		15
MAY	10 27	2	18
JUN	12 29	4	20
JUL	15	7	23
AUG	1 17	9	25
SEP	3 19	11	27
OCT	6 22	14	30
NOV	8 24	16	
DEC	11 27	19	2

TABLE M67

	CRITICALS	HIGHS	LOWS
JAN	17	25	9
FEB	2 19	27	10
MAR	6 23	31	14
APR	8 25		16
MAY	11 28	3	19
JUN	13 30	5	21
JUL	16	8	24
AUG	2 19	10	26
SEP	4 20	12	28
OCT	7 23	15	31
NOV	9 25	17	
DEC	12 28	20	3

TABLE M68

	CRITICALS	HIGHS	LOWS
JAN	1 18	26	9
FEB	3 20	28	11
MAR	7 24		15
APR	9 26	1	17
MAY	12 29	4	20
JUN	14	6	22
JUL	1 17	9	25
AUG	3 19	11	27
SEP	5 21	13	29
OCT	8 24	16	
NOV	10 26	18	1
DEC	13 29	21	4

TABLE M69

	CRITICALS	HIGHS	LOWS
JAN	2 19	27	10
FEB	4 21	29	12
MAR	8 25		16
APR	10 27	2	18
MAY	13 30	5	21
JUN	15	7	23
JUL	2 18	10	26
AUG	4 20	12	28
SEP	6 22	14	30
OCT	9 25	17	
NOV	11 27	19	2
DEC	14 30	22	5

TABLE M70

	CRITICALS	HIGHS	LOWS
JAN	3 20	28	11
FEB	5 22		13
MAR	9 26	1	17
APR	11 28	3	19
MAY	14 31	6	22
JUN	16	8	24
JUL	3 19	11	27
AUG	5 21	13	29
SEP	7 23	15	
OCT	10 26	18	1
NOV	12 28	20	3
DEC	15 31	23	6

TABLE M71

	CRITICALS	HIGHS	LOWS
JAN	4 21	29	12
FEB	6 23		14
MAR	10 27	2	18
APR	12 29	4	20
MAY	15	7	23
JUN	1 17	9	25
JUL	4 20	12	28
AUG	6 22	14	30
SEP	8 24	16	
OCT	11 27	19	2
NOV	13 29	21	4
DEC	16	24	7

TABLE M72

	CRITICALS	HIGHS	LOWS
JAN	5 22	30	13
FEB	7 24		15
MAR	11 28	3	19
APR	13 30	5	21
MAY	16	8	24
JUN	2 18	10	26
JUL	5 21	13	29
AUG	7 23	15	31
SEP	9 25	17	
OCT	12 28	20	3
NOV	14 30	22	5
DEC	17	25	8

TABLE M73

	CRITICALS	HIGHS	LOWS
JAN	6 23	31	14
FEB	8 25		16
MAR	12 29	4	20
APR	14	6	22
MAY	1 17	9	25
JUN	3 19	11	27
JUL	6 22	14	30
AUG	8 24	16	
SEP	10 26	18	1
OCT	13 29	21	4
NOV	15	23	6
DEC	1 18	26	9

TABLE M74

	CRITICALS	HIGHS	LOWS
JAN	7 24		15
FEB	9 26	1	17
MAR	13 30	5	21
APR	15	7	23
MAY	2 18	10	26
JUN	4 20	12	28
JUL	7 23	15	31
AUG	9 25	17	
SEP	11 27	19	2
OCT	14 30	22	5
NOV	16	24	7
DEC	2 19	27	10

TABLE M75

	CRITICALS	HIGHS	LOWS
JAN	8 25		16
FEB	10 27	2	18
MAR	14 31	6	22
APR	16	8	24
MAY	3 19	11	27
JUN	5 21	13	29
JUL	8 24	16	
AUG	10 26	18	1
SEP	12 28	20	3
OCT	15 31	23	6
NOV	17	25	8
DEC	3 20	28	11

TABLE M76

	CRITICALS	HIGHS	LOWS
JAN	9 26	1	17
FEB	11 28	3	19
MAR	15	7	23
APR	1 17	9	25
MAY	4 20	12	28
JUN	6 22	14	30
JUL	9 25	17	
AUG	11 27	19	2
SEP	13 29	21	4
OCT	16	24	7
NOV	1 18	26	9
DEC	4 21	29	12

TABLE M77

	CRITICALS	HIGHS	LOWS
JAN	10 27	2	18
FEB	12 29	4	20
MAR	16	8	24
APR	2 18	10	26
MAY	5 21	13	29
JUN	7 23	15	
JUL	10 26	18	1
AUG	12 28	20	3
SEP	14 30	22	5
OCT	17	25	8
NOV	2 19	27	10
DEC	5 22	30	13

TABLE M78

	CRITICALS	HIGHS	LOWS
JAN	11 28	3	19
FEB	13	5	21
MAR	1 17	9	25
APR	3 19	11	27
MAY	6 22	14	30
JUN	8 24	16	
JUL	11 27	19	2
AUG	13 29	21	4
SEP	15	23	6
OCT	1 18	26	9
NOV	3 20	28	11
DEC	6 23	31	14

TABLE M79

	CRITICALS	HIGHS	LOWS
JAN	12 29		20
FEB	14	6	22
MAR	2 18	10	26
APR	4 20	12	28
MAY	7 23	15	31
JUN	9 25	17	
JUL	12 28	20	3
AUG	14 30	22	5
SEP	16	24	7
OCT	2 19	27	10
NOV	4 21	29	12
DEC	7 24		15

TABLE M80

	CRITICALS	HIGHS	LOWS
JAN	13 30	5	21
FEB	15	7	23
MAR	3 19	11	27
APR	5 21	13	29
MAY	8 24	16	
JUN	10 26	18	1
JUL	13 29	21	4
AUG	15 31	23	6
SEP	17	25	8
OCT	3 20	28	11
NOV	5 22	30	13
DEC	8 25		16

TABLE M81

	CRITICALS	HIGHS	LOWS
JAN	14 31	6	22
FEB	16	8	24
MAR	4 20	12	28
APR	6 22	14	30
MAY	9 25	17	
JUN	11 27	19	2
JUL	14 30	22	5
AUG	16	24	7
SEP	1 18	26	9
OCT	4 21	29	12
NOV	6 23		14
DEC	9 26	1	17

TABLE M82

	CRITICALS	HIGHS	LOWS
JAN	15	7	23
FEB	1 17	9	25
MAR	5 21	13	29
APR	7 23	15	
MAY	10 26	18	1
JUN	12 28	20	3
JUL	15 31	23	6
AUG	17	25	8
SEP	2 19	27	10
OCT	5 22	30	13
NOV	7 24		15
DEC	10 27	2	18

TABLE M83

	CRITICALS	HIGHS	LOWS
JAN	16	8	24
FEB	2 18	10	26
MAR	6 22	14	30
APR	8 24	16	
MAY	11 27	19	2
JUN	13 29	21	4
JUL	16	24	7
AUG	1 18	26	9
SEP	3 20	28	11
OCT	6 23	31	14
NOV	8 25		16
DEC	11 28	3	19

APPENDIX B:
NAMES AND BIRTHDATES
FOR READER INVESTIGATION

Aaron, Hank	February 15, 1934
Abdul-Jabbar, Kareem	April 16, 1947
Abernathy, Ralph	March 11, 1926
Abzug, Bella	July 24, 1920
Adderley, Julian "Cannonball"	September 15, 1928
Agee, Tommy	August 9, 1942
Agnew, Spiro	November 9, 1918
Ailey, Alvin	January 5, 1931
Albee, Edward	March 12, 1928
Albert, Carl	May 10, 1908
Aldrin, Edwin "Buzz"	January 20, 1930
Ali, Muhammad	January 18, 1942
Alioto, Joseph	February 12, 1916
Allen, Dick	March 8, 1942
Allen, Steve	December 26, 1921
Allen, Woody	December 1, 1935
Allison, Mose	November 11, 1927
Alpert, Herb	March 31, 1935
Altman, Robert	February 20, 1925
Amsterdam, Morey	December 14, 1914
Anderson, Eric	February 14, 1943
Anderson, Jack	October 19, 1922
Andersson, Bibi	November 11, 1935
Andress, Ursula	March 19, 1938
Andretti, Mario	February 28, 1940
Andrews, Julie	October 1, 1935
Angelou, Maya	April 4, 1928
Anka, Paul	July 30, 1941
Ann-Margret	April 28, 1941
Antonioni, Michelangelo	September 29, 1912
Archibald, Nate	April 18, 1948
Arkin, Alan	March 26, 1934
Armstrong, Louis	July 4, 1900
Armstrong, Neil	August 5, 1930
Arnaz, Desi	March 2, 1917
Arness, James	May 26, 1923
Arnold, Eddy	May 15, 1918
Arthur, Beatrice	May 13, 1926
Ashe, Arthur Jr.	July 10, 1943
Ashley, Elizabeth	August 30, 1940
Asimov, Isaac	January 2, 1920
Asner, Edward	November 15, 1929
Astaire, Fred	May 10, 1899
Atkins, Chet	June 20, 1924

Auerbach, Arnold "Red"	September 20, 1917
Avedon, Richard	May 15, 1923
Bacall, Lauren	September 16, 1924
Bacharach, Burt	September 16, 1924
Backus, Jim	February 25, 1913
Baez, Joan	January 9, 1941
Bailey, Pearl	March 29, 1918
Bailyn, Bernard	September 10, 1922
Baker, Carroll	May 28, 1935
Baker, Josephine	June 3, 1906
Baker, Russell	August 8, 1925
Balanchine, George	January 22, 1904
Baldwin, Billy	May 30, 1903
Baldwin, James	August 2, 1924
Ball, Lucille	August 6, 1911
Ballard, Kaye	November 20, 1926
Bancroft, Anne	September 17, 1931
Banks, Ernie	January 31, 1931
Bardot, Brigitte	September 28, 1935
Barnes, Clive	May 13, 1927
Barry, Gene	June 4, 1922
Barry, Rick	March 28, 1944
Basie, Count	August 21, 1904
Bayh, Birch	January 22, 1928
Beame, Abraham	March 20, 1906
Bean, Orson	July 22, 1928
Beard, James	May 5, 1903
Beatles, The	
Harrison, George	February 25, 1943
Lennon, John	October 9, 1940
McCartney, Paul	June 18, 1942
Starr, Ringo	July 7, 1940
Beaton, Cecil	January 14, 1904
Beatty, Warren	March 30, 1937
Beckett, Samuel	April 13, 1906
Beene, Geoffrey	August 30, 1927
Belafonte, Harry	March 1, 1927
Bellamy, Ralph	June 17, 1904
Belli, Melvin	July 29, 1907
Bellow, Saul	July 10, 1915
Belmondo, Jean-Paul	April 9, 1933
Bench, Johnny	December 7, 1947
Benjamin, Richard	May 22, 1938
Benny, Jack	February 14, 1894
Bentley, Eric	September 14, 1916
Bergen, Candice	May 9, 1946
Bergman, Ingmar	July 14, 1918

Bergman, Ingrid	August 29, 1915
Berle, Milton	July 12, 1908
Berlin, Irving	May 11, 1888
Bernardi, Herschel	October 20, 1923
Bernstein, Carl	February 14, 1944
Bernstein, Leonard	August 25, 1918
Berra, Yogi	May 12, 1925
Berry, Chuck	October 18, 1926
Bikel, Theodore	May 2, 1924
Bing, Dave	November 24, 1943
Bishop, Joey	February 3, 1918
Black, Karen	July 1, 1942
Black, Shirley Temple	April 23, 1928
Blake, Robert	September 18, 1938
Blakey, Art	October 11, 1919
Blanc, Mel	May 30, 1908
Blass, Bill	June 22, 1922
Blondell, Joan	August 30, 1909
Bloom, Claire	February 15, 1931
Blue, Vida	July 28, 1949
Bly, Robert	December 23, 1926
Blyden, Larry	June 13, 1925
Bond, Julian	January 14, 1940
Bono, Sonny	February 16, 1940
Boone, Pat	June 1, 1934
Booth, Shirley	August 30, 1909
Borg, Björn	June 6, 1956
Borge, Victor	January 3, 1909
Borman, Frank	March 14, 1928
Bosley, Tom	October 1, 1927
Boyd, Stephen	July 4, 1928
Boyer, Charles	August 28, 1899
Bradley, Bill	July 28, 1943
Brando, Marlon	April 3, 1924
Braun, Wernher von	March 23, 1912
Brennan, Walter	July 25, 1894
Brennan, William, Jr.	April 25, 1906
Breslin, Jimmy	October 17, 1930
Bridges, Beau	December 9, 1941
Brinkley, David	July 10, 1920
Brock, Lou	June 18, 1938
Bronson, Charles	November 3, 1922
Brooke, Edward	October 26, 1919
Brothers, Dr. Joyce	October 20, 1928
Brown, Helen Gurley	February 18, 1922
Brown, James	May 3, 1934
Brown, Jim	September 19, 1947
Brubeck, Dave	December 6, 1920
Bruhn, Eric	October 3, 1928
Buchwald, Art	October 20, 1925
Buckley, William F.	November 24, 1925
Burger, Warren	September 17, 1907
Burnett, Carol	April 26, 1933
Burns, Arthur F.	April 27, 1904
Burns, George	January 20, 1896

Burr, Raymond	May 21, 1917
Burroughs, William	February 5, 1914
Burton, Richard	November 10, 1925
Butz, Earl	July 3, 1909
Caesar, Sid	September 8, 1922
Cagney, James	July 17, 1900
Caine, Michael	March 14, 1933
Calder, Alexander	July 22, 1898
Calloway, Cab	December 25, 1907
Cambridge, Godfrey	February 26, 1933
Campanella, Roy	November 19, 1921
Campbell, Glen	April 22, 1938
Cantrell, Lana	August 7, 1944
Capote, Truman	September 30, 1924
Capra, Frank	May 19, 1897
Cardin, Pierre	July 7, 1922
Carmichael, Hoagy	November 22, 1899
Carney, Art	November 4, 1918
Carroll, Diahann	July 17, 1935
Carson, Johnny	October 23, 1925
Carter, Jimmy	October 1, 1924
Casals, Pablo	December 29, 1876
Cash, Johnny	February 26, 1932
Casper, Billy	June 24, 1931
Cassavetes, John	December 9, 1929
Cassidy, David	April 12, 1950
Cavett, Dick	November 19, 1936
Chagall, Marc	July 7, 1887
Chamberlain, Richard	March 31, 1935
Chamberlain, Wilt	August 21, 1936
Chancellor, John	July 14, 1927
Channing, Carol	January 31, 1923
Chaplin, Charlie	April 16, 1889
Charles, Ray	September 23, 1932
Chavez, César	March 31, 1927
Cher	May 20, 1946
Child, Julia	August 15, 1912
Chisholm, Shirley	November 30, 1924
Christie, Agatha	September 15, 1890
Christie, Julie	April 14, 1941
Church, Frank	July 25, 1924
Clark, Kenneth B.	July 24, 1914
Clark, Petula	November 15, 1934
Clark, Ramsey	December 18, 1927
Cleaver, Eldridge	August 31, 1935
Cliburn, Van	July 12, 1934
Cobb, Lee J.	December 9, 1911
Coburn, James	August 31, 1928
Coco, James	March 21, 1929
Collins, Judy	May 1, 1939
Connally, John B.	February 28, 1917
Connery, Sean	August 25, 1930
Connors, Jimmy	September 2, 1952
Connors, Mike	August 15, 1925
Conrad, William	September 27, 1920

Considine, Bob	November 4, 1906
Cooke, Alistair	November 20, 1908
Cooper, Alice	February 4, 1948
Copland, Aaron	November 14, 1900
Cosby, Bill	July 12, 1937
Cosell, Howard	March 25, 1929
Costellano, Richard	September 3, 1934
Cousins, Norman	June 24, 1915
Cousteau, Jacques	June 11, 1910
Cox, Archibald	May 17, 1912
Cronkite, Walter	November 4, 1916
Crosby, Bing	May 2, 1904
Crosby, Gary	June 27, 1933
Cugat, Xavier	January 1, 1900
Cullen, Bill	February 18, 1920
Culp, Robert	August 16, 1931
Curtis, Tony	June 3, 1925
Dahl, Arlene	August 11, 1928
Daley, Richard	May 15, 1902
Dali, Salvador	May 11, 1904
Davis, Angela	January 26, 1944
Davis, Bette	April 5, 1908
Davis, Miles	May 25, 1926
Davis, Ossie	December 18, 1917
Davis, Sammy Jr.	December 8, 1925
Day, Doris	April 3, 1924
Dayan, Moshe	May 20, 1915
DeBusschere, Dave	October 16, 1940
Dee, Ruby	October 27, 1924
Delon, Alain	November 8, 1935
Deneuve, Catherine	October 22, 1943
Dennis, Sandy	April 27, 1937
Denver, John	December 31, 1943
Dewhurst, Colleen	June 3, 1926
Diamond, Neil	January 24, 1941
Dickey, James	February 2, 1923
Dickinson, Angie	September 30, 1932
Dietrich, Marlene	December 27, 1904
DiGregorio, Ernie	January 15, 1951
Diller, Phyllis	July 17, 1917
DiMaggio, Joe	November 25, 1914
Douglas, Kirk	December 9, 1916
Douglas, William O.	October 16, 1898
Dubinsky, David	February 22, 1892
Duke, Patty	December 14, 1946
Dunaway, Faye	January 14, 1941
Durocher, Leo	July 27, 1906
Dylan, Bob	May 24, 1941
Eagleton, Thomas	September 4, 1929
Eastwood, Clint	May 31, 1930
Eggar, Samantha	March 5, 1939
Ehrlichman, John	March 20, 1925
Eisenhower, Dwight	October 14, 1890
Eisenhower, Mamie	November 14, 1891
Ellington, Duke	April 29, 1899
Elliot, Cass	February 19, 1941

Ernst, Max	April 2, 1891
Ervin, Sam	September 27, 1896
Evers, Charles	September 14, 1923
Evert, Chris	December 21, 1954
Fairbanks, Douglas Jr.	December 9, 1909
Falk, Peter	September 16, 1927
Farrow, Mia	February 9, 1945
Feliciano, José	September 10, 1945
Fellini, Federico	January 20, 1920
Ferrer, José	January 8, 1912
Fiedler, Arthur	December 17, 1894
Finney, Albert	May 9, 1936
Fischer, Bobby	March 9, 1943
Fitzgerald, Ella	April 25, 1918
Fonda, Henry	May 16, 1905
Fonda, Jane	December 21, 1937
Fonda, Peter	February 23, 1939
Fonteyn, Dame Margot	May 18, 1919
Ford, Gerald	July 14, 1913
Ford, Glenn	May 1, 1916
Ford, Henry, I	July 30, 1863
Ford, Henry, II	September 4, 1919
Foreman, George	January 10, 1949
Foxx, Redd	December 9, 1922
Foyt, A. J.	January 16, 1935
Franklin, Aretha	March 25, 1942
Frazier, Joe	January 17, 1944
Frazier, Walt	March 29, 1945
Friedan, Betty	February 4, 1921
Friedkin, William	August 29, 1939
Frost, David	April 7, 1939
Fulbright, J. William	April 9, 1905
Fuller, Buckminster	July 12, 1895
Furstenberg, Diane Von	December 31, 1946
Gabor, Zsa Zsa	February 6, 1923
Galbraith, John Kenneth	October 15, 1908
Gallup, George	November 18, 1901
Garbo, Greta	September 18, 1905
Garner, Erroll	June 15, 1921
Garner, James	April 17, 1928
Gazzara, Ben	August 28, 1930
Gehrig, Lou	June 19, 1903
Getty, J. Paul	December 15, 1892
Gibson, Kenneth	May 15, 1932
Gielgud, Sir John	April 14, 1904
Gillespie, Dizzy	October 21, 1917
Ginsberg, Allen	June 3, 1926
Gleason, Jackie	February 26, 1916
Glenn, John	July 15, 1921
Goldsboro, Bobby	January 11, 1941
Goldwater, Barry	January 1, 1909
Gonzales, Pancho	May 9, 1928

Goodman, Benny	May 30, 1909	Huntley, Chet	December 10, 1911
Goodrich, Gail	April 23, 1943	Hurok, Sol	April 9, 1888
Goolagong, Evonne	July 31, 1951	Huston, John	August 5, 1906
Gordon, Ruth	October 30, 1896	Jackson, Glenda	May 9, 1936
Goren, Charles	March 4, 1901	Jackson, Henry	May 3, 1912
Gormé, Eydie	August 16, 1931	Jackson, Rev. Jesse	October 8, 1941
Gould, Elliott	August 29, 1938	Javits, Jacob K.	May 18, 1904
Graham, Billy	November 7, 1918	John, Elton	March 25, 1947
Graham, Martha	May 11, 1894	Johnson, Lyndon	August 27, 1908
Granger, Stewart	May 6, 1913	Johnson, Rafer	August 18, 1935
Grant, Cary	January 18, 1904	Johnson, Virginia	February 11, 1925
Graves, Peter	March 18, 1936	Jones, James	November 6, 1921
Graziano, Rocky	June 7, 1922	Jones, James Earl	January 17, 1931
Greene, Graham	October 4, 1904	Jones, Tom	June 7, 1940
Greene, Lorne	February 12, 1915	Kaye, Danny	January 18, 1913
Gregory, Dick	October 12, 1932	Kazan, Lainie	May 16, 1940
Grey, Joel	April 11, 1932	Keach, Stacy	June 2, 1941
Griffin, Merv	July 6, 1925	Kelley, Clarence	October 24, 1911
Griffith, Andy	June 1, 1926	Ken, Walter	July 8, 1913
Guccione, Bob	December 17, 1930	Kennedy, Edward	February 22, 1932
Guinness, Alec	April 2, 1914	Kennedy, Ethel	
Guthrie, Arlo	July 10, 1947	(Skakel)	April 11, 1928
Hackman, Gene	January 30, 1931	Kennedy, John F.	May 29, 1917
Haig, Alexander	December 2, 1924	Kennedy, Robert	November 20, 1925
Hall, Monty	August 25, 1923	Kennedy, Rose	July 22, 1890
Hamill, Pete	June 24, 1935	King, Billie Jean	November 22, 1943
Hampton, Lionel	April 20, 1913	King, Coretta	May 27, 1927
Harris, Julie	December 2, 1925	King, Dr. Martin	
Harris, Richard	October 1, 1933	Luther Jr.	January 15, 1929
Harrison, Rex	March 5, 1908	Kissinger, Henry	May 27, 1923
Hawn, Goldie	November 2, 1945	Koufax, Sandy	December 30, 1935
Hayes, Elvin	November 17, 1945	Kubrick, Stanley	July 26, 1928
Hayes, Helen	October 10, 1900	Kuhn, Bowie	October 28, 1926
Hayes, Isaac	August 20, 1942	Kunstler, William	July 7, 1919
Hefner, Hugh	April 9, 1926	Lancaster, Burt	November 2, 1913
Hellman, Lillian	June 20, 1905	Landers, Ann	July 4, 1918
Hepburn, Audrey	May 4, 1929	Lansbury, Angela	October 16, 1925
Hepburn, Katharine	November 8, 1909	Laver, Rod	August 9, 1938
Herblock	October 13, 1909	Lawrence, Steve	July 8, 1935
Heston, Charlton	October 4, 1923	Leary, Timothy	October 22, 1920
Heyerdahl, Thor	October 6, 1914	Lee, Peggy	May 26, 1920
Hitchcock, Alfred	August 13, 1899	Lemmon, Jack	February 8, 1925
Hoffa, James	February 14, 1913	Lerner, Max	December 20, 1902
Hoffman, Abbie	November 30, 1936	Lewis, Jerry	March 16, 1926
Hoffman, Dustin	August 8, 1937	Liberace	May 16, 1919
Hope, Bob	May 29, 1903	Lindbergh, Charles	February 4, 1902
Hopkins, Anthony	December 31, 1937	Linden, Hal	March 20, 1931
Horne, Lena	June 17, 1917	Lindsey, John V.	November 24, 1921
Horowitz, Vladimir	October 1, 1904	Lopez, Trini	May 15, 1937
Hudson, Rock	November 17, 1925	Lord, Jack	December 30, 1928
Hughes, Harold	February 10, 1922	Loren, Sophia	September 20, 1934
Hughes, Howard	December 24, 1910	MacGraw, Ali	April 1, 1939
Humperdinck,		MacLaine, Shirley	April 24, 1934
Engelbert	May 2, 1936	Mailer, Norman	January 21, 1923
Humphrey,		Makarova, Natalia	November 21, 1940
Hubert H.	May 27, 1911	Malden, Karl	March 22, 1913

Mansfield, Mike	March 16, 1903
Mao Tse-tung	December 26, 1893
Maravich, Pete	June 22, 1948
Marshall, Thurgood	July 2, 1908
Martin, Dean	June 17, 1917
Martin, Mary	December 1, 1913
Marvin, Lee	February 19, 1924
Marx, Groucho	October 2, 1895
Masters, William	December 27, 1915
Mathis, Johnny	September 30, 1935
Matthau, Walter	October 1, 1920
May, Elaine	April 21, 1932
Mays, Willie	May 6, 1931
McAdoo, Bob	September 25, 1951
McCarthy, Eugene	March 29, 1916
McGovern, George	July 19, 1922
McMahon, Ed	March 6, 1923
McNamara, Robert S.	June 9, 1916
Mead, Margaret	December 16, 1901
Meany, George	August 16, 1894
Meir, Golda	March 3, 1898
Menotti, Gian-Carlo	July 7, 1911
Menuhin, Yehudi	April 22, 1916
Miller, Arthur	October 17, 1915
Miller, Henry	December 26, 1891
Miller, Johnny	April 29, 1947
Mills, Wilbur	May 24, 1909
Minnelli, Liza	March 12, 1946
Mitchell, John	September 15, 1913
Mitchell, Joni	November 7, 1943
Mitchum, Robert	August 6, 1917
Mondale, Walter	January 5, 1928
Monroe, Earl	November 21, 1944
Monroe, Marilyn	June 1, 1926
Morse, Robert	May 18, 1931
Mostel, Zero	February 18, 1915
Moynihan, Daniel Patrick	March 16, 1927
Muskie, Edmund	March 28, 1914
Nabokov, Vladimir	April 23, 1899
Nader, Ralph	February 27, 1934
Namath, Joe	May 31, 1943
Nero, Peter	May 22, 1934
Newcombe, John	May 23, 1943
Newhart, Bob	September 5, 1929
Newley, Anthony	September 24, 1931
Newlin, Mike	January 2, 1949
Newman, Paul	May 22, 1934
Nicholson, Jack	April 22, 1937
Nicklaus, Jack	January 21, 1940
Niven, David	March 1, 1910
Nixon, Pat	March 16, 1912
Nixon, Richard	January 9, 1913
Nureyev, Rudolf	March 17, 1938
Oates, Joyce Carol	June 16, 1938
Odetta	December 31, 1930

Olivier, Lord Laurence	May 22, 1907
Onassis, Aristotle	January 20, 1906
Onassis, Jackie Kennedy	July 28, 1929
O'Neal, Ryan	April 20, 1941
Ono, Yoko	February 18, 1933
Orr, Bobby	March 20, 1948
O'Toole, Peter	August 2, 1933
Pacino, Al	April 25, 1940
Palmer, Arnold	September 10, 1929
Patterson, Floyd	January 4, 1935
Peck, Gregory	April 5, 1916
Peppard, George	October 1, 1928
Percy, Charles H.	September 27, 1919
Perkins, Tony	April 14, 1932
Petty, Richard	July 2, 1937
Plimpton, George	March 18, 1927
Poitier, Sidney	February 20, 1924
Polanski, Roman	August 18, 1933
Presley, Elvis	January 8, 1935
Price, Vincent	May 27, 1911
Proxmire, William	January 11, 1915
Quayle, Anthony	September 7, 1913
Queen, Ellery	October 20, 1905
Quinn, Anthony	April 21, 1916
Rawls, Lou	December 1, 1935
Reagan, Ronald	February 6, 1911
Reasoner, Harry	April 17, 1923
Redford, Robert	August 18, 1937
Redgrave, Sir Michael	March 20, 1908
Redgrave, Vanessa	January 30, 1937
Reed, Oliver	February 13, 1938
Reed, Rex	October 2, 1939
Reiner, Carl	March 20, 1922
Reston, James	November 3, 1909
Revson, Peter	February 27, 1939
Reynolds, Burt	February 11, 1936
Reynolds, Debbie	April 1, 1932
Rhodes, John	September 18, 1916
Richardson, Elliot	July 20, 1921
Rickles, Don	May 8, 1926
Rigg, Diana	July 20, 1938
Rivera, Geraldo	July 4, 1943
Robards, Jason Jr.	July 26, 1922
Robbins, Harold	May 21, 1916
Robeson, Paul	April 9, 1898
Robinson, Brooks	May 18, 1937
Robinson, Sugar Ray	May 3, 1920
Rockefeller, David	June 12, 1915
Rockefeller, Laurance S.	May 26, 1910
Rockefeller, Nelson	July 8, 1908
Rockwell, Norman	February 3, 1894
Rodgers, Richard	June 28, 1902

Rolling Stones, The	
Jagger, Mick	July 26, 1943
Richard, Keith	December 18, 1943
Watts, Charles	June 2, 1941
Wyman, Bill	October 24, 1941
Ross, Diana	March 26, 1944
Roth, Philip	March 19, 1933
Roundtree, Richard	September 7, 1942
Rozelle, Pete	March 1, 1926
Rubinstein, Artur	January 28, 1887
Ruckelshaus, William	July 24, 1932
Russell, Bill	February 12, 1934
Ruth, Babe	February 6, 1895
Ryun, Jim	April 29, 1947
St. John, Jill	August 19, 1940
Salinger, J. D.	January 1, 1919
Salinger, Pierre	June 14, 1925
Salk, Jonas	October 28, 1914
Schlesinger, Arthur, Jr.	October 15, 1917
Schlesinger, James	February 15, 1929
Schmidt, Mike	September 27, 1949
Scott, George C.	October 18, 1927
Scott, Hugh	November 11, 1900
Scranton, William	July 19, 1917
Seaver, Tom	November 17, 1944
Seeger, Pete	May 3, 1919
Sevareid, Eric	November 26, 1912
Shankar, Ravi	April 7, 1920
Sharif, Omar	April 10, 1932
Sheen, Bishop Fulton J.	May 8, 1895
Shirer, William L.	February 23, 1904
Shoemaker, Willie	August 19, 1931
Shore, Dinah	March 1, 1917
Shriver, Sargent	November 9, 1915
Sills, Beverly	May 25, 1929
Simon, Carly	June 25, 1945
Simon, Paul	October 13, 1940
Simone, Nina	February 21, 1933
Sinatra, Frank	December 12, 1915
Slick, Grace	October 30, 1941
Smothers Brothers,	
Dick	November 20, 1938
Tom	February 2, 1937
Sommer, Elke	November 5, 1942
Sorenson, Theodore	May 8, 1928
Spitz, Mark	February 10, 1950
Spock, Dr. Benjamin	May 2, 1903
Stapleton, Maureen	June 21, 1925
Stassen, Harold	April 13, 1907
Staubach, Roger	February 5, 1942
Steiger, Rod	April 14, 1925
Steinem, Gloria	March 25, 1936
Stengel, Casey	July 30, 1891
Stewart, Jimmy	May 20, 1908
Stokes, Carl	June 21, 1927
Stone, Irving	July 14, 1903
Streisand, Barbra	April 24, 1942
Struthers, Sally	July 28, 1948
Susskind, David	December 19, 1920
Sutherland, David	July 17, 1934
Symington, Stuart	June 26, 1901
Taft, Robert Jr.	February 26, 1917
Tarkenton, Fran	February 3, 1940
Taylor, Elizabeth	February 27, 1932
Taylor, James	March 12, 1948
Taylor, Rod	January 11, 1930
Thomas, Lowell	April 6, 1892
Thomas, Marlo	November 21, 1938
Thurmond, Strom J.	December 5, 1902
Tolkien, J. R. R.	January 3, 1892
Tomjanovich, Rudy	November 24, 1948
Tomlin, Lily	September 1, 1939
Trevino, Lee	December 1, 1939
Truffaut, François	February 6, 1932
Truman, Margaret	February 17, 1924
Tunney, John V.	June 26, 1934
Turner, Lana	February 8, 1920
Twiggy (Leslie Hornby)	September 19, 1949
Udall, Morris K.	June 15, 1922
Udall, Stewart	January 31, 1920
Uggams, Leslie	May 25, 1943
Ullman, Liv	December 16, 1939
Unitas, Johnnie	May 7, 1933
Unseld, Wes	March 14, 1946
Unser, Al	May 29, 1939
Unser, Bobby	February 20, 1924
Untermeyer, Louis	October 1, 1885
Updike, John	March 18, 1932
Uris, Leon	August 3, 1924
Ustinov, Peter	April 16, 1921
Vaccaro, Brenda	November 18, 1939
Van Buren, Abigail (Dear Abby)	July 4, 1918
Van Dyke, Dick	December 13, 1925
Van Horne, Harriet	May 17, 1920
Vaughan, Sarah	March 27, 1924
Vereen, Ben	October 10, 1946
Vidal, Gore	October 3, 1925
Voight, Jon	December 29, 1938
Vonnegut, Kurt Jr.	November 11, 1922
Wagner, Robert F.	April 20, 1910
Wallace, George	August 25, 1919
Wallace, Irving	March 19, 1916
Wallach, Eli	December 7, 1915
Walters, Barbara	September 25, 1931
Walton, Bill	November 5, 1952
Warhol, Andy	August 7, 1927
Warwicke, Dionne	December 12, 1940
Wayne, John	May 26, 1907

Name	Birthdate	Name	Birthdate
Weicker, Lowell	May 16, 1931	Winters, Jonathan	November 11, 1925
Weinberger, Casper	August 18, 1917	Winters, Shelley	August 18, 1922
Welch, Racquel	September 5, 1940	Wood, Natalie	July 20, 1938
Welk, Lawrence	March 11, 1903	Woodcock, Leonard	February 14, 1911
Welles, Orson	May 6, 1915	Woodward, Joanne	February 27, 1930
West, Mae	August 17, 1892	Wouk, Herman	May 27, 1915
Westmoreland,		Wyeth, Andrew	July 12, 1917
William	March 26, 1914	Wyeth, James	July 6, 1946
White, Byron	June 8, 1917	Yastrzemski, Carl	August 22, 1939
Wicker, Tom	June 18, 1926	York, Susannah	January 9, 1942
Wilkins, Roy	August 30, 1901	Yorty, Sam	October 1, 1909
Williams, Ted	August 30, 1918	Young, Robert	February 22, 1907
Williams, Tennessee	March 26, 1914	Ziegler, Ronald	May 12, 1939
Wilson, Earl	May 3, 1907		

Name	Birthdate	Name	Birthdate

Name	Birthdate	Name	Birthdate

APPENDIX C:
ANNUAL RECORD FORMS

Your Own Step-by-Step
Biorhythm Record for the Current Year

Step 1. Your Date of Birth _____	Look up your own Birth Codes; insert each in the appropriate box.
Birth Codes	

Physical	*Emotional*	*Intellectual*

Step 2. Select Inquiry Year _____	Select the current year as your Inquiry Year. Look opposite the Inquiry Year and under your Birth Codes to find the Inquiry Year Codes; insert each in the appropriate box.

Inquiry Year	*Inquiry Year Codes*		
	Physical	*Emotional*	*Intellectual*

Step 3. Copy your biorhythm data for the year.

Month	Physical Crit. High Low			Emotional Crit. High Low			Intellectual Crit. High Low			
Jan.										Look under the Inquiry Year Codes in the Biorhythm Data Tables and insert your bio-rhythm data for each month.
Feb.										
March										
April										
May										
June										
July										
Aug.										
Sept.										
Oct.										
Nov.										
Dec.										

Your Own Step-by-Step
Biorhythm Record for the Current Year

Step 1. Your Date of Birth _____	Look up your own Birth Codes; insert each in the appropriate box.
Birth Codes *Physical Emotional Intellectual*	

Step 2. Select Inquiry Year _____	Select the current year as your Inquiry Year. Look opposite the Inquiry Year and under your Birth Codes to find the Inquiry Year Codes; insert each in the appropriate box.
Inquiry Year Inquiry Year Codes *Physical Emotional Intellectual*	

Step 3.   Copy your biorhythm data for the year.

Month	Physical Crit.	High	Low	Emotional Crit.	High	Low	Intellectual Crit.	High	Low	
Jan.										
Feb.										Look under the Inquiry Year Codes in the Biorhythm Data Tables and insert your biorhythm data for each month.
March										
April										
May										
June										
July										
Aug.										
Sept.										
Oct.										
Nov.										
Dec.										

Your Own Step-by-Step
Biorhythm Record for the Current Year

Step 1. Your Date of Birth ———————	Look up your own Birth Codes; insert each in the appropriate box.
Birth Codes *Physical Emotional Intellectual* 	

Step 2. Select Inquiry Year ———————	Select the current year as your Inquiry Year. Look opposite the Inquiry Year and under your Birth Codes to find the Inquiry Year Codes; insert each in the appropriate box.
Inquiry Year · *Inquiry Year Codes* *Physical Emotional Intellectual*	

Step 3.   Copy your biorhythm data for the year.

Month	Physical Crit.	High	Low	Emotional Crit.	High	Low	Intellectual Crit.	High	Low	
Jan.										Look under the Inquiry Year Codes in the Biorhythm Data Tables and insert your bio- rhythm data for each month.
Feb.										
March										
April										
May										
June										
July										
Aug.										
Sept.										
Oct.										
Nov.										
Dec.										

Your Own Step-by-Step
Biorhythm Record for the Current Year

Step 1. Your Date of Birth _____	Look up your own Birth Codes; insert each in the appropriate box.
Birth Codes	
Physical *Emotional* *Intellectual*	

Step 2. Select Inquiry Year _____	Select the current year as your Inquiry Year. Look opposite the Inquiry Year and under your Birth Codes to find the Inquiry Year Codes; insert each in the appropriate box.
Inquiry *Inquiry Year Codes*	
Year *Physical* *Emotional* *Intellectual*	

Step 3. Copy your biorhythm data for the year.

Month	Physical Crit.	High	Low	Emotional Crit.	High	Low	Intellectual Crit.	High	Low	
Jan.										
Feb.										Look under the Inquiry Year
March										Codes in the Biorhythm Data
April										Tables and
May										insert your bio-
June										rhythm data for
July										each month.
Aug.										
Sept.										
Oct.										
Nov.										
Dec.										

BIBLIOGRAPHY

Books

Cohen, Daniel. *Biorhythms in Your Life*. Greenwich, Conn.: Fawcett, 1976.

Gittelson, Bernard. *Biorhythms: A Personal Science*, 2nd ed. New York: Arco, 1976.

Huff, Darrell. *Cycles in Your Life*. New York: W. W. Norton, 1964.

Luce, Gay Gaer. *Biological Rhythms in Human and Animal Physiology*. New York: Dover, 1971.

Luce, Gay Gaer, and Segal, J. *Sleep*. New York: Coward McCann, 1966.

Mallardi, Vincent. *Biorhythms and Your Behavior*. Bethlehem, Pa.: Media America, Inc., 1975.

O'Neil, Barbara, and Phillips, Richard. *Biorhythms: How to Live with Your Life Cycles*. Pasadena, Ca.: Ward Ritchie Press, 1975.

Rosenfeld, Edward. *The Book of Highs*. New York: Quadrangle, 1973.

Shealy, C. Norman, M.D., with Freese, Arthur S. *Occult Medicine Can Save Your Life*. New York: Dial Press, 1975.

Sill, Henry. *Of Time Tides and Inner Clocks*. New York: Pyramid, 1975.

Strughold, Hubertus. *Your Body Clock*. New York: Scribners, 1971.

Thommen, George S. *Is This Your Day?*, rev. ed. New York: Crown, 1973.

Ward, Ritchie. *The Living Clocks*. New York: Knopf, 1971.

Wernli, Hans J. *Biorhythm: A Scientific Exploration into the Life Cycles of the Individual*. Translated by Rosemary Colmers. New York: Crown, 1960.

Articles

Bigham, Barbara. "How to Put Your Rhythms to Work on Your Blues." *Writer's Digest*, August, 1975.

"Biorhythms in the Sky." *Science Digest*, June, 1975, pp. 16-17.

"Biorhythm Theory Claims Ability to Spot Accident-Prone Periods." *Aviation Week*, January 23, 1961, pp. 101-102.

"Clocks Within the Body." *Current Research on Sleep and Dreams*, U.S. Department of Health, Education, and Welfare, Public Health Service Publication No. 1389.

Deyo, M. E. "Is Today for the Birds—or the Birdies?" *The Professional Golfer*, April, 1976, pp. 18-26.

Hersey, Rex B. *"Emotional Cycles in Man."* Journal of Mental Science, January, 1931, pp. 151-169.

Hirsh, Tom. "Biorhythm—Or, Is It a Critical Day?" *National Safety News*, February, 1976.

Mackenzie, Jean. "How Biorhythms Affect Your Life." *Science Digest*, August, 1973.

"Moonstruck Scientists." *Time*, January 10, 1972, p. 48.

Nelson, Ed. "Death and Biorhythm." *Science Digest*, May, 1976.

Richelle, Marc. "Biological Clocks." *Psychology Today*, May, 1970.

Zito, Tom. "Pilots' Biorhythm Cycles Are Studied as Factors in Crashes." *The Washington Post*, February 2, 1975.

About the Author

Dr. Robert E. Smith is both a computer analyst and author. He has written over thirty books on computer science, including the COMPUTER EXPLORER SERIES, a number of titles relating computer applications to such diverse popular subjects as: astrology, palmistry, gamesmanship, handwriting analysis, gambling, numerology, as well as more classical computer applications. He has more than twenty years of experience as a former teacher and Professor of Mathematics, and was Chairman of the Mathematics Department at Duquesne University. In recent years, Dr. Smith has inaugurated a unique program of classes on special applications of computer terminology and numerical analysis.